CHICAGO PUBLIC LIBRARY
HAROLD WASHINGTON LIBRARY CENTER

R0002769799

REF
QH
545
.05
N43

Neff, Jerry M.

Response of marine animals to petroleum and specific petroleum hydrocarbons

DATE

© THE BAKER & TAYLOR CO.

RESPONSE OF MARINE ANIMALS TO PETROLEUM AND SPECIFIC PETROLEUM HYDROCARBONS

RESPONSE OF MARINE ANIMALS TO PETROLEUM AND SPECIFIC PETROLEUM HYDROCARBONS

JERRY M. NEFF

*Battelle New England Marine Research Laboratory,
397 Washington Street, Duxbury,
Massachusetts 02332, USA*

and

JACK W. ANDERSON

*Battelle Pacific North West Division,
Marine Research Laboratory, Route 5,
Box 1000, Sequim, Washington 98382, USA*

APPLIED SCIENCE PUBLISHERS LTD, LONDON

HALSTED PRESS DIVISION
JOHN WILEY & SONS, NEW YORK & TORONTO

First published in 1981 in Great Britain by
Applied Science Publishers Ltd

British Library Cataloguing in Publication Data

Neff, Jerry M.
 Response of marine animals to petroleum and
 specific petroleum hydrocarbons.
 1. Marine ecology   2. Marine pollution
 3. Petroleum pollution of water
 I. Title   II. Anderson, Jack W.
 574.5   GC1085

ISBN 0-85334-953-3 (APPLIED SCIENCE PUBLISHERS)
ISBN 0-470-27215-5 (HALSTED)

WITH 35 TABLES AND 52 ILLUSTRATIONS

© APPLIED SCIENCE PUBLISHERS LTD 1981

Published in the U.S.A. and Canada by
Halsted Press, a Division of
John Wiley & Sons, Inc.,
New York

All rights reserved. No part of this publication may be reproduced, stored in a retrieval system, or transmitted in any form or by any means, electronic, mechanical, photocopying, recording, or otherwise, without the prior written permission of the publishers, Applied Science Publishers Ltd, Ripple Road, Barking, Essex, England

Printed in Great Britain by Galliard (Printers) Ltd, Great Yarmouth

Acknowledgments

Most of the research reported here was performed under contracts No. 0S20C and 206-76 from the American Petroleum Institute, Environmental Affairs Department to J. W. Anderson and J. M. Neff through the Center for Marine Resources, Texas A&M University.

This book represents the product of the cooperative effort of several individuals at different levels of training. We wish to express our appreciation to all those who contributed their research, secretarial, or editorial expertise to the completion of this research project. We also gratefully acknowledge the valuable contributions of Dr J. S. Warner, Battelle Columbus Laboratory, who performed many gas chromatographic analyses of oil, water and tissue samples for us and of Dr R. J. Pancirov, Exxon Research and Engineering, Linden, N.J., who determined the chemical and physical characteristics of the four API reference oils.

Preface

During the past decade, a very large volume of research has been performed dealing with the fate and effects of petroleum in the marine environment. Major interest in and financial support for research in this area followed the wreck of the tanker 'Torrey Canyon' off Cornwall, Great Britain on 18 March 1967 and the smaller oil spill resulting from the wreck of the oil barge 'Florida' off Falmouth, Massachusetts, USA on 16 September 1969. Both spills were near major marine research laboratories and both were studied intensively by leading marine scientists. From these investigations has come the increased realization that deleterious impacts of oil on the marine ecosystem may persist long after the visible oil pollution has been cleaned up or washed away. In the decade since these spills, the volume of petroleum transported by sea has increased substantially and there has been a large increase in exploration for and production of offshore oil deposits.

Between 1972 and 1977 our laboratory, with financial support from the American Petroleum Institute, Environmental Affairs Department, conducted extensive laboratory and field research dealing with the toxicity and sublethal biological effects of petroleum to marine organisms. Two earlier reports have dealt with some aspects of this research. Results of laboratory studies conducted during the first two and a half years of this project are summarized in API publication No. 4249: *Laboratory Studies on the Effects of Oil on Marine Organisms: An Overview*. Results of the field studies are summarized in API publication No. 4291: *Effects of Oilfield Brine Effluent on Benthic Organisms in Trinity Bay, Texas*.

In this book we present the major results and conclusions of the laboratory research performed in this research program and discuss these results in relation to major published information on the subject from other laboratories.

JERRY M. NEFF
JACK W. ANDERSON

Contents

	Acknowledgments	v
	Preface	vii
1	SOURCES AND DISTRIBUTION OF HYDROCARBONS IN THE MARINE ENVIRONMENT	1
2	TOXICITY OF PETROLEUM AND SPECIFIC PETROLEUM HYDROCARBONS	5
	2.1 Introduction	5
	2.2 Materials and Methods	6
	2.2.1 Petroleum and Petroleum Hydrocarbons	6
	2.2.2 Hydrocarbon Analysis	8
	2.2.3 Test Animals	8
	2.2.4 Bioassay Procedures	9
	2.3 Results and Discussion	10
	2.3.1 Characteristics of API Reference Oils	10
	2.3.2 Comparative Toxicity of the Four Oils	17
	2.3.3 Comparative Toxicity of the WSF and OWD	20
	2.3.4 Comparative Toxicity of Oil to Different Species	22
	2.3.5 Comparative Toxicity of Oil to Marine Larvae	23
	2.3.6 Comparative Toxicity of Petroleum-derived Aromatic Hydrocarbons	28
	2.3.7 Effect of Salinity on Acute Toxicity of Aromatic Hydrocarbons	32
3	REPRODUCTION, DEVELOPMENT AND GROWTH	35
	3.1 Introduction	35
	3.2 Materials and Methods	39
	3.3 Results and Discussion	42
	3.3.1 Embryonic and Larval Development: Molluscs	42
	3.3.2 Embryonic Development: Fish	45
	3.3.3 Larval Development and Growth	62
	3.3.4 Reproductive Success: Polychaetes	72
	3.3.5 Photosynthetic Carbon Fixation by Coral Zooxanthellae	77
	3.3.6 Calcium Deposition by Reef Corals	80
	3.3.7 Calcium Deposition in Sea Urchin Spines	89

4	ACCUMULATION AND RELEASE OF PETROLEUM HYDROCARBONS	93
	4.1 Introduction	93
	4.2 Materials and Methods	99
	4.3 Results and Discussion	100
	4.3.1 Uptake and Release of Petroleum Hydrocarbons from Water: Laboratory Studies	100
	4.3.2 Uptake and Release of Petroleum Hydrocarbons: Field Studies	121
	4.3.3 Effects of Salinity and Temperature on Hydrocarbon Uptake	123
	4.3.4 Accumulation of Hydrocarbons from Sediment and Food	130
	4.3.5 Metabolism of ^{14}C-Naphthalene by *Neanthes arenaceodentata*	135
5	EFFECTS OF PETROLEUM ON THE BEHAVIOR OF MARINE ANIMALS	143
	5.1 Introduction	143
	5.2 Materials and Methods	146
	5.3 Results and Discussion	148
	REFERENCES	155
	GLOSSARY OF SELECTED TECHNICAL TERMS	171
	Index	175

CHAPTER 1

Sources and Distribution of Hydrocarbons in the Marine Environment

As a result of the rapid increase in the volume of petroleum shipped by sea and the exploitation of offshore oil and gas deposits in recent years, there is a growing concern over the impacts of spilled petroleum on marine organisms and the marine environment as a whole. Several attempts have been made to estimate the annual influx of petroleum to the marine environment from natural and anthropogenic sources. The best estimate of natural marine oil seepage is 0·6 million metric tons per year (mta) (Wilson *et al.*, 1974), but it does not include oil entering the ocean from terrestrial oil seeps and erosion of tar sands, oil-bearing shales, etc. The most widely quoted estimate of annual petroleum influx to the marine environment from all sources, 6·2 mta, is by the National Academy of Sciences (1975). Grossling (1976) more recently approximated the amount of oil entering the oceans each year to be between 29 and 71 million barrels (4–10 mta). The 2·5-fold range of his estimate was due primarily to uncertainty about the magnitude of natural submarine oil seepage.

The most important anthropogenic sources of petroleum entering the marine environment are those associated with marine transportation (2·2 mta) and surface runoff from land (1·6 mta) (National Academy of Sciences, 1975). Pollution due to marine transportation includes losses during normal ship operations, oil spills resulting from accidents at sea, and spills occurring during terminal operations. Offshore oil drilling and production activities contribute only 0·1 mta to the total, most of which results from blow-outs and accidental large spills (50 barrels or more).

Assessment of the environmental impact of this influx of petroleum to the oceans is complicated by the fact that crude oils and most refined products are extremely complex mixtures of organic compounds. Hydrocarbons are most abundant, usually representing more than 75 per cent of the oil with the remainder being made up primarily of various sulfur, oxygen, and nitrogen-containing organic compounds (Speers and Whitehead, 1969; Kallio, 1976). Most evaluations of environmental levels of petroleum are based on analyses of total, or selected classes of, hydrocarbons (Farrington et al., 1976).

However, petroleum is not the sole source of hydrocarbons present in marine waters, sediments, and organisms. Koons and Monaghan (1976) estimated the input of hydrocarbons from biochemical synthesis in the oceans to be 6 mta, a figure similar to the estimated total input of hydrocarbons from all petroleum sources. Alkenes, n-alkanes, and branched alkanes, in that order, are the predominant classes of biogenic hydrocarbons, whereas the most abundant in petroleum are cycloalkanes, branched alkanes, n-alkanes, and aromatics.

Aromatic hydrocarbons, and in particular, polycyclic aromatic hydrocarbons (PAH) may enter the marine environment from sources other than petroleum (Neff, 1979). A major source of PAH in the environment is the combustion or pyrolysis of organic materials including fossil fuels. Youngblood and Blumer (1975) identified a complex mixture of PAH in recent marine sediments and suggested that they were derived from particulates formed during natural fires. Municipal incinerators produce PAH which may be released to the environment in ash waste, stack gases, and wastewaters (Davies et al., 1976). Suess (1976) estimated anthropogenic emissions of the PAH benzo[a]pyrene (BaP) to the atmosphere to be more than 5000 metric tons/year. The major sources of BaP were coal burning and coke production. Much BaP may reach the marine environment in surface runoff and atmospheric fallout. Biosynthesis of PAH by microorganisms, phytoplankton, algae, and higher plants has been reported (ZoBell, 1971; Suess, 1976). However, Hase and Hites (1976) provided evidence against synthesis of PAH in recent marine sediments by anaerobic bacteria.

Neff (1979) recently estimated that about 230 000 metric tons of PAH enter the aquatic environment each year from all sources. The largest portions of the total come from petroleum spillage (170 000 mta) and fallout and rainout from the air (50 000 mta). However, for the higher molecular weight four- through six-ring PAH, which include the major known PAH carcinogens like benzo[a]pyrene, the pattern is different. Neff estimated that nearly 700 metric tons of BaP enter the aquatic environment each year from all sources. The major sources are fallout and rainout from air (500 mta) and surface runoff from land (118 mta) of BaP derived primarily from combustion of fuels.

Because of the diversity of sources of hydrocarbons in the marine environment, the presence of hydrocarbons in a marine sample is not by itself an indication of pollution from urban, industrial, or petroleum sources. However, high levels, particularly of aromatic and high molecular weight aliphatic hydrocarbons, are often indicative of petroleum pollution (Hites, 1976).

McAuliffe (1976) reviewed available literature on levels of hydrocarbons in seawater and marine sediments. Low molecular weight C_2–C_4 hydrocarbons are found at all depths in the oceans at a mean total concentration of 45 ng/liter (parts per trillion). In surface waters near the natural oil seep at Coal Oil Point on the Santa Barbara Channel, C_3–C_8 hydrocarbons were present at a concentration of 290 ng/liter (Koons and Brandon, 1975). Total dissolved C_1–C_{10} hydrocarbons under an oil slick produced by a major oil spill in the Gulf of Mexico reached maximum concentrations of 1–200 µg/liter (parts per billion) (McAuliffe et al., 1975). In surface waters of the coastal northwestern Gulf of Mexico, an area of extensive oil exploration and production, concentrations of total C_6–C_{14} n-alkanes ranged from 7 to 50 ng/liter and of total volatile aromatics (benzene, toluene, ethylbenzene and xylenes) from 20 to 450 ng/liter (Sauer et al., 1978). At most stations, 80–90 per cent of the total liquid hydrocarbons in surface water were aromatic. These volatile hydrocarbons represented a few per cent at most of the total solvent-extractable hydrocarbons in seawater.

As higher molecular weight hydrocarbons are generally restricted to the upper 10 m of the water column, they may be particulate in form (McAuliffe, 1976). The surface microlayer (0–0·3 mm) generally contains the highest concentrations, which range from 0·2 to 155 µg/liter. Below this surface layer, concentrations are generally in the low parts per billion range and decrease with depth. Marty and Saliot (1976) measured hydrocarbons in surface microlayer and subsurface water samples from several stations in the eastern North Atlantic and western Mediterranean. Concentrations of dissolved and particulate hydrocarbons in the surface microlayer ranged from 15·0 to 114 µg/liter and 3·34 to 1214 µg/liter, respectively. In subsurface water, concentrations of dissolved and particulate hydrocarbons ranged from 0·11 to 5·6 µg/liter and 0·28 to 7·2 µg/liter, respectively. The authors identified much of the hydrocarbon as biogenic, though petroleum hydrocarbons were identified in some samples.

Concentrations of high molecular weight hydrocarbons are generally three to four orders of magnitude higher in sediments than in the water column and range from about 40 mg/kg (parts per million) on the continental shelf to 1–5 mg/kg in deep water. In coastal areas subjected to petroleum and industrial waste inputs, sediment hydrocarbon concentrations greater than 500 mg/kg have been reported (McAuliffe, 1976).

Concentrations of hydrocarbons in sediments in and near the New York Bight dump site off the eastern United States were 500–3000 µg/g dry weight and were identified as primarily of fossil fuel origin (Farrington and Tripp, 1977). Sediment hydrocarbon concentration decreased with distance offshore and away from the dump site and reached a level of 1–7 µg/g dry weight on the North Atlantic abyssal plain. No more than 1 µg/g fossil fuel hydrocarbons could be identified in the latter samples. Two months after the wreck of the oil tanker 'Argo Merchant' on the Nantucket Shoals off Massachusetts USA, sediments in the immediate vicinity of the wreck contained up to 122 µg/g dry weight of petroleum hydrocarbons, primarily of a fine particulate nature (Hoffman and Quinn, 1978, 1979). Background concentrations of sediment hydrocarbons in this region were less than 0·5 µg/g.

A major research effort has been initiated in this country and abroad to assess the impacts of hydrocarbon inputs on marine organisms and ecosystems. Much laboratory research to date has concentrated on acute toxicity bioassays with one or a few oils, or petrochemicals, and selected species of marine organisms. Such bioassays, while important, are of limited value in judging the potential impact of long term, low level petroleum contamination of marine ecosystems. Here, studies on sublethal biological effects of hydrocarbons are potentially of more use (Wilson, 1975; Perkins, 1979). The laboratory at Texas A&M University, under contract to the American Petroleum Institute since 1972, has conducted extensive laboratory and field research concerning the effects of petroleum and specific petroleum hydrocarbons on marine organisms and ecosystems. The purpose of this volume is to summarize and review lethal and sublethal effects of oil on marine organisms, based on results of laboratory studies completed since the publication of an earlier report (Anderson, 1975).

CHAPTER 2

Toxicity of Petroleum and Specific Petroleum Hydrocarbons

2.1 INTRODUCTION

A considerable volume of published information on acute toxicity of different crude and refined petroleum products to marine organisms is available. However, comparisons of relative toxicities of different petroleum products and the sensitivity to oil of different marine species are difficult because a variety of methods were used to introduce petroleum into the water; in many cases the actual concentration of hydrocarbons in the aqueous phase of the exposure medium was not measured (Moore and Dwyer, 1974). Petroleum is an extremely complex mixture of thousands of different hydrocarbons and related hetero-compounds. Different crude and refined oils vary tremendously in the relative concentrations of different components and as a result show substantial variability in solubility, dispersibility, and persistence in seawater (Boylan and Tripp, 1971; Anderson et al., 1974a; Lee et al., 1974). In addition, weathering of petroleum may substantially alter its aqueous solubility and dispersibility and will result in a change in the composition of the water-accommodated fraction (Burwood and Speers, 1974; Lysyj and Russell, 1974; Young and Sethi, 1975; MacKay and Shiu, 1976). As a result, bioassays in which results are reported in terms of 'oil added' are of little comparative value. For laboratory bioassay and biological effects studies, the simplest reliable techniques for analysis of petroleum hydrocarbons in seawater are infrared (IR) analysis for total hydrocarbons (American Petroleum Institute, 1958; Carlberg and Skarstedt, 1972), or a method for measuring soluble aromatics such as ultraviolet (UV) spectrophotometry (Neff and Anderson, 1975) or fluorescence spectroscopy (Gordon et al., 1974).

When these parameters are taken into consideration, several trends emerge. Crude petroleums are generally less toxic to marine organisms than refined products. However, crude oils from different sources vary tremendously in their toxicity and some may be more toxic than certain highly refined light distillate products such as JP-8 jet fuel (Allen, 1971). Recent evidence strongly indicates that in most cases acute toxicity of a petroleum product is directly correlated to its content of soluble aromatic derivatives (Moore and Dwyer, 1974; Anderson *et al.*, 1974*a*). These include benzene, naphthalene, phenanthrene, and their alkyl homologs.

Marine animals vary tremendously in their sensitivity to oil. Hyland and Schneider (1976) reviewed much of the relevant literature and concluded that for the adult stages of a wide variety of marine animals, lethal effects from water-soluble fractions of petroleum and petrochemicals occur in the 1–100 ppm (parts per million) range. Larval and juvenile life stages are usually more sensitive to oil pollutants with lethal concentrations often in the 0·1–1·0 ppm range.

Most bioassays conducted to date have been with coastal and estuarine species. It has been assumed by some researchers that these species, because of natural adaptation to environmental stresses (i.e., variable salinity, temperature, and oxygen availability), would be more resistant to pollutant stress than oceanic species which live under more constant and thus less stressful temperature and salinity regimes. Little comparative information is available in this area and also on effects of temperature and salinity stress on the toxicity of oil to marine animals.

The laboratory at Texas A&M University has tried to address several of these problems. Some of the major results and conclusions of these studies are discussed below.

2.2 MATERIALS AND METHODS

2.2.1 Petroleum and Petroleum Hydrocarbons

Two crude and two refined oils, supplied in 55-gal barrels by API, were used in these investigations: Kuwait crude (API reference oil No. I), south Louisiana crude (API reference oil No. II), No. 2 fuel oil (API reference oil No. III, approximately 38 per cent aromatics), and Venezuelan bunker C residual oil (API reference oil No. IV). The contents of each 55-gal barrel were transferred to hexane-rinsed 1-gal amber glass jars, residual jar space flushed with purified nitrogen, and the jars sealed with aluminium foil or Teflon-lined caps.

Oil–water mixtures for use in bioassays were prepared by two methods as described by Anderson *et al.* (1974*a*). Oil-in-water dispersions (OWD) were prepared by adding measured volumes of oil to artificial seawater (prepared by mixing distilled water with Instant Ocean synthetic sea salts (Aquarium Systems, Eastlake, Ohio)), and shaking the mixtures vigorously for 5 min at approximately 200 cycles/min on a shaker platform. In most cases, dispersions were prepared in bioassay containers which were sealed with Teflon-lined caps during mixing. The oil and water layers were allowed to separate for 30–60 min before test animals were added or the aqueous layer sampled for hydrocarbon analysis.

Water-soluble fractions (WSF) of oil were prepared by placing one part oil over nine parts artificial seawater in 5-gal, or smaller, Pyrex bottles and slowly stirring the mixtures with magnetic stirrers for a 20 h period at room temperature ($20 \pm 2\,°C$). Stirring speeds were adjusted so that the vortices did not extend more than 25 per cent of the distance to the bottom of the container. After mixing, oil and water phases were allowed to separate for 1–2 h before the water phase was siphoned off and used immediately in experiments. Different concentrations of WSF were prepared by diluting stock WSF (100 per cent WSF) by volume with seawater immediately before use. Since bunker C oil had a density similar to that of seawater and tended to form persistent droplets in the water phase, it was necessary to filter the WSF of this oil through Nitex nylon filter pads before use. Water-soluble fractions prepared in this way showed a high degree of inter-batch uniformity in initial aqueous hydrocarbon concentration and composition.

Experiments with some pure hydrocarbons were also performed. Specific hydrocarbons used were of the highest purity available (usually 99 + per cent) and were obtained from Aldrich, Pfaltz and Bauer, or Chemical Samples Co. Stock solutions of the hydrocarbons were prepared in nanograde acetone. Exposure mixtures were prepared by adding specific volumes of the acetone stock to seawater with vigorous mixing to ensure thorough dispersion and dissolution of the hydrocarbon. The concentration of the stock solution was adjusted such that no more than 1 ml acetone was added to each liter of exposure water. This concentration of acetone was found to be completely non-toxic to all the test species. It is recognized that this procedure of preparing exposure solutions of specific hydrocarbons is highly artificial. Acetone, even at non-toxic concentrations, could influence the bioavailability and toxicity of hydrocarbons to marine animals. However, this method of introducing sparingly soluble compounds into aquatic exposure systems is widely accepted because of the ease with which reproducible hydrocarbon concentrations in water can be achieved.

2.2.2 Hydrocarbon Analysis

Detailed analyses of the hydrocarbon composition of the four API reference oils were performed by Dr J. S. Warner, Battelle Memorial Laboratory, Columbus, Ohio and Dr R. J. Pancirov, Exxon Research and Engineering Co., Linden, New Jersey. Details of the analytical methodology are described by Pancirov (1974), Pancirov and Brown (1975), and Warner (1976). Concentrations of petroleum hydrocarbons present in the aqueous phase of OWD, WSF, or specific hydrocarbon solutions were determined by IR, UV, or gas chromatographic methods. Aliquots of water were carefully siphoned from below the surface of the OWD in such a way as to avoid contamination of the water sample by any surface oil slick that might be present. A 200–400 ml aliquot of each stock WSF was taken for analysis before the WSF was diluted for bioassays. In addition, similar aliquots of diluted WSF exposure mixtures were taken for analysis at the beginning of the bioassay and at several times during the bioassay so that the rate of decrease in concentration of hydrocarbons in exposure media could be assessed. Aqueous samples of pure hydrocarbon exposure media were handled in a similar fashion.

Total petroleum hydrocarbons in water samples (carbon tetrachloride-extractable oily material) were determined by API method No. 733-58 (American Petroleum Institute, 1958) using a Miran I single beam IR analyzer (Wilkes Scientific Instruments, South Norwalk, Conn.). Naphthalene and alkyl naphthalenes were determined by the UV spectrophotometric method of Neff and Anderson (1975) with a Pye–Unicam SP1800 recording UV spectrophotometer. Selected samples were analyzed by Dr J. S. Warner (Battelle Memorial Laboratory, Columbus, Ohio) using a gas chromatographic method (Warner, 1976) for n-paraffins and mono-, di- and tricyclic aromatic hydrocarbons. The three analytical methods used differ in their sensitivity and specificity. The IR method is specific for compounds, extractable with CCl_4 from acidified water, which contain CH, CH_2, or CH_3 groups, and is therefore subject to interference by many organic compounds. Gas chromatographic and UV spectrophotometric techniques have been compared for both tissue and water samples (Neff and Anderson, 1975). The former has greater precision and sensitivity, but both techniques yield comparable values in the types of studies reported here.

2.2.3 Test Animals

The estuarine species utilized in this study were collected from the

Galveston Bay System, Texas and especially in the vicinity of Galveston Island. Oceanic species were collected in the vicinity of the Buccaneer gas production field approximately 35 miles south of Galveston, Texas. Animals were transported to the College Station, Texas laboratory in large styrofoam coolers with aeration. Before experimentation, animals were maintained in large tanks for 5 to 7 days at $20 \pm 2\,°C$ and a salinity similar to that at the collection site. Polychaete worms *Neanthes arenaceodentata* (Moore) and *Capitella capitata* (O. Fabricius) were reared in the laboratory from populations originally cultured by Dr D. J. Reish (California State University, Long Beach), using culture methods described by Reish and Richards (1966). The worms were maintained at a salinity of $32 \pm 1‰$, temperature of $22 \pm 2\,°C$, and dissolved oxygen concentration of 7 ± 0.2 mg/liter. They were fed dried alfalfa powder and the green alga *Enteromorpha*.

All specimens were kept is air conditioned rooms and in adequately filtered and aerated tanks and aquaria. Seawater solutions were prepared with distilled water and artificial sea salts (Instant Ocean, Aquarium Systems, Eastlake, Ohio). The salinity of the seawater was determined with an optical refractometer (American Optical Co.).

2.2.4 Bioassay Procedures

Methods used in this study are similar to those recommended by Sprague (1969) and LaRoche *et al.* (1970). Bioassay containers varied in size depending on the size of the test organism. Generally, for larvae and polychaete worms, 135 ml Erlenmeyer flasks containing 50 ml exposure medium were used. For larger animals, 1-3 liter glass containers with 500 or 750 ml of exposure medium were used. Aeration was supplied so that a bubble was produced from the tip of a disposable Pasteur pipet approximately three times/second. The number of individuals/container varied according to the size of the species and the volume of exposure medium used. Generally, 5-6 concentrations of the test mixture plus a control were used in each bioassay. A minimum of 10 animals were exposed to each concentration in all cases.

At approximately 12 h intervals over the 96-h test period, containers were examined, mortalities recorded, and dead individuals removed. Whenever possible, LC_{50} values (concentration of toxicant causing 50 per cent mortality in the specified time), their 95 per cent confidence intervals, and slope functions were calculated by the graphic method of Litchfield and Wilcoxon (1949). In those cases where the nature or distribution of data points precluded computation by this method, LC_{50} values were computed on semilogarithmic paper by the standard APHA method (American Public Health Association, 1971).

Some of the bioassays described below required slight modifications of these methods; variations in procedure are described in the text.

2.3 RESULTS AND DISCUSSION

2.3.1 Characteristics of the API Reference Oils

The four reference oils have been characterized both chemically and physically (Anderson et al., 1974b; Pancirov, 1974; Pancirov and Brown, 1975; Youngblood and Blumer, 1975). The two crude oils are light petroleums (API gravity > 30) (Table 1). Kuwait crude is a relatively high sulphur petroleum while south Louisiana crude is low in sulfur. Nickel and vanadium in crude oils are nearly always associated with fossil porphyrins indigenous to the oil and therefore show limited bioavailability to aquatic organisms. Kuwait crude oil contains a relatively high concentration of vanadium and bunker C residual oil contains high concentrations of both nickel and vanadium. The two refined oil products, No. 2 fuel oil and bunker C residual oil, have a higher total aromatic concentration than the crudes. Bunker C also contains significant amounts of polar compounds (probably primarily phenols) and insoluble materials.

TABLE 1
COMPOSITION OF API REFERENCE OILS USED IN BIOLOGICAL STUDIES (FROM PANCIROV, 1974)

	Kuwait crude	South Louisiana crude	No. 2 fuel oil	Bunker C residual oil
API gravity at 60°C	31·4	34·5	31·6	7·3
Sulfur (weight per cent)	2·44	0·25	0·32	1·46
Nitrogen (weight per cent)	0·14	0·069	0·024	0·94
Nickel (ppm)	7·7	2·2	0·5	89
Vanadium (ppm)	28	1·9	1·5	73
Naphtha (up to 400°F, weight per cent)	22·7	18·6	—	0·0
Saturates (weight per cent)	34·0	56·3	61·8	21·1
Aromatics (weight per cent)	21·9	16·5	38·2	34·2
Polars (weight per cent)	17·9	8·4	—	30·3
Insolubles (weight per cent)	3·5	0·2	—	14·4

TABLE 2
CONCENTRATIONS OF C_{12}–C_{24} n-PARAFFINS AND AROMATIC HYDROCARBONS IN THE API REFERENCE OILS USED IN BIOLOGICAL STUDIES (PARAFFINS · AND NAPHTHALENE–FLUORENES FROM ANDERSON et al., 1974b; BENZENES FROM PANCIROV, 1974; PHENANTHRENE–BENZO[ghi]PERYLENE FROM PANCIROV AND BROWN, 1975.) CONCENTRATIONS ARE IN mg/kg (PARTS PER MILLION)

Compound	Kuwait crude	South Louisiana crude	No. 2 fuel oil	Bunker C residual oil
Total C_{12}–C_{24} n-paraffins	40 000	39 800	73 800	12 600
Benzenes[a]	80 000	92·000	222 000	60 000
Naphthalene	400	400	4 000	1 000
1-Methylnaphthalene	500	800	8 200	2 800
2-Methylnaphthalene	700	900	18 900	4 700
Dimethylnaphthalenes	2 000	3 600	31 100	12 300
Trimethylnaphthalenes	1 900	2 400	18 400	8 800
Biphenyls[b]	<100	<100	1 600	<100
Fluorenes[b]	<100	200	3 600	2 400
Phenanthrene	26	70	429	482
1-Methylphenanthrene	—	111	173	43
2-Methylphenanthrene	89	144	7 677	828
Fluoranthene	2·9	5·0	37	240
Pyrene	4·5	3·5	41	23
Benz[a]anthracene	2·3	1·7	1·2	90
Chrysene	6·9	17·56	2·2	196
Triphenylene	2·8	10	1·4	31
Benzo[ghi]fluoranthene	<1	1	—	—
Benzo[b]fluoranthene	<1	<0·5	—	—
Benzo[j]fluoranthene	<1	<0·9	—	—
Benzo[k]fluoranthene	<1	<1·3	—	—
Benzo[a]pyrene	2·8	0·75	0·6	44
Benzo[e]pyrene	0·5	2·5	0·1	10
Perylene	<0·1	34·8	—	22
Benzo[ghi]perylene	<1	1·6	—	—
Total hydrocarbons analyzed	125 637·7	140 503·4	389 962·5	106 609

[a] Includes benzene, alkylbenzenes, indans, tetralins and dinaphthenobenzenes.
[b] Includes the parent compounds and their alkyl homologs.

The hydrocarbon compositions of the four reference oils are quite different (Table 2), however, some trends emerge. All but bunker C are relatively rich in low molecular weight n-paraffins. There is a very conspicuous trend in all four oils for the concentration of specific aromatic hydrocarbon types to decrease substantially with increasing ring number. However, south Louisiana crude oil has unusually high concentrations of chrysene and perylene. Number 2 fuel oil contains a high concentration of 2-methylphenanthrene, while bunker C residual oil contains high concentrations of 2-methylphenanthrene, fluoranthene, chrysene, benzo[a]pyrene, and perylene. Typically, the alkyl homologs are present at higher concentrations than the parent PAH in crude and refined oils (Pancirov and Brown, 1975; Youngblood and Blumer, 1975; Blumer, 1976). This is reflected in the relative concentrations of naphthalene and C_1-C_3 naphthalenes in the four API reference oils. Although more than 60 hydrocarbons were identified and quantified in the analyses summarized in Table 2, their total concentrations represented only 10–40 per cent of the total weight of the oils. This is further indication of the extreme compositional complexity of crude and refined oils.

The chemical and physical characteristics of the oils affect their behavior in water. The relationship between the amount of oil added to seawater and the concentration of total hydrocarbons in the aqueous phase of OWD, prepared as described above, is different for each oil (Fig. 1). For

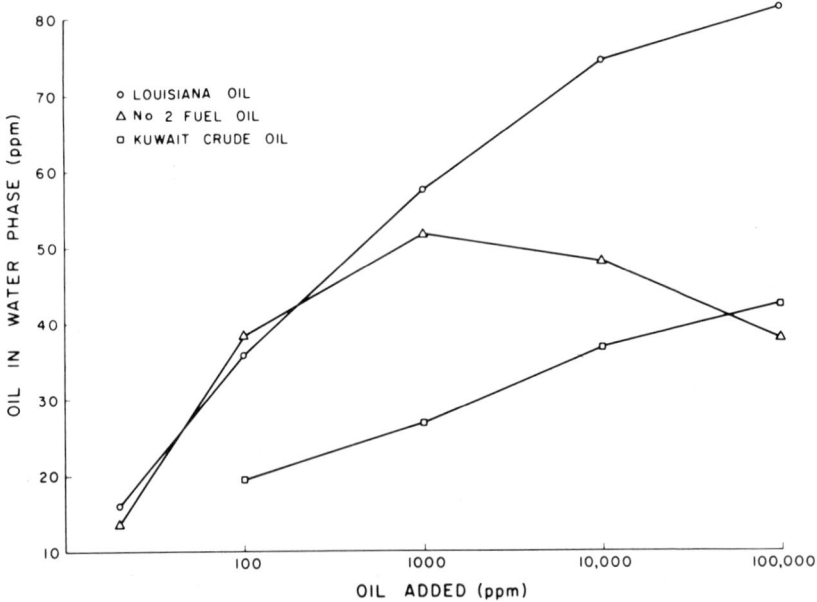

FIG. 1. Relationship between the amount of oil added and the concentration of total hydrocarbons (as determined by IR) in the aqueous phase of the oil-in-water dispersion (OWD) of three API reference oils (from Anderson et al., 1974a, with permission of Springer–Verlag).

the two crude oils, there is a nearly linear relationship between the amount of oil added and the concentration of total hydrocarbons in the aqueous phase after 1 h. However, south Louisiana crude oil produces a much more concentrated OWD than does Kuwait crude oil at all levels of oil added. The No. 2 fuel oil has a different behaviour. Between 20 and 1000 ppm oil added, there is a nearly linear increase in the concentration of total hydrocarbons in the aqueous phase. When more oil is added, there is a drop in the concentration of hydrocarbons in the aqueous phase.

It is clear from these results that a relatively small fraction of the total oil added to water becomes dispersed as fine droplets in the water column even after relatively violent mixing. Violent mixing of 100 000 ppm (v:v) south Louisiana crude oil (the most dispersible oil tested) with seawater produces an OWD containing only 81·2 ppm total hydrocarbons. The unusual behavior of No. 2 fuel oil in water can be explained as follows: with an increase in the amount of oil added and thus an increase in the concentration of oil droplets in the aqueous phase, the rate of droplet coalescence increases (Smoluchowski equation). A dispersion with dissimilar sized droplets is less stable than one with fewer but similar sized droplets. Thus, droplet coalescence increases the rate at which oil returns to the surface slick, resulting in lower hydrocarbon concentrations in the aqueous phase. This phenomenon is not observed in the crude oils due to the presence in them of indigenous surface-active agents which tend to decrease the incidence of droplet coalescence. South Louisiana crude oil produces a more concentrated OWD than Kuwait crude at all levels of added oil because of its lower viscosity, allowing droplets to be more readily generated for the same mixing energy.

Oil-in-water dispersions are unstable under the bioassay conditions used in this investigation. The concentration of total hydrocarbons in the aqueous phase of the OWD drops rapidly when it is gently aerated. Approximately 10 per cent of the hydrocarbons originally present in south Louisiana crude oil OWD remain after 24-h aeration. Concentrations of aliphatic hydrocarbons in the OWD decrease more rapidly than concentrations of aromatic hydrocarbons (Table 3), due to the much lower aqueous solubility of the former.

Oil-in-water dispersions of bunker C residual oil cannot be prepared reproducibly because of the physical characteristics of this oil. Because of its high viscosity and high specific gravity (s.g. = 1·019 compared to 1·022 for 32‰ S seawater at 20 °C), it tends to sink in the water and form macro oil droplets. As a result, a true aqueous phase cannot be separated from the oil–water mixture.

TABLE 3

HYDROCARBON COMPOSITION OF AQUEOUS PHASE OF 1 000 ppm OF OIL-IN-WATER DISPERSIONS (OWD) AND EFFECTS OF 24-h AERATION. ALL VALUES IN μg/liter (ppb) (FROM ANDERSON et al., 1974a, WITH PERMISSION OF SPRINGER–VERLAG)

Hydrocarbon	Kuwait Initial	Kuwait Final	South Louisiana Initial	South Louisiana Final	No. 2 fuel oil Initial	No. 2 fuel oil Final
C_{12}–C_{24} n-paraffins	1 320	71	1 988	64	4 234	820
Tri- and tetramethylbenzenes	260	70	135	99	547	79
Naphthalene	19	15	64	53	671	292
1-Methylnaphthalene	12	1	40	24	416	132
2-Methylnaphthalene	16	17	46	25	646	159
Dimethylnaphthalenes	33	4	108	32	1 430	179
Trimethylnaphthalenes	19	3	56	6	872	105
Phenanthrene	2	2	34	2	214	65
Methylphenanthrenes	2	2	20	2	152	46
Total n-paraffins	1 320	71	1 988	64	4 234	820
Per cent decrease		95		97		81
Total aromatics	359	110	506	241	4 950	1 057
Per cent decrease		69		52		79
Total hydrocarbons measured	1 679	181	2 494	305	9 184	1 877
Per cent decrease		89		88		80

The hydrocarbon composition of the OWD can be expected to resemble that of the parent oil since most of the hydrocarbons are still present in dispersed oil droplets. The situation is quite different in the water-soluble fraction (WSF) of oil. Since aromatic hydrocarbons are more soluble in water than alkanes of similar molecular weight (McAuliffe, 1966), they will tend to be enriched in the WSF compared to the alkanes.

The hydrocarbon composition and concentrations in the WSF prepared from the four reference oils have been determined by Dr J. S. Warner (Table 4). The WSF of the crude oils are richer in light aliphatics (ethane to methylcyclohexane) and light aromatics (benzene to xylenes) than are the WSF of the refined oil products. On the other hand, the WSF of the refined oil products contain much higher concentrations of naphthalene and alkyl naphthalenes than the crudes. The WSF of all four oils have very low concentrations of C_{12}–C_{14} n-paraffins, reflecting the low water solubility of these compounds.

The concentrations of total hydrocarbons in the 100 per cent WSF of the four reference oils as determined by IR spectrometry are listed in Table 5.

TABLE 4
CONCENTRATIONS OF SPECIFIC HYDROCARBONS IN THE 100 PER CENT WATER-SOLUBLE FRACTION OF THE FOUR API REFERENCE OILS. CONCENTRATIONS ARE IN mg/liter (ppm) (FROM ANDERSON et al., 1974a, WITH PERMISSION OF SPRINGER–VERLAG)

Compound	Kuwait crude oil	South Louisiana crude oil	No. 2 fuel oil	Bunker C residual oil
Alkanes				
Ethane	0·23	0·54		
Propane	3·30	3·01		
Butane	3·66	2·36	0·39 [a]	0·05 [a]
Isobutane	0·90	1·69		
Pentane	1·31	0·49		
Isopentane	0·98	0·70		
Cyclopentane + 2-methylpentane	0·59	0·38	0·02	0·005
Methylcyclopentane	0·19	0·23	0·019	0·004
Hexane	0·29	0·09	0·014	0·004
Methylcyclohexane	0·08	0·22	0·03	0·002
Heptane	0·09	0·06	0·02	0·004
Total C_{12}–C_{24} n-paraffins	0·004	0·089	0·047	0·012
Aromatics				
Benzene	3·36	6·75	0·55	0·04
Toluene	3·62	4·13	1·04	0·08
Ethylbenzene + m- + p-xylenes	1·58	1·56	0·95	0·09
O-Xylene	0·67	0·40	0·32	0·03
Trimethylbenzenes	0·73	0·76	0·97	0·11
Naphthalene	0·02	0·12	0·84	0·21
1-Methylnaphthalene	0·02	0·06	0·34	0·19
2-Methylnaphthalene	0·008	0·05	0·48	0·20
Dimethylnaphthalenes	0·02	0·06	0·24	0·20
Trimethylnaphthalenes	0·003	0·008	0·03	0·10
Biphenyl	0·001	0·001	0·011	0·001
Methylbiphenyls	0·001	0·001	0·014	0·001
Dimethylbiphenyls	0·001	0·001	0·003	0·001
Fluorene	0·001	0·001	0·009	0·005
Methylfluorenes	0·001	0·001	0·009	0·004
Dimethylfluorenes	0·001	0·001	0·002	0·002
Dibenzothiophene	0·001	0·001	0·004	0·001
Phenanthrene	0·001	0·001	0·010	0·009
Methylphenanthrenes	0·001	0·002	0·007	0·011
Dimethylphenanthrenes	0·001	0·001	0·003	0·003
Total saturates	11·62	9·86	0·56	0·081
Total aromatics	10·04	13·91	5·83	1·288
Total dissolved hydrocarbons measured	21·66	23·77	6·39	1·369

[a] Unresolved GC peaks, probably includes some olefins.

TABLE 5
CONCENTRATION OF TOTAL DISSOLVED HYDROCARBONS IN THE 100 PER CENT WSF OF THE FOUR API REFERENCE OILS AS DETERMINED BY IR SPECTROMETRIC METHOD NO. 733-48 (AMERICAN PETROLEUM INSTITUTE, 1958) (FROM ANDERSON et al., 1974a, WITH PERMISSION OF SPRINGER–VERLAG)

API reference oil number	Type of oil	Total dissolved hydrocarbons (ppm)
I	Kuwait crude	10·4
II	South Louisiana crude	19·8
III	No. 2 fuel oil	8·7
IV	Bunker C residual oil	6·3

Hydrocarbon concentrations vary between 6·3 ppm in the WSF of bunker C residual oil to 19·8 ppm in the WSF of south Louisiana crude oil. The values obtained by IR are lower for the two crudes and higher for the two distillate oil products than values obtained by GC (Table 3). Both methods rank the oils in the same order with respect to hydrocarbon concentration in the WSF.

When oil is equilibrated with water, as is done in the preparation of the WSF, the aqueous phase never becomes saturated with hydrocarbons. This is due to the fact that oil–water partition coefficients of hydrocarbons favor their retention in the oil phase. Thus, although naphthalene has a solubility of about 20 ppm in seawater (Rossi and Neff, 1978) and No. 2 fuel oil contains 4000 ppm naphthalene, the WSF prepared from No. 2 fuel oil contains only 0·84 ppm naphthalene. Because the oil–water partition coefficients of hydrocarbons increase with increasing molecular weight, one can predict that the concentration of a five-ring PAH such as the carcinogen benzo[a]pyrene in a WSF would be less than 0·001 μg/liter (ppb).

Loss of naphthalenes from bioassay containers of the WSF of No. 2 fuel oil follows first-order kinetics described by the equation:

$$\log_{10} Y_t = \log_{10} Y_0 - k't$$

where Y_t = concentration of total naphthalenes at time t
  Y_0 = initial concentration of total naphthalenes
  t = time
  $-k'$ = 0·4343k, negative here because loss from solution occurs, k is a constant.

Salinity has little effect on the rate of loss, but temperature has a marked effect (Laughlin et al., 1979). Half-times for loss of naphthalenes from the 15 per cent WSF vary from 3–5 days at 20 °C to 0·2–0·4 days at 30 °C.

It should be recognized that the WSF is a purely artificial mixture and is not meant to precisely simulate the conditions of hydrocarbon composition and concentration in the water column under oil spilled in the marine environment. Equilibration conditions in nature are quite different from those used to produce the WSF. The WSF represents a compromise. It is a means of generating a highly reproducible oil-in-water mixture and therefore is extremely useful for comparing the relative toxicity of different crude and refined petroleums to marine organisms.

2.3.2 Comparative Toxicity of the Four Oils

The results of bioassays with two polychaete worms, *Neanthes arenaceodentata* and *Capitella capitata*, and the WSF of the four test oils are summarized in Table 6. Hydrocarbon concentrations shown are extrapolated from analyses of total hydrocarbons in each 100 per cent WSF. For both species and all four oils, no control group mortality was observed. The No. 2 fuel oil WSF was most toxic to *N. arenaceodentata*, with a 48-h LC_{50} value of 3·2 ppm. The bunker C WSF was only slightly less toxic (48-h LC_{50} = 4·6 ppm). In bioassays with *C. capitata* the reverse was seen as bunker C was significantly more toxic than No. 2 fuel oil (48-h LC_{50} = 1·1 ppm and 3·5 ppm, respectively). Since the bunker C WSF is known to contain higher concentrations of phenols than the other WSF (J. S. Warner, personal communication), *C. capitata* may well be especially sensitive to this class of compound.

For both species, the south Louisiana crude oil WSF was substantially less toxic than either No. 2 fuel oil or bunker C residual oil. However, this WSF was more toxic than anticipated from earlier studies (Anderson et al., 1974a). In earlier studies, the WSF were aerated due to the high oxygen requirements of test organisms. Since polychaetes generally require less oxygen (Mangum, 1970) and there was a high surface area to volume ratio in the test containers, no aeration was supplied during these bioassays. Oxygen analyses of the exposure media with a YSI Model 51A oxygen meter showed that at all times during the bioassays dissolved oxygen concentrations were at or near saturation. It is possible that this lack of aeration produced the unexpected higher mortality by causing prolonged retention in solution of toxic benzenes and naphthalenes. To test this hypothesis, a bioassay was conducted using a 100 per cent WSF that had been lightly aerated for 4 h. Toxicity of the WSF was markedly reduced by

TABLE 6

ACUTE TOXICITY OF THE WSF OF FOUR PETROLEUM PRODUCTS TO THE MARINE POLYCHAETE WORMS *Neanthes arenaceodentata* AND *Capitella capitata*. VALUES ARE EXPRESSED AS LC$_{50}$ VALUES (CONCENTRATION IN ppm TO CAUSE 50 PER CENT MORTALITY IN THE SPECIFIED TIME). THE 95 PER CENT CONFIDENCE INTERVALS (95 PER CENT C.I.) AND SLOPE FUNCTIONS (S.F.) WERE COMPUTED FOR EACH LC$_{50}$ VALUE BY THE METHOD OF LITCHFIELD AND WILCOXON (1949) (FROM ROSSI et al., 1976)

Oil type	Statistical value	Neanthes arenaceodentata 24	48	96	Capitella capitata 24	48	96
No. 2 fuel oil WSF	LC$_{50}$	>8.7	3.2	2.7	>8.7	3.5	2.3
	95 per cent C.I.	—	1.8–5.7	1.4–5.4	—	2.1–5.9	1.1–4.8
	S.F.	—	1.8	2.0	—	0.8	0.7
Bunker C WSF	LC$_{50}$	>6.3	4.6	3.6	>6.3	1.1	0.9
	95 per cent C.I.	—	2.7–8.0	2.5–5.3	—	2.5–4.8	0.3–2.4
	S.F.	—	1.8	1.5	—	0.55	0.6
South Louisiana crude WSF	LC$_{50}$	18.0	13.9	12.5	>19.8	16.2	12.0
	95 per cent C.I.	15.1–21.4	11.9–16.1	10.7–14.6	—	9.6–27.4	8.1–17.7
	S.F.	1.2	1.2	1.17	—	0.9	0.8
Kuwait crude WSF	LC$_{50}$	>10.4	>10.4	>10.4	>10.4	>10.4	9.8
	95 per cent C.I.	—	—	—	—	—	—
	S.F.	—	—	—	—	—	0.9

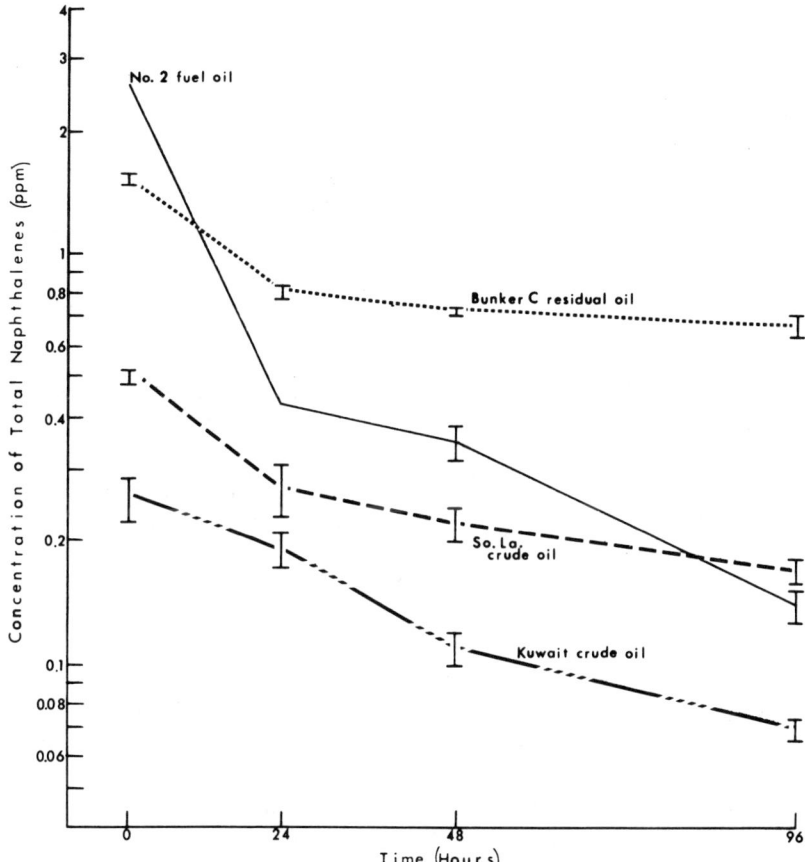

FIG. 2. Rate of decrease in the concentration of total naphthalenes in the WSF of the four API reference oils during static bioassays. Mean values for four samples are given with the corresponding standard deviations represented by vertical lines. The WSF was not aerated (from Rossi et al., 1976).

aeration. It is suggested that this brief period of aeration removed a major portion of the more toxic, volatile components of the WSF.

The Kuwait crude oil WSF was the least toxic of the tested solutions (Table 6). At full strength (10·4 ppm hydrocarbons) the WSF failed to produce 50 per cent or greater mortality in all cases but one. The 96 h LC_{50} for *C. capitata* was 9·8 ppm (94 per cent WSF).

Figure 2 shows initial concentration of total naphthalenes (naphthalene, methylnaphthalenes, and dimethylnaphthalenes) in the four 100 per cent

WSF and rate of decrease during the course of the 96-h bioassays. Initial total naphthalenes concentrations decreased in this order: No. 2 fuel oil, bunker C residual oil, south Louisiana crude oil, and Kuwait crude oil. This is the same order as the relative toxicity of the WSF of the four oils. It has been suggested previously that naphthalenes may be responsible for a major portion of the toxicity produced by WSF of these oils (Anderson et al., 1974a).

It appears that high initial concentrations of these aromatic compounds in the No. 2 fuel oil WSF contribute to its being most toxic. In addition to naphthalenes, the bunker C WSF contains high concentrations of phenols which may approach the toxicity of naphthalenes to some species. The toxicity of bunker C may therefore be the result of the combined action of naphthalenes and phenols. Low toxicity of the two crude oils may be attributable to lower levels of naphthalenes in their WSF. All four WSF showed a substantial drop in concentration of naphthalenes during the course of the 96-h bioassay. This decrease varied from about 90 per cent for the No. 2 fuel oil WSF to about 50 per cent for the two crude oil WSF.

2.3.3 Comparative Toxicity of the WSF and OWD

The comparative acute toxicity of the WSF and OWD of three oils to three species of crustaceans and three species of fish is summarized in Table 7. When LC_{50} is measured in terms of initial concentration of total hydrocarbons in the exposure water, the WSF is more toxic than the OWD in all cases. This is undoubtedly due to the fact that the WSF is enriched in toxic aromatic hydrocarbons in comparison to the oil, whereas the hydrocarbon composition of the OWD, at least initially, resembles that of the oil from which it was prepared. In addition, a significant portion of the hydrocarbons in the OWD is present as micro oil droplets. These hydrocarbons are probably less bioavailable than hydrocarbons in solution and therefore contribute less than the latter to the toxicity of the oil.

Following a spill of nearly 400 000 gal of bunker C residual oil and a No. 2 fuel oil into the lower Houston Ship Channel on 9 March 1973, large mortalities of the supralittoral isopod *Ligia exotica* (Roux) were observed in regions where the oil came ashore (Dillon et al., 1978). Little damage to other marine species in the area was noted. In an effort to determine whether this ecologically important species was particularly sensitive to oil, bioassays were conducted with the WSF and OWD of No. 2 fuel oil using animals collected from a nearby unimpacted area. Animals were exposed individually in 125 ml Erlenmeyer flasks containing 7 ml of exposure medium in 30‰ S artificial seawater. The volume of exposure medium used

TABLE 7

COMPARATIVE ACUTE TOXICITY, MEASURED AS 48-h LC_{50}, OF THE WATER-SOLUBLE FRACTIONS (WSF) AND OIL-IN-WATER DISPERSIONS (OWD) OF THREE OILS TO SEVERAL SPECIES OF MARINE ANIMALS. ALL VALUES ARE EXPRESSED AS CONCENTRATION IN mg/liter (ppm) OF TOTAL HYDROCARBONS IN THE WATER AT THE BEGINNING OF THE BIOASSAY (FROM ANDERSON et al., 1974a, WITH PERMISSION OF SPRINGER–VERLAG)

Species	South Louisiana crude oil WSF	OWD	Kuwait crude oil WSF	OWD	No. 2 fuel oil WSF	OWD
Crustacea						
Mysidopsis almyra	8·7	25	6·6	18	0·9	1·3
Palaemonetes pugio	>19·8	62	>10·2	36	4·1	3·4
Penaeus aztecus	>19·8	>58	—	—	5·0	9·4
Fish						
Cyprinodon variegatus	>19·8	78	—	>42	>6·9	43
Menidia beryllina	8·7	69	6·6	38	5·2	40
Fundulus similis	16·8	71	>10·4	38	4·7	23

was sufficient to allow continuous contact while not inducing forced swimming by this semiterrestrial invertebrate.

Bioassay results indicated that *L. exotica* was very tolerant to No. 2 fuel oil prepared as either a WSF or OWD (Table 8). The 100 per cent WSF (8·7 ppm total hydrocarbons, 2·5 ppm total naphthalenes) failed to produce 50 per cent mortalities in 96 h. LC_{50} values for the OWD were 73 ppm oil added (=5·4 ppm total naphthalenes) at 24 and 48 h and 36·5 ppm oil added (=2·0 ppm total naphthalenes) at 96 h.

TABLE 8

ACUTE TOXICITY TO *Ligia exotica* OF NO. 2 FUEL OIL AS EITHER A WATER-SOLUBLE FRACTION (WSF) OR AN OIL IN WATER DISPERSION (OWD). LC_{50} = CONCENTRATION PRODUCING 50 PER CENT MORTALITY AT THE TIME SPECIFIED (FROM DILLON et al., 1978, WITH PERMISSION OF SPRINGER–VERLAG)

	Median tolerance limit (LC_{50}), in hours		
	24 h	48 h	96 h
WSF	>100 per cent (8·7 ppm)	>100 per cent	>100 per cent
OWD	73·0 ppm	73·0 ppm	36·5 ppm

Ligia exotica samples were also collected from the spill site and were found to be heavily contaminated with petroleum hydrocarbons. A pooled sample of several animals (approximately 10 g wet weight) contained approximately 1600 ppm total hydrocarbons. Especially noteworthy were the relatively high concentrations of sulfur-containing aromatics (mainly dibenzothiophenes) found in the tissues. Concentrations of the various dibenzothiophenes determined by GC coupled to a sulfur-specific flame photometric detector were 10 ppm dibenzothiophene, 50 ppm methyldibenzothiophenes, 60 ppm C_2-dibenzothiophenes, 25 ppm C_3-dibenzothiophenes, and 15 ppm C_4-dibenzothiophenes.

The high mortality of *L. exotica* at the spill site may have been due to bunker C oil. This is supported by GC data. The gas chromatogram of Fraction 3 of a bunker C residual oil (API reference oil No. IV) closely resembled qualitatively the gas chromatogram of Fraction 3 of *L. exotica* tissue sample (Warner, 1975), the main difference between the two being that the *L. exotica* sample was greatly enriched in dibenzothiophenes, suggesting selective accumulation of these compounds. Practically nothing is known about uptake kinetics and acute toxicity of dibenzothiophenes in marine animals.

2.3.4 Comparative Toxicity of Oil to Different Species

Table 9 summarizes data on the toxicity of south Louisiana crude oil WSF to 10 species of marine animals representing three phyla. LC_{50} values are based on concentrations of total hydrocarbons in the exposure media as measured by IR analysis. Polychaete worms tested were relatively tolerant. The oceanic pelagic species *Platynereis dumerili* (Andouin and Milne Edwards) was somewhat more sensitive than two benthic species, *Neanthes arenaceodentata* and *Capitella capitata*, which showed similar sensitivities to oil. Crustaceans studied showed substantial differences in sensitivity to the south Louisiana crude oil WSF. The opossum shrimp *Mysidopsis almyra* (Bowman), an estuarine species, was more sensitive than the oceanic prawn *Leander tenuicornis* (Say). Two other estuarine shrimp, *Palaemonetes pugio* (Holthuis) and *Penaeus aztecus* (Ives), were extremely tolerant.

The three fish species tested are common sympatric estuarine species from the Texas Gulf coast, yet the silverside minnow *Menidia beryllina* (Cope) was substantially more sensitive than either the Gulf killifish *Fundulus similis* (Baird and Girard) or the sheepshead minnow *Cyprinodon variegatus* (Lacépède). Among the 10 species tested, there was no relationship between phylogenetic position or habitat and relative

TABLE 9
CONCENTRATION OF THE WSF OF SOUTH LOUISIANA CRUDE OIL (ppm TOTAL HYDROCARBONS AND WSF IN PARENTHESES) LETHAL TO 50 PER CENT OF THE TEST ANIMALS IN 48 OR 96 h (FROM NEFF et al., 1976a, WITH PERMISSION OF THE AMERICAN INSTITUTE OF BIOLOGICAL SCIENCES)

Species	48-h LC_{50}	96-h LC_{50}
Polychaeta		
Sargassum worm, *Platynereis dumerilii*	12·3(62)	9·5(48)
Errant benthic worm, *Neanthes arenaceodentata*	13·9(70)	12·5(63)
Sedentary benthic worm, *Capitella capitata*	16·2(82)	12·0(61)
Crustacea		
Opossum shrimp, *Mysidopsis almyra*	8·7(44)	—
Prawn, *Leander tenuicornis*	10·2(52)	6·0(30)
Grass shrimp, *Palaemonetes pugio*	>16·8(85)	>16·8(85)
Brown shrimp postlarvae, *Penaeus aztecus*	>19·8(100)	>19·8(100)
Teleosti		
Silverside minnow, *Menidia beryllina*	8·7(44)	5·5(28)
Gulf killifish, *Fundulus similis*	16·8(85)	16·8(85)
Sheepshead minnow, *Cyprinodon variegatus*	>19·8(100)	>19·8(100)

sensitivity to oil. These data and those of Hyland and Schneider (1976) indicate that we cannot predict *a priori* the relative sensitivity of an animal to oil based on its phylogenetic position or its habitat.

2.3.5 Comparative Toxicity of Oil to Marine Larvae

To test the widely stated hypothesis that larvae of marine animals are more sensitive to oil than are juveniles and adults (Hyland and Schneider, 1976), bioassays were conducted with several life stages of four species of marine crustaceans and the polychaete worm *Neanthes arenaceodentata*. The comparative crustacean bioassays are summarized in Table 10. Postlarvae of the commercially important brown shrimp *Penaeus aztecus* were significantly more tolerant to the WSF of No. 2 fuel oil than were either the early or late juvenile stages. However, postlarvae and juveniles of the closely related white shrimp *Penaeus setiferus* (Linn.) did not show differential sensitivity and both stages were more sensitive than *P. aztecus*. The estuarine opossum shrimp *Mysidopsis almyra* behaved similarly to *P. aztecus*; 1-day- and 7-day-old postlarvae were significantly more tolerant to the WSF than were adults. Adult *M. almyra* were the most sensitive animals tested in the laboratory to date with a 96-h LC_{50} to No. 2 fuel oil WSF of 0·65 ppm total hydrocarbons. The grass shrimp *Palaemonetes*

TABLE 10
COMPARATIVE TOXICITY OF THE WSF OF NO. 2 FUEL OIL TO DIFFERENT LIFE STAGES OF FOUR MARINE CRUSTACEANS (CONCENTRATIONS IN ppm IN TOTAL HYDROCARBONS AND IN PER CENT WSF IN PARENTHESES) (FROM NEFF et al., 1976a, WITH PERMISSION OF THE AMERICAN INSTITUTE OF BIOLOGICAL SCIENCES)

Species	Description	96-h LC_{50} (ppm)	95 per cent C.I.
Penaeus aztecus (brown shrimp)	Postlarvae (2·23 mg dry weight)	6·6 (76)	6·1–6·9
	Early juveniles (29·2 mg dry weight)	3·7 (43)	3·0–5·1
	Late juveniles (225·8 mg dry weight)	2·9 (33)	2·2–3·8
Penaeus setiferus (white shrimp)	Postlarvae (0·33 mg dry weight)	1·4 (16)	0·9–2·1
	Juveniles (77·8 mg dry weight)	1·0 (11)	0·8–1·2
Palaemonetes pugio (grass shrimp)	Larvae (<3-week-old)	1·2 (14)	1·0–1·5
	Postlarvae (4–5-week-old)	2·4 (28)	2·1–2·8
	Adults (>5-week-old)	3·5 (40)	2·4–4·9
Mysidopsis almyra (opossum shrimp)	Postlarvae (1-day-old)	1·75 (20)	—
	Postlarvae (7-day-old)	1·75 (20)	—
	Adults (20-day-old)	0·65 (7)	—

pugio showed an opposite trend. Larvae were significantly more sensitive to the WSF of No. 2 fuel oil than were either the postlarvae or adults.

Bioassays were performed with the WSF of No. 2 fuel oil and south Louisiana crude oil on the following life stages of *Neanthes arenaceodentata*: 4-, 18-, 32- and 40-segment juveniles, as well as 60-segment mature adult worms (Rossi and Anderson, 1976). Juvenile stages were obtained from mature worm couples (one gravid female and one male), previously isolated from the laboratory population. For bioassays with juveniles, 10 animals were placed in uncovered 100 ml culture dishes containing 50 ml WSF. Adults were held individually in unstoppered 125 ml Erlenmeyer flasks containing 50 ml of exposure medium. Each bioassay was repeated four times so that $n \geq 40$ for each WSF concentration tested.

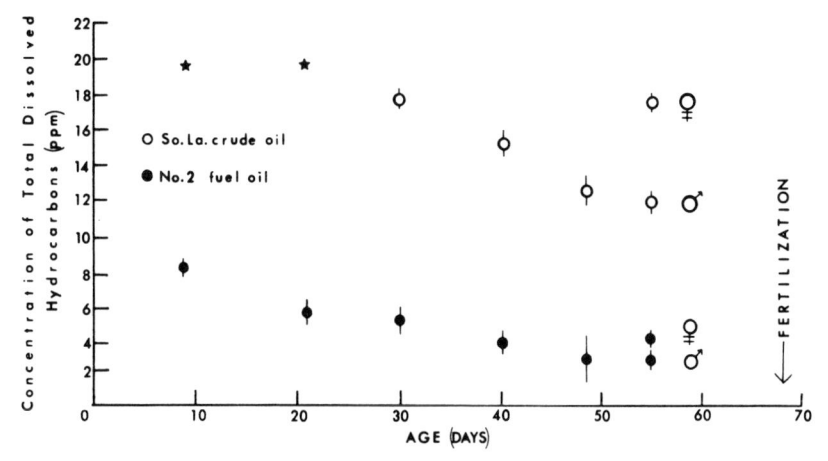

FIG. 3. Median lethal concentrations at 96 h (96-h LC$_{50}$) of the WSF of south Louisiana crude oil (open circles) and No. 2 fuel oil (dark circles) to six life stages of the marine polychaete worm *Neanthes arenaceodentata*. Vertical lines represent 95 per cent confidence intervals. LC$_{50}$ values are given in ppm total aqueous hydrocarbons as determined by IR. Mature male and female worms (55-day-old animals) had different sensitivities to the WSF. ★, LC$_{50}$ > 19·8 ppm (from Rossi and Anderson, 1976, with permission of Springer–Verlag).

Results of bioassays with seven life stages of *N. arenaceodentata* are summarized in Fig. 3. As expected, the WSF of south Louisiana crude oil was significantly less toxic than that of No. 2 fuel oil to all life stages of the worms. However, toxicity of the WSF of both oils increased as the worms grew older and increased in size. Differences in sensitivity to the two oils were greatest between the youngest juvenile stage tested (9-day-old animals) and adults (54-day-old), except in the case of gravid females whose sensitivity was similar to that of medium-sized (30-day-old) juveniles. Bioassays performed on immature adult worms failed to show a significant difference in sensitivity between male and female animals at this stage of development.

Greater resistance to WSF of the youngest juvenile stages and gravid females may be related to their high content of lipid-rich yolk material. It is possible that the more toxic petroleum hydrocarbons are being sequestered into yolk material, to the extent that their effect on other tissues is reduced. Holden (1962) and Fabacher and Chambers (1971) have proposed sequestration of chlorinated hydrocarbons into lipid-containing tissues or organs as the mechanism for the development of insecticide resistance in brown trout and mosquito fish. As a logical sequel to these experiments,

TABLE 11

RESULTS OF SENSITIVITY BIOASSAYS WITH NO. 2 FUEL OIL AND SOUTH LOUISIANA CRUDE OIL (SO. LA.) ON THREE LIFE STAGES FROM EACH OF THREE SUCCESSIVE GENERATIONS OF POLYCHAETES (*Neanthes arenaceodentata*) CHRONICALLY EXPOSED TO SUBLETHAL CONCENTRATIONS OF NO. 2 FUEL OIL WSF. (FROM ROSSI AND ANDERSON, 1978b, WITH PERMISSION OF SPRINGER–VERLAG)

Generation	Life stage	\multicolumn{8}{c}{Chronic exposure concentration (per cent WSF No. 2 fuel oil)}
		\multicolumn{2}{c}{0}	\multicolumn{2}{c}{2.5}	\multicolumn{2}{c}{5}	\multicolumn{2}{c}{10}
		\multicolumn{8}{c}{96-h LC$_{50}$ during exposure to No. 2 fuel oil or south Louisiana crude oil WSF (ppm total hydrocarbons)}
		No. 2	So. La.	No. 2	So. La.	No. 2	So. La.	No. 2	So. La.
F$_1$	Larvae	8.4±0.4 1.0	a	8.3±0.5 1.2	a	8.6±0.3 1	a	8.1±0.2 1.05	a
	Juveniles	5.7±0.4 1.2	a	4.9±0.6 1.2	16.1±0.3 1.1	4.5±0.8 1.3	12.5±1.4 1.3	5.8±0.2 1	15.4±0.5 1
	Adults	2.7±1.1 2.0	12.5±1 1.2	7.1±0.2 1.1	a	7.2±0.3 1	a	6.9±0.5 1.1	a
F$_2$	Larvae			8.0±0.6 1.2	a	8.6±0.2 1.1	a	7.8±0.5 1.1	a
	Juveniles			5.6±0.3 1.1	18.2±0.7 1.05	6.0±0.2 1	17.6±1.1 1.1	6.0±0.2 1.05	16.5±2.0 1.2
	Adults			7.2±0.3 1.1	a	6.7±0.6 1.1	a	7.6±0.5 1.1	a
F$_3$	Larvae			6.5±1.0 1.7	a	8.7±0.8 1.2	a	8.7±0.3 1.2	a
	Juveniles			1.7±0.3 1.1	16.8±1.7 1.1	5.8±0.4 1.1	17.8±1.0 1.1	6.3±0.4 1.1	16.9±0.4 1.2
	Adults			6.7±0.5 1.3	a	7.0±0.5 1.2	a	5.8±1.2 1.7	a

96-h LC$_{50}$ values ±95 per cent confidence intervals and corresponding slope functions were calculated according to Litchfield and Wilcoxon (1949). Values represent mean results of at least two bioassays, with 10 animals used per each of six WSF concentrations (including a control group in each).
a = No mortality in a 100 per cent WSF after 96 h.
100 per cent WSF No. 2 fuel oil and south Louisiana crude oil = 8.7 ppm and 19.8 ppm total dissolved hydrocarbons, respectively.
For 0 per cent chronic exposure concentration (control) results, F$_2$ and F$_3$ sensitivities were identical to those stated (F$_1$).

additional investigations were conducted to determine if chronic exposure to sublethal concentrations of the WSF would result in development in subsequent generations of *N. arenaceodentata* of an increased hydrocarbon resistance.

Groups of *N. arenaceodentata* were exposed continuously to selected concentrations of No. 2 fuel oil WSF for three generations (Rossi and Anderson, 1978*b*). These studies were initiated by exposing male/female pairs (parental generation) of immature adult polychaetes to WSF during brood tube formation, oogenesis (the formation of eggs), and subsequent fertilization of spawned eggs. Three WSF concentrations were used: 2·5, 5, and 10 per cent (0·21, 0·44, and 0·87 ppm total hydrocarbons), and compared to control responses. Twenty couples were used at each concentration. Pairs were held individually in 300 ml culture dishes containing 150 ml unaerated WSF/dish; WSF were replaced daily. The offspring (F_1 generation) were interbred and exposed according to the same regime. This was repeated for the subsequent two generations (F_2 and F_3 generations). Larval, juvenile and immature adult worms were harvested from each generation and subjected to bioassays with the WSF of No. 2 oil and south Louisiana crude oil. To test the lability of hydrocarbon resistance, selected groups of immature adult worms were removed from exposure either 7 or 14 days prior to testing.

Resistance of each generation of oil-exposed worms to No. 2 fuel oil and south Louisiana crude oil WSF was compared to that of unexposed (= susceptible = control) animals (Table 11). Larvae (3-segment worms or metatrochophores) from all chronic exposure concentrations in each generation were quite similar in sensitivity to the WSF of both test oils. Therefore, induction of hydrocarbon resistance was not evident at this early stage (9–10-day-old) of development.

Unlike larvae, exposed adults were considerably more resistant (approximately twice as much) to the two oils than were unexposed adults. Increased resistance in adults was not related to chronic exposure concentration, nor was there a significant increase in resistance in successive generations over that attained in the F_1 generation. For all chronic exposure concentrations, F_1 adults were just as resistant as were F_3 adults.

Unexpectedly, exposed juveniles were less resistant than their unexposed counterparts. Juveniles within each generation exhibited similar sensitivities in that LC_{50} values were not significantly related to exposure concentration. There was, however, a distinct trend toward increasing resistance in later generations. F_3 juveniles were more resistant than F_1 juveniles in bioassays with No. 2 fuel oil. Among unexposed control

TABLE 12

RESULTS OF BIOASSAYS ON GROUPS OF ADULT *Neanthes arenaceodentata* REMOVED FROM CHRONIC EXPOSURE TO 5 PER CENT WSF OF NO. 2 FUEL OIL EITHER 7 OR 14 DAYS PRIOR TO RESISTANCE TESTING. 96-h LC$_{50}$ VALUES AND THEIR RESPECTIVE ±95 PER CENT CONFIDENCE INTERVALS (C.I.) AND SLOPE FUNCTIONS (S.F.) ARE GIVEN. LC$_{50}$ VALUES EXPRESSED IN ppm TOTAL DISSOLVED HYDROCARBONS INITIALLY PRESENT IN THE CHALLENGE OIL (NO. 2 FUEL OIL). MEAN VALUES FROM TWO BIOASSAYS ARE GIVEN, WITH 10 ANIMALS FOR EACH OF SIX CONCENTRATIONS (INCLUDING A CONTROL GROUP) USED PER BIOASSAY (FROM ROSSI AND ANDERSON, 1978b, WITH PERMISSION OF SPRINGER–VERLAG)

Generation	Prior removal time (days)	96-h LC$_{50}$ ppm (per cent WSF)	C.I.	S.F.
Control		2·7(31)	3·8–1·9	2·0
	0	7·2(83)	7·5–6·9	1
F$_1$	7	7·1(82)	7·5–6·7	1·2
	14	4·9(56)	5·4–4·4	1·5
	0	6·7(77)	7·3–6·1	1·1
F$_2$	7	7·1(82)	7·7–6·5	1·3
	14	6·4(74)	6·9–5·9	1·3
	0	7·0(80)	7·5–6·5	1·2
F$_3$	7	7·0(80)	7·5–6·5	1·2
	14	7·2(83)	7·6–6·8	1·1

N. arenaceodentata, the order of sensitivity to both oils proceeds as adults > juveniles > larvae, whereas for chronically exposed worms one observes the sensitivity order of juveniles > adults > larvae.

Results of bioassays performed with groups of adult *N. arenaceodentata* removed from chronic exposure prior to bioassays are summarized in Table 12. Worms removed from exposure 7 days before testing were just as resistant as constantly exposed animals for all three generations. Only first generation adults removed 14 days prior to testing were significantly less resistant than constantly exposed adults. However, F$_1$ worms removed 14 days before testing were still more resistant than unexposed animals. These data are evidence of the rapidity of onset and stability over time of induced petroleum hydrocarbon resistance.

2.3.6 Comparative Toxicity of Petroleum-derived Aromatic Hydrocarbons

Bioassays were performed with immature, young adult polychaetes (*Neanthes arenaceodentata*) as described above (Rossi and Neff, 1978). Ten polycyclic aromatic hydrocarbons (PAH) were tested: naphthalene; 2,6-dimethylnaphthalene; 2,3,6-trimethylnaphthalene; phenanthrene; 1-methylphenanthrene; fluorene; fluoranthene; chrysene; benzo[a]pyrene;

TABLE 13

RELATIVE SOLUBILITIES OF SELECTED POLYCYCLIC AROMATIC HYDROCARBONS IN 32‰ S SEAWATER AT 22 °C. EQUILIBRATION, 24 h MEAN VALUES ± S.D. FOR THREE DETERMINATIONS ARE LISTED (FROM ROSSI AND NEFF, 1978, WITH PERMISSION OF PERGAMON PRESS)

PAH	Molecular weight	Relative solubility, ppm (μg/g seawater)
Naphthalene	138	20 ± 2
Dimethylnaphthalene	156	2·4 ± 0·5
Trimethylnaphthalene	170	1·7 ± 0·6
Fluorene	166	0·8 ± 0·2
Phenanthrene	178	0·6 ± 0·1
Methylphenanthrene	192	0·3 ± 0·1
Fluoranthene	202	0·1 ± 0·06
Chrysene	228	<0·005, >0·001
Benzo[a]pyrene	252	<0·010, >0·005
Dibenzanthracene	278	<0·010, >0·005

and dibenz[ah]anthracene. Replicate flasks without animals were set up to monitor decreases in PAH concentration during the test period. In addition, experiments were performed to determine relative solubility of the PAH in seawater. An excess of the PAH was mixed with 0·45 μ millipore-filtered seawater for 24 h. The aqueous phase was filtered three times through a 1 × 30 cm column filled with lightly packed glass wool to remove suspended PAH particulates. PAH were extracted from aqueous solution with n-hexane and their concentrations determined spectrophotometrically using the method of Neff and Anderson (1975) by reference to UV spectra of authentic standards in hexane (Friedel and Orchin, 1951; Schwarz and Wasik, 1976).

Relative solubilities in seawater of 10 PAH are listed in Table 13. Solubility was closely related to molecular weight (i.e., solubility varied inversely with molecular weight) and molar volume. The lighter diaromatic hydrocarbons naphthalene, dimethylnaphthalene, and trimethylnaphthalene, exhibited seawater solubilities in the low ppm range. The triaromatic compounds fluorene, phenanthrene, and methylphenanthrene had solubilities approximately one-tenth of those of the diaromatics. High molecular weight hydrocarbons such as chrysene, benzo[a]pyrene, and dibenzanthracene were minimally soluble in seawater, exhibiting variable solubilities in the low ppb range. The threshold of detection for UV spectrophotometric procedures employed here was 1 ppb. It seems likely that the latter three PAH require more than 24 h to equilibrate with seawater. This could account for variation observed in their relative solubilities. Published solubilities in distilled water of these three PAH are in the range of 0·2–6 µg/liter (Neff, 1979).

Little published information is available concerning solubility in seawater of PAH. Schwarz and Wasik (1976) reported maximum solubilities in freshwater at 25 °C for naphthalene, fluoranthene, and benzo[a]pyrene of 22, 0·236, and 0·004 ppm. These values are similar to those reported above. Differences are accounted for by the slightly lower solubility of aromatic hydrocarbons in seawater than in freshwater (Eganhouse and Calder, 1976; May et al., 1978).

McAuliffe (1966) showed that for homologous series of monocyclic aromatic hydrocarbons, the logarithm of solubility in water is a linear function of hydrocarbon molar volume. The relationship between solubility and molar volume becomes more complex for PAH. For example, isopropylbenzene, with a molar volume of 139·5 ml/mol at 20 °C, has an aqueous solubility of 50 ppm (McAuliffe, 1966), while naphthalene, with a molar volume of 133·2 ml/mol at 25 °C (Bohon and Claussen, 1951), has an aqueous solubility of only 22 ppm. This value is somewhat lower than would be predicted from a strict linear relationship between log solubility and molar volume. As shown by McAuliffe (1966) each homologous series of hydrocarbons has a slightly different solubility/molar volume relationship.

Results of bioassays with *Neanthes arenaceodentata* and specific PAH are summarized in Table 14. Only 96-h LC_{50} values are given since mortalities greater than 50 per cent were not observed after 24- and 48-h exposure. PAH toxicity was also closely related to molecular weight, i.e., a trend towards increasing toxicity with increasing molecular weight was observed in the PAH series naphthalene through fluoranthene. 1-Methylphenanthrene was the most toxic PAH tested; its toxicity was approximately an order of magnitude greater than those of the naphthalenes. PAH with previously established very low seawater solubilities proved least toxic, to the extent that 50 per cent mortality was not observed in solutions containing chrysene, benzo[a]pyrene, or dibenzanthracene at nominal concentration up to 1 ppm.

Concentrations of representative di-, tri-, tetra-, and pentacyclic aromatic hydrocarbons in the exposure media decreased substantially during the 96-h bioassays. Observed decreases were presumably due to volatilization of naphthalene and/or photo-oxidation of hydrocarbons in test solutions (Acheson et al., 1976). Residence times of PAH in test

TABLE 14

TOXICITY OF SELECTED POLYCYCLIC AROMATIC HYDROCARBONS (PAH) TO *Neanthes arenaceodentata*. 96-h LC_{50} VALUES, 95 PER CENT CONFIDENCE INTERVAL (C.I.) AND CORRESPONDING SLOPE FUNCTIONS (S.F.) WERE CALCULATED BY THE METHOD OF LITCHFIELD AND WILCOXON (1949). LC_{50} VALUES ARE EXPRESSED AS CONCENTRATION (ppm) OF PAH INITIALLY PRESENT IN TEST SOLUTION AS DETERMINED BY UV SPECTROPHOTOMETRIC ANALYSIS. VALUES FOR CHRYSENE, BENZO[a]PYRENE, AND DIBENZANTHRACENE WERE NOT CALCULATED SINCE THESE HYDROCARBONS WERE NOT LETHAL AT THE HIGHEST CONCENTRATION TESTED (1 ppm)(FROM ROSSI AND NEFF, 1978, WITH PERMISSION OF PERGAMON PRESS)

PAH	96-h LC_{50}	95 per cent C.I.	S.F.
Naphthalene	3·8	4·1–3·5	1·5
Dimethylnaphthalene	2·6	2·9–2·3	1·4
Trimethylnaphthalene	2·0	2·4–1·6	1·2
Fluorene	1·0	1·3–0·7	1·4
Phenanthrene	0·6	0·8–0·4	1·7
Methylphenanthrene	0·3	0·4–0·2	1·1
Fluoranthene	0·5	0·7–0·3	1·2
Chrysene	—	—	—
Benzo[a]pyrene	—	—	—
Dibenzanthracene	—	—	—

solutions appeared to be directly related to molecular weight, i.e., decreases in naphthalene concentrations were relatively greater than decreases in higher molecular weight hydrocarbons, fluoranthene and benzo[a]pyrene.

Virtually no other published information is available concerning toxicity to marine animals of non-chlorinated aromatic hydrocarbons with molecular weights exceeding those of naphthalene and its alkyl-derivatives (Neff, 1979). Diaromatic hydrocarbons are toxic to shrimp *Penaeus aztecus* and *Palaemonetes pugio* at concentrations of approximately 1 ppm, which corresponds well with findings presented here (Anderson et al., 1974b). These two shrimp species are also quite similar to *Neanthes arenaceodentata* in sensitivity to whole oil WSF (Rossi et al., 1976). Triaromatic PAH were quite toxic, their toxicity to *N. arenaceodentata* was an order of magnitude greater than that of the naphthalenes. Surprisingly, the higher molecular weight PAH, chrysene, benzo[a]pyrene, and dibenzanthracene, all of which are active mammalian carcinogens (International Agency for Research on Cancer, 1973), exhibited little acute toxicity. This finding may have been related to low relative solubility of these compounds. These higher molecular weight hydrocarbons may not have formed true solutions with seawater, thereby affecting their mobility across biological membranes and hence modifying their apparent toxicity. It seems likely that these hydrocarbons could have formed biologically inactive micelles with dissolved organic matter in test solution, in the manner suggested by Boehm and Quinn (1976).

These data support the hypothesis that alkylation (e.g., methylation) can substantially modify chemical and biological characteristics of aromatic hydrocarbons (Anderson et al., 1974b; Neff, 1979). For example, in addition to being much less soluble than their parent compounds, di- and trimethylnaphthalene, and 1-methylphenanthrene proved significantly more toxic than naphthalene and phenanthrene, respectively.

2.3.7 Effect of Salinity on the Acute Toxicity of Aromatic Hydrocarbons

Acute toxicity bioassays were performed with phenanthrene on grass shrimp *Palaemonetes pugio* and white shrimp *Penaeus setiferus* (Young, 1977). Effects of acclimation to different salinities and of salinity shock on the time required to produce 50 per cent mortality (ET_{50}) for different concentrations of phenanthrene was determined.

Grass shrimp *P. pugio* were acclimated in the laboratory for at least two weeks to salinities of 2, 17, and 32‰. White shrimp *P. setiferus* were acclimated for two weeks in the laboratory to 8, 20, and 32‰ S. Mean size of the grass shrimp was 2 cm and that of white shrimp, 5 cm. Acclimations and bioassays were conducted at room temperature, $23 \pm 1\,°C$. ET_{50} values were computed according to the graphical technique of Litchfield (1949).

The following bioassays were conducted:

1. *Palaemonetes pugio*—acclimated to 17‰ S and exposed to 0·2, 0·3, 0·6, 0·8, and 1·0 ppm phenanthrene
2. *Palaemonetes pugio*—acclimated to 2, 16, and 32‰ S and exposed to 0·5 ppm phenanthrene at the acclimation salinity
3. *Palaemonetes pugio*—same as 2 except that phenanthrene concentration was 2·0 ppm
4. *Palaemonetes pugio*—transferred from 17‰ S to 2, 17, and 32‰ S and immediately exposed to 2·0 ppm phenanthrene (salinity shock)
5. *Penaeus setiferus*—acclimated to 8, 20, and 32‰ S and exposed to 0·5 ppm phenanthrene at the acclimation salinity.

The results of these experiments are summarized in Table 15. ET_{50} values (time in hours to 50 per cent mortality) for *P. pugio* acclimated and exposed at 17‰ S to several concentrations of phenanthrene were low, indicating that the 24-h LC_{50} for *P. pugio* at this salinity was approximately 0·5 ppm.

TABLE 15

EFFECT OF SALINITY ACCLIMATION AND SALINITY SHOCK ON THE TOXICITY OF PHENANTHRENE TO *Palaemonetes pugio* AND *Penaeus setiferus*. ET_{50} VALUES (MEDIAN TIME IN HOURS TO 50 PER CENT MORTALITY), THEIR 95 PER CENT CONFIDENCE INTERVALS C.I. AND SLOPE FUNCTIONS S.F. WERE COMPUTED BY THE METHOD OF LITCHFIELD (1949) (FROM YOUNG, 1977)

	Phenanthrene concentration (ppb)	Salinity (‰)	ET_{50} (hours)	95 per cent C.I.	S.F.
1. *Palaemonetes pugio* (salinity acclimated)	600 800 1 000	17	12·5 6·8 6·8	11·3–13·8 6·2–7·4 6·2–7·4	1·18 1·15 1·15
2. *Palaemonetes pugio* (salinity acclimated)	500	2 17 32	22·0 13·0 12·8	19·8–24·4 11·7–14·4 11·6–14·1	1·17 1·18 1·17
3. *Palaemonetes pugio* (salinity acclimated)	2 000	2 17 32	3·6 4·2 2·8	3·3–3·9 3·8–4·7 2·4–3·2	1·15 1·18 1·27
4. *Palaemonetes pugio* (salinity shock)	2 000	2 17 32	3·5 3·3 2·6	3·0–4·1 2·4–4·5 2·0–3·3	1·28 1·66 1·49
5. *Penaeus setiferus* (salinity shock)	500	8 20 32	19·5 29·0 15·2	14·0–27·1 22·7–37·1 11·6–19·8	1·68 1·48 1·54

Shrimp acclimated and exposed at 17 and 32‰ S to 0·5 ppm phenanthrene had similar ET_{50}s. Those acclimated to 2‰ S were more tolerant.

The experiment was repeated using 2 ppm phenanthrene. At this concentration, shrimp acclimated and tested at 17‰ S were slightly more tolerant than those at 2 and 32‰ S. Salinity shock had little effect on the ET_{50} of *P. pugio* exposed to 2 ppm phenanthrene. *Peneaus setiferus*, acclimated and tested at 8, 20, and 32‰ S and 0·5 ppm phenanthrene were more tolerant than similarly exposed *P. pugio*. Tolerance to phenanthrene was greater at 20‰ S than at either 8 or 32‰ S.

In this and an earlier similar study with naphthalene, salinity stressed shrimp seem to be slightly more sensitive to aromatic hydrocarbons than unstressed individuals. This may be related to salinity-induced changes in integumentary permeability. Phenanthrene had no effect on the rate of water flux across the body wall of *P. pugio* (Young, 1977). However, the rate of phenanthrene accumulation by this species was lower at low and high salinities than at the intermediate salinity (See pages 123–5). Thus, the increased sensitivity of *P. pugio* to phenanthrene at high and low salinities is due to the added stress of non-optimal salinities and not to the effect of salinity on the rate of phenanthrene accumulation by the shrimp.

CHAPTER 3

Reproduction, Development and Growth

3.1 INTRODUCTION

Acute toxicity of petroleum and specific petroleum hydrocarbons to different life stages of marine animals was reviewed in the previous section. While such studies are of value in comparing relative sensitivities to oil of various life stages of an organism, they tend to omit possible subtle sublethal responses which may have a deleterious effect on long-term success of a population. Reproductive processes are characterized by large-scale biochemical, proliferative, and morphogenic events, some of which may be highly sensitive to pollutant stress. Any environmental factor which produces alterations in normal reproductive events and patterns will have a long term effect on population structure and dynamics of the species and may ultimately result in the demise of the population in an impacted area (Thorson, 1966; Mileikovsky, 1970; Vernberg et al., 1978). Due to this consideration, increased emphasis has been placed in recent years upon studies of the effects of chronic low-level pollutant exposure on parameters such as reproductive success, fecundity, embryonic and larval development rate, growth rate of larvae and juveniles, and production of developmental abnormalities (teratogenesis).

Little work has been done concerning effects of chronic low-level oil exposure on reproductive success and fecundity of marine animals. Linden (1976a) reported that chronic exposure to low concentrations (0·3–0·4 ppm measured aqueous hydrocarbons) of a light Venezuelan crude oil decreased brood numbers (fecundity) of female amphipods *Gammarus oceanicus*. The frequency with which the male and female entered the precopulation stage, which is necessary for successful fertilization, was also significantly decreased by oil exposure. Sublethal concentrations of oil and aromatic hydrocarbons also decreased fecundity of estuarine copepods (Berdugo *et al.*, 1977; Ott *et al.*, 1978; Ustach, 1979). When copepods *Eurytemora affinis* were exposed to a WSF of aromatic heating oil (0·52 mg/liter total hydrocarbons), feeding rate was depressed by about 38 per cent (Berdugo *et al.*, 1977). Female copepods exposed to a 3 mg/liter WSF of the oil for 80 min or more and then returned to clean seawater showed a significant reduction in subsequent length of life, total number of eggs produced, mean brood size, and rate of egg production. Chronic exposure of female *E. affinis* to about 10 μg/liter naphthalene or alkyl naphthalenes for the duration of their adult life resulted in significant reductions in length of life, total number of nauplii produced, and mean brood size (Ott *et al.*, 1978). Egg production rate of naphthalene-exposed copepods was about 50 per cent of that of control animals. Brood size of the herpacticoid copepod *Nitocra affinis* was decreased by about 40 per cent by chronic exposure to a dilute WSF of Louisiana crude oil (Ustach, 1979).

Struhsaker (1977) found that when female Pacific herring *Clupea harengus pallasi* were exposed to 800 ppb benzene for 48 h prior to spawning, a significant reduction occurred in survival of ovarian eggs and the resultant embryos and larvae through yolk absorption. Exposure to benzene also resulted in premature spawning of the females. Recently, Loya and Rinkevich (1979) reported that the Red Sea coral *Stylophora pistillata* extruded its planulae (larvae) immediately upon exposure to 0·1–10 ml/liter of the WSF of Iranian crude oil.

Several recent publications deal with the influence of oil and petroleum hydrocarbons on larval development and growth. Lucas and LeRoux (1975) reported a decreased growth rate and increased incidence of developmental abnormalities in larvae of the marine mussel *Mytilus edulis* exposed to four crude oils. Byrne and Calder (1977) studied effects of WSF of three crude and three refined oils on survival and growth of larvae of the hard shell clam *Mercenaria* sp. Forty-eight-hour LC_{50} values for embryos varied from 12 ppm for Kuwait crude oil (API reference oil No. I) to 0·04 ppm for a used crankcase motor oil. Larvae were somewhat less sensitive with 48-h LC_{50} values between >25 ppm Kuwait crude oil and 0·10 ppm used motor oil. Interestingly, one of the crude oils tested, Florida Jay crude from the Jay field near Destin, Florida, was significantly more toxic to both embryos and larvae than were two refined oil products, No. 2 fuel and bunker C residual oil (API reference oils Nos. III and IV, respectively). All the oil WSF reduced growth rate of the larvae during continuous exposure. Concentrations of oil causing a 50 per cent reduction of growth in 10 days as compared to controls (EC_{50}) varied from 4·2 ppm for Kuwait crude oil to 0·22 ppm for Florida Jay crude oil.

Katz (1973) reported that the WSF of a light Venezuelan crude oil retarded molting of larvae of the crab *Neopanope texana*. The first zoeal stage was more sensitive than subsequent larval stages. Linden (1976b) exposed newly-hatched larvae of the amphipod *Gammarus oceanicus* to an aqueous extract of a light Venezuelan crude oil (0·36 ppm mean aqueous hydrocarbon concentration) continuously for 60 days. A significant decrease in the growth rate of exposed larvae as compared to control larvae was reported. First-stage larvae of the American lobster *Homarus americanus* were more sensitive (96-h $LC_{50} = 0.86$ ppm) than third- and fourth-stage larvae (96-h $LC_{50} = 4.9$ ppm) to dispersions of Venezuelan Tia Juana crude oil (Wells and Sprague, 1976). The minimum oil concentration resulting in retardation of development rate was 0·14 ppm. Supernumerary zoeae were observed in several exposure groups but not among the controls. Caldwell *et al.* (1977) observed that chronic exposure to the WSF of Cook Inlet (Alaska) crude oil (0·22 ppm as total dissolved aromatics) or to naphthalene (0·13 ppm) resulted in a decreased molting rate of larvae of the Dungeness crab *Cancer magister*. These concentrations, however, were without effect on the size of surviving fifth-stage zoeae. Larvae of the brachyuran crab *Hyas araneus* were quite tolerant to the WSF of Ekofisk crude oil (Christiansen and Stormer, 1978). Only the highest concentration tested (3·0 ppm) significantly decreased survival to the megalops stage. However, there was a graded increase in the duration of larval development from hatching to the megalops stage with increasing WSF concentration from a mean of 38·8 days among controls to 64·6 days at 3·0 ppm oil.

The WSF of No. 2 fuel oil (API reference oil No. III) at concentrations as low as 0·6 ppm depressed respiration, mobility of sperm, interfered with fertilization and cleavage, and retarded larval development of the sand dollar *Melitta quinquiesperforata* (Nicol *et al.*, 1977). Kuwait crude oil (API reference oil No. I) was much less toxic. Lonning and Hagstrom (1975) reported that dispersions of Kuwait and Ekofisk crude oil alone or in combination with the dispersant Corexit 8666 had little effect on fertilization or early development of embryos of the sea urchins *Psammechinus miliaris* and *Paracentrotus lividus*. However, the oils interfered seriously with later development and differentiation of embryos. Falk-Petersen (1979) reported similar responses to oil in developing embryos of the sea urchins *Strongylocentrotus droebachiensis* and *S. pallidus*. Gasoline and kerosine had a relatively low toxicity, Ekofisk crude oil was intermediate, and gas oil, Ekofisk crude oil residue and No. 6 fuel oil were highly toxic. DeAngelis and Giordano (1974) studied the influence of benzo[a]pyrene and 7,12-dimethylbenz[a]anthracene on early development of the sea urchin *Paracentrotus lividus*. Benzo[a]pyrene had no effect but dimethylbenz[a]anthracene caused significant damage to developing mesenchymal cells of the embryos, interfering with development of the larval skeleton.

Kuhnhold (1974) exposed developing eggs of the cod *Gadus morhua* to seawater extracts of three crude oils—Venezuelan Tia Juana, Iranian Agha Jari, and Libyan Sarir—and made observations on survival, development rate, and incidence of developmental abnormalities in embryos and larvae. Venezuelan crude oil was the most acutely toxic, but Iranian crude produced a significant retardation of development rate at all exposure concentrations, especially during and after gastrulation. A wide variety of developmental abnormalities were observed in embryos exposed to Iranian crude oil extracts. Linden (1975) reported similar developmental abnormalities as well as fin rot disease in larvae of Baltic herring *Clupea harengus membras* exposed to Venezuelan crude oil and crude oil–dispersant (BP1100X and Finasol OSR2) mixtures. Subsequently, Linden (1978) reported that up to 10 ppm crude oil (Russian Tuimaza crude and Venezuelan crude) in water or 5 ppm light fuel oil in water had no effect on fertilization of Baltic herring eggs. However, if embryos were continuously exposed during development to 3·1–8·9 ppm light fuel oil or 3·3–11·9 ppm crude oil, there were significant alterations in embryonic activity, decreases in heart beat rate, and increased incidence of premature or delayed hatching. The majority of larvae that hatched following exposure to oil were malformed and most were shorter than control larvae. Similar responses were reported in Pacific herring *Clupea harengus pallasi* exposed to Prudhoe Bay crude oil (Smith and Cameron, 1979).

Recently, Edwards (1978) reported reduced growth rate of juvenile shrimp *Crangon crangon* during continuous exposure to the WSF of North Sea Brent Field crude oil (0·95–6·3 ppm total hydrocarbons). Growth retardation, which was accompanied by a reduction in respiration rate, was most marked at the highest temperature tested, 20 °C. Feeding rates of juvenile hard clams *Mercenaria mercenaria* were significantly depressed during continuous exposure for six weeks to WSF of Nigerian crude oil containing 0·06–7·0 ppm total hydrocarbons (Keck *et al.*, 1978). Growth rate of unexposed control clams was three times those of the WSF exposure groups.

Over the past several years, the laboratory at Texas A&M University has conducted extensive studies concerning the effects of oil on various reproductive parameters in marine invertebrates and fish. The results of these studies are discussed in the following sections.

3.2 MATERIALS AND METHODS

The four API reference oils described earlier were used in these studies. Most of the research involved south Louisiana crude and No. 2 fuel oils. Water-soluble fractions (WSF) and oil-in-water dispersions (OWD) were prepared by methods described in the previous section. Concentrations of oil in the exposure media were monitored by UV spectrophotometry (Neff and Anderson, 1975) or IR spectrometry (American Petroleum Institute, 1958).

Most of the gravid female specimens of each species or their freshly laid eggs were obtained from the coastal bays of Texas, particularly Galveston Bay and the vicinity of Port Aransas, Texas. Ovigerous mud crabs *Rhithropanopeus harrisii* were collected from Sykes Creek, Brevard County, Florida. Eggs and larvae of the polychaete *Neanthes arenaceodentata* were taken from cultures maintained in the laboratory (cultures originally obtained from Dr D. Reish, California State University, Long Beach). Adult killifish *Fundulus heteroclitus* (Wolbaum) were obtained from a wild population at Beaufort, North Carolina, and were shipped by air freight to Texas. They were maintained in the laboratory in 550-liter recirculating marine aquaria (Living Stream, Frigid Units, Inc., Toledo, Ohio) under conditions shown by Boyd and Simmonds (1974) to favor continuous production of fertile eggs: temperature $20 \pm 2\,°C$; salinity $20\%_{00}$; 16 h light/8 h dark. All test animals were maintained in glass aquaria containing artificial seawater (Instant Ocean, Aquarium Systems, Eastlake, Ohio) at a temperature of $19-22\,°C$.

In most cases, exposure of eggs, embryos, and larvae to oil–water mixtures was performed in glass finger bowls or 125 ml Erlenmeyer flasks containing measured volumes of the oil–water mixtures. Except where noted otherwise, exposure media were changed daily to ensure a continuous exposure of the organisms to oil. Aeration was not deemed necessary because of the small size of the test animals and large surface to volume ratio of the exposure containers. Periodic monitoring of the exposure media revealed oxygen concentrations always in excess of 80 per cent saturation.

Several experiments were performed at Texas A&M University and at the Bermuda Biological Station for Research, St Georges West, Bermuda to determine the effects of petroleum and specific petroleum hydrocarbons on photosynthesis of coral zooxanthellae, and skeletal calcium deposition by reef corals and the sea urchin *Diadema antillarum*.

Five species of reef corals were used in the investigations conducted in

Bermuda. These were four scleractinean corals, the encrusting coral *Madracis decatis* (Lyman), the ivory bush coral *Oculina diffusa* (Lamarck), the common star coral *Montastrea annularis* (Ellis and Solander), and the star coral *Favia fragum* (Esper); and one hydrocoral, the stinging coral *Millepora* sp. Coral specimens were collected by hand using scuba equipment in approximately 6 m of water off the northeast coast of Bermuda. They were returned to the Bermuda Biological Station where they were maintained in glass aquaria with flowing natural seawater at 20 °C and 34‰ S. Coral reef sea urchins *Diadema antillarum* (Phillipii) were also collected from the coral reefs and maintained under similar conditions in the laboratory.

The ivory bush coral *Oculina diffusa* (Lamarck) was used in all investigations performed at Texas A&M University. Coral specimens were collected in approximately 6 m of water from rock jetties at Port Aransas, Texas. They were maintained in the laboratory in large glass aquaria containing artificial seawater (Instant Ocean, Aquarium Systems, Eastlake, Ohio) at a salinity of 28–30‰ and a temperature of 20–23 °C.

Groups of corals were exposed for 72 h to the WSF of south Louisiana crude oil or No. 2 fuel oil (API reference oils Nos. II and III, respectively). Corals were exposed to 0, 10, 20 and 30 per cent dilutions of the WSF in natural seawater in 20 cm glass finger bowls containing 1 liter of exposure medium. Mean initial concentrations of total hydrocarbons, as measured by IR spectrometry (American Petroleum Institute, 1958), in the 10, 20, and 30 per cent WSF of south Louisiana crude oil were 2·0, 4·0 and 5·9 mg/liter, respectively. In the 10, 20, and 30 per cent WSF of No. 2 fuel oil, the initial mean total hydrocarbon concentrations were 0·9, 1·7, and 2·6 ppm, respectively. The hydrocoral *Millepora* sp. was also exposed for up to 96 h to phenanthrene at concentrations of 25, 100 and 500 μg/liter (ppb).

At the end of the exposure period, pieces of coral were transferred to glass finger bowls containing 500 ml of clean filtered seawater to which had been added 50 μl of sodium (^{14}C) bicarbonate (specific activity 60·5 mCi/mmol, Amersham Searle). The corals were incubated in the ^{14}C-bicarbonate solution for 6 h in natural light. They were then rinsed in distilled water, placed in test tubes containing 1 N NaOH, and incubated in a 42 °C water bath overnight to digest coral tissue. Aliquots of the NaOH digest were added to Aquasol II liquid scintillation cocktail and ^{14}C activity was measured in a Packard TriCarb liquid scintillation spectrometer and corrected for quench. Protein concentration in the digests was measured by the biuret method using albumen as a protein standard (Gornall *et al.*, 1949). Photosynthetic ^{14}C fixation is expressed as counts per minute per mg protein nitrogen (CPM/mg protein-N).

For recovery experiments, corals were transferred to finger bowls containing natural seawater after the 72 h exposure period. The water was changed daily over the 72 h recovery period. Photosynthetic ^{14}C fixation was then measured during a 6 h incubation period in natural daylight as described above.

The majority of experiments on the effects of oil on calcium carbonate deposition in the skeleton of corals was performed in Bermuda. Preliminary experiments were performed at Texas A&M University with *Oculina diffusa* and the WSF of No. 2 fuel oil. Pieces of living coral were placed in exposure chambers (600 ml glass jars containing 300 ml seawater) and allowed to acclimate for 24 h. The seawater was then replaced with freshly prepared WSF of No. 2 fuel oil or south Louisiana crude oil. Six coral fragments were exposed to each WSF concentration for 24–72 h. At the end of the exposure period, exposure media were replaced with 300 ml freshly prepared WSF of the same concentration containing 10 μl $^{45}CaCl_2$ (specific activity, 0·2 mCi/ml, Amersham Searle). Exposure was continued for 3–6 h.

The coral fragments were then removed and rinsed in distilled water to remove unincorporated ^{45}Ca. The coral tissues were removed from the skeleton by incubating the coral fragments in 0·5 N NaOH at 40 °C overnight. Protein concentration in the NaOH digest was determined by the Lowry procedure (Lowry *et al.*, 1951). The Lowry and biuret procedures yielded similar results with these samples. The cleaned skeleton was dissolved in a minimum volume of 1 N HCl and neutralized with 1 N NaOH. An aliquot of the dissolved skeletal material was added to Aquasol II liquid scintillation cocktail and ^{45}Ca activity was measured with a Packard TriCarb liquid scintillation spectrometer.

Additional experiments were performed in Bermuda on the effects of exposure to the WSF of a No. 2 fuel oil or pure petroleum hydrocarbons on the rate of mineral deposition in the spines of the sea urchin *Diadema antillarum*. The methods for studying spine regeneration were essentially similar to those of Heatfield (1970). Aboral spines of intact healthy urchins were trimmed with iridectomy scissors to within 1 cm of the test. Several spines were left intact to prevent abrasion of the tips of those which were clipped. The animals were then returned to the seawater tanks for a 48-h recovery period. Heatfield (1970) showed that, in *Strongylocentrotus purpuratus*, at least 2 days are required for wound healing and tissue reorganization at the site of fracture before regeneration begins.

After the recovery period, previously clipped spines were removed from the urchins by severing the basal ligaments and muscles with a scalpel. Spines in groups of 4–5 were placed in 4-cm finger bowls containing 15 ml of exposure medium. Several replicate bowls were used at each exposure concentration and the experiments were duplicated.

The regenerating spines were exposed to 25, 100, and 500 ppb (μg/liter) naphthalene, phenanthrene, and dibenzothiophene, or to 10, 20, and 30 per cent dilutions of the WSF of a diesel fuel oil in natural seawater. A control series in hydrocarbon-free seawater was run simultaneously with each experimental exposure.

The three aromatic hydrocarbons were obtained from the Aldrich Chemical Co. and had a certified purity of better than 95 per cent. The diesel fuel oil was obtained from the Bermuda Biological Station for Research and is the fuel used in their oceanographic vessel. The WSF of this oil initially contained a mean of about 10 ppm total hydrocarbons as measured by IR analysis (American Petroleum Institute, 1958) and 1 ppm total naphthalenes as measured by UV spectrophotometry (Neff and Anderson, 1975). It is estimated that the initial concentrations of total hydrocarbons in the 10, 20, and 30% WSF were 1, 2, and 3 ppm, respectively.

Each exposure solution was labeled with 10 μl of $^{45}CaCl_2$ (stock concentration = 0·2 mCi/ml, Amersham Searle) at the beginning of each exposure. All exposures were continued for 24 h, after which spines were removed from the exposure medium and rinsed with distilled water. The distal 5 mm of each spine was clipped and placed in Clorox for several hours to remove all tissue and unincorporated ^{45}Ca. Individual spines were placed in scintillation vials and just enough concentrated HCl was added to dissolve the mineral. The digests were then assayed for radioactivity in Aquasol II liquid scintillation cocktail using a Packard TriCarb liquid scintillation spectrometer. Radioactive counts per minute (CPM), after quench correction, were converted to the quantity of calcium deposited in each spine.

All radioisotope uptake data in the photosynthesis and calcification experiments were analyzed statistically by analysis of variance (Barr *et al.*, 1976).

3.3 RESULTS AND DISCUSSION

3.3.1 Embryonic and Larval Development: Molluscs

Adult specimens of the oyster drill *Thais haemostoma* (Conrad), a gastropod mollusc, were collected from Aransas Bay. One cluster of egg capsules was obtained from these snails. This egg cluster was carefully

TABLE 16
MORTALITY OF *Thais haemostoma* EGG CAPSULES EXPOSED TO VARIOUS DILUTIONS OF SOUTH LOUISIANA CRUDE OIL WSF

WSF concentration per cent WSF and (ppm hydrocarbons)	Number of dead capsules	Day died
0 (0)	0	—
2·5 (0·5)	1	5
5 (1·0)	2	8, 12
10 (2·0)	0	—
25 (4·95)	3	2, 10, 14
50 (9·9)	4	12(2), 14, 18
75 (14·85)	5	3, 4, 10, 13, 15, 18
100 (19·8)	5	9(3), 15, 18

removed from the aquarium and the egg capsules divided into groups of five. Each capsule contained numerous embryos in the form of actively dividing blastula-like balls of cells which developed into trochophore larvae. While in the trochophore stage, the egg capsules were subjected to various dilutions of south Louisiana crude oil WSF. The exposure regime consisted of control (seawater only), 2·5, 5, 10, 25, 50, 75, and 100 per cent WSF (0·5–19·8 ppm total hydrocarbons). One group of five capsules was exposed at each concentration.

Tables 16 and 17 show mortality and hatching rates of egg capsules exposed to the various WSF concentrations. There were increasing mortality rates and decreasing hatching rates with progressive increases in

TABLE 17
HATCHING OF *Thais haemostoma* EGG CAPSULES EXPOSED TO VARIOUS DILUTIONS OF SOUTH LOUISIANA CRUDE OIL WSF

WSF concentration per cent WSF and (ppm hydrocarbons)	Number of capsules hatched	Day hatched
0 (0)	5	9, 23(3), 26
2·5 (0·5)	4	14, 22, 26(2)
5 (1·0)	3	12, 15, 26
10 (2·0)	4	11, 21, 22(2)
25 (4·95)	2	8, 27
50 (9·9)	0	—
75 (14·85)	0	—
100 (19·8)	0	—

WSF concentration. All trochophores died in the 75 and 100 per cent WSF and four in the 50 per cent WSF. Development was also severely retarded in the 50 per cent and higher WSF while the larvae were still viable.

On the first day of exposure, all embryos were developing from blastula-like cell masses into trochophore larvae. By the second day, all capsules contained crawling trochophore larvae. During the latter part of the first week of exposure trochophore larvae in the 25 per cent and weaker WSF dilutions began metamorphosing into veliger larvae. Designation of the veliger stage was based on the observation that the larvae had developed a shell and were actively crawling around in their capsules. On day 8 of exposure, all larvae in the control and 25 per cent and weaker WSF exposures developed a conspicuous dark gland-like organ located under the visceral mass inside the shell. The exact identity of this gland is not known; it was very conspicuous being black in color while all other parts of the veligers were colorless or near transparent. Capsules began hatching in the control and 25 per cent and weaker WSF exposures on days 8 and 9, and by day 26 all capsules in these exposure concentrations had hatched except two, which contained very well developed and actively crawling veligers.

Such staggered hatching is apparently normal. Although hatching was staggered, larval development appeared to proceed at the same rate and stage in all capsules. Those capsules which hatched on days 8 and 9, however, produced weak, non-actively swimming veligers. Capsules which hatched at later dates produced stronger, actively swimming veligers.

The development of larvae in egg capsules exposed to the three highest WSF concentrations varied from development in lower concentrations. While the blastula-like egg masses developed into trochophore larvae by day 2 of exposure, development apparently ceased after this stage. Four of the five egg capsules in the 50 per cent WSF exposure died by day 18, and at the time of death none of the larvae had developed beyond the trochophore stage. One capsule was still viable on day 28. In this capsule, the dark organ below the visceral mass was first noticed on day 22 of exposure; it was not very distinct and comprised only about one-half of the larval body. Shells of the veligers also appeared to be poorly developed. By day 25, almost all of the larvae had developed the dark gland but the shells still appeared incompletely formed. During the entire development, larvae were actively crawling in their capsules.

Larval development in capsules exposed to the 75 and 100 per cent WSF apparently ceased almost entirely on day 4 of exposure. From day 4 through day 13 of exposure the only movement of trochophore larvae observed in viable capsules was an occasional, very slight turn of one or two larvae. On day 13 of exposure, capsules which were still viable (one in 75 per cent WSF and two in 100 per cent WSF) were transferred to clean seawater. On day 15, the capsule originally exposed to 75 per cent WSF died and one exposed to 100 per cent WSF also died. The one remaining viable capsule originally exposed to 100 per cent WSF appeared to resume development after 2 days in clean seawater and larvae began slowly crawling about. This continued for 3 days after the capsule was placed in clean seawater after which larvae in this capsule also expired.

Results of this experiment would tend to indicate that development of *T. haemostoma* egg capsules proceeds normally at WSF concentrations as high as 25 per cent. It should be noted, however, that total naphthalenes in a 25 per cent WSF drop from 0·1 ppm at the beginning of exposure to 0·04 ppm after 24 h. In a 50 per cent WSF, development is severely retarded in capsules which are able to survive this concentration. After 13 days exposure to 75 and 100 per cent WSF, some capsules were still alive although development had nearly ceased. When placed in clean seawater, development in the capsules resumed—the most interesting aspect of this experiment. These results indicate that *T. haemostoma* embryos may be able to recover completely from exposure to potentially lethal WSF concentrations and resume normal development if exposure is terminated before 15 days.

3.3.2 Embryonic Development: Fish

Sperm and eggs of the estuarine fish *Cyprinodon variegatus*, *Fundulus heteroclitus*, and *F. similis* were obtained by stripping adult fish (Trinkaus, 1967). Embryos resulting from the fertilized ova of each species were maintained in 15-cm glass finger bowls at room temperature ($21 \pm 1\,°C$) prior to use in experiments. The period of development at which exposure to the WSF of oil was initiated varied from immediately after fertilization to 130 h of development. In each test, hatching rates of embryos and per cent of successful hatchings were recorded. At the developmental stage in which the heart was visible and active, daily measurements of heart beat rate were determined for control and WSF-exposed embryos (Anderson *et al.*, 1977*a*).

To determine permeability of the chorionic membrane of the embryos, two different tests were conducted. First, the chorion of a group of *Cyprinodon variegatus* embryos was surgically removed using watchmaker forceps. Tolerance of these embryos was compared to that of embryos of the same age with intact chorionic membranes. In addition, concentration of specific hydrocarbons (naphthalenes) present in *F. similis* eggs exposed

TABLE 18
THE HATCHING SUCCESS OF EMBRYOS OF THREE FISH SPECIES EXPOSED TO WSF OF NO. 2 FUEL OIL (NO. 2 F.O.) AND SOUTH LOUISIANA (SO. LA.) CRUDE OIL (FROM ANDERSON et al., 1977a, WITH PERMISSION OF ACADEMIC PRESS)

Species	Oil	Condition	Concentration (per cent WSF)
			0 10 25 30 50 70 75 90 100
Fundulus	No. 2 F.O.	No. dead	0 0 2 5
heteroclitus		No. hatched	5 5 3 0
		Per cent hatched	100 100 60 0
Cyprinodon	No. 2 F.O.	No. dead	0 1 3 8
variegatus	test 1	No. hatched	7 7 5 0
(with chorion)		Per cent hatched	100 88 62 0
	test 2	No. dead	0 1 3 5
		No. hatched	5 4 2 0
		Per cent hatched	100 80 40 0
(chorion	No. 2 F.O.	No. dead	0 4 4 10
removed)		No. hatched	9 6 6 0
		Per cent hatched	100 60 60 0
(with	No. 2 F.O.	No. dead	3 1 25 25 25 25
chorion)	renewed	No. hatched	22 24 0 0 0 0
	daily	Per cent hatched	88 82 0 0 0 0
Fundulus	So. La.	No. dead	2 7 1 7 24 25 25
similis	renewed	No. hatched	23 18 24 18 1 0 0
	daily	Per cent hatched	92 72 96 72 4 0 0

to three concentrations of WSF of south Louisiana crude oil was determined.

The experimental results for hatching success of embryos of all three species are summarized in Table 18. Some interesting comparisons can be made between results of tests involving different treatment conditions. In those studies where the exposure media were not renewed during the test period, which was generally about 8 days, the 100 per cent WSF of No. 2 fuel oil was the only concentration producing 100 per cent mortality. Survival to hatching for both *F. heteroclitus* and *C. variegatus* was approximately 50 per cent at the 50 per cent WSF concentration. At 25 per cent WSF, 60–100 per cent of the embryos of both species survived to hatching. It should be noted that *C. variegatus* embryos which lacked the chorionic membrane from 130 h to hatching (approximately 11 days) survived at about the same level as normal embryos.

The effect of daily renewals of WSF is evident from the decreased survival of *C. variegatus* embryos in 30 and 50 per cent WSF of No. 2 fuel oil.

Without renewal, survival and hatching of *C. variegatus* was 40–62 per cent in 50 per cent WSF and 60–100 per cent in 25 per cent WSF. Since survival was reduced to 0 per cent at a concentration of 30 per cent WSF, the daily replenishment of WSF significantly reduced hatching success. When south Louisiana crude oil WSF was renewed daily in tests with *F. similis* the median lethal concentration was shown to lie between 50 and 70 per cent WSF. It is apparent, though not certain, from these tests that the tolerances of embryos of all three species of estuarine fish are approximately equal.

Adult *F. similis* and *C. variegatus* showed similar tolerances to WSF of No. 2 fuel oil and Venezuelan bunker C oil (Anderson *et al.*, 1974*a*). It is therefore likely that the significantly greater sensitivity exhibited by *C. variegatus* embryos in 30 per cent WSF of No. 2 fuel oil (100 per cent mortality), as compared with *F. similis* in 30 per cent WSF of south Louisiana crude oil (96 per cent hatching), is the result of differences in the hydrocarbon composition of the extracts. Since Tests 1 and 2 (with chorion) on *C. variegatus* and the test with *F. heteroclitus* were identically conducted, it would appear that embryos of these two species are very similar in tolerance to oil WSF and more tolerant than the adults (Anderson *et al.*, 1974*a*).

As noted above, those embryos of *C. variegatus* which lacked the chorionic membrane responded to the various hydrocarbon solutions very much like normal embryos. These findings would indicate that this outer membrane is not a barrier to hydrocarbons. However, analyses of normal and dechorionated embryos for hydrocarbon content would be more direct evidence. In a later experiment with *F. similis*, embryos which were so obviously deformed that hatching was impossible were sacrificed for naphthalenes analyses by the method of Neff and Anderson (1975). Results of these analyses are summarized in Table 19. Dimethylnaphthalenes were

TABLE 19

NAPHTHALENE (N), METHYLNAPHTHALENES (MN), DIMETHYLNAPHTHALENES (DMN) AND TOTOAL NAPHTHALENES (TN) CONCENTRATIONS IN EMBRYOS OF *Fundulus similis* EXPOSED TO 70, 90 AND 100 PER CENT WSF OF SOUTH LOUISIANA CRUDE OIL (FROM ANDERSON *et al.*, 1977*a*, WITH PERMISSION OF ACADEMIC PRESS)

Sampling time (days)	Exposure concentration, per cent WSF and (ppm total naphthalenes)	Concentration (ppm)			
		N	MN	DMN	TN
28	70 (0·21)	0·88	1·59	2·50	4·97
27	90 (0·27)	4·01	4·47	7·60	16·08
25	100 (0·30)	2·02	3·21	4·22	9·45

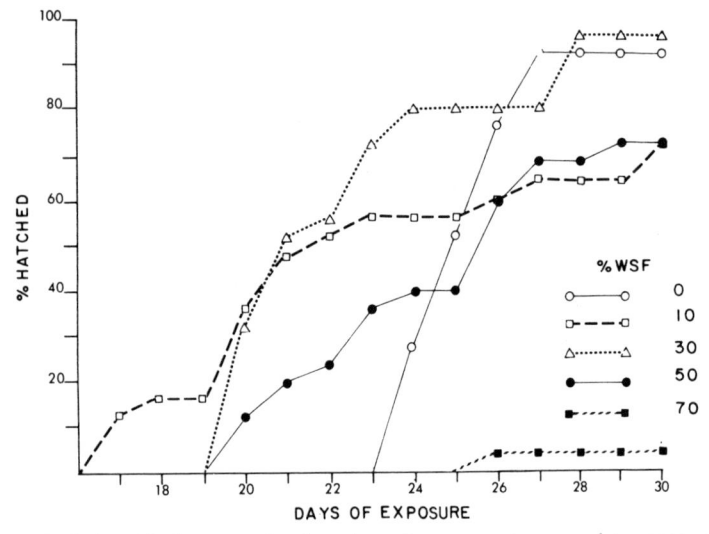

FIG. 4. Relationship between the duration of exposure to several concentrations of the WSF of south Louisiana crude oil and the per cent cumulative hatch of embryos of the killifish *Fundulus similis*. Twenty embryos were exposed continuously during development to each WSF concentration (from Anderson *et al.*, 1977a, with permission of Academic Press).

present in highest concentration in all embryos tested and concentrations of total naphthalenes ranged from about 5 to 16 ppm. It should be noted that 4 per cent of the embryos exposed to 70 per cent WSF of south Louisiana crude oil survived to hatching (Table 18), although they contained approximately 5 ppm of total naphthalenes.

Rate of hatching by *F. similis* embryos was also recorded and the results are quite interesting. As shown in Fig. 4, exposure to hydrocarbons apparently stimulated hatching since those embryos in 10 per cent WSF of south Louisiana crude oil began hatching on the 17th day. By the time control embryos began to hatch (day 24), oil exposed organisms exhibited hatching percentages of 80, 56, and 40 at 30, 10, and 50 per cent WSF, respectively. Final hatching percentages at these latter concentrations and for control animals ranged from 72 to 96. While the hatching rate of oil-exposed embryos was rather irregular and took place over a period of about 10 days, control fish all hatched within a period of 4 days.

The final parameter measured in these studies was rate of embryonic heart beat. In *C. variegatus*, the heart developed sufficiently so that pulsations were observed by 130 h, while heart beats of *F. heteroclitus*

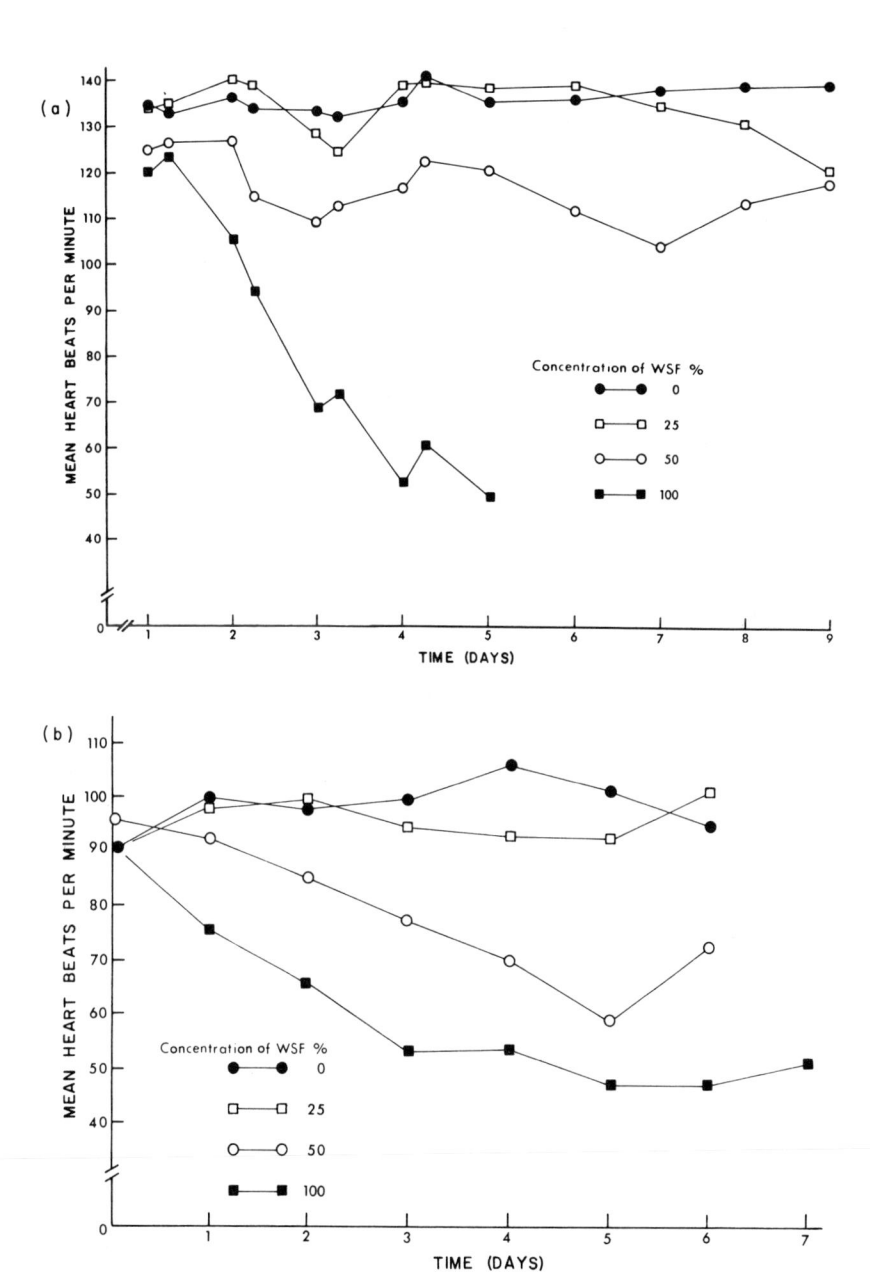

FIG. 5. Effect of continuous exposure to the WSF of No. 2 fuel oil on the heart beat rate of embryos of (a) the mummichog *Fundulus heteroclitus* and (b) the sheepshead minnow *Cyprinodon variegatus*. The WSF were not renewed during the exposure period (from Anderson *et al.*, 1977a, with permission of Academic Press).

embryos could not be measured until the 11th day of development. Mean heart beat rate for control *F. heteroclitus* embryos and those exposed to a single dose of 25 per cent WSF of No. 2 fuel oil was approximately 135 beats/min (Fig. 5a); *C. variegatus* embryos under the same conditions exhibited a mean of about 100 beats/min (Fig. 5b). In both species, heart beat rates of embryos subjected to 50 per cent WSF were significantly lower. In *C. variegatus* embryos, this suppression increased gradually with time. An even more drastic reduction in heart beat rates was exhibited by embryos of both species subjected to the 100 per cent WSF of No. 2 fuel oil. For both species, suppression began immediately after the beginning of exposure and the decrease continued until death of the embryos. Further studies with *C. variegatus* verified the above findings.

Studies on the effects of WSF of south Louisiana crude oil on *F. similis* embryonic heart beat are summarized in Fig. 6. Since there were daily variations in the mean rates of heart beat in control, 10, 30, and 50 per cent WSF, the zone of variation has been enclosed by shading. No relationship between heart beat rate and concentration of WSF at exposure

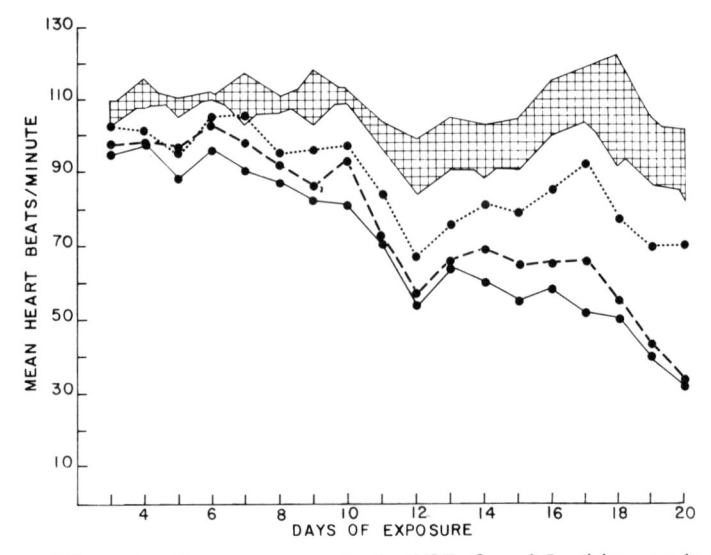

FIG. 6. Effect of continuous exposure to the WSF of south Louisiana crude oil on the heart beat rate of embryos of the gulf killifish *Fundulus similis*. The shaded zone represents the range of all means from embryos exposed to concentrations ranging from 0 to 50 WSF. ●——●, 70 per cent WSF; ●····●, 90 per cent WSF; ●---●, 100 per cent WSF (from Anderson *et al.*, 1977a, with permission of Academic Press).

concentrations of 0 to 50 per cent WSF was evident. However, at concentrations of 70, 90, and 100 per cent WSF heart beat rates were lower from day 4 to hatching on day 20. Embryonic heart beat rates of specimens retained in 90 and 100 per cent WSF decreased during the course of exposure. Heart beat rates in embryos exposed to 70 per cent WSF decreased to a low on day 12, but then increased and finally leveled off at approximately the day 12 heart beat rate. As shown in Table 18, none of the embryos hatched in concentrations of 90 to 100 per cent WSF, but 4 per cent in the 70 per cent WSF did hatch.

Several additional experiments were performed to determine more precisely the effect of exposure length on toxicity of WSF to *F. heteroclitus* embryos and whether embryos could recover from acute exposure to oil (Sharp et al., 1979). In the first experiment, eggs were stripped from 10 females and each group of eggs was kept separately from the others. Egg groups were fertilized with sperm from one of 10 males and then exposed to a 25 per cent WSF of No. 2 fuel oil (2·1 ppm total hydrocarbons) which was renewed daily. Control eggs ($n = 241$) were collected from five females and each group of these eggs was fertilized by one of five males. For this experiment, five sets of the exposed eggs ($n = 257$) were transferred to hydrocarbon-free seawater after 12-day exposure to No. 2 fuel oil WSF. Exposure to the 25 per cent WSF was continued for the other five sets of eggs ($n = 301$) until all had hatched or died. Upon hatching, fry were transferred to clean artificial seawater.

Eggs in the second experiment were exposed after fertilization to a WSF of the No. 2 fuel oil at concentrations of 0 (control), 10, and 20 per cent WSF (0, 0·87, and 1·7 ppm total hydrocarbons) renewed daily. The total number of eggs used was 313 in the control, 208 in the 10 per cent WSF mixture, and 283 in the 20 per cent WSF. No attempt was made to keep these eggs separated according to parentage. Several replicates of each exposure were used. Fry were continuously exposed during development and after hatching.

Another experiment utilized fertilized eggs exposed to a 25 per cent WSF of No. 2 fuel oil, which was renewed daily. Groups of these eggs were exposed for periods of 0, 4, 8, and 12 days before being transferred to clean water. One group of eggs was continuously exposed for the duration of the experiment. Approximately 65 eggs were used in each test group.

In Experiment 1, in which *Fundulus heteroclitus* eggs were exposed to 25 per cent WSF, survival of exposed eggs was much less than that of controls (Fig. 7). Control eggs first began hatching after 14 days and after 48 days, 192 of a total 241 eggs (82 per cent) had hatched. Of the eggs that hatched,

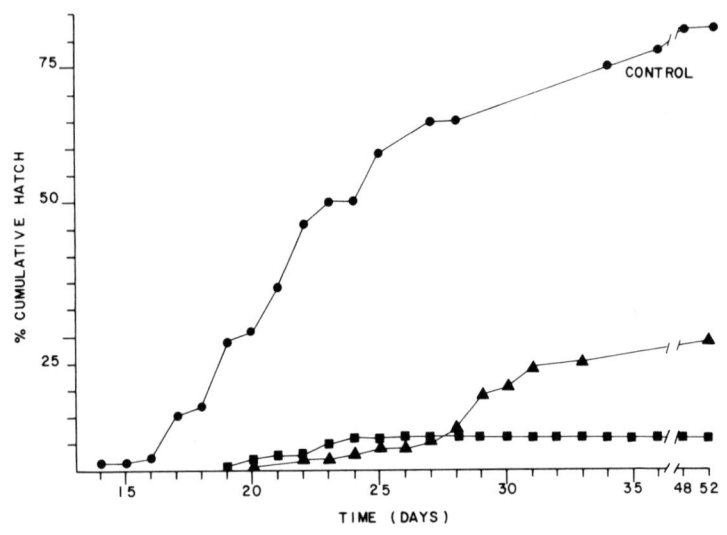

FIG. 7. The per cent cumulative hatch of embryos of mummichogs *Fundulus heteroclitus* exposed to a 25 per cent WSF of No. 2 fuel oil, renewed daily. One group of eggs was exposed continuously throughout the experiment (■──■) while another groups was exposed to the WSF for only the first 12 days of development (▲──▲).

89 per cent of the fry survived. Hatching was delayed in both groups of exposed eggs. The first hatch among eggs in which exposure to fuel oil was terminated after 12 days occurred on the 20th day after fertilization. Likewise, the first hatch among eggs continuously exposed to No. 2 fuel oil was observed on the 19th day after fertilization. Hatching rate was slightly enhanced in those eggs continuously exposed as compared to the 'partially' exposed eggs, but tapered off after 5 days with only 6 per cent of these eggs eventually hatching. In contrast, partially exposed eggs continued hatching through the 52nd day with a total hatch of 22 per cent.

Survival of fry was also much less in exposed fish even though fry were transferred to clean water upon hatching. Of the controls, 175 of 192 emergent fry survived representing an 89 per cent survival rate. By comparison, 78 per cent of the partially exposed fry survived (37 of 47), while none of the 'totally' exposed fry lived for more than a few days after hatching.

Since there was an almost total mortality when *F. heteroclitus* eggs were exposed to 25 per cent WSF, the second experiment was designed to investigate effects of exposure to more dilute WSF on hatching success.

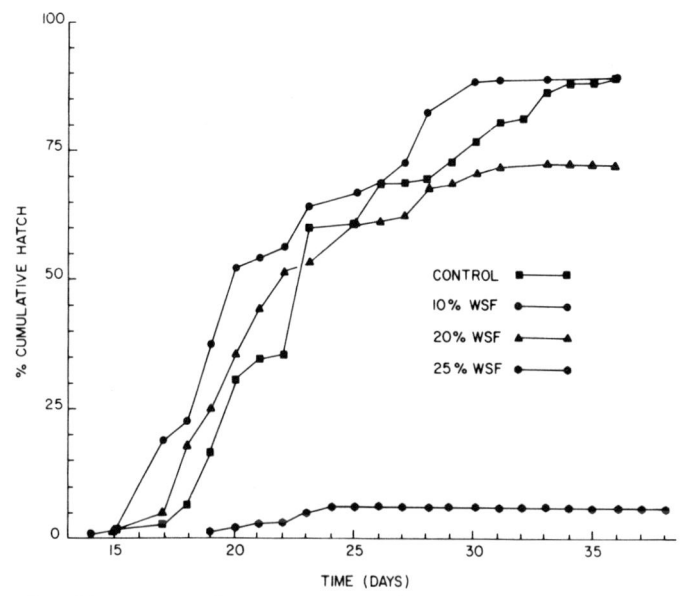

FIG. 8. Hatching rate and per cent cumulative hatch of embryos of the mummichog *Fundulus heteroclitus* continuously exposed to several concentrations of the WSF of No. 2 fuel oil, renewed daily (from Sharp *et al.*, 1979, with permission of Academic Press).

Figure 8 shows the per cent cumulative hatch of eggs exposed to 10 and 20 per cent WSF of No. 2 fuel oil. First hatches were noted on day 14 at each of the exposure concentrations.

The first control hatch occurred on the 15th day after fertilization. In the control and 10 per cent WSF exposure, final hatching success was approximately the same with 91 and 92 per cent of the eggs hatching in the two groups. Hatching success was somewhat less in the 20 per cent WSF exposure with 74 per cent of the eggs hatching. Interestingly, hatching rate was somewhat enhanced in the 10 per cent WSF exposure as compared to the control hatching rate.

Fry in the test described above were continuously exposed to WSF after hatching. Survival was greatly reduced among fry hatched from eggs that had been exposed to the WSF. Approximately 97 per cent of the control fry survived as compared to 88 per cent of the fish exposed to 10 per cent WSF and only 61 per cent of 20 per cent WSF-exposed fish.

Based on recoveries seen after 12-day exposure to the WSF in the first experiment, the third experiment was designed to determine the length of

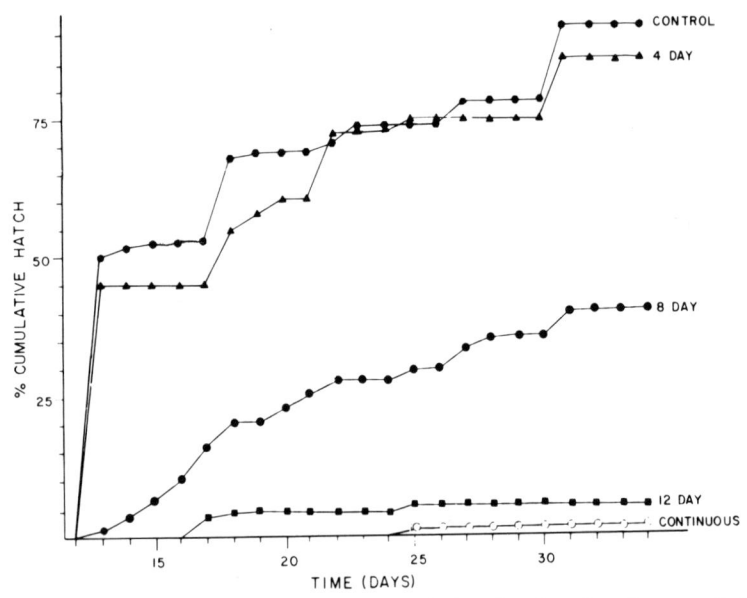

FIG. 9. Hatching rate and per cent cumulative hatch of embryos of the mummichog *Fundulus heteroclitus* continuously exposed from fertilization for different lengths of time to a 25 per cent WSF of No. 2 fuel oil, renewed daily (from Sharp et al., 1979, with permission of Academic Press).

exposure to the oil that could be tolerated by the embryos before irreversible effects were noted. The per cent cumulative hatch in each egg group exposed for different lengths of time to 25 per cent WSF is presented in Fig. 9. Hatching success of control and 4-day-exposed eggs was similar with hatches approximating 92 and 87 per cent, respectively. In contrast, only 40 per cent of the 8-day-exposed eggs, 6 per cent of the 12-day-exposed eggs, and 1·5 per cent of the continuously exposed eggs eventually hatched.

Analysis of variance revealed that duration of exposure had a highly significant effect on hatching success of the embryos ($p > F = 0.0001$). The Waller–Dunkan K-ratio T-test for variable hatch revealed that there were three exposure groups showing significantly different hatching rates and success. These were the controls and 4-day-exposed embryos (group 1), the 8-day-exposed embryos (group 2), and the 12-day- and continuously exposed embryos (group 3).

Initial hatches were recorded on the 13th day after fertilization in the control and 4-day exposures, and on day 14 for eggs exposed for an 8 day period. First hatches were noted on the 17th and 25th days in 12-day- and

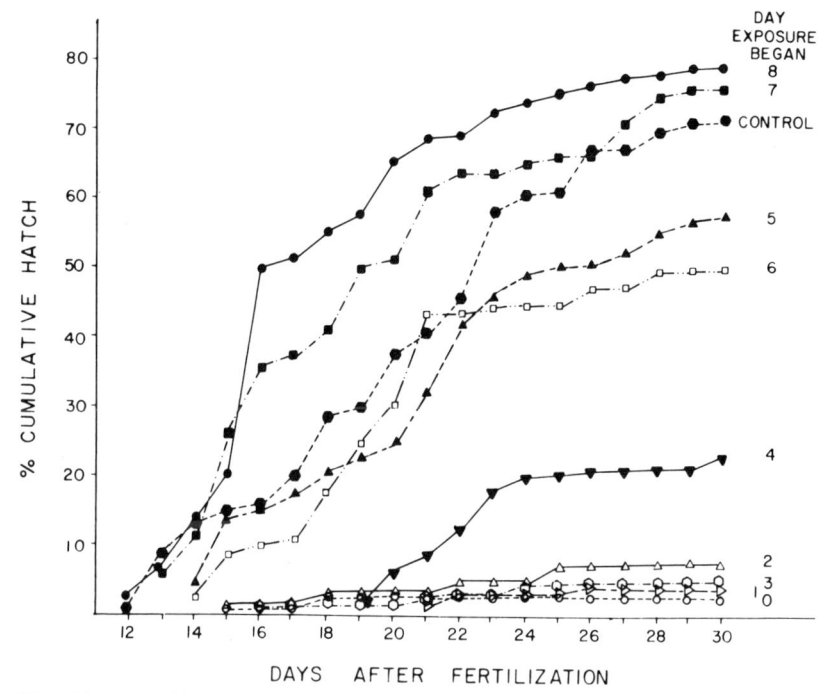

FIG. 10. Hatching rate and per cent cumulative hatch of *Fundulus heteroclitus* embryos when continuous exposure to the 25 per cent WSF of No. 2 fuel oil was initiated after different periods of development in hydrocarbon-free seawater (from Sharp *et al.*, 1979, with permission of Academic Press).

continuously exposed groups, respectively. No records were kept on the number of fry surviving after hatching.

When exposure to the 25 per cent WSF was delayed for different lengths of time after fertilization of the eggs, a similar graded hatching success response was noted (Fig. 10). Four distinct subgroupings of exposure groups were evident. Hatching success of the controls and of embryos first exposed to the 25 per cent WSF after 7 or 8 days of development in hydrocarbon-free seawater was similar, 72·5, 76·3 and 78·8 per cent, respectively. Hatching rates of the day-7- and day-8-exposure groups appeared to be somewhat higher than that of the controls. Embryos first exposed on days 5 and 6 of development had a similar hatching success of 57·5 and 50 per cent, respectively and also had similar hatching rates. Hatching rate and success (22·5 per cent) were greatly reduced in embryos first exposed to the WSF on day 4 of development. The fourth subgroup consisted of embryos first exposed on days 3, 2, 1 and 0 of development. Hatching success of these groups was 3·75, 7·5, 3·75 and 2·5 per cent, respectively. An analysis of variance of the hatching success data revealed that the day on which exposure to the WSF was initiated had a highly significant effect on hatching success ($p > F = 0.0001$). Duncan's multiple range test for variable hatch indicated four subgroups with significantly different hatching rates. These were the controls, 5-, 7- and 8-day embryos (group 1), the 5- and 6-day embryos (group 2), the 4-day embryos (group 3), and the 0-, 1-, 2- and 3-day embryos (group 4).

The onset of hatching also was related to the time at which exposure to the WSF was initiated. However, there was not a clear-cut dose × time response relationship for this developmental parameter. The first hatch of controls and 8-day embryos occurred on day 12 of development. The 7-day embryos began hatching on day 13, the 5- and 6-day embryos on day 14, the 0- and 2-day embryos on day 15, the 3-day embryos on day 16, the 4-day embryos on day 19 and the 1-day embryos on day 21.

The results discussed above indicate that the period of embryonic development most sensitive to oil was from about the third to the sixth day of development. After this period, embryos became progressively more tolerant to oil. In an attempt to ascertain if this change in sensitivity to oil was due to changes during embryogenesis in the rate of hydrocarbon accumulation by the embryos, pulse labeling experiments were performed with ^{14}C-naphthalene and ^{3}H-water. After different periods of development in clean seawater, groups of embryos were placed in seawater containing both radioisotopes and incubated at room temperature for 2 h. Uptake by the embryos of ^{14}C and ^{3}H was measured by liquid scintillation counter. The instantaneous naphthalene uptake rate decreased markedly from 2530 DPM ^{14}C/embryo × 2 h on day 2 of development in hydrocarbon-free seawater to 390 DPM ^{14}C/embryo × 2 h on day 10 of development (Fig. 11). The sharpest drops in the apparent permeability of the embryos to naphthalene occurred between days 2 and 4 and days 8 and 10. In contrast, instantaneous water uptake rates increased slightly in a nearly linear fashion from 220 DPM ^{3}H/embryo × 2 h on day 2 to 460 DPM ^{3}H/embryo × 2 h on day 10. Standard deviations were never more than 10 per cent of the means. Thus, the apparent permeability of the embryos to naphthalene and water changed in different directions during the time course of development.

The heart beat rates of control animals increased slightly but not significantly from 127 ± 10 beats/min (bpm) on day 9 of development to 135 ± 7.8 bpm on day 11 (Fig. 12). Embryos first exposed to the WSF on

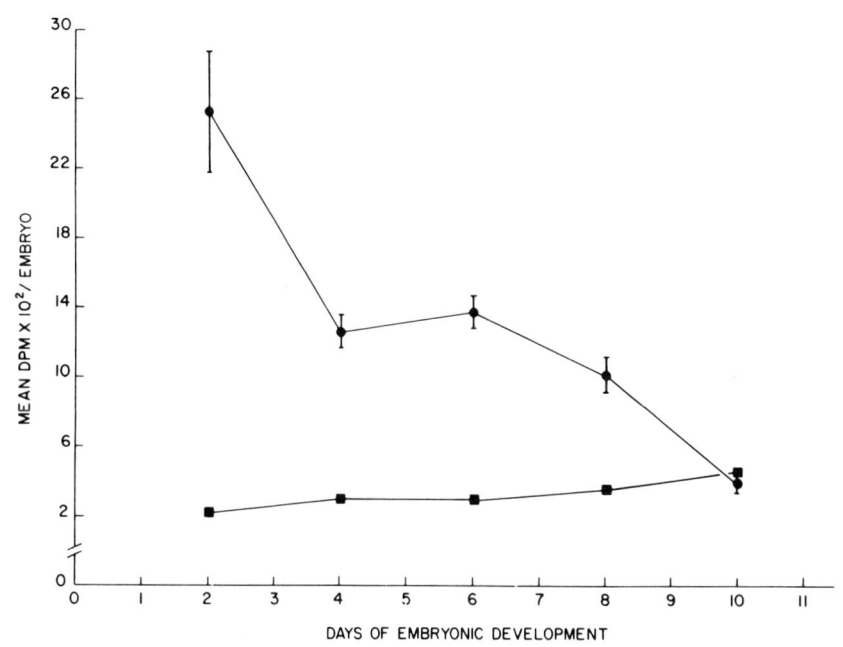

FIG. 11. 'Instantaneous' uptake rates of ^{14}C-naphthalene (●——●) and ^{3}H-water (■——■) by *Fundulus heteroclitus* embryos at different times during development in hydrocarbon-free seawater. Data points represent radioactivity accumulated during 2-h incubation in seawater containing ^{14}C-naphthalene and ^{3}H-water. Vertical lines represent standard deviations (from Sharp *et al.*, 1979, with permission of Academic Press).

day 8 of development had heart beat rates essentially identical to those of controls at all sampling times. Embryos first exposed to the WSF on day 6 had slightly depressed heart beat rates ranging from 113 ± 5.7 bpm on day 9 to 122 ± 9.6 bpm on day 11. Embryos first exposed to the WSF on day 2 or 4 of development had heart beat rates that were significantly depressed below those of controls on days 10 and 11. In addition, heart beat rates showed a decreasing trend between days 9 and 11. The heart beat rates of day-2 embryos dropped from 107 ± 11 bpm on day 9 to 94.4 ± 7 bpm on day 11. The relatively large standard deviations in the heart beat rates measured at any given time of development seem to be typical for embryos of this species (Glaser, 1929).

Salinity and temperature had a marked effect on the sensitivity of *Fundulus heteroclitus* embryos to the WSF of No. 2 fuel oil. During exposure to a 15 per cent WSF (0·28 ppm total naphthalenes), survival to

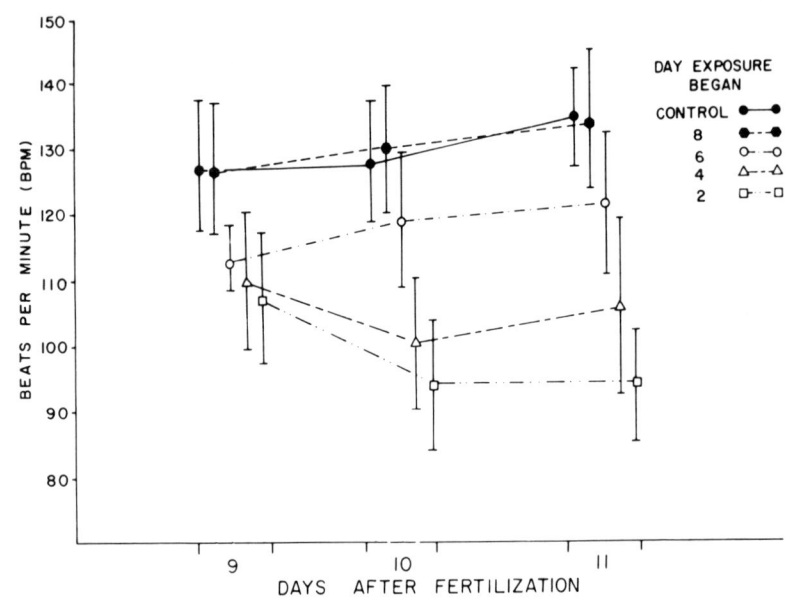

FIG. 12. Mean heart beat rates of *Fundulus heteroclitus* embryos on days 9, 10 and 11 of development during continuous exposure to the 25 per cent WSF of No. 2 fuel oil initiated at different times after fertilization (from Sharp *et al.*, 1979, with permission of Academic Press).

hatching of embryos was highest at 20‰ S and 25 °C, and decreased significantly at all higher and lower salinities and temperature (Linden *et al.*, 1979). Survival was lowest at 10‰ S, 20 °C and 30‰ S, 30 °C. Duration of embryogenesis from fertilization to hatching tended to decrease with increasing WSF exposure concentration at all temperature–salinity combinations. However, temperature had a greater effect than salinity or exposure to oil on development rate.

Fish fry hatching from eggs that had been exposed continuously to WSF during embryonic development were significantly shorter in overall length than control fry (Linden *et al.*, 1980). Temperature also exerted an effect on growth at low but not high WSF concentrations. Fry from oil-exposed embryos had larger yolk diameters than did controls, indicating a decreased utilization of yolk material during development. Animals formerly exposed to oil had a decreased number of vertebrae and exhibited a variety of spinal deformities such as scoliosis and lordosis.

Developmental anomalies and histopathological lesions, were characterized in *Fundulus grandis* (Baird and Girard) embryos and fry exposed continuously during development to the WSF of No. 2 fuel oil (Ernst *et al.*, 1977). At different times after fertilization, embryos and fry were either fixed in 10 per cent neutral formalin and embedded in paraffin or they were fixed in 2 per cent glutaraldehyde, postfixed in 1 per cent OsO_4 and embedded in Epon 812. Tissue blocks were sectioned, mounted on glass slides, and stained using standard histological procedures.

Observations were made through the complete embryonic period. According to the terminology of Balon (1975), the embryonic period of this and many other species of fish can be divided into three distinct phases: cleavage egg, embryo, and eleutheroembryo. The cleavage phase extends from the time development commences until organogenesis begins. The embryonic phase extends from the beginning of organogenesis until hatching is completed, and the eleutheroembryonic phase commences with hatching and lasts until the yolk is completely absorbed and the fish begins external feeding.

Mortalities of embryos and eleutheroembryos were recorded during the exposure period. All embryos in 12·5 per cent WSF hatched but some eleutheroembryos later died. A number of embryos in 26 per cent WSF hatched but these eleutheroembryos subsequently died. No embryos in 50 per cent WSF survived to hatching.

Embryos exposed to 12·5 per cent WSF hatched earlier than controls. At 12·5 per cent WSF all the fish had hatched by day 22, whereas hatching in control groups was not complete until day 28. Eleutheroembryos from 12·5 per cent WSF exposure appeared normal. Kuptsis (1902, as cited by Kuhnhold, 1974) also observed early hatching of fish *Salmo* and *Coregonus* embryos due to exposure to crude oils. Similar stimulatory responses to exposure to low sublethal oil concentrations were observed in several of the experiments reported earlier in this section.

No macroscopic evidence for the cause of death was observed in animals treated with 12·5 per cent WSF except that one eleutheroembryo was not able to come completely out of its chorion and remained alive in this condition for 2 days. Embryos treated with 25 per cent WSF took longer to develop to a given stage than did controls and appeared to produce less melanin. Embryos treated with 50 per cent WSF were even slower in development, were much smaller in size, and also appeared to have less melanin. Some of these embryos were macroscopically deformed, even to the extent of being merely an aggregation of living cells with no recognizable structure. The most striking aspect noticed was shrinkage of the yolk. In normal embryos, the growing embryo and yolk sac almost filled the chorion. In embryos treated with 50 per cent WSF, these structures occupied only about one-quarter of this space. After fixation, space around embryos and yolk contained a serous fluid, probably fixative and seawater, while space around shrunken embryos was jelled, indicating a loss of protein by embryos and yolk.

Fundulus grandis eggs contained many small lipid droplets that were visible on the yolk surface. These lipid droplets in control and 12·5 per cent WSF treated embryos appeared to be normal, while in those exposed to 25 and 50 per cent WSF the lipid droplets tended to coalesce. This coalescence was more pronounced in 50 per cent WSF treated embryos indicating a physical alteration of the lipid in the droplets due to the WSF.

Most of this lipid is not used by the embryo in the initial stages of development as it is still present upon hatching. In teleosts, these lipid droplets contain glycerides and are not metabolized until shortly before the eleutheroembryo starts to feed (Smith, 1957). Therefore, pollution studies which only describe hatching in fish do not accurately describe the total effect of pollutant if it reacts mainly with neutral fats. Pathological effects from such pollutants may not necessarily be demonstrated just after hatching, as eleutheroembryos have not absorbed all of the yolk contents at that time.

Only normal tissues were observed in histological studies of embryos and eleutheroembryos of controls and animals treated with 12·5 per cent WSF. Embryos treated with 25 per cent WSF varied in degree of histopathology from normal to very abnormal. Development in general was slower in these animals. Whether certain organs or tissues were differentially slower in development was very difficult to ascertain. The liver usually contained necrotic areas and in some cases few to no blood sinuses. Most kidneys were under-developed with no lumina. Some had necrotic cells and in one no renal tissue was distinguishable. Other animals were without recognizable mucus cells in the gut; a few contained entirely abnormal gut epithelia while others had gut tissues which appeared normal. The lenses of most animals were visiculated. Occasionally necrotic muscle tissues were observed. Embryos treated with 50 per cent WSF also varied in the degree of histopathology from almost totally undifferentiated and necrotic cells to organized but altered tissues.

These experiments have shown that embryos of estuarine gastropod molluscs and fish are quite tolerant to petroleum hydrocarbons. Relatively high concentrations of petroleum hydrocarbons are required (compared to concentrations normally detected in the water column in the vicinity of an oil spill or a chronic hydrocarbon input source) to produce significant mortality or delay in development. Deleterious effects of petroleum were often reversible if the animals were returned to hydrocarbon-free seawater before mortality occurred.

Rice *et al.* (1975) reported that eggs of the pink salmon *Oncorhynchus gorbuscha* were extremely tolerant to 96-h exposures to Prudhoe Bay crude oil. Generally, developing embryos enclosed within protective egg envelopes are more tolerant than newly hatched larvae. Increased sensitivity in larvae has been attributed to loss of protection afforded by the relatively impermeable chorion or analogous envelope which surrounds the embryo (Wilson, 1976). However, dechorionated eggs of *C. variegatus* were not significantly more sensitive to oil WSF than were normal eggs, indicating that the chorion does not constitute an effective barrier against hydrocarbon pollutants. Hydrocarbons and other lipophilic pollutants tend to accumulate selectively in lipid deposits of the yolk. Hydrocarbons may then alter lipid structure, as indicated in histological studies reported here, and interfere with lipid mobilization and catabolism which occurs after the embryo hatches (Smith, 1957). This hypothesis may partially explain the increased sensitivity to oil and other pollutants of newly-emergent fry (eleutheroembryos) and mortality of fry that have been exposed to oil during embryonic development.

It would appear that early cleavage stages of newly fertilized eggs are somewhat less sensitive to pollutants than gastrula and later morphogenic stages. Thus, *Thais haemostoma* eggs developed normally through the blastula stage but further development of the gastrula was inhibited by high WSF concentrations. Embryos of *F. heteroclitus* could tolerate a 25 per cent WSF of No. 2 fuel oil for 4 days without a significant reduction in hatching success. With exposure extended to 8 days a significant decrease in hatching success was observed. According to Armstrong and Child (1965) the period between 4 and 8 days of embryonic development in this species is characterized by major morphogenic events leading to the production of major organ system primordia. It would appear that these organogenic processes are particularly sensitive to pollutants, an assumption supported by histopathological evidence of increased incidence of lesions in embryonic organ systems with exposure to high WSF concentrations.

These conclusions tend to be supported by observations of other investigators. Kuhnhold (1972, 1974) observed that cod eggs were most sensitive to petroleum at the gastrula stage and immediately afterward. Eggs that were exposed to Iranian crude oil immediately after fertilization showed normal development for 2–3 days after which development rate slowed in comparison to that of controls. Embryos initially exposed at day 10, when the heart was fully differentiated, were also much less sensitive than those exposed during gastrulation and early morphogenesis. The majority of harmful effects of oil and dispersants on development of sea urchin embryos were observed after gastrulation and during differentiation (Lonning and Hagstrom, 1975). In addition, several oil dispersants have been shown to cause abnormal cell division and differentiation in fish embryos (Linden, 1974; Wilson, 1976).

In the present investigation, heart beat rates were depressed in *Cyprinodon variegatus*, *Fundulus similis*, and *F. heteroclitus* embryos by exposure to concentrations of WSF approaching acutely toxic levels. Similarly, depressed heart beat rates have been observed in fish embryos exposed to sublethal concentrations of oil spill dispersants (Linden, 1974; Wilson, 1976). These depressant effects may be attributed to the anesthetic effect of lipophilic hydrocarbons.

An important observation in these studies with gastropod and fish embryos is that if exposure to potentially lethal concentrations of oil is terminated early enough, embryos may recover and resume normal development. This is undoubtedly related to the ability of embryos to rapidly release petroleum hydrocarbons from their tissues when returned to oil-free seawater.

3.3.3 Larval Development and Growth

Juvenile polychaete worms *Neanthes arenaceodentata* (mean initial wet weight, 20 mg/worm) were exposed continuously for 28 days in a flow-through system to three concentrations of No. 2 fuel oil WSF (Rossi and Anderson, 1978a). Mean exposure concentrations were 60, 95, and 180 ppb (μg/liter) total naphthalenes measured, corresponding to 180, 310, and 600 ppb total aqueous hydrocarbons, respectively. These concentrations are all well below the 96-h LC_{50} value of 2·7 ppm total hydrocarbons for this species (Rossi *et al.*, 1976).

Growth rates were suppressed by all three WSF concentrations. This effect was most noticeable after 28-day exposure (Fig. 13). Analysis of variance verified that, just prior to exposure, worms were equal in weight ($p = 0.05$). After 14-days exposure, only those juveniles exposed to the two highest WSF concentrations were growing at rates significantly slower than that of control animals ($p = 0.05$). Following 28-day exposure, final weights of worms grown in all three WSF concentrations were significantly lower than controls ($p = 0.05$).

The ecological significance of these results is difficult to assess, especially in light of the observation that equal numbers of females just becoming gravid were present in the control and WSF exposure aquaria on the 28th

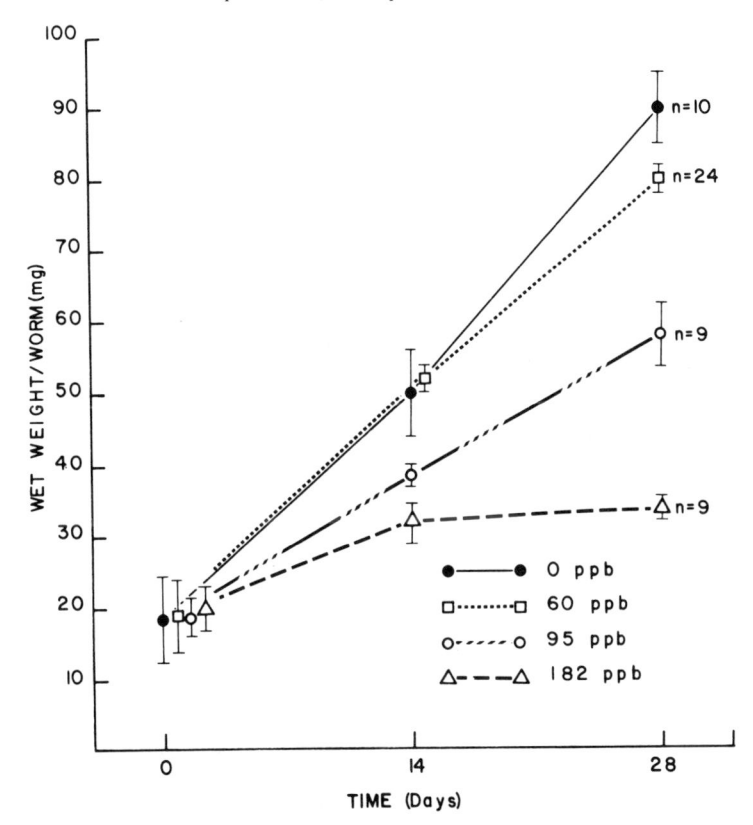

FIG. 13. Effect of several concentrations of the WSF of No. 2 fuel oil on the growth of juvenile polychaete worms *Neanthes arenaceodentata*. Exposure concentrations are given in µg/liter (ppb) total naphthalenes. Each finger bowl contained 20 worms, and *n* indicates the number of bowls in each exposure group. Vertical lines are standard deviations of the means (from Rossi and Anderson, 1978a, with permission of D. Reidel Publishing Co.).

exposure day. Thus, although presence of the WSF apparently retarded growth, it did not seem to retard the development of sexual maturity or egg maturation.

In another experiment, the effect of No. 2 fuel oil WSF (270–340 ppb total naphthalenes, 520–850 ppb total hydrocarbons) on the growth of grass shrimp *Palaemonetes pugio* larvae was investigated (Tatem, 1977). Exposure and control groups were maintained in petri dishes containing the appropriate exposure medium. Exposure media were changed and larvae

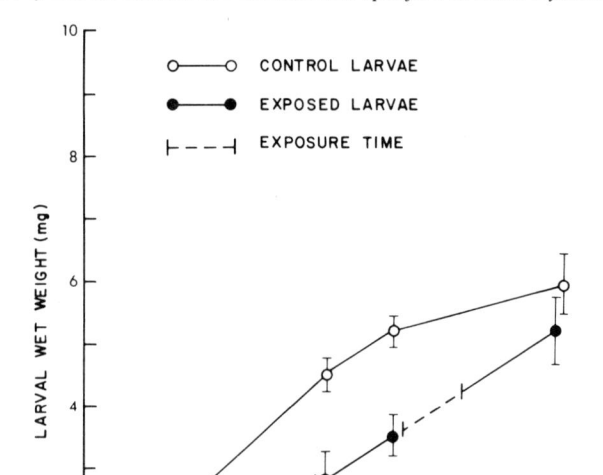

FIG. 14. Effects of exposure to a dilute WSF of No. 2 fuel oil (0·5–0·8 ppm total hydrocarbons and 0·3 ppm total naphthalenes) on the growth of larvae of the grass shrimp *Palaemonetes pugio*. Larvae were exposed to the WSF only during periods indicated by the dashed lines. Each value represents the mean wet weight of 10–12 larvae. Vertical lines represent standard errors of the mean (from Tatem, 1977, with permission of Pergamon Press).

fed newly hatched brine shrimp *Artemia salina* daily. After 12 days, WSF-exposed larvae showed significantly less growth than controls ($p < 0.001$) (Fig. 14). On the 12th day of the experiment, larvae exposed to the WSF were transferred to petri dishes containing oil-free seawater and allowed to depurate for 5 days. During this time, growth rates of the experimentals were similar to those of the controls. Experimental shrimp were then exposed again to the WSF for 2 days and returned to oil-free seawater for 10 days before final weight determinations. It was concluded from the data obtained that the exposed larvae recovered from the effects of exposure and grew at an accelerated rate, since final weights of control and experimental animals were not significantly different ($p > 0.05$). These results can be explained in part by the observation, stated earlier in this review, that young larvae of this species are very sensitive to oil and sensitivity decreases as larvae grow to juvenile and adult life stages (Table 8).

TABLE 20
Palaemonetes pugio, SURVIVAL AND DURATION OF ZOEAL STAGES OF LARVAE EXPOSED TO VARIOUS CONCENTRATIONS OF PHENANTHRENE (FROM YOUNG, 1977)

Phenanthrene concentration, µg/liter (ppb)	Duration of first zoeal stage (days)	Duration of total zoeal stages (days)	Survival[a] (per cent)
Control	3·31 ± 0·3	11·14 ± 0·1	77·8 ± 22·5
25	3·41 ± 0·3	11·38 ± 0·1	93·3 ± 10·3
50	3·38 ± 0·3	11·19 ± 0·1	93·3 ± 10·3
100	3·19 ± 0·3	11·27 ± 0·1	86·7 ± 24·2

[a] Means ± standard deviation from six replicates per exposure concentration.

Newly hatched larvae of *P. pugio* were exposed continuously for 30 days to phenanthrene (0–100 ppb) (Young, 1977). All surviving larvae appeared to develop normally to the postlarval stage. Chronic phenanthrene exposure did not significantly affect duration of the first larval instar, total duration of larval development, or survival of larvae to metamorphosis (Table 20). By the eighth day of development, larvae exposed to 100 ppb phenanthrene were observed to be less active than those in control and other exposure groups. Approximately one week after all animals reached the postlarval stage, shrimp in the 100 ppb group were significantly smaller in comparison to the other groups. Also, they were less active and appeared to consume fewer *Artemia salina* nauplii. About one-third of the shrimp in this 100 ppb exposure group developed an abnormal carapace which projected laterally away from the body. The condition disappeared completely within a few days when shrimp were transferred to clean seawater. Preliminary observations in the laboratory at Texas A&M University with other crustacean species indicate that phenanthrene interferes with the normal hardening of the crustacean exoskeleton after the molt. Body length of the shrimp was measured after 30-day exposure to phenanthrene and again 40 days after return to hydrocarbon-free seawater (Fig. 15). The mean length of shrimp in all exposure groups was less than that of controls, with the greatest growth inhibition occurring at 100 ppb phenanthrene. However, this latter group experienced accelerated growth when returned to clean seawater so that after 40 days all exposure groups had mean sizes similar to that of the controls.

To better define the effects of oil on crustacean growth, mud crabs *Rhithropanopeus harrisii* (Gould) were exposed continuously from hatching for six months to the WSF of No. 2 fuel oil (Laughlin *et al.*, 1978).

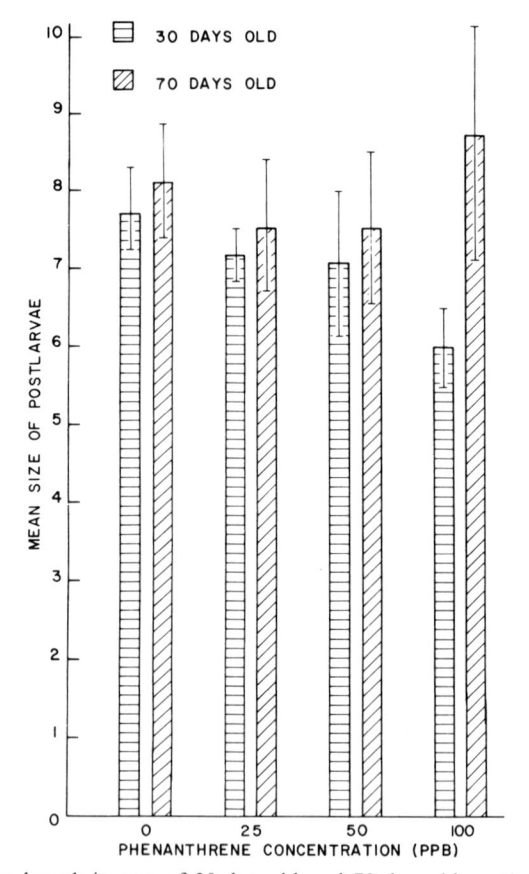

FIG. 15. Mean length in mm of 30-day-old and 70-day-old postlarvae of grass shrimp *Palaemonetes pugio*. Shrimp were exposed continuously to the phenanthrene concentrations indicated during the first 30 days only and were then returned to hydrocarbon-free seawater for the remaining 40 days. Vertical lines represent standard deviations of the means (from Young, 1977).

Observations were made on mortality, molting rate, and final weight. This growth study was conducted at a temperature of 25 °C and salinity of 15‰. Each exposure group (0, 0·16, 0·31, 0·63, 0·94, and 1·26 ppm total aqueous hydrocarbons, corresponding to 0, 2·5, 5, 10, 15 and 20 per cent WSF, respectively) consisted of 100 larvae. Animals were censused, fed *Artemia salina* nauplii, and exposure mixtures changed daily.

Larval development of *R. harrisii* is characterized by four zoeal stages followed by a megalops stage. The megalops then molts to the first crab

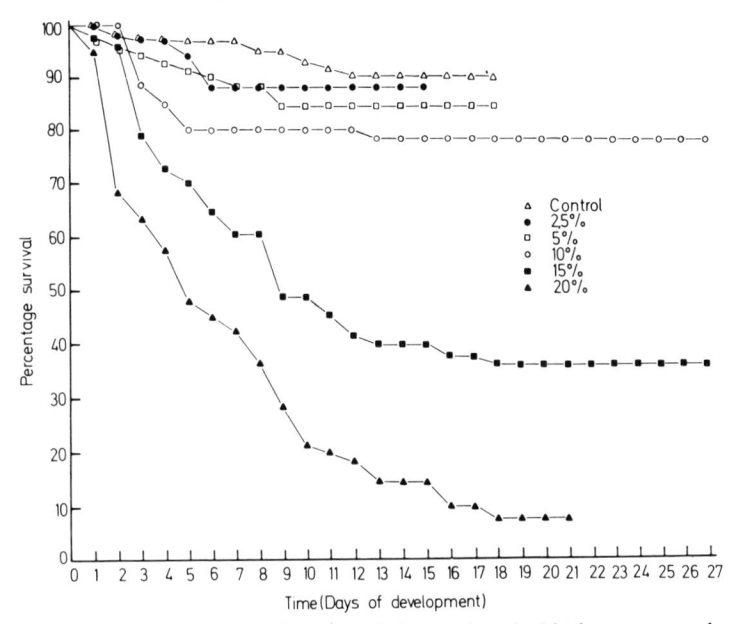

FIG. 16. Survival of the zoeal stages of the mud crab *Rhithropanopeus harrisii* during continuous exposure to several concentrations of the WSF of No. 2 fuel oil. Lines continue until the last surviving zoea metamorphosed to the megalops stage or died (from Laughlin *et al.*, 1978, with permission of Springer–Verlag).

stage. Control, 0·16, and 0·31 ppm exposure groups had similar numbers of survivors to the megalops stage (∼90 per cent) (Fig. 16). In the 0·63, 0·94, and 1·26 ppm exposure groups, survival to the megalops stage was 76, 30, and 6 per cent respectively. Most mortalities occurred between days 4 and 6 of exposure, the approximate time of the first zoeal molt. The time required to reach the megalops stage was different for each exposure group (Fig. 17). At lower WSF concentrations, the mean time of molting to the megalops was decreased relative to that of controls. At higher concentrations, mean time to the megalopal molt exceeded that of the controls. Zoeal molting was highly synchronized in the control and lower-concentration exposure groups. At the higher exposure concentrations, this synchrony was partially abolished and resulted in wide variations in intermolt periods among individuals in these groups.

Relatively few mortalities occurred among the megalops and crab stages at any exposure concentration. In addition, the mean duration of the megalops and first crab stages was not greatly affected by exposure to

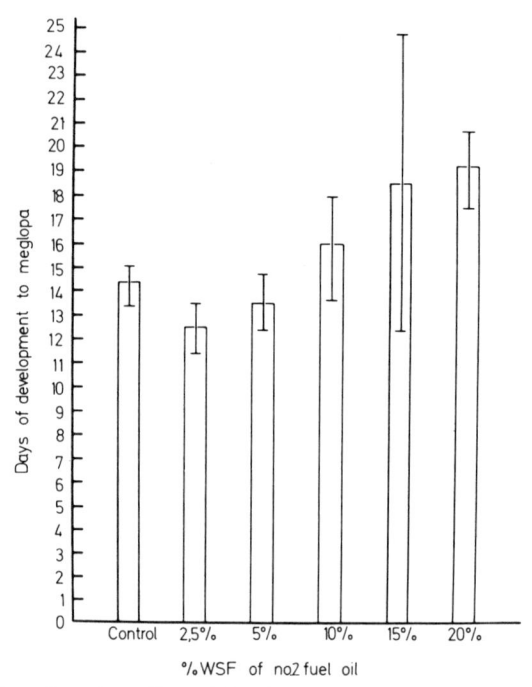

FIG. 17. Mean duration of larval development of the mud crab *Rhithropanopeus harrisii* from hatching to the molt to the megalops during continuous exposure to several concentrations of the WSF of No. 2 fuel oil. Vertical lines represent standard deviations (from Laughlin et al., 1978, with permission of Springer–Verlag).

No. 2 fuel oil WSF (Fig. 18). There was a tendency for the mean duration of the megalops stage to increase with increasing WSF concentration between 0 and 0·63 ppm WSF, but the variations were not significant. At 0·94 and 1·26 ppm WSF, mean duration of the megalops was only slightly longer than that of controls. Mean duration of the first crab stage exceeded that of the controls at all exposure concentrations. However, there was no correlation between exposure concentration and duration of the first crab stage and only in the 0·16 and 0·63 ppm exposure groups were deviations from the control significant.

The mean size of crabs in different instars after six months' continuous exposure to the WSF was determined (Table 21). The mean size of controls at any stage was consistently greater than that of animals exposed to 0·16, 0·31, or 0·63 ppm WSF. All surviving crabs, with the exception of the ninth stage individuals, exposed to 0·94 and 1·26 ppm WSF were larger than the

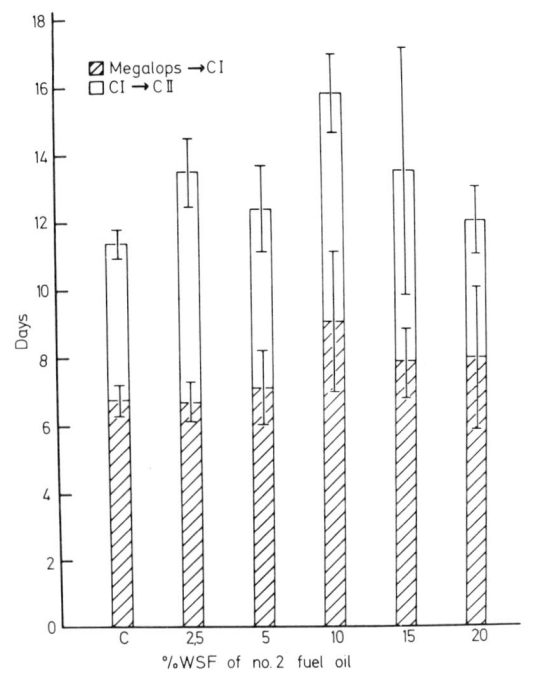

FIG. 18. Effect of the WSF of No. 2 fuel oil on the mean duration of the megalops (Megalops → CI) and first crab (CI → CII) stage of the mud crab *Rhithropanopeus harrisii*. Vertical lines represent standard deviations (from Laughlin *et al.*, 1978, with permission of Springer–Verlag).

corresponding controls. This may be due in part to the small sample size in these two exposure groups. However, the largest individuals in these groups were larger than any in other exposure groups or the controls.

After six months' continuous exposure to the WSF, surviving crabs were returned to oil-free seawater and observed for an additional five months. During this time one-quarter to one-third of the crabs from control and all exposure groups except the 1·26 ppm group were observed to produce eggs. However, only three individuals comprised this latter group. Zoeae were frequently seen in the aquaria indicating that the eggs were viable.

In this study, the first zoeal stage of *R. harrisii* was the stage most sensitive to oil. Later larval stages, the megalops and the crab stages, were much less sensitive to the No. 2 fuel oil WSF. This may in part be an artifact of the experimental design, since the most sensitive individuals would be eliminated early in the exposure period. Thus, only the heartiest, most

TABLE 21
THE MEAN SIZE OF EACH INSTAR OF *Rhithropanopeus harrisii* AFTER SIX MONTHS OF EXPOSURE TO VARIOUS CONCENTRATIONS OF WSF OF NO. 2 FUEL OIL (FROM LAUGHLIN *et al.*, 1978, WITH PERMISSION OF SPRINGER–VERLAG)

Crab stage	Concentration of WSF No. 2 fuel oil (per cent WSF and in parentheses ppm total hydrocarbons) Mean size of stage in mm plus standard deviation					
	0	2·5 (0·16)	5 (0·31)	10 (0·63)	15 (0·94)	20 (1·26)
4	—	3·07 ± 0·47	4·15 ± 0·35	2·60 ± 0·14	—	—
5	5·63 ± 1·23	3·69 ± 0·61	4·54 ± 1·01	3·91 ± 0·68	5·9 ± 0·85	—
6	5·23 ± 0·57	4·12 ± 0·51	4·83 ± 0·54	4·42 ± 0·79	6·63 ± 0·75	—
7	5·53 ± 0·90	4·97 ± 0·81	5·12 ± 0·50	4·59 ± 0·84	7·41 ± 0·96	6·97 ± 1·61
8	6·04 ± 0·77	5·09 ± 0·69	5·91 ± 0·69	5·1 ± 0·54	6·62 ± 0·88	—
9	6·87 ± 0·84	6·0[a]	6·13 ± 0·29	5·8[a]	6·57 ± 0·25	6·8[a]
10	—	4·8[a]	—	7·0[a]	—	—

[a] Absence of a standard deviation indicates only one animal was present at this stage.

resistant individuals survived long enough to be exposed at the megalops and later molt stages. However, Katz (1973) exposed each zoeal stage of the closely related crab *Neopanope texana* separately to a WSF of light Venezuelan crude oil and showed that larvae exposed initially at the first zoeal stage had a much higher mortality than groups exposed initially at later zoeal stages. Recently, Cucci and Epifanio (1979) reported that the 48-h LC_{50} of the WSF of Kuwait crude oil (API reference oil No. I) to first and second zoeae of the mud crab *Eurypanopeus depressus* was 10 ppm and 17 ppm total dissolved hydrocarbons, respectively. Mortality of larvae initially exposed to the WSF at zoeal stage III or IV was not significantly higher than that of controls. Thus we can conclude that the first zoeal stage of mud crabs is the most sensitive to oil and tolerance increases with age and larval instar.

Katz (1973) and Cucci and Epifanio (1979) also reported delayed molting among those crabs that survived oil exposure. Cucci and Epifanio (1979) reported that more than 15 per cent of the megalops stage larvae surviving continuous exposure to a 40 per cent WSF of Kuwait crude oil molted to an extra morphologically abnormal megalops stage. In the experiments reported here, zoeal molting rate and thus time to the megalops stage was decreased by lower and increased by higher concentrations of the WSF. Recently, we showed that exposure to sublethal concentrations of phenanthrene decreased the rate of development of *R. harrisii* larvae while exposure to sublethal concentrations of naphthalene increase larval development rate (Laughlin and Neff, 1979). It would appear, as suggested by Epifanio (1971) that changes in larval molting rate may be a sensitive index of sublethal pollutant stress in decapod crustaceans. Later molt stages and intermolt periods were less sensitive to pollutant-induced perturbation. After six months' exposure, mean sizes of crabs exposed to lower concentrations of WSF were less than those of controls, and mean sizes of crabs exposed to the two highest concentrations were greater than the average weights of controls. The small variations observed are of uncertain ecological significance.

Several important conclusions can be drawn from the development and growth studies reported here. Concentrations of aqueous petroleum hydrocarbons which elicited significant sublethal development and growth responses are close to acutely toxic concentrations, for the species studied. Growth perturbations appear, in some cases, to decrease in magnitude during the course of long term exposure to the WSF of oil. In growth experiments with *Palaemonetes pugio*, sublethal responses rapidly diminished when shrimp were returned to oil-free seawater. Thomas and Rice (1975) and Rice *et al.* (1977) observed transitory changes in the breathing and coughing rates (brief reversal of water flow in the opercular cavity) of pink salmon *Oncorhynchus gorbuscha* fry during exposure to concentrations of the WSF of Prudhoe Bay crude oil equivalent to 20 per cent of the 96-h LC_{50} concentration. Breathing and coughing rates rose to a maximum in 3–6 h after the beginning of exposure and then began to drop back toward the control rate. When fish were returned to oil-free seawater breathing and coughing rates rapidly returned to normal.

These findings all strongly suggest that marine animals have some ability to acclimate to and recover from exposure to sublethal concentrations of oil. Acclimation may involve induction or increase in activity of hydrocarbon-metabolizing enzymes in response to hydrocarbon exposure (see recent reviews by Varanasi and Malins, 1977; Bend and James, 1978; Neff, 1979).

The enzyme system responsible for the metabolism of aromatic hydrocarbons and several related classes of lipophilic organic compounds is the microsomal cytochrome P-450 mixed-function oxygenase system. This enzyme system has been detected in tissues of marine annelids, crustaceans, elasmobranchs and teleost fish (Neff, 1979). Its presence in marine molluscs is uncertain (Anderson, 1978; Vandermeulen and Penrose, 1978; Payne and May, 1979). Exposure to petroleum or specific aromatic hydrocarbons results in induction of mixed-function oxygenase activity in polychaete worms (Lee et al., 1977, 1979; Rossi and Anderson, 1978b), copepods (Walter et al., 1979) and possibly some bivalve molluscs (Anderson, 1978), but apparently not in the other species of marine invertebrates examined to date (Singer and Lee, 1977; Payne and May, 1979). Several investigators have demonstrated induction of the mixed-function oxygenase system in tissues of marine fish as a result of exposure to oil in the laboratory and field (Payne and Penrose, 1975; Payne, 1976; Gruger et al., 1977; Kurelec et al., 1977; Stegeman, 1978; Walton et al., 1978; Chambers, 1979). Induction of petroleum-metabolizing enzyme systems may be an important factor in the development of resistance adaptation to oils by marine animals.

3.3.4 Reproductive Success: Polychaetes

To determine the effects of chronic petroleum exposure on the reproductive biology of the polychaete worm *Neanthes arenaceodentata*, groups of worms were exposed continuously to several concentrations of the WSF of No. 2 fuel oil for three successive generations (Rossi and Anderson, 1978a). These studies were initiated by exposing 20 male/female pairs of immature adult worms to each of four WSF concentrations, with a control, during brood tube formation, oogenesis, and subsequent fertilization of spawned eggs. Powdered alfalfa (20 mg/couple) was added every other day for feeding and tube construction. Oocyte development rate was measured by noting the time period between appearance of oocytes within female coeloms and fertilization. Fecundity measurements were made by counting the number of zygotes produced per couple. Exposure was continued throughout subsequent development of zygotes into larvae, and eventually juvenile polychaetes. Larval development rate was determined by noting the time required for newly hatched metatrochophore larvae to develop into the feeding juvenile stage. The number of 32-segment juveniles produced per brood (clutch of zygotes) was taken as a relative index of brood mortality among members of this first (F_1) generation of WSF-exposed polychaetes. Exposure and feeding was continued as juveniles developed into immature adult polychaetes. Some of these were harvested for interbreeding of males and females ($F_1 \times F_1$) by the same techniques used to produce F_1 larvae from parental generation matings ($P \times P$). The above procedures were repeated for three successive generations (F_1, F_2, F_3).

Exposure concentrations chosen were 2·5, 5, 10, and 25 per cent WSF, corresponding to mean initial measured hydrocarbons concentrations of 0·25, 0·5, 0·95, and 2·1 ppm. During the 24 h between media changes, exposure concentrations dropped by approximately one order of magnitude (Fig. 19).

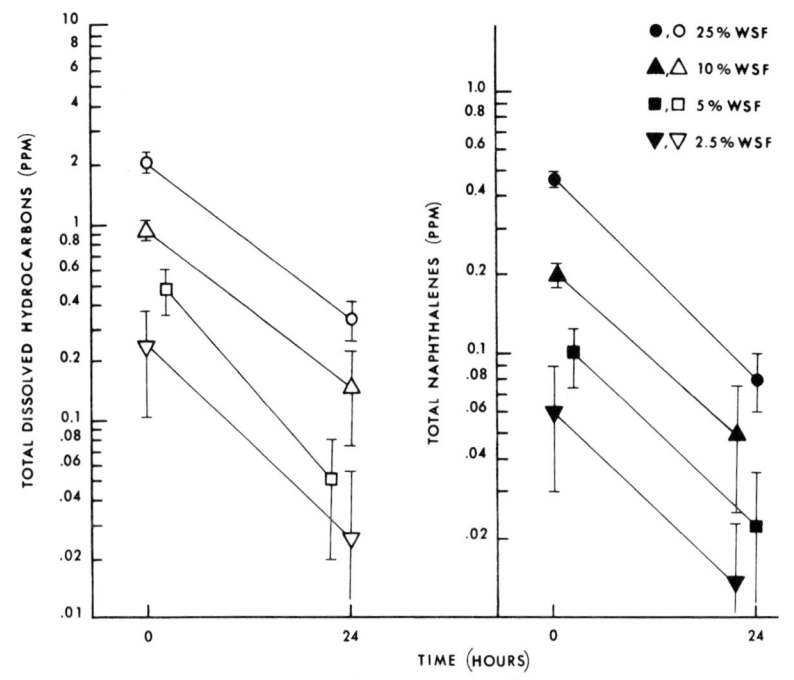

FIG. 19. Mean concentrations of total naphthalenes and total dissolved hydrocarbons in the WSF of No. 2 fuel oil used for exposures of successive generations of reproducing polychaete worms *Neanthes arenaceodentata*. Values are the means of 20 analyses ± standard deviations (vertical lines) (from Rossi and Anderson, 1978a, with permission of D. Reidel Publishing Co.).

Sublethal concentrations of No. 2 fuel oil WSF failed to affect the time required for successive generations to mature into the feeding juvenile stage (18-segment or 21-day-old animals) (Table 22). None of the concentrations tested, over all three generations, resulted in larvae growing at rates different from those of unexposed controls.

Unlike larval development rate, fecundity in exposed couples, for each generation, was affected by all WSF concentrations (Table 22). A decrease in the mean number of zygotes produced/couple was noted in the parental (P × P) and subsequent first generation ($F_1 \times F_1$) matings in 5, 10, and 25 per cent WSF. Only after a complete generation of exposure did couples in 2·5 per cent WSF produce zygotes in significantly reduced numbers.

There was little evidence that continued exposure reduced fecundity beyond the initial repression observed in P × P pairings. Fecundity among second generation couples which were inbred ($F_2 \times F_2$) was quite similar to

TABLE 22

ZYGOTE PRODUCTION AND SUBSEQUENT LARVAL DEVELOPMENT RATE IN SUCCESSIVE GENERATIONS OF POLYCHAETES (*Neanthes arenaceodentata*) CHRONICALLY EXPOSED TO SUBLETHAL CONCENTRATIONS OF NO. 2 FUEL OIL WSF. FECUNDITY IS EXPRESSED AS THE MEAN NUMBER OF ZYGOTES ($\pm$ S.D.) PRODUCED PER COUPLE FOR n NUMBER OF COUPLES (IN PARENTHESES) AT EACH CONCENTRATION. LARVAL DEVELOPMENT RATE WAS THE TIME (DAYS) REQUIRED FOR ZYGOTES TO DEVELOP INTO FEEDING STAGE JUVENILES (FROM ROSSI AND ANDERSON, 1978a, WITH PERMISSION OF D. REIDEL PUBLISHING CO.)

| | Generation | \multicolumn{6}{c}{Exposure concentration (per cent WSF and ppm total hydrocarbons in parentheses)} |
		0	2·5 (0·25)	5 (0·5)	10 (0·95)	25 (2·1)
Fecundity (number of zygotes)	P × P	240 ± 20 (20)	220 ± 20 (16)	159 ± 22 (16)	138 ± 23 (12)	86 ± 26 (14)
	F$_1$ × F$_1$	235 ± 21 (20)	123 ± 34 (34)	125 ± 40 (15)	79 ± 28 (13)	83 ± 40 (4)
	F$_2$ × F$_2$	242 ± 14 (10)	145 ± 36 (17)	150 ± 26 (18)	92 ± 40 (17)	82 ± 22 (10)
Larval development rate (days)	F$_1$	21 ± 1 (10)	19 ± 2 (16)	19 ± 1 (10)	20 ± 3 (10)	18 ± 2 (6)
	F$_2$	21 ± 0·5 (10)	21 ± 3 (15)	21 ± 3 (14)	23 ± 2 (10)	20 ± 3 (3)
	F$_3$	21 ± 0·4 (10)	21 ± 1 (17)	20 ± 2 (18)	20 ± 1 (14)	21 ± 1 (7)

that in original pairings of unexposed adults (P × P). A trend developed for fecundity to be inversely related to exposure concentrations among P × P couples. This trend was not seen in the $F_1 \times F_1$ pairings. Likewise, fecundity among $F_2 \times F_2$ couples was not significantly related to exposure concentration, except in comparing responses of worms in 25 per cent WSF to those in 2·5 per cent WSF. Nevertheless, the similarity in fecundity for the 2·5 and 5 per cent WSF-exposed couples, and the 10 and 25 per cent WSF-exposed pairs in both first and second generation matings, suggests the presence of a threshold value for concentration-dependent fecundity suppression. This threshold falls somewhere between 5 and 10 per cent WSF which is equivalent to 0·1 and 0·2 ppm total naphthalenes initially present, respectively.

Brood mortality among F_1 generation juveniles was strongly influenced by WSF concentration (Fig. 20). Per cent survival was inversely related to exposure concentration. Juvenile mortality data for F_2 and F_3 generations indicated that the animals have some ability to adapt to chronic hydrocarbon exposure. This adaptation is best illustrated by the increased survival of F_2 and F_3 juveniles exposed to 25 per cent WSF. In addition, the per cent survival among juveniles exposed to 5 and 10 per cent WSF increased to control (unexposed) values in the F_2 and F_3 generations.

Perhaps the most dramatic effect of chronic exposure to No. 2 fuel oil WSF was the concentration-dependent reduction in fecundity among *N. arenaceodentata* couples. This reduction in fecundity was due to reduction in the actual number of eggs produced per female since greater than 95 per cent of the mature oocytes developed into viable metatrochophore larvae. The failure of eleocytes (nurse cells) to transfer available energy reserves to developing secondary oocytes was shown to account for the reduction of fecundity in *N. arenaceodentata* exposed to hypoxic conditions (Davis and Reish, 1975). The effect of the WSF on fecundity is probably indirect. Mated pairs consumed food in an inverse relationship to WSF concentration. Oil-exposed couples probably possessed reduced energy reserves owing to their comparatively poor nutritional condition, and consequently produced fewer zygotes. No evidence of reduced oocyte diameter was seen in these studies, as has been noted in polychaetes exposed to other environmental stresses (Bellan *et al.*, 1971; Davis and Reish, 1975).

Decreases in oocyte maturation time for all exposure concentrations through successive generations were not expected (Fig. 20), since chronic exposure failed to affect the timecourse of larval maturation. Increases in oocyte maturation time were observed in the polychaete *Capitella capitata*

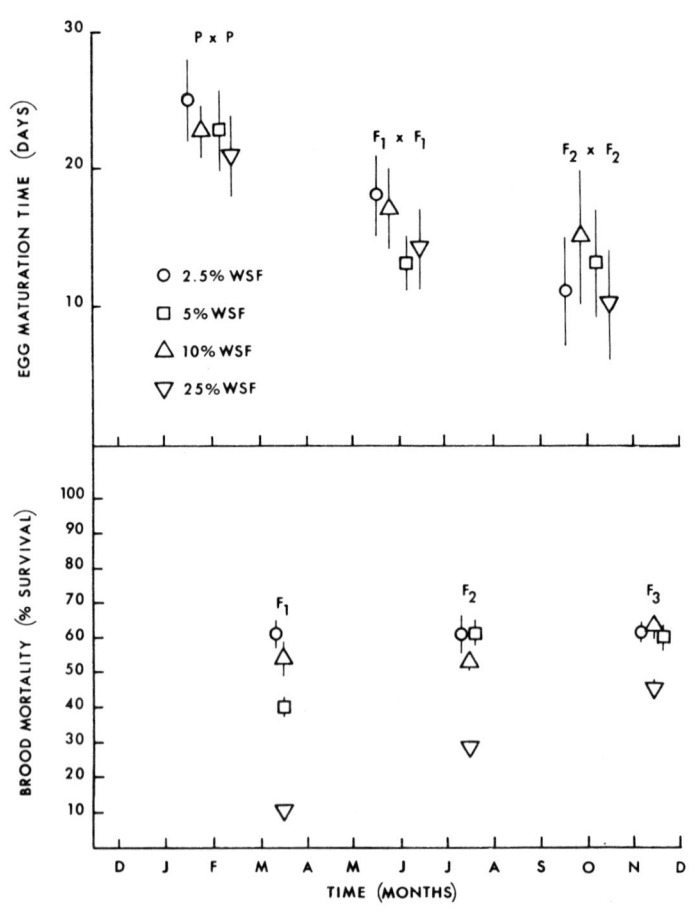

FIG. 20. Egg maturation rate and subsequent brood mortality in successive generations of the polychaete worm *Neanthes arenaceodentata* continuously exposed to sublethal concentrations of the WSF of No. 2 fuel oil. Symbols represent mean values for data from 10 polychaete couples $\pm$ standard deviations (vertical lines). In unexposed (control) worms egg maturation requires 22 ± 2 days, and the per cent survival to the 32-segment feeding stage is 64 ± 4 per cent (from Rossi and Anderson, 1978a, with permission of D. Reidel Publishing Co.).

during exposure to the detergent, polyethyleneglycol fatty acid (Bellan et al., 1972). Disruptions in timing of reproductive functions is commonly observed in studies with various environmental pollutants (Sprague and McLeese, 1968; Anderson et al., 1974b; Peakall, 1975). One can postulate that observed increases in oogenesis rate represent part of a more generalized response to biochemical stress, as is seen in recent work with chlorinated hydrocarbons and the horseshoe crab *Limulus polyphemus* (Neff and Giam, 1977).

Nutritional stress probably could account for differences in brood mortality in oil-exposed animals. Oil exposure apparently produced concentration-dependent decreases in the ability of juvenile worms to utilize available food and substrate. Most of the mortality among feeding juveniles was due to cannibalism of smaller, weaker animals. Consequently, experimental conditions produced strong pressure for selection of juveniles least impaired by oil exposure. Evidence of strong selection pressure was seen in the improved vigor (decreased brood mortality) among F_2 and F_3 juveniles, especially among those exposed to 25 per cent WSF. Induction of adaptive biochemical mechanisms may play a compensatory role here. Thus, exposure to high (10 and 25 per cent) WSF concentrations decreased the reproductive potential in the first generation *Neanthes arenaceodentata*. However, increased survival among F_2 and F_3 broods reduced the net effect of chronic exposure on reproductive potential.

3.3.5 Photosynthetic Carbon Fixation by Coral Zooxanthellae

Exposure to the WSF of No. 2 fuel oil for 72 h resulted in a highly significant ($p = 0.002$) decrease in the rate of photosynthetic carbon fixation by zooxanthellae of fire coral *Millepora* sp. (Fig. 21). ^{14}Carbon uptake rate decreased in a linear fashion with increasing WSF concentration. Exposure to the 30 per cent WSF (2·6 ppm total hydrocarbons) resulted in an approximately 50 per cent decrease in ^{14}C fixation rate. However, when corals were allowed to recover for 72 h, no statistically significant differences in ^{14}C fixation rate could be detected between control and previously exposed specimens. In all groups, ^{14}C fixation rates were significantly higher after a 72-h recovery period than at the end of the exposure period.

The reverse was seen in *Madracis decatis*. No differences were observed in ^{14}C fixation rates of control and WSF-exposed corals immediately after exposure to the No. 2 fuel oil WSF was terminated (Fig. 22). However, after 72 h of recovery in hydrocarbon-free seawater there was a statistically significant ($p = 0.017$) inverse relationship between the previous exposure

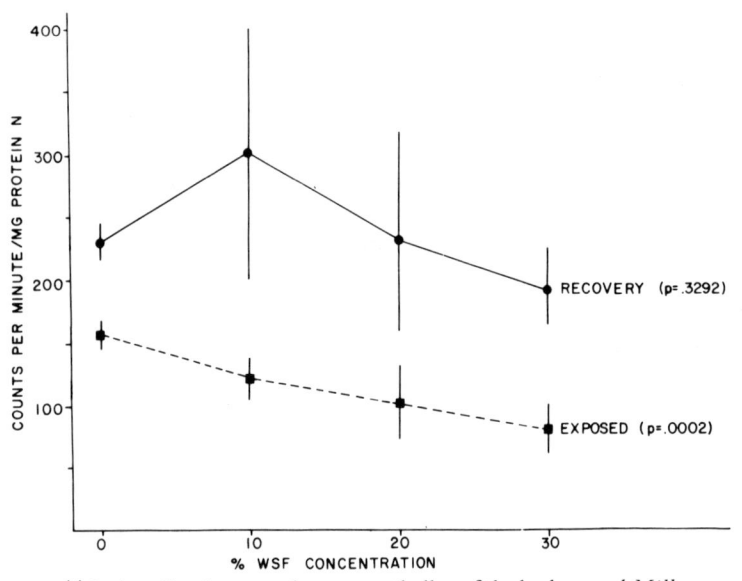

FIG. 21. ^{14}Carbon fixation rates by zooxanthellae of the hydrocoral *Millepora* sp. immediately after 72-h exposure (exposed) to WSF of No. 2 fuel oil and after 72-h recovery (recovery) in hydrocarbon-free seawater. Vertical bars represent standard deviations. Results of ANOVA are included for each curve.

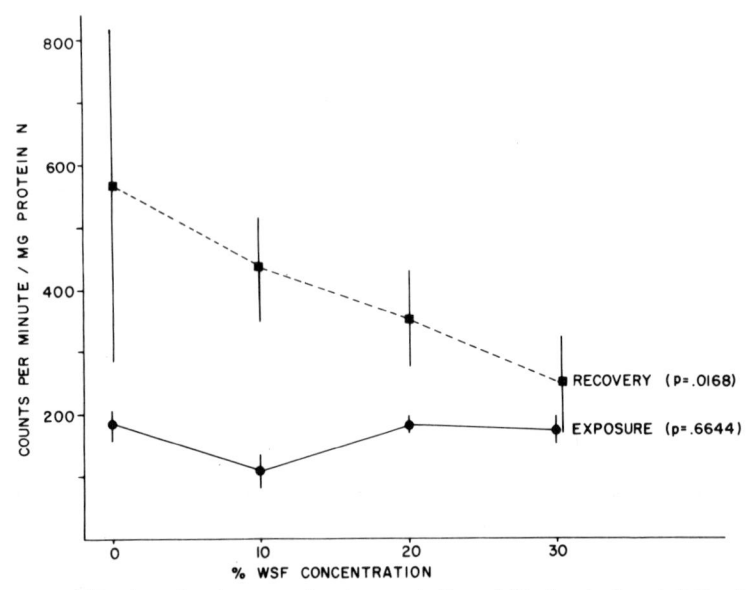

FIG. 22. ^{14}Carbon fixation rates by zooxanthellae of *Madracis decatis* following exposure to WSF of No. 2 fuel oil for 72 h and following recovery in clean seawater for 72 h.

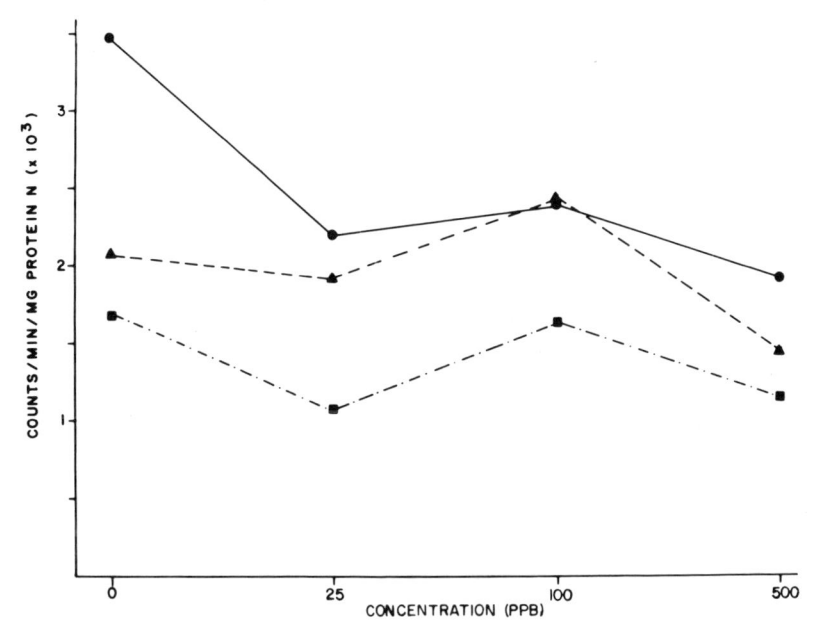

FIG. 23. ^{14}Carbon fixation rates by zooxanthellae of *Millepora* sp. following exposure to phenanthrene for 24–96 h. ●——● 24 h ($p = 0.0168$), ▲---▲ 48 h ($p = 0.0889$), ■—·—■ 96 h ($p = 0.1289$).

concentration and ^{14}C fixation rate by the coral zooxanthellae. Previous exposure to the 30 per cent WSF produced a 60 per cent decrease in the mean photosynthetic carbon fixation rate. In all groups, mean ^{14}C fixation rates were again higher in the animals allowed to recover for 72 h than in those sampled immediately after exposure.

Photosynthetic carbon fixation rates of zooxanthellae of *Favia fragum* and *Montastrea annularis* were unaffected by 24-h exposure to the WSF of No. 2 fuel oil and south Louisiana crude oil, respectively.

Exposure to phenanthrene for 24 h significantly inhibited ^{14}C fixation by zooxanthellae of *Millepora* sp. (Fig. 23). The inhibition was dose-dependent and reached a maximum of 54 per cent at a phenanthrene concentration of 500 ppb. After 48 h, phenanthrene exposure produced only a marginally significant ($p = 0.089$) decrease in ^{14}C fixation rates. Photosynthetic carbon fixation rates were unaffected by 96-h exposure of the corals to phenanthrene. The lack of a highly significant difference with exposure to phenanthrene at 48 and 96 h may have been due to a general decline in health of the control animals. Rate of ^{14}C fixation by the controls dropped from a mean of 3491 CPM/mg protein-N at 24 h to 1168 CPM/mg protein-N at 96 h. This decline in control values tended to obscure any effects due to phenanthrene exposure. At all three sampling times, corals exposed to 100 ppb phenanthrene accumulated ^{14}C at a slightly higher rate than those exposed to 25 ppb phenanthrene.

We can conclude from these experiments that photosynthesis by coral zooxanthellae is not greatly affected by acute exposure to concentrations of oil WSF that are sublethal to the host. Free-living marine algae are quite sensitive to oil and particularly to petroleum-derived aromatic hydrocarbons such as naphthalene (see review of Vandermeulen and Ahern, 1976). In some species, photosynthesis and growth are stimulated by low concentrations of petroleum hydrocarbons and inhibited by high concentrations. The greater tolerance of coral zooxanthellae may be attributed to the very limited ability of the coral host to accumulate petroleum hydrocarbons from the ambient medium (see Chapter 4).

3.3.6 Calcium Deposition by Reef Corals

Preliminary experiments were performed at Texas A&M University to determine the effect of the WSF of No. 2 fuel oil on the rate of calcium deposition in the skeleton of the coral *Oculina diffusa*. The results of this experiment are summarized in Table 23.

The rate of calcium deposition was quite variable at all WSF-exposure concentrations. However, there was a definite trend toward a decrease in the rate of calcium deposition with increasing concentrations of the WSF. The mean amount of ^{45}Ca deposited in 3 h by control corals was 6.77×10^{-5} µg/mg protein-N. Corals exposed to 20 per cent WSF (approximately 1·7 ppm total hydrocarbons) for 24 h deposited skeletal ^{45}Ca at a mean rate of 3.01×10^{-5} µg/mg protein-N. If the single extremely anomalous result in this group (sample 20-1) is excluded, the mean rate is reduced to 1.65×10^{-5} µg ^{45}Ca/mg protein-N. A one-way analysis of variance (ANOVA) of the ^{45}Ca deposition data revealed that regardless of whether the anomalous value is included, there was a significant difference in the amount of ^{45}Ca deposited by control and WSF-exposed corals ($p < 0.01$). Exposure to 5 and 10 per cent WSF caused an approximately 60 per cent reduction in the amount of ^{45}Ca deposited by *O. diffusa* in 3 h.

At all WSF concentrations the coral polyps were extended on all coral fragments during some part of the exposure period. However, it was apparent that polyp extension was more pronounced in control corals than in WSF-exposed fragments.

More detailed experiments were performed with five species of Bermuda reef corals and the WSF of No. 2 fuel oil and south Louisiana crude oil.

TABLE 23

EFFECTS OF EXPOSURE FOR 24 h TO THE WATER-SOLUBLE FRACTION (WSF) OF NO. 2 FUEL OIL ON THE RATE OF ^{45}Ca DEPOSITION IN THE SKELETON OF THE CORAL *Oculina diffusa* FROM TEXAS. ^{45}CALCIUM DEPOSITION WAS MEASURED OVER A 3-h PERIOD. THE RATE OF CALCIUM DEPOSITION BY WSF-EXPOSED CORALS IS SIGNIFICANTLY LOWER THAN THAT BY CONTROLS AT THE $p < 0.01$ LEVEL

Exposure concentration, per cent WSF and (ppm total hydrocarbons)	Sample No.	Total protein-N (mg)	^{45}Ca deposition/3 h (μg ^{45}Ca/mg protein-N $\times 10^{-5}$ ($\pm$ standard deviation))
0 (0)	0-1	11.0	5.41
	0-2	13.0	3.38
	0-3	4.1	7.52 $\bar{x} = 6.77 \pm 3.73$
	0-4	12.7	2.38
	0-5	2.9	10.10
	0-6	1.1	11.80
5 (0.45)	5-1	7.5	2.18
	5-2	9.0	2.12
	5-3	10.0	3.42 $\bar{x} = 2.15 \pm 1.12$
	5-4	13.1	1.60
	5-5	11.9	3.21
	5-6	2.9	0.36
10 (0.87)	10-1	12.7	1.80
	10-2	4.7	3.03
	10-3	7.5	4.04 $\bar{x} = 2.08 \pm 1.37$
	10-4	8.9	1.61
	10-5	1.0	[a]
	10-6	1.7	1.99
20 (1.74)	20-1	4.3	9.84
	20-2	4.4	2.82
	20-3	1.1	1.92 $\bar{x} = 3.01 \pm 3.64$[b]
	20-4	11.5	3.50 $\bar{x} = 1.65 \pm 1.61$[c]
	20-5	9.7	[a]
	20-6	3.4	[a]

[a] CPM below background, ^{45}Ca deposition assumed to be zero.
[b] Includes sample 20-1.
[c] Excludes sample 20-1.

The effects of exposure to No. 2 fuel oil and south Louisiana crude oil WSF on the rate of calcium deposition in the coral skeleton were quite variable. Calcium deposition data for experiments with five species of corals are summarized in Table 24. *Millepora* sp. was little affected by exposure to the WSF of either No. 2 fuel oil or south Louisiana crude oil. Immediately after 72-h exposure to the WSF, or following 72 h of recovery in clean

TABLE 24
SUMMARY OF THE EFFECTS OF EXPOSURE TO THE WSF OF NO. 2 FUEL OIL (NO. 2 F.O.) AND SOUTH LOUISIANA CRUDE OIL (SO. LA.) ON ^{45}Ca DEPOSITION RATE BY BERMUDA REEF CORALS

Species	WSF	Test of significance	Comments[a]
Millepora sp.	No. 2 F.O.	n.s.[b]	+
	recover	n.s.	+
	So. La.	n.s.	+
	recovery	n.s.	+
Madracis decatis	No. 2 F.O.	1 significant	+ significant
		2 n.s.	−n.s.
	recovery	1 significant	all +
		2 n.s.	
	So. La.	significant	+
	recovery	significant	+
Montastrea annularis	No. 2 F.O.	n.s.	+
	recovery	significant	+
Oculina diffusa	No. 2 F.O.	significant	−(at 20 and 30 per cent WSF)
	recovery	significant	+(at 30 per cent WSF)
Favia fragum	No. 2 F.O.		
	(24 h)	n.s.	variable

[a] +, Increased ^{45}Ca deposition rate with increasing exposure concentration.
−, Decreased ^{45}Ca deposition rate with increasing exposure concentration.
[b] n.s., Not statistically significant at the $p < 0.1$ level by analysis of variance.

seawater, there was a trend for ^{45}Ca deposition rates to increase with increasing exposure concentrations. None of the rates showed significant differences from those of controls. *Favia fragum* was unaffected by 24-h exposure to the WSF of No. 2 fuel oil.

In three experiments, *Madracis decatis* showed highly variable responses to exposure to No. 2 fuel oil WSF (Fig. 24). The rate of ^{45}Ca deposition at all exposure concentrations in each experiment was quite variable, resulting in large standard deviations. Only one of the experiments showed deposition rates of WSF-exposed animals to be significantly different from those of the controls ($p = 0.08$), and in this case ^{45}Ca uptake increased with increasing exposure concentration. A second experiment showed a declining trend with increasing WSF concentration, but the differences were not statistically significant. In the recovery experiment, all three experimental groups showed trends toward increasing rates of ^{45}Ca deposition corresponding to increasing exposure concentrations, but only

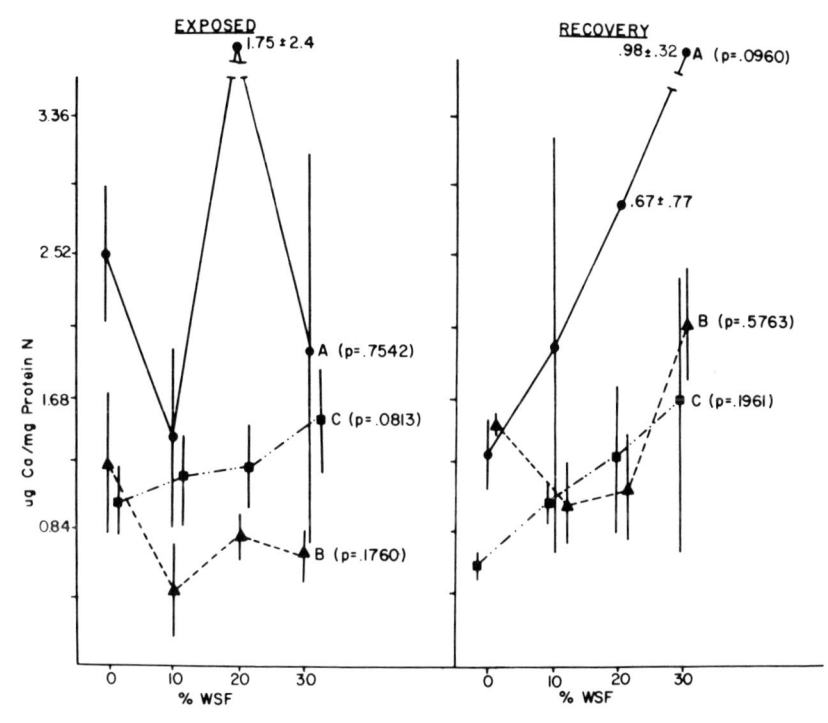

FIG. 24. ^{45}Calcium deposition rates by *Madracis decatis* in three experiments following exposure to WSF of No. 2 fuel oil for 72 h and following recovery for 72 h in hydrocarbon-free seawater.

one of these was marginally significant ($p = 0.096$). There was little difference between mean ^{45}Ca deposition rates of animals immediately after exposure to the WSF and corresponding groups after 72-h recovery.

Only one experiment was performed with *Madracis decatis* and south Louisiana crude oil WSF. The rate of ^{45}Ca deposition increased significantly ($p = 0.053$) with increasing WSF-exposure concentration immediately after 72 h of exposure to oil (Fig. 25). The same trend was seen following 72 h of recovery in clean seawater. The rate of ^{45}Ca deposition was similar in the three groups of animals formerly exposed to different concentrations of the WSF and was significantly higher than that of the controls ($p = 0.081$). In each exposure group, there was little difference in deposition rate before and after 72 h of recovery.

Oculina diffusa responded quite differently to exposure to No. 2 fuel oil WSF. ^{45}Calcium deposition rates were similar in the controls and 10 and 20

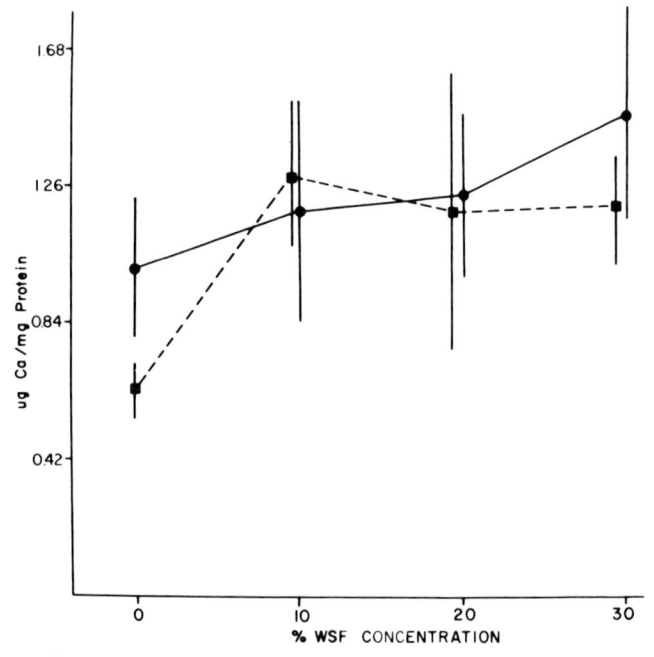

FIG. 25. ^{45}Calcium deposition rates of *Madracis decatis* following exposure for 72 h (●——●, $p = 0.0528$) to WSF of south Louisiana crude oil and following recovery for 72 h (■---■, $p = 0.0810$) in clean seawater.

per cent WSF-exposure groups, but were significantly ($p = 0.062$) depressed in the 30 per cent WSF-exposure group (Fig. 26). Controls and animals exposed to 10 per cent WSF and then permitted to recover for 72 h showed greatly enhanced mean ^{45}Ca deposition rates, as compared to animals sampled immediately after the exposure period. However, the mean rates for these groups had large standard deviations. Following exposure to 20 or 30 per cent WSF, ^{45}Ca deposition rates of recovered animals were similar to or slightly higher than those of the same groups immediately after exposure. Because of the large standard deviations, the rate of mineral deposition of recovered animals was only marginally affected by WSF exposure ($p = 0.108$).

^{45}Calcium deposition rates of *Montastrea annularis* increased slightly but not significantly compared to the rates of control animals immediately after 72 h of exposure to No. 2 fuel oil WSF (Fig. 27). Following 72 h of recovery in clean seawater, ^{45}Ca deposition rates increased significantly ($p = 0.094$) with increasing WSF exposure concentration. There was little

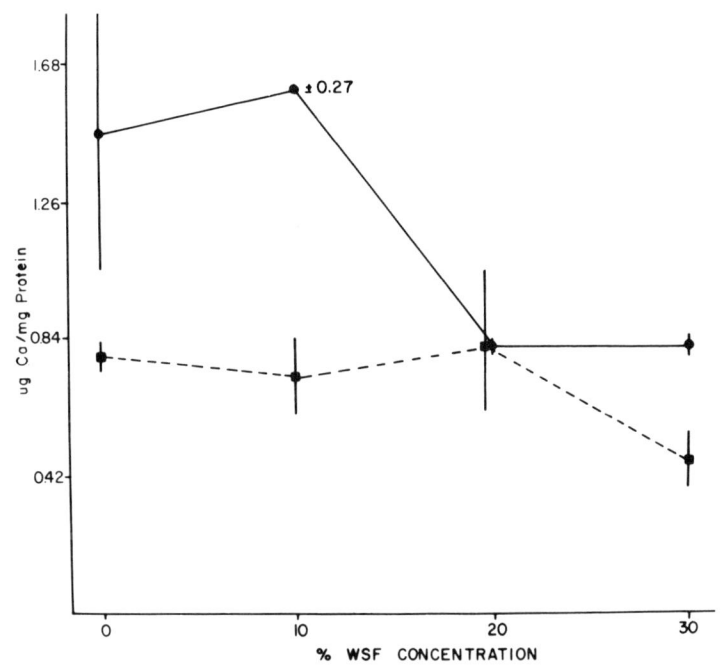

FIG. 26. ^{45}Calcium deposition rates of *Oculina diffusa* following exposure for 72 h (●——●, $p = 0.1077$) to WSF of No. 2 fuel oil and following recovery for 72 h (■---■, $p = 0.0624$) in clean seawater.

difference in the ^{45}Ca uptake rate of controls at the end of the exposure period and 72 h later. However, at all WSF concentrations, ^{45}Ca deposition rates were much higher following 72 h of recovery than immediately after exposure.

Exposure to phenanthrene had variable effects on the rate of ^{45}Ca deposition by *Millepora* sp. (Fig. 28). Control rates were similar at the 24-, 48-, and 96-h sampling periods. ^{45}Calcium uptake was greatly enhanced following a 24-h exposure to 25 ppb phenanthrene but decreased slightly with exposure to 100 and 500 ppb phenanthrene. Because of the variability of responses, the effect of phenanthrene on deposition rates was significant at only 24 and 96 h.

During exposure to the WSF of No. 2 fuel oil, polyps of *Madracis decatis* and *Montastrea annularis* remained partially or totally retracted. Polyps of control corals were fully expanded during all or most of the exposure period. At the two highest WSF concentrations, some pieces of *M. annularis* extruded their zooxanthellae and *M. decatis* polyps took on a

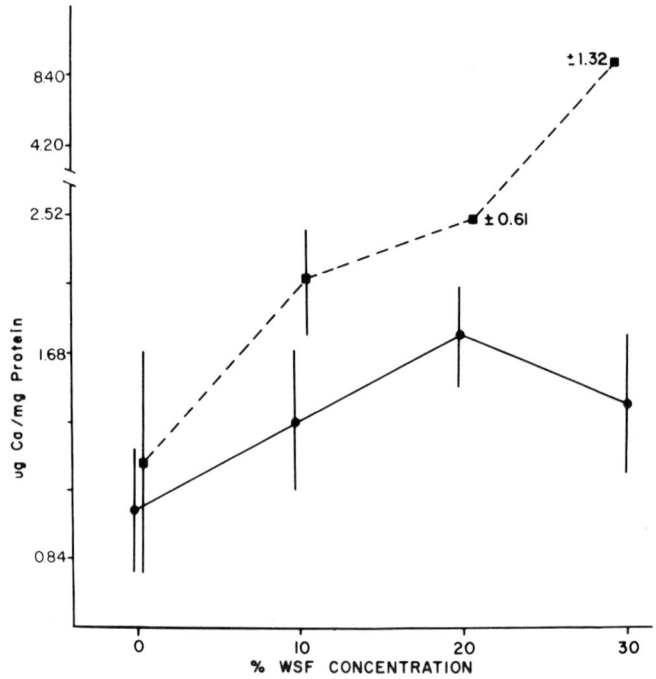

FIG. 27. ^{45}Calcium deposition rates of *Montastrea annularis* following exposure for 72 h (●——●, $p = 0.2807$) to WSF of No. 2 fuel oil and following recovery for 72 h (■---■, $p = 0.0094$) in clean seawater.

slightly bleached appearance. In contrast, the corals *Oculina diffusa* and *Millepora* sp. appeared normal at all WSF concentrations. Polyp behaviour and appearance in all species appeared to be unaffected by exposure to the WSF of the south Louisiana crude oil.

The apparent rate of ^{45}Ca deposition proved to be a sensitive indicator of sublethal pollutant stress in corals. Ability to quantify even small amounts of ^{45}Ca in skeletal material made it possible to detect small differences in calcium deposition rates between control and exposed animals. The ease of handling ^{45}Ca and the short period of exposure to the isotope required to obtain results made it possible to assay large numbers of animals.

A major problem with this approach is the large variation in ^{45}Ca uptake rates by corals. Large variations in ^{45}Ca deposition rates between different coral heads or even different regions of the same colony under both field and laboratory conditions have been reported by several other investigators (Goreau, 1959; Goreau and Goreau, 1959; Drew, 1973; Clausen and Roth,

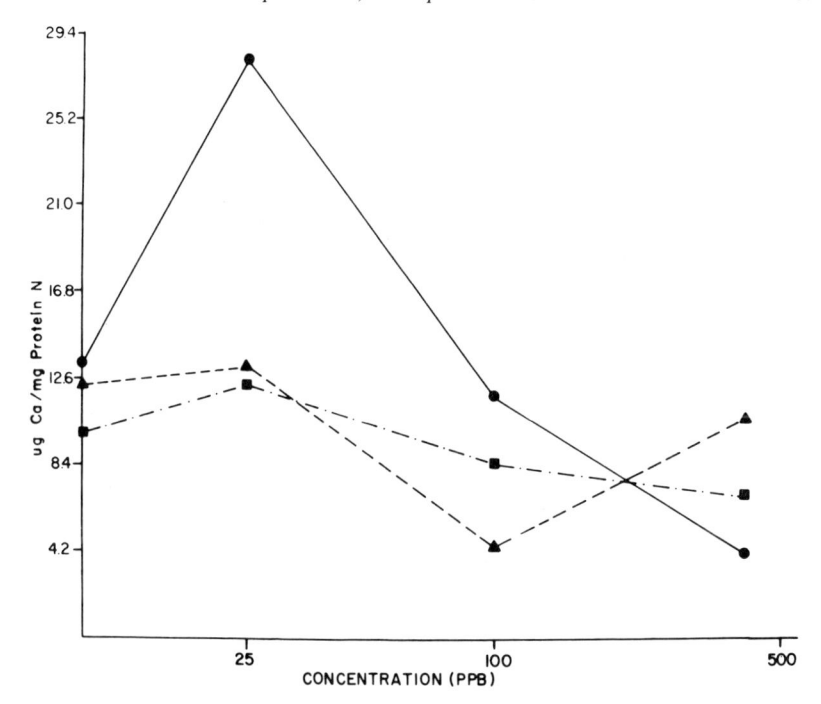

FIG. 28. ^{45}Calcium deposition rates by *Millepora* sp. following exposure to phenanthrene for 24 to 96 h. ●——● 24 h ($p = 0.0148$), ▲---▲ 48 h ($p = 0.8359$), ■-·-·-■ 96 h ($p = 0.0495$).

1975; Stromgren, 1976). The calcification process may be discontinuous and individual polyps within the colony may be at different phases of the calcification cycle at any given time (Goreau and Goreau, 1959; Stromgren, 1976; Sweeney, 1976; Chalker and Taylor, 1978). In the experiments reported here, between sample variability was greater in hydrocarbon-exposed animals than in controls. This trend was also observed by Birkeland *et al.* (1976), who reported that the mean growth rates of individual branches of coral heads exposed to bunker oil differed significantly while growth rates of individual control branches showed much less variability. They suggested that this was due to individual variation between colonies or parts of colonies in sensitivity to oil. Use of a sufficiently large sample size, with similar sized coral pieces from the same region of colonies under uniform light conditions should partially ameliorate this problem and allow for accurate assessment of sublethal responses of corals to pollutants (Barnes and Crossland, 1977).

Some general trends emerged from the experiments reported here. In some experiments, the rate of ^{45}Ca deposition increased with increasing oil exposure concentrations. In other experiments the rate of ^{45}Ca deposition decreased with increasing oil exposure concentration.

It is quite possible that different results would have been obtained after a longer recovery period. This is indicated by those experiments in which changes in ^{45}Ca and ^{14}C uptake rates were observed after the 72-h recovery period but not immediately after the exposure period. Birkeland et al. (1976) reported that differences in coral survival between controls and corals exposed to marine diesel oil were not detectable for as long as one month after exposure was terminated. Delayed mortality of oil-exposed corals occurred up to 61 days after the end of the exposure period. In addition, corals surviving for 61 days after oil exposure had significantly lower growth rates than controls.

Although exposure to phenanthrene caused a reduction in uptake of both ^{45}Ca and ^{14}C by *Millepora* sp., in *Madracis decatis*, exposure to the No. 2 fuel oil WSF caused an increase in apparent ^{45}Ca deposition rate and a decrease in ^{14}C fixation rate. These results are surprising for it is well established that the rate of calcification by hermatipic corals is dependent on the rate of photosynthesis by their zooxanthellae. The rate of calcium deposition in the skeleton of hermatipic corals is much higher in the light than in the dark (Goreau, 1959; Goreau and Goreau, 1959; Pearse and Muscatine, 1971; Clausen and Roth, 1975). Specific inhibitors of photosynthesis also inhibit calcification (Vandermeulen et al., 1972; Chalker and Taylor, 1976). Thus, it is apparent that the hydrocarbon-induced enhancement of calcification rate observed in most of the experiments reported here is not due to any effect of the hydrocarbons on photosynthetic rate of the coral zooxanthellae.

^{45}Calcium in seawater readily exchanges with calcium adsorbed to the surface of coral skeletons in direct contact with seawater or separated from seawater by dead tissue. Passive exchange of ^{45}Ca is relatively low when the skeleton is separated from the seawater by living tissue (Goreau, 1959; Goreau and Goreau, 1960; Clausen and Roth, 1975; Barnes and Crossland, 1977). Exposure to petroleum hydrocarbons may increase the relative permeability of living coral polyp tissues to Ca^{2+} thereby increasing the rate of passive ^{45}Ca exchange between the medium and the coral skeleton. Phenanthrene, an aromatic component of the oil WSF (Anderson et al., 1974a), apparently was not responsible for this effect as it caused a decrease in calcification rate in *Millepora* sp.

Coral species apparently vary substantially in their sensitivity to oil. That

Madracis decatis and *Montastrea annularis* were severely stressed by the No. 2 fuel oil WSF was indicated by their aberrant behaviour in the presence of oil. Retraction of polyps and especially extrusion of zooxanthellae and bleaching are signs of severe stress in corals. These are the species that showed the most marked increase in ^{45}Ca deposition rate following exposure to the WSF. In contrast, *Oculina diffusa* and *Millipora* sp. which were behaviorally unaffected by exposure to the WSF, showed a slight decrease in calcification rate following exposure to the No. 2 fuel oil WSF or phenanthrene.

3.3.7 Calcium Deposition in Sea Urchin Spines

Spines of similar size from a single specimen of the sea urchin *Diadema antillarum* had similar rates of mineral regeneration as measured by ^{45}Ca incorporation. However, the rate of ^{45}Ca incorporation into regenerating spines of different sea urchin specimens was quite variable and was not correlated with urchin size, color morph, or other recognizable physical characteristics. Mean rates of regeneration of spines from different individuals under control conditions varied from 0·6 µg Ca/spine/day to 4·6 µg Ca/spine/day. Because of this variability and because 60 or more spines could be harvested from a single sea urchin, each experiment was performed with spines from a single individual.

The effects of the three petroleum hydrocarbons—naphthalene, phenanthrene and dibenzothiophene—on the rate of spine regeneration were roughly similar (Figs. 29, 30, 31). The highest exposure concentration used for all three hydrocarbons, 500 ppb, significantly decreased the rate of calcium incorporation into regenerating spines in comparison to control rates with one exception (one experiment with dibenzothiophene). In the replicate dibenzothiophene experiment, 500 ppb dibenzothiophene caused an approximately 70 per cent decrease in the rate of mineral regeneration. Phenanthrene exposure also gave variable results. In one experiment, exposure to 500 ppb phenanthrene resulted in an approximately 50 per cent reduction in regeneration rate, but in the second experiment the same concentration resulted in only a 20 per cent decrease in spine regeneration rate. Naphthalene at a concentration of 500 ppb caused a 60 per cent decrease in calcium deposition rate. Exposure to 25 or 100 ppb of the three hydrocarbons produced variable results. In three cases, naphthalene at 25 ppb and phenanthrene and dibenzothiophene at 100 ppb, the exposed spines incorporated calcium at a slightly faster rate, though not significantly so, than did the control spines,

Exposure to the WSF of diesel fuel oil produced more consistent results

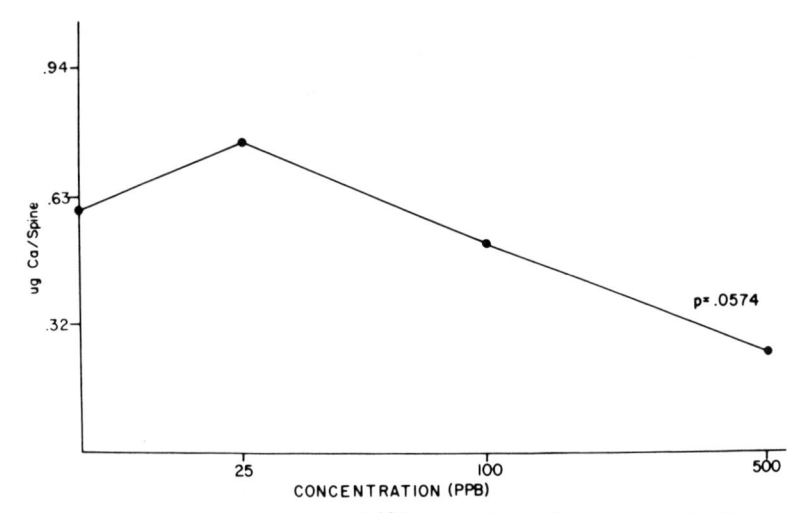

90 *Response of Marine Animals to Petroleum and Specific Petroleum Hydrocarbons*

FIG. 29. The rate of deposition of ^{45}Ca by spines of the sea urchin *Diadema antillarum* during exposure to naphthalene for 24 h.

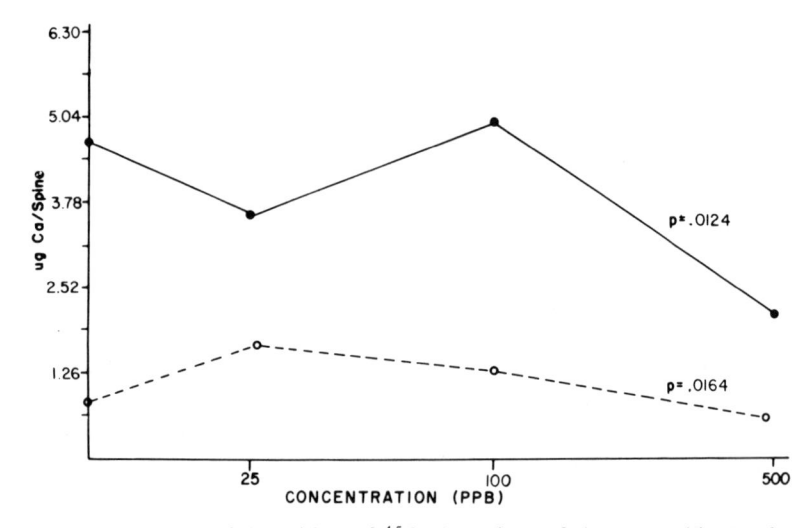

FIG. 30. The rate of deposition of ^{45}Ca by spines of the sea urchin *Diadema antillarum* during exposure to phenanthrene for 24 h. Results from two experiments are shown.

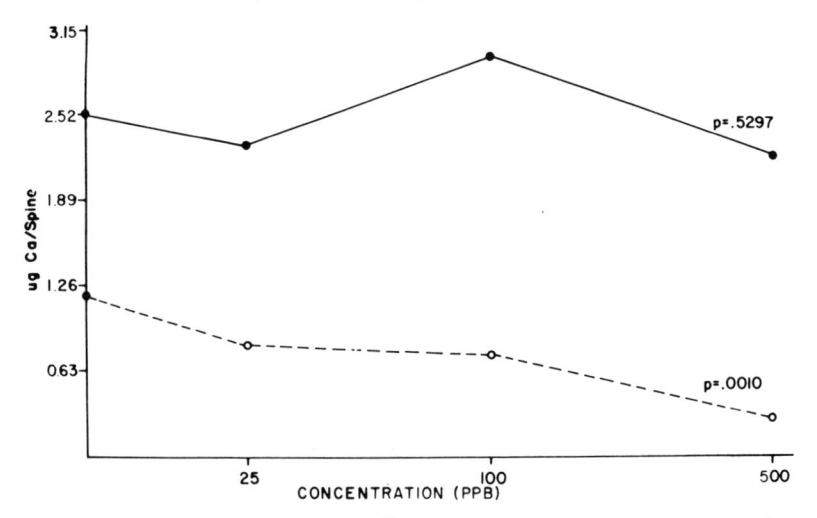

FIG. 31. The rate of deposition of ^{45}Ca by spines of the sea urchin *Diadema antillarum* during exposure to dibenzothiophene for 24 h. Results from two experiments are shown.

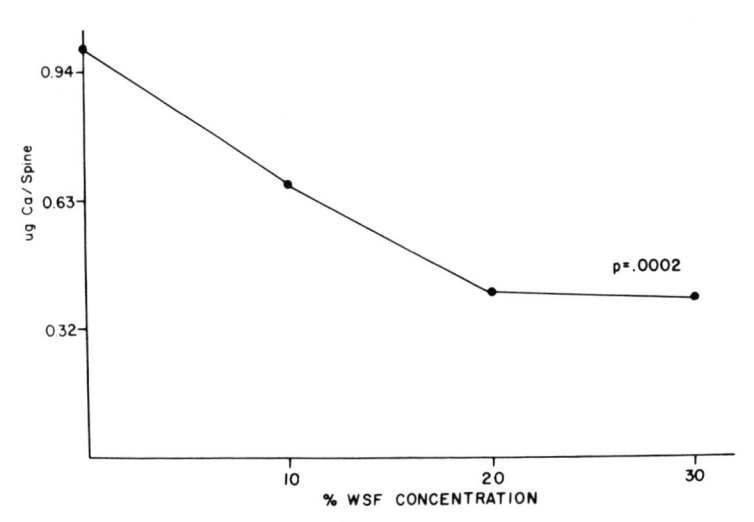

FIG. 32. The rate of deposition of ^{45}Ca by spines of the sea urchin *Diadema antillarum* during exposure to WSF of diesel fuel for 24 h.

(Fig. 32). The rate of spine regeneration was significantly inhibited, in comparison to the control rate, at all WSF concentrations used. Maximum inhibition was approximately 70 per cent and occurred in 20 and 30 per cent WSF (2–3 ppm total hydrocarbons). The 10 per cent WSF produced a 30 per cent inhibition of spine regeneration rate.

The rate of ^{45}Ca deposition in *in vitro* spine preparations from the sea urchin *Diadema antillarum* proved to be a useful indicator of pollutant stress. The main advantage of this system is that a large number of spines can be harvested from a single individual and exposed *in vitro* to a variety of pollutants and pollutant concentrations. Thus, regeneration rates of spines from a single individual, which vary less than those of spines from several sea urchins, can be compared under a variety of exposure conditions.

The process underlying spine regeneration in echinoids is poorly understood. Heatfield (1971a) suggested that the tissue responsible for spine regeneration is derived by migration of cells from proximal regions of relatively high mitotic activity within the shaft near the level of the milled ring to more distal regions. The migration of cells could account for the 48-h time lag between fracture of a spine and initiation of spine regeneration. The rate of spine regeneration was shown to be nearly linear for at least 4 days *in vitro* (Heatfield, 1970) and for at least 40 days *in vivo* (Heatfield, 1971b; Davies *et al.*, 1972). Both *in vivo* and *in vitro* regeneration rates were temperature dependent. Ultrastructure and mineralogy of the regenerated calcium carbonate as well as ultrastructure of the cell types associated with mineral regeneration have been described (Heatfield, 1971b; Davies *et al.*, 1972; Heatfield and Travis, 1975a, b). These papers reveal that spine regeneration is a complex physiological process under precise cellular control. Therefore, any environmental stress which interferes with normal cellular functions can be expected to adversely affect spine regeneration.

CHAPTER 4

Accumulation and Release of Petroleum Hydrocarbons

4.1 INTRODUCTION

Data on the presence, distribution, and possible sources of hydrocarbons in the marine environment were reviewed in Chapter 1. Because of their persistence (particularly the polycyclic aromatic hydrocarbons, PAH) and lipophilic nature, it is not surprising that hydrocarbons have frequently been detected in the tissues of fish and shellfish from a wide variety of different polluted and unpolluted habitats. Patterns of accumulation and release of petroleum hydrocarbons by marine animals were reviewed recently by Varanasi and Malins (1977) and Neff (1979). Ehrhardt (1972) detected a total of 236 ppm hydrocarbons in the tissues of oysters *Crassostrea virginica* from a region of Galveston Bay, heavily contaminated with industrial and petroleum pollution. Concentrations of aromatic hydrocarbons (mainly mono-, di-, and tricyclic aromatic hydrocarbons, 134 ppm) were higher than those of alkanes (102 ppm).

Cahnmann and Kuratsune (1957) detected several PAH, including pyrene, fluoranthrene, and benzo[*a*]pyrene, in the tissues of oysters from the vicinity of Norfolk, Virginia. Fazio (1971) obtained similar results with oysters from several polluted and unpolluted stations in Galveston Bay and from relatively unpolluted Aransas Bay, Texas. Total PAH burdens in oysters collected from several stations in Galveston Bay in November 1969 and January 1971 varied from 11 to 237 ppb. Several PAH, but not benzo[*a*]pyrene, were detected in oysters from both contaminated and uncontaminated stations. PAH present at highest concentrations in the tissues of oysters from contaminated stations (those closed to commercial fishing) were fluoranthrene (7·8 ppb), pyrene (6·5 ppb), benzo[*b*]fluoranthene (2·2 ppb), and benzo[*e*]pyrene (2·1 ppb). Oysters from uncontaminated

Aransas Bay contained lower concentrations of these PAH, ranging from 0·2 ppb (benzo[*e*]pyrene) to 1·7 ppb (fluoranthene).

Concentrations of the carcinogenic PAH, benzo[*a*]pyrene, in sediments and mussels *Mytilus edulis* and *M. californianus* from the vicinity of Vancouver, British Columbia were studied by Dunn and Stich (1975; 1976*a*). Benzo[*a*]pyrene (B*a*P) concentrations were highest near major sources of pollution and decreased with distance away from the pollutant source. Sediment B*a*P concentrations varied from 121 µg B*a*P/kg dry sediment (ppb) near the Iona Island sewage outfall to about 0·4 ppb at the stations furthest downstream. Mussels were heavily contaminated with B*a*P in the immediate vicinity of a powerboat marina. Highest levels were found in animals collected from creosoted pilings (up to 215 ppb) which indicated creosote to be an important source of B*a*P contamination. Further support for this conclusion comes from the observation of Dunn and Fee (1979) that lobsters *Homarus americanus* which had been kept in a commercial tidal pond constructed of creosoted timber contained very high concentrations of benzo[*a*]pyrene and several other PAH. The digestive glands of contaminated lobsters contained up to 2300 ng/g (2·3 ppm) benzo[*a*]pyrene, the highest value ever reported for any foodstuff. B*a*P concentrations in mussels from southern California coastal waters were at or near zero, except in areas of intense industrial, shipping or recreational activity (Dunn and Young, 1976). Bivalve molluscs from several bays along the Oregon, USA coast contained highly variable body burdens of B*a*P (Mix *et al.*, 1977; Mix and Schaffer, 1979). Concentrations of B*a*P in the mollusc tissues ranged from less than 0·1 ng/g (ppb) to as high as 170 ng/g, with a majority of values less than about 5 ng/g. The authors concluded that point sources such as creosoted pilings, marinas, fish processing factories and boat traffic contributed to the B*a*P contamination. The majority of marine fin fish analyzed to date contained very low or undetectable levels of PAH (Pancirov and Brown, 1977; Dunn and Fee, 1979; Neff, 1979).

The presence of pollutant hydrocarbons in the tissues of natural populations of marine animals suggests that these organisms are able to accumulate hydrocarbons from water, food, or sediments. Several laboratory and field studies have lent support to this hypothesis.

Following a spill of $65-70 \times 10^4$ liters of No. 2 fuel oil in Buzzards Bay near Falmouth, Massachusetts on 16 September 1969, Blumer *et al.* (1970) studied the incorporation of petroleum hydrocarbons into bottom sediments and bivalve molluscs in the vicinity of the spill site. They reported that because of gale-force winds at the time of the spill, oil was incorporated into sediments lying under as much as 10 m of water. The GC profile of the hydrocarbon fraction of the sediment resembled that of the spilled oil, except that there was a general decrease in the low molecular weight hydrocarbons in the sediment. Tissue samples of oysters *Crassostrea virginica* and scallops *Aequipecten irradians* collected from Wild Harbor and West Falmouth Harbor two months after the spill were contaminated with petroleum hydrocarbons. The tissue samples appeared to be enriched in the higher molecular weight alkanes and in an unresolved background identified as the cyclic-aromatic fraction of the oil. However only three scallop samples and one oyster sample from oil contaminated stations were analyzed. Total hydrocarbon concentrations in these samples, determined gravimetrically, ranged from 7·4 to 69 mg/kg wet weight (ppm), compared to 2·3 and 5·5 ppm total hydrocarbons in two scallop samples from an uncontaminated station. Based on this limited data, the authors concluded that the more toxic aromatic hydrocarbons are retained in the mollusc tissues for long periods of time after removal of the source of pollution.

More recent studies of aromatic petroleum hydrocarbons in sediments from the spill site (Youngblood and Blumer, 1975; Teal *et al.*, 1978) reveal that many of the aromatics, and in particular the highly alkylated naphthalenes and phenanthrenes, were retained in the sediments for periods in excess of six years (although at greatly reduced concentrations). These sediment hydrocarbons undoubtedly provide a source of low-level chronic hydrocarbon contamination to the resident benthic fauna. Thus, although Blumer *et al.* (1970) did demonstrate that benthic marine animals are able to accumulate a wide spectrum of hydrocarbons from oil-contaminated water and sediments, they did not prove that once accumulated the aromatic hydrocarbon fraction would persist indefinitely in the tissues of the animals.

Subsequent studies have shown that in most cases marine animals are able to release petroleum hydrocarbons from their tissues rapidly when they are returned to an oil-free environment. Lee *et al.* (1972*a,b*) showed that the marine mussel *Mytilus edulis* and several species of fish were capable of accumulating aliphatic and aromatic hydrocarbons from dilute solution in seawater. The hydrocarbons were released rapidly when the animals were returned to hydrocarbon-free seawater. Oysters *Crassostrea virginica* accumulated hydrocarbons to high concentrations in their tissues when they were chronically exposed to oil-contaminated seawater. The maximum amount of hydrocarbons accumulated was directly correlated to tissue lipid levels. When returned to a low hydrocarbon medium, oysters released most of the hydrocarbons from their tissues (Stegeman and Teal, 1973). When mussels *Mytilus edulis* and *M. californianus* were transferred from clean seawater stations to polluted stations in San Francisco, California they accumulated hydrocarbons in their tissues (DiSalvo et al., 1975). When returned to clean water stations, their tissue hydrocarbon levels returned to near clean water baseline levels.

The mussel *Mytilus edulis* rapidly accumulated radiolabeled hydrocarbons from seawater solution (Hansen et al., 1978). Following exposure for 8 h, bioaccumulation factors (concentration in tissues/concentration in water) for toluene, naphthalene and phenanthrene were 4, 30, and more than 70, respectively. When the mussels were returned to clean seawater, toluene and naphthalene were released very rapidly and phenanthrene was released more slowly. Hexadecane was accumulated and released very slowly by the molluscs. Oysters *Crassostrea virginica* were suspended at a depth of 7 m in a controlled ecosystem enclosure (c 2 m diameter and 15 m deep containing 60 000 liters seawater) to which was added Prudhoe Bay crude oil spiked with several aromatic hydrocarbons (Lee et al., 1978). The oysters rapidly accumulated aromatic hydrocarbons, particularly naphthalene and alkyl naphthalenes. When oysters were transferred to clean seawater they released hydrocarbons to non-detectable levels within 23 days. Bieri and Stamoudis (1977) obtained similar results when they exposed oysters *Crassostrea virginica* and hard shell clams *Mercenaria mercenaria* to a spill of No. 2 fuel oil in an estuarine environment. Accumulation and release of aromatic hydrocarbons were rapid.

Recently Boehm and Quinn (1977) reported that hard shell clams *Mercenaria mercenaria* chronically exposed to petroleum hydrocarbons accumulated hydrocarbons to high levels in their tissues, but failed to release them when returned to clean seawater. These authors suggested that when molluscs are exposed to pollutant hydrocarbons for long enough a major proportion of the accumulated hydrocarbons enters a more stable tissue compartment (for example, depot lipid). Hydrocarbon release is then very slow as long as the molluscs are in positive nutritional balance. In contrast, Dunn and Stich (1976b) showed that mussels presumed to be chronically contaminated with BaP from nearby creosote pilings released the accumulated PAH when placed in a clean flowing seawater system. The mussels initially contained a mean concentration of 45 ppb BaP; tissue BaP declined exponentially over the six weeks of the experiment, with an overall half-life of 16 days.

Lee et al. (1976) showed that the blue crab *Callinectes sapidus* could accumulate paraffinic and aromatic hydrocarbons from both water and food. Hydrocarbons or their metabolites were rapidly eliminated, primarily in the feces. All hydrocarbons studied were readily metabolized to more polar metabolites in the hepatopancreas.

Rice *et al.* (1977) measured the accumulation of paraffinic and aromatic hydrocarbons by pink salmon fry *Oncorhynchus gorbuscha* during a 96-h exposure to the WSF of Cook Inlet crude oil and hydrocarbon release following return to oil-free seawater. Only the gills showed a net increase in the concentration of paraffins during the exposure period, with the maximum concentration being reached after 10 h. Aromatic hydrocarbons, benzene through trimethylnaphthalene, were accumulated by the three tissues studied: gill, gut, and muscle. These hydrocarbons again reached maximum concentrations after 10-h exposure then decreased during the remainder of the exposure period. Methylnaphthalenes were accumulated to highest concentrations in all tissues. Depuration to background levels was accomplished in 20 h or less. Interestingly, the pattern of aromatic hydrocarbon uptake and release correlated well with the sublethal responses measured (breathing and coughing rates). These responses reached a maximum after 5–10-h exposure and then decreased toward normal rates for the remainder of the exposure period.

Codfish *Gadus morhua* fed a diet containing Kuwait crude oil accumulated high concentrations of n-alkanes in the liver, but not in muscle tissue (Hardy *et al.*, 1974). The liver selectively retained n-alkanes with chain lengths of C_{24}–C_{28} rather than those with shorter chain lengths, although the crude oil was particularly rich in the latter. Korn *et al.* (1976) measured uptake and release of ^{14}C-benzene in different tissues of the northern anchovy *Engraulis mordax* and the striped bass *Marone saxatilis* during 48-h exposure to low concentrations of this compound in seawater. The anchovy accumulated benzene to higher concentrations in all tissues than did the striped bass. Maximum tissue concentrations were reached in all tissues but the gall bladder and intestine in 6–24 h. These latter two tissues continued to accumulate ^{14}C activity for the full 48-h exposure period. Gall bladder and intestinal tissues also contained the highest concentrations of ^{14}C activity at each sampling time. Radioactivity reached undetectable levels in all tissues of striped bass except gall bladder and liver within 7 days after return to benzene-free seawater. No attempt was made to identify benzene metabolites.

The studies discussed above all demonstrate that marine animals are able to accumulate hydrocarbons from either dilute solution or dispersion in seawater. Molluscs tend to accumulate hydrocarbons to higher levels and retain them much longer than do the other marine taxa so far investigated. This is probably due to the apparent inability of molluscs to metabolize hydrocarbons to more water-soluble and easily excreted polar metabolites (Lee et al., 1972a), since molluscs do not contain significantly higher concentrations of tissue lipids than other marine animals. Thus, hydrocarbon concentrations in mollusc tissues tend to reflect ambient levels of these materials in the environment. Consequently, several investigators have recommended use of molluscs to monitor the degree of environmental contamination by hydrocarbons and other pollutant materials (Fossato and Siviero, 1974; DiSalvo et al., 1975; Goldberg, 1975; Dunn and Stich, 1976a).

There is much less evidence for accumulation of petroleum hydrocarbons by marine animals from contaminated food and sediments (see recent review of Neff, 1979). The marine copepod *Colanus helgolaneicus* accumulated ^{14}C-labeled naphthalene from its food (Corner et al., 1976b; Harris et al., 1977). Uptake from food was much more efficient than uptake from solution. When juvenile blue crabs *Callinectes sapidus* were fed food containing radiolabeled aromatic hydrocarbons, between 7 and 10 per cent of the radioisotope was transferred from the stomach to other body tissues (Lee et al., 1976). Fish generally accumulated very little hydrocarbon from petroleum hydrocarbon-contaminated food (Corner et al., 1976b; Roubal et al., 1977; Whittle et al., 1977). This may reflect low absorption of hydrocarbons from the gut or rapid metabolism and excretion of accumulated hydrocarbons.

Bioavailability to benthic marine animals of hydrocarbons from petroleum contaminated sediments appears to be low. In several experiments, the sipunculid worm *Phascolosoma agassizii* and two species of bivalve molluscs, *Macoma inquinata* and *Protothaca staminea*, failed to accumulate significant amounts of aromatic hydrocarbons from sediments heavily contaminated with crude oil (Anderson et al., 1977b; Roesijadi et al., 1978a,b). Similar results were obtained when marine polychaete worms *Arenicola marina* were exposed to sediments contaminated with ^{14}C-naphthalene (Lyes, 1979). In most cases hydrocarbon concentrations in the tissues of sediment-exposed animals were similar to or lower than hydrocarbon concentrations in the sediment. Much of the uptake could be accounted for by accumulation of soluble hydrocarbons desorbed from the sediments. When English sole *Parophrys vetulus* were maintained for four months in aquaria containing sediments heavily contaminated with Alaskan crude oil, they rapidly accumulated alkane and aromatic hydrocarbons in skin, muscle and liver (McCain et al., 1978). However, during chronic exposure, tissue hydrocarbon concentrations decreased so that after 27 days' continuous exposure only the liver contained detectable levels of hydrocarbons.

Many questions remain unanswered concerning patterns of accumulation and release of petroleum hydrocarbons, in particular the more highly toxic and potentially carcinogenic aromatic hydrocarbons, by different species of marine animals. In order to better understand health hazards that might arise from consumption by man of petroleum-contaminated seafoods, the laboratory at Texas A&M University has investigated patterns of accumulation, body distribution, and release of petroleum-derived aromatic hydrocarbons by marine molluscs, worms, crustaceans, and fish. Results of these investigations are summarized below.

4.2 MATERIALS AND METHODS

Oils used in these experiments were API reference oils. Except where stated otherwise, water-soluble fractions (WSF) and oil-in-water dispersions (OWD) were prepared by methods described earlier (Anderson *et al.*, 1974*a*). For specific hydrocarbon uptake studies, the pure hydrocarbon was dissolved in nanograde acetone and aliquots of this acetone stock solution were added to artificial seawater to prepare exposure media of the appropriate concentration. The maximum amount of acetone used to prepare these solutions was 1 mg/liter, a concentration found to be completely non-toxic to the species under investigation.

Uptake and release experiments, except field studies, were conducted in static recirculating, or flow-through seawater systems. Exposure media were monitored periodically for hydrocarbon residues but were not renewed during the exposure phase of the experiments. After exposure, animals were returned to aquaria containing hydrocarbon-free artificial seawater which was continuously recirculated through activated charcoal filters to remove aqueous hydrocarbons.

At various times during the exposure period and following return to oil-free seawater, animals were sacrificed for hydrocarbon analysis. Specimens were rinsed several times in distilled water, blotted dry, and weighed. If not analyzed immediately, tissues were placed in all-glass containers with Teflon-lined caps and stored at $-76\,°C$ until analysis. In most cases, samples were analyzed for naphthalenes (naphthalene, methylnaphthalenes, and dimethylnaphthalenes) by the UV spectrophotometric technique developed by Neff and Anderson (1975). The same technique was used in several of the studies on accumulation of specific PAH.

A general outline of the procedure is as follows: tissues were homogenized at 0 °C in spectrophotometric grade n-hexane (Burdick & Jackson) with a Sorval Omnimixer; the homogenate was centrifuged, the supernatant recovered and treated with activated Florisil to remove UV-absorbing biogenic materials. Hexane extracts were then scanned with a Pye-Unicam SP 1800 recording spectrophotometer between the wavelengths of 210 and 240 nm. Concentrations in the extracts of naphthalene, methylnaphthalenes, and dimethylnaphthalenes were then computed by simultaneous equations constructed from UV spectra of authentic standards in hexane.

In several experiments, tissue hydrocarbons were analyzed by GC by Dr J. S. Warner, Battelle Columbus Laboratories (Warner, 1976). Tissues were packed in dry ice and air-shipped to Columbus, Ohio for this purpose.

4.3 RESULTS AND DISCUSSION

4.3.1 Uptake and Release of Petroleum Hydrocarbons from Water: Laboratory Studies

Oysters *Crassostrea virginica* (Gmelin) exposed in a static system to a concentrated No. 2 fuel oil OWD (1 per cent in water) for 4 days accumulated a wide spectrum of petroleum hydrocarbons in tissues (Table 25) (Neff *et al.*, 1976*b*). Petroleum hydrocarbons analyzed by GC included C_{12}–C_{25} n-paraffins and aromatic hydrocarbons in the boiling point range from naphthalene to dimethylphenanthrene. n-Paraffins accounted for about 3·2 per cent of the total 96·7 ppm petroleum hydrocarbons accumulated by the oysters. Naphthalenes were accumulated to the highest concentrations. Particularly noteworthy was the presence of 35·6 ppm mixed isomers of dimethylnaphthalene. Among the other aromatic hydrocarbons analyzed, monomethyl analogs were usually present at higher concentration in oyster tissues than were either dimethyl analogs or parent compounds. Dibenzothiophene, the only heterocyclic compound analyzed, was present at a relatively low concentration.

A group of clams *Rangia cuneata* (Gray) and oysters *Crassostrea virginica* was exposed to No. 2 fuel oil in a flow-through system (Neff *et al.*, 1976*b*). Oysters were collected from Galveston Bay and clams from Trinity Bay just prior to the experiment. Oyster tissues contained low background concentrations of naphthalenes and those of the clams a small amount of methylphenanthrenes (Tables 26 and 27, exposure time 0). The total volume of the exposure tank was 46 liters; flow rate was adjusted to

TABLE 25
OIL-DERIVED HYDROCARBONS IN THE TISSUES OF OYSTERS *Crassostrea virginica* FOLLOWING EXPOSURE TO A 1 PER CENT OWD OF NO. 2 FUEL OIL FOR 4 DAYS. ALL CONCENTRATIONS ARE IN ppm (μg/g WET WEIGHT OF TISSUE) (FROM NEFF *et al.*, 1976*b*, WITH PERMISSION OF SPRINGER–VERLAG)

Oil hydrocarbon	Tissue concentration (ppm)	Total (ppm)
Paraffins		
C_{12}–C_{25} n-paraffins		3·1
Naphthalenes		
naphthalene	6·3	
1-methylnaphthalene	9·4	
2-methylnaphthalene	15·4	
dimethylnaphthalene	35·6	
trimethylnaphthalene	17·4	84·1
Other aromatics and heterocyclics		
biphenyl	0·2	
methylbiphenyl	1·0	
dimethylbiphenyl	0·7	
fluorene	1·2	
methylfluorene	1·7	
dimethylfluorene	0·5	
dibenzothiophene	0·6	
phenanthrene	1·7	
methylphenanthrene	1·7	
dimethylphenanthrene	0·2	9·5
Total oil-derived hydrocarbons analyzed		96·7

15·6–16·5 liters/h. Oil was continuously injected into the inflow stream before it reached the centrifugal metering pump. Following 8-h exposure in the flow-through system, clams and oysters were returned to clean filtered seawater.

The flow-through exposure system was highly efficient in dispersing oil into the aqueous phase of the exposure tank. Concentration of dispersed aqueous petroleum hydrocarbons in the exposure tank, as measured by IR analysis, rose rapidly to a maximum of 470 ppm at 4 h. Concentration then dropped gradually, reaching 302 ppm at 8 h.

Following 8-h exposure to the concentrated OWD oysters were heavily contaminated with petroleum hydrocarbons (Table 26). A total of 312 ppm petroleum hydrocarbons were detected in the tissues; n-paraffins were present at the highest concentration. The depuration pattern was extremely complex and variable, undoubtedly reflecting the great variability in the

TABLE 26

ACCUMULATION OF PETROLEUM HYDROCARBONS BY THE OYSTER *Crassostrea virginica* DURING EXPOSURE TO DISPERSED NO. 2 FUEL OIL IN A FLOW-THROUGH SYSTEM AND SUBSEQUENT RELEASE OF HYDROCARBONS WHEN THE OYSTERS WERE RETURNED TO OIL-FREE SEAWATER (FROM NEFF *et al.*, 1976*b*, WITH PERMISSION OF SPRINGER–VERLAG)

Time (hours)	n-p[a]	N	1-MN	2-MN	DMN	TMN	B	MB	F	MF	DBT	P	MP	DMP	Total
Exposure														
0	—	0·2	0·1	0·3	1·0	0·8	—	—	—	—	—	—	—	—	2·4
8	235	14·7	8·7	15·0	21·8	9·1	0·3	0·5	1·0	1·2	0·3	1·9	1·9	0·3	312
Depuration														
3	156	12·0	8·4	12·0	22·7	10·8	0·3	0·4	0·7	0·7	0·3	1·3	1·3	0·2	228
6	68	7·3	5·1	7·3	13·2	5·7	0·1	0·2	0·4	0·2	0·1	0·6	0·6	0·1	109
24	18	6·5	5·7	7·6	14·8	9·5	0·2	0·2	0·5	0·7	0·2	1·2	1·3	0·3	67
120	10	8·2	4·7	6·8	13·4	4·9	0·1	0·1	0·2	0·1	0·1	0·4	0·4	0·2	54
672	—	—	—	0·1	0·5	0·9	—	—	—	—	—	—	—	—	1·5

[a] n-p, n-Paraffins; N, naphthalene; 1-MN, 1-methylnaphthalene; 2-MN, 2-methylnaphthalene; DMN, dimethylnaphthalenes; TMN, trimethylnaphthalenes; B, biphenyl; MB, methylbiphenyls; F, fluorene; MF, methylfluorenes; DBT, dibenzothiophene; P, phenanthrene; MP, methylphenanthrenes; DMP, dimethylphenanthrenes.

TABLE 27
ACCUMULATION OF PETROLEUM HYDROCARBONS BY THE CLAM *Rangia cuneata* DURING EXPOSURE TO DISPERSED NO. 2 FUEL OIL IN A FLOW-THROUGH SYSTEM AND SUBSEQUENT RELEASE OF HYDROCARBONS WHEN THE CLAMS WERE RETURNED TO OIL-FREE SEAWATER
(FROM NEFF et al., 1976b, WITH PERMISSION OF SPRINGER–VERLAG)

Time (hours)	n-p[a]	N	1-MN	2-MN	DMN	TMN	B	MB	F	MF	DBT	P	MP	DMP	Total
Exposure															
0	—	—	—	—	—	—	—	—	—	—	—	—	—	—	0·2
8	66	3·8	2·6	3·9	7·4	3·6	0·1	0·1	0·3	0·3	0·1	0·5	0·8	0·1	89
Depuration															
3	23	2·2	1·7	2·6	4·2	1·8	—	—	0·1	—	—	0·2	0·2	—	36
6	1·1	2·0	1·9	2·9	4·9	1·9	—	0·1	0·1	0·2	0·1	0·3	0·5	—	16
24	1·0	0·6	0·7	1·1	2·5	1·3	—	—	0·1	0·2	—	0·3	0·3	—	8·1
120	—	—	0·1	0·1	0·6	0·6	—	—	—	—	—	—	—	—	1·4
672	—	—	—	—	—	—	—	—	—	—	—	—	—	—	0

[a] n-p, n-Paraffins; N, naphthalene; 1-MN, 1-methylnaphthalene; 2-MN, 2-methylnaphthalene; DMN, dimethylnaphthalenes; TMN, trimethylnaphthalenes; B, biphenyl; MB, methylbiphenyls; F, fluorene; MF, methylfluorenes; DBT, dibenzothiophene; P, phenanthrene; MP, methylphenanthrenes; DMP, dimethylphenanthrenes.

TABLE 28

ACCUMULATION, RELEASE AND BIOACCUMULATION FACTORS (TISSUE CONCENTRATION/EXPOSURE WATER CONCENTRATION) OF NAPHTHALENES BY THE CLAM *Rangia cuneata* EXPOSED TO THE WATER-SOLUBLE FRACTION OF NO. 2 FUEL OIL (FROM NEFF et al., 1976b, WITH PERMISSION OF SPRINGER–VERLAG)

Petroleum hydrocarbon	Concentration in exposure water (ppm)	Tissue concentration after 24-h exposure (ppm)	Bioaccumulation factor	Tissue concentration after 24-h depuration (ppm)	Per cent released in 24 h
Naphthalene	0·84	1·9	2·3	0·4	79
1-Methylnaphthalene	0·34	2·9	8·5	1·9	34
2-Methylnaphthalene	0·48	3·9	8·1	1·9	51
Dimethylnaphthalenes	0·24	4·1	17·1	2·8	32
Trimethylnaphthalenes	0·03	0·8	26·7	0·4	50

response of oysters during and following exposure to high concentrations of oil. In general, release of accumulated hydrocarbons was extremely rapid. More than 90 per cent n-paraffins were released in 24 h. n-Paraffins remaining were lost more slowly. After 120 h (5 days) in oil-free seawater, 10 ppm n-paraffins remained; after 672 h (28 days) no petroleum-derived n-paraffins could be detected in oyster tissues. Aromatic hydrocarbons, though present in tissues at much lower concentrations than n-paraffins after 8-h exposure, were lost more slowly. All aromatic hydrocarbons showed roughly similar behavior.

Concentrations of different aromatics in oyster tissues remained essentially constant during the first 120 h post-exposure. However, after 672 h, concentration of aromatic hydrocarbons had reached background levels. As observed in unexposed control oysters, di- and trimethylnaphthalenes were the aromatics remaining in highest concentrations following 28 days in oil-free seawater.

Clams accumulated much smaller amounts of petroleum hydrocarbons in their tissues than did oysters (Table 27). After 8-h exposure, clam tissues contained a total of 89 ppm petroleum hydrocarbons, only 28·5 per cent of the amount found in oysters at the same time. Again, n-paraffins were the petroleum hydrocarbons present in highest concentrations at the end of the exposure period. However, more than 98 per cent of the n-paraffins were released within 6 h. The remaining n-paraffins were lost more slowly, but reached undetectable levels in the tissues by 120 h. Aromatic hydrocarbons were released more slowly. Concentrations of different aromatic hydrocarbons dropped by 30–60 per cent during the first 3 h of depuration and remained essentially constant during the next 3 h. Only alkyl naphthalenes remained in tissues after 120 h, di- and trimethylnaphthalenes being the compounds present at highest concentration. No petroleum hydrocarbons were detected in clam tissues at or above the 0·1 ppm level after 672 h in oil-free seawater.

Because naphthalenes appeared in most cases to be the petroleum hydrocarbons accumulated to the greatest extent by marine animals and retained for the longest period following exposure, further experiments were conducted to better assess patterns of accumulation and release of these aromatic hydrocarbons. Clams *Rangia cuneata* exposed for 24 h to the WSF of No. 2 fuel oil (total dissolved hydrocarbons 6·28 ppm) accumulated 13·6 ppm total naphthalenes in their tissues (Table 28) (Neff et al., 1976b). Methyl- and dimethylnaphthalenes were the analogs reaching highest concentrations. Bioaccumulation factors (tissue concentration/exposure water concentration) increased in direct relation to the molecular weight of the naphthalene analog, from 2·3 for naphthalene to 26·7 for trimethylnaphthalenes. When clams were returned to oil-free seawater and allowed to depurate for 24 h, concentrations of tissue naphthalenes dropped; naphthalene was released most rapidly. Alkyl naphthalenes were released more slowly, with 32–51 per cent released in 24 h.

In an effort to determine whether uptake and retention of individual hydrocarbons is influenced by the presence of other hydrocarbons in the exposure medium (the WSF) and whether the direct relationship between bioaccumulation factors and molecular weight holds true over a wider molecular weight range, uptake experiments were conducted with pure polycyclic aromatic hydrocarbons (Neff et al., 1976a). Groups of 10 clams were exposed to each hydrocarbon for 24 h. Five clams were then sacrificed for analysis and the remaining five clams were returned to hydrocarbon-free seawater for 24 h before being sacrificed. Aqueous and tissue PAH concentrations were measured by UV spectrophotometry. Phenanthrene was accumulated most rapidly and released most slowly (Table 29). Chrysene and benzo[a]pyrene were accumulated more slowly and also released relatively slowly. Rapid release of naphthalene from clam tissues probably masked a similarly rapid uptake during exposure, since both influx and efflux of this compound undoubtedly occur simultaneously.

These observations may be used to partially explain the relatively high acute toxicity of phenanthrene observed earlier. Uptake/release kinetics for

TABLE 29

ACCUMULATION AND RELEASE OF DIFFERENT AROMATIC HYDROCARBONS BY THE CLAM *Rangia cuneata* (FROM NEFF et al., 1976a, WITH PERMISSION OF THE AMERICAN INSTITUTE OF BIOLOGICAL SCIENCES)

Aromatic hydrocarbon	Naphthalene	Phenanthrene	Chrysene	Benzo[a]pyrene
Exposure concentration (ppm)	0·071	0·089	0·066	0·052
Tissue concentration after 24-h exposure (ppm)	0·43 ± 0·1	2·85 ± 1·1	0·54 ± 0·2	0·45 ± 0·1
Bioaccumulation factor [tissue]/[water]	6·1	32·0	8·2	8·7
Tissue concentration after 24-h depuration (ppm)	0·15 ± 0·02	2·47 ± 1·2	0·40 ± 0·15	0·38[a]
Per cent released in 24 h	66	13	26	16

[a] Only one sample analyzed.

this compound favor its rapid accumulation in tissues of marine animals. Chrysene and benzo[a]pyrene, because they are accumulated much more slowly, have a relatively low acute toxicity. However, their slow release from tissues of exposed animals indicates a potential for high chronic toxicity. Lower toxicity of naphthalene may be related to the rapidity with which it is released from tissues of exposed animals.

The lower bioaccumulation factors for naphthalene uptake from the WSF of No. 2 fuel oil (2·3) (Table 28) in comparison to that for uptake from pure solution (6·1) (Table 29) may imply that bioaccumulation of a particular hydrocarbon is influenced by the presence of other aromatic hydrocarbons in the exposure medium. However, bioaccumulation factors of most pollutants tend to increase as exposure concentrations decrease, so that the 10-fold difference in exposure concentrations in the two experiments may account for some of the difference in bioaccumulation factors observed. There is a definite trend for bioaccumulation factors to increase with molecular weight between naphthalene and phenanthrene. Higher molecular weight PAH, chrysene and benzo[a]pyrene, do not follow the trend. This is due primarily to a lower than expected uptake rate, since these hydrocarbons are released relatively slowly. Since aqueous concentrations of these poorly soluble PAH were higher than published values for their aqueous solubilities (Neff, 1979), they may have been present in the exposure water in miceller form rather than in true solution, decreasing their uptake rate by clams.

A group of clams was exposed for 24 h to a dilute OWD of No. 2 fuel oil (Fig. 33) (Neff *et al.*, 1976*b*). The clams rapidly accumulated naphthalenes in their tissues. Concentrations of naphthalene, methylnaphthalenes, and dimethylnaphthalenes in tissues, as determined by UV spectrophotometry, reached 10·9, 17·4, and 7·6 ppm, respectively, in 24 h.

When clams were returned to oil-free seawater they immediately began to release accumulated naphthalenes. Naphthalene and methylnaphthalenes were released at approximately the same rate and more rapidly than dimethylnaphthalenes. After 200 h in oil-free seawater tissue naphthalenes levels had dropped to approximately pre-exposure baseline levels. Similar experiments were conducted with the dilute OWD of south Louisiana crude oil and bunker C residual oil. Essentially the same patterns of uptake and release of naphthalenes were seen in these experiments. In the south Louisiana crude oil exposure, tissue concentrations of naphthalene, methylnaphthalenes, and dimethylnaphthalenes reached levels of 3·0, 3·6, and 0·8 ppm, respectively, in 24 h. Upon return to oil-free seawater, naphthalenes were depurated to background levels in 220 h. In the bunker

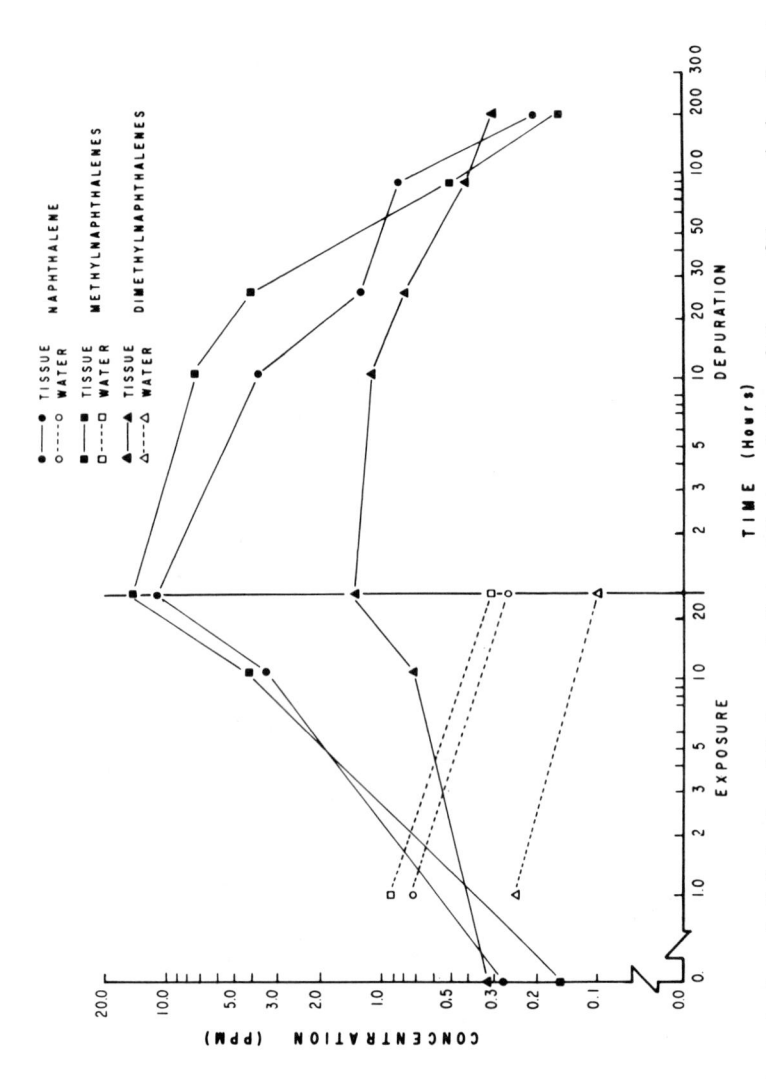

FIG. 33. Concentration of oil-derived naphthalene and alkylnaphthalenes in the soft tissues of the marsh clam *Rangia cuneata* during exposure to a dilute OWD of No. 2 fuel oil (384 ppm oil added), and following return to oil-free seawater. Each data point represents the mean concentration of the hydrocarbon in the tissues of five clams (from Neff *et al.*, 1976b, with permission of Springer–Verlag).

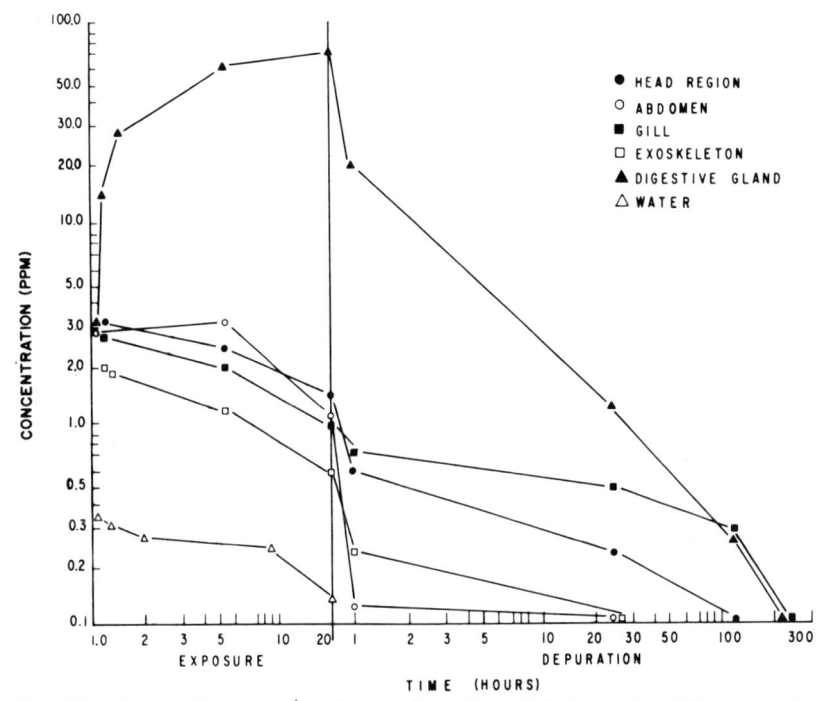

FIG. 34. Accumulation and release of total naphthalenes by different body regions of juvenile brown shrimp *Penaeus aztecus* exposed to a 20 per cent WSF of No. 2 fuel oil. Body regions from three shrimp were pooled for each data point, and background absorbance was subtracted (from Neff *et al.*, 1976b, with permission of Springer–Verlag).

C residual oil exposure, tissue concentrations of naphthalene, methylnaphthalenes, and dimethylnaphthalenes following 24-h exposure were 2·7, 6·2, and 0·75 ppm, respectively. Depuration to background levels in hydrocarbon-free seawater required 400 h. The OWD of these oils contained lower concentrations of naphthalenes than dispersions of No. 2 fuel oil, accounting for the lesser accumulation of naphthalenes in these experiments.

A group of juvenile brown shrimp *Penaeus aztecus* was exposed to a 20 per cent WSF of No. 2 fuel oil for 20 h and returned to oil-free seawater (Neff *et al.*, 1976b). Animals were sacrificed at different times during and following exposure and dissected into different body regions. Each body region was analyzed separately for total naphthalenes by the UV spectrophotometric technique (Fig. 34). Total naphthalenes in the head

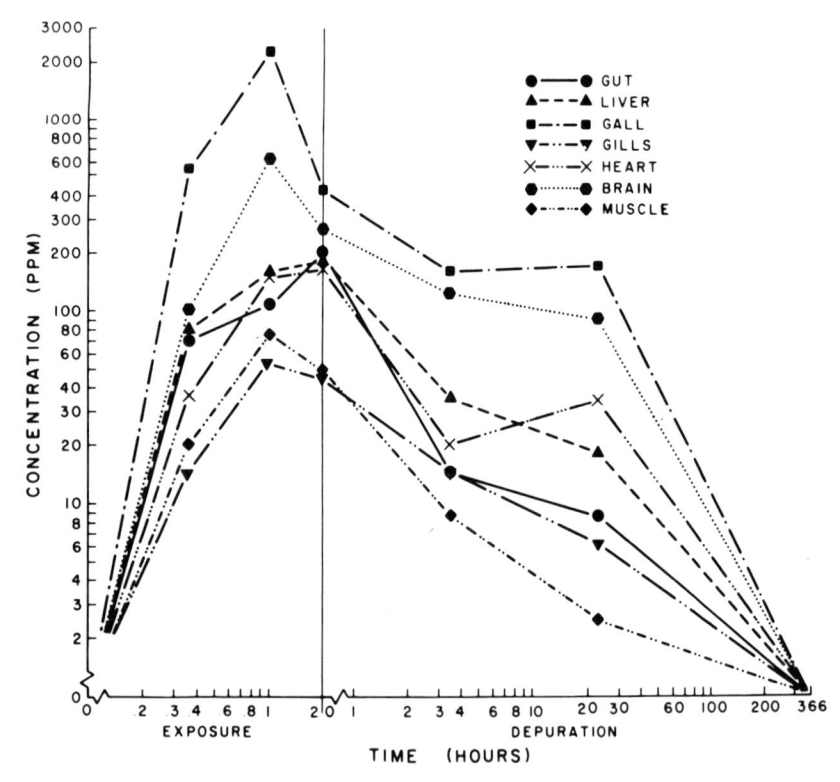

FIG. 35. Distribution of total naphthalenes in several tissues of the gulf killifish *Fundulus similis* during exposure to a WSF of No. 2 fuel oil and at different times following return to oil-free seawater. The WSF contained approximately 2 ppm total naphthalenes (from Neff *et al.*, 1976b, with permission of Springer–Verlag).

region, abdomen, gill, and exoskeleton reached maximum concentrations after only 1 h of exposure then dropped gradually during the remainder of the exposure period. The digestive gland continued to accumulate naphthalenes during the entire 20-h exposure period. Concentration of total naphthalenes in this organ reached 70 ppm in 20 h. When shrimp were returned to oil-free seawater, initial rapid release of naphthalenes from all body regions occurred followed by a long phase of more gradual release of tissue naphthalenes. Naphthalenes were released most rapidly from abdominal muscle and exoskeleton and could not be detected (limits of detection, 0·1 ppm) in any body region of shrimp maintained in oil-free seawater for 250 h following exposure.

A similar experiment was conducted using the Gulf killifish *Fundulus similis* (Fig. 35) (Neff *et al.*, 1976*b*). All organ systems examined rapidly accumulated naphthalenes during the 2-h exposure period. Maximum levels of naphthalenes were reached in different tissues after either 1 or 2 h of exposure. The gall bladder contained approximately 2300 ppm total naphthalenes after 1 h. Particularly noteworthy, however, was the presence of approximately 620 ppm naphthalenes in the brain at the same exposure time. When fish were returned to oil-free seawater, all organ systems immediately began to release accumulated naphthalenes. Somatic muscles released naphthalenes most rapidly, whereas the gall bladder and brain released naphthalenes much more slowly. Other tissues examined released naphthalenes at an intermediate rate. Naphthalenes dropped to undetectable concentrations in all organ systems of fish maintained in oil-free seawater for 366 h.

Mature male and gravid female polychaete worms *Neanthes arenaceodentata* were exposed for 24 h to a 25 per cent dilution of the No. 2 fuel oil WSF ($\sim$2 ppm total aqueous hydrocarbons) (Rossi and Anderson, 1977). Several times during exposure and following return to oil-free seawater, groups of worms were sacrificed and analyzed spectrophotometrically for naphthalenes. Both sexes accumulated naphthalenes from solution very rapidly (Figs. 36, 37). Maximum accumulation occurred in 1–3 h in most cases. Concentrations of naphthalenes then remained relatively constant in both sexes during the remainder of the exposure period. After 24-h exposure, both sexes had accumulated naphthalene, methylnaphthalenes, and dimethylnaphthalenes to approximately equal concentrations.

When returned to oil-free seawater, male worms rapidly released accumulated naphthalenes (Fig. 36). Half-time for release of naphthalenes (time required to release 50 per cent of accumulated naphthalenes) was approximately 20 h. Tissue naphthalenes had reached the limit of detection by the analytical technique ($\sim$0·1 ppm) in 400 h.

The pattern of hydrocarbon release by gravid females was quite different (Fig. 37). Little release of accumulated naphthalenes took place during the initial 300 h in oil-free seawater. During this time, tissue total naphthalenes concentration dropped from approximately 15 ppm to approximately 10 ppm. Females spawned between 300 and 500 h; those analyzed immediately after spawning contained a mean 0·2 ppm total naphthalenes. Zygotes and trochophore larvae produced by exposed worms contained an average 18 ppm total naphthalenes. Forty days after the beginning of the depuration period, 18-segment juveniles still contained residual amounts of

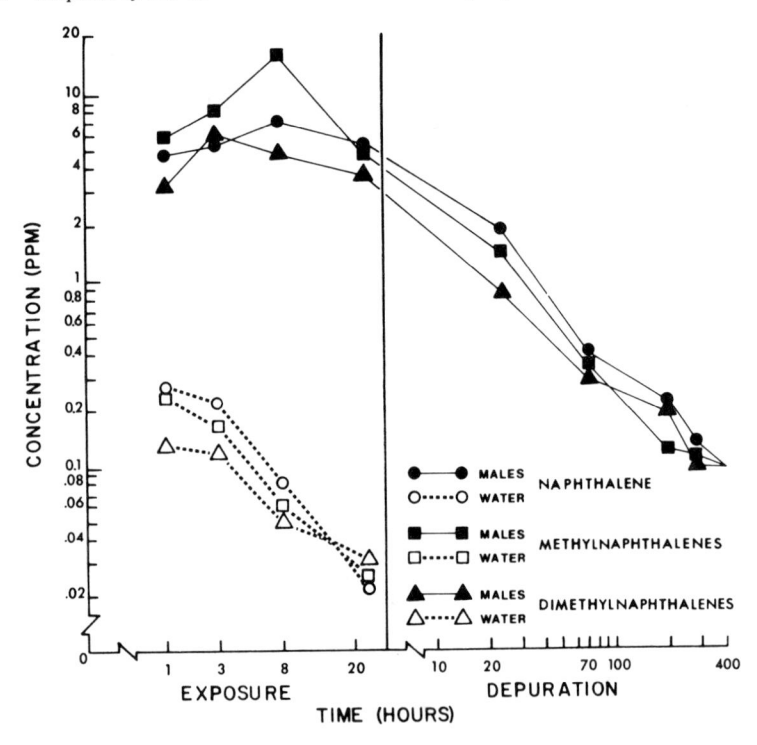

FIG. 36. Concentration of naphthalene and alkyl naphthalenes in whole body tissue samples of mature male polychaete worms *Neanthes arenaceodentata* during exposure to a 25 per cent WSF of No. 2 fuel oil and at different times following return to oil-free seawater. Tissue samples were composed of 15 worms (0·8–1·0 g wet weight sample). Each data point represents the mean value for four samples. All standard deviations are ≤ ±0·05 ppm (from Rossi and Anderson, 1977, with permission of Springer–Verlag).

naphthalenes (0·4 ppm). Five days later, when worms had developed to the 32-segment stage, naphthalenes were at undetectable levels in tissues.

These data indicate that naphthalenes accumulated by gravid female worms were passed to offspring through storage in yolk material. Developing oocytes of nereid polychaetes are rich in triglycerides, glyceryl ether diesters, and other lipids (Pocock *et al.*, 1971). Thus, lipid-rich eggs represent a hydrophobic compartment in which large amounts of hydrocarbons can be sequestered. During the time that oocytes are developing and are retained within the female, they represent an extremely stable compartment in that lipids stored in them are metabolically

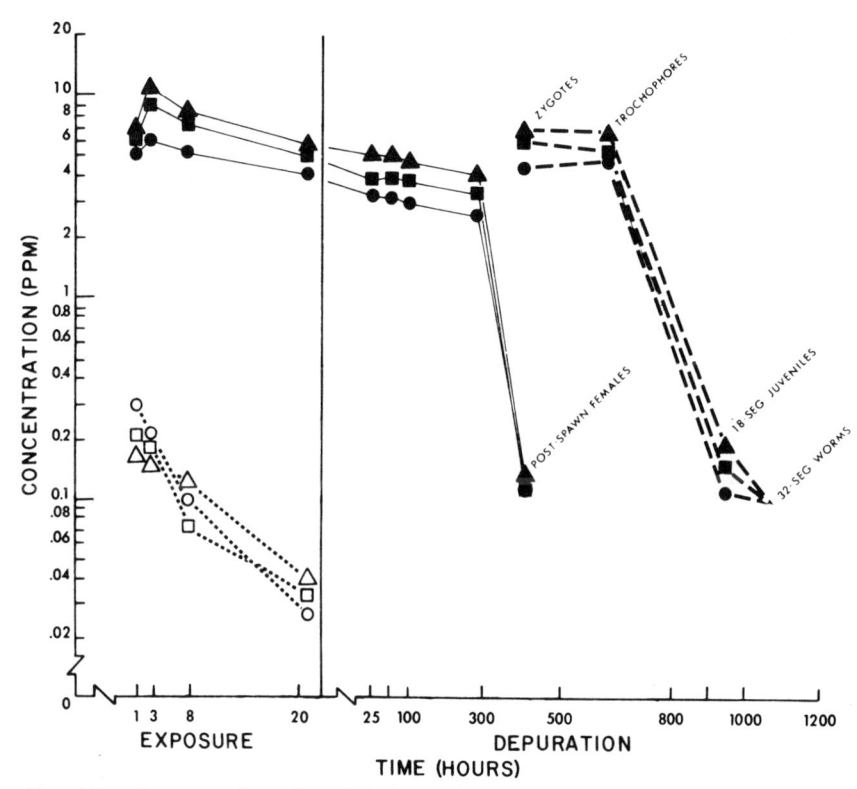

FIG. 37. Concentration of naphthalene and alkyl naphthalenes in whole body tissue samples of adult female polychaete worms *Neanthes arenaceodentata* during exposure to a 25 per cent WSF of No. 2 fuel oil, and at different times after return to oil-free seawater. Included are naphthalenes concentrations in offspring which developed from exposed gravid females during the depuration period. Each data point is the mean value for four samples each composed of 10–200 individuals. All standard deviations are ≤0·06 ppm. SEG: segment (from Rossi and Anderson, 1977, with permission of Springer–Verlag).

	Females	Offspring	Water
Naphthalenes	●——●	●--●	○·····○
Methylnaphthalenes	■——■	■--■	□·····□
Dimethylnaphthalenes	▲——▲	▲--▲	△·····△

unavailable to the female. As suggested earlier in this review and by other investigators (Stegeman and Teal, 1973; Hamelink and Spacie, 1977), the equilibrium capacity of a particular tissue compartment for hydrophobic pollutants and the rate of their release from this compartment are related to the turnover rate of the hydrophobic tissue components (mainly lipids) in it. Thus, we can differentiate between compartments characterized by rapid lipid turnover (such as plasma membranes) in which hydrocarbon uptake and release can be expected to be rapid, and compartments characterized by slow lipid turnover or a net increase with time in lipid concentration (such as depot lipid stores in animals in positive nutrient balance) in which hydrocarbons may be sequestered at high concentrations for long periods of time.

Groups of worms were exposed continuously for three generations to several concentrations of the WSF of No. 2 fuel oil. Naphthalenes concentrations were determined in pooled samples (usually two males and two females/sample) from each generation (Rossi and Anderson, 1978b). Exposure concentrations were 2·5, 5, 10, and 25 per cent WSF, corresponding to 0·06, 0·1, 0·2, and 0·46 ppm total naphthalenes, respectively. Tissue total and specific naphthalenes concentrations declined with each successive generation at all exposure concentrations (Fig. 38). Concentrations of naphthalenes in F_1 worms were greater than those in the exposure media. Concentrations in F_3 animals were quite similar to final (24-h-old WSF) concentrations in the exposure media. These data indicate the presence in polychaetes of an exposure-induced adaptation to hydrocarbons resulting in either a decrease in hydrocarbon uptake rate or an increase in depuration rate. An alternative hypothesis is that progeny of chronically exposed worms contained smaller lipid reserves and therefore a lower hydrocarbon equilibrium capacity.

The reef corals *Montastrea annularis*, *Acropora cervicornis* and *A. palmata* from Carysfort reef, Florida were exposed to a surface oil slick of south Louisiana crude oil for up to 3 days. The corals did not come in direct contact with the surface slick but were exposed to solubilized and dispersed oil in the water column. Concentrations of total naphthalenes in the aqueous phase of the flow-through exposure tanks containing a surface layer of south Louisiana crude oil fluctuated between 0·006 and 0·29 ppm.

The corals did not accumulate significant amounts of naphthalenes from the slick. A few of the coral heads contained concentrations of naphthalenes just above the detection limits of the UV technique (0·01–0·02 ppm). Aromatic hydrocarbons could not be detected by GC at or above the 0·1 ppm level in the tissue of corals exposed to the crude oil.

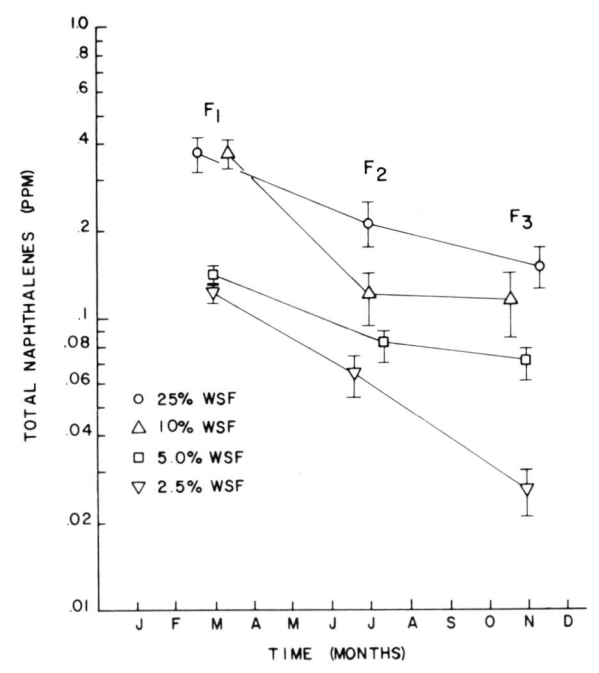

Fig. 38. Concentration of total naphthalenes in tissues of immature adult polychaete worms *Neanthes arenaceodentata* from three generations continuously exposed to sublethal concentrations of the WSF of No. 2 fuel oil. Data points are mean values for six or more samples, each composed of two male and two female worms. Vertical lines are standard deviations (from Rossi and Anderson, 1978a, with permission of D. Reidel Publishing Co.).

The exposed corals contained slightly higher concentrations of several n-paraffins than did the controls (Table 30). J. S. Warner (personal communication) suggested that some of the aliphatic hydrocarbons detected in the coral samples might be due to paraffin wax contamination from the aluminium foil in which the specimens were wrapped. However, corals and the mucus they secrete contain significant amounts of wax esters and other hydrocarbon-like compounds. The most abundant wax ester in the species studied to date is cetyl palmitate which contains 32 carbon atoms (Young *et al.*, 1971; Benson and Muscatine, 1974). This compound may correspond to the unidentified RT_{33} hydrocarbon which was present at high concentration in both control and oil-exposed corals.

All the hydrocarbons quantified, except RT_{31}, were present at higher concentrations in oil-exposed corals than in controls. The controls contained

TABLE 30
CONCENTRATIONS OF n-PARAFFINS AND BIOGENIC OLEFINS ANALYZED BY GC IN CONTROL CORALS AND CORALS THAT HAD BEEN EXPOSED TO A SURFACE SLICK OF SOUTH LOUISIANA CRUDE OIL FOR 3 DAYS[a]

Sample	\multicolumn{10}{c}{Hydrocarbon (ppm)}
	C_{15}	C_{16}	C_{17}	C_{18}	C_{19}	C_{20}	C_{21}	C_{22}	C_{23}	C_{24}	C_{25}
Control	1·7	0·4	5·4	0·3	1·0	0·6	0·4	0·5	0·5	0·4	0·4
Exposed	4·2	5·0	11·9	4·4	4·9	4·4	2·6	2·5	3·5	2·7	2·5
Difference	2·5	4·6	6·5	4·1	3·9	3·8	2·2	2·0	3·0	2·3	2·1

Sample	\multicolumn{10}{c}{Hydrocarbon (ppm)}
	C_{26}	C_{27}	C_{28}	C_{29}	C_{30}	C_{31}	BO_{20}[b]	BO_{28}[b]	RT_{31}[c]	RT_{33}[c]	RT_{35}[c]
Control	0·4	0·3	0·4	0·3	0·4	0·4	12·5	3·0	30	1 120	380
Exposed	2·6	1·6	1·3	1·0	0·9	0·9	14·3	7·7	28	1 420	433
Difference	2·2	13	0·9	0·7	0·5	0·5	1·8	4·7	−2	300	53

[a] No aromatic hydrocarbons were detected. All concentrations are in mg/kg wet tissue (ppm). Difference between control and exposed values is presumed to represent petroleum hydrocarbons accumulated by corals or hydrocarbon-like substances (e.g., waxes and olefins) synthesized by corals in response to oil exposure.
[b] BO_{20} and BO_{28} are biogenic olefins with GC retention times similar to C_{20} and C_{28} n-paraffins, respectively.
[c] RT, unknown with GC retention time near that given for n-paraffin.

a total of 1559·3 ppm hydrocarbons and the oil-exposed animals contained 1959·9 ppm. The difference can be assumed to represent n-paraffins accumulated from the oil and hydrocarbon-like substances (e.g., waxes and olefins) synthesized by the corals in response to oil exposure. In the n-paraffin series, the pattern of relative 'accumulated' hydrocarbon concentrations (the difference between control and exposed values) resembles that of the oil. The concentration of individual n-paraffins in south Louisiana crude oil increases to a maximum at C_{17} and then decreases gradually with increasing chain length (Pancirov, 1974). The same pattern is seen in the 'accumulated' n-paraffins in the corals. The olefins and unidentified hydrocarbon-like materials eluting between C_{31} and C_{35} are probably biogenic. They represent more than 95 per cent of the total hydrocarbon in both control and oil-exposed animals. It is interesting to note that the concentration of these compounds, with the exception of

RT_{31}, were also higher in oil-exposed rather than control corals. This may be due to oil-induced stimulation of the production of wax-rich mucus by the corals.

Mucus production by *Montastrea annularis* was stimulated by exposure to the WSF of south Louisiana crude oil. No attempt was made to quantify the amount of mucus produced. However, the mucus produced by the corals after they were returned to oil-free seawater was collected and analyzed for naphthalenes by UV spectrophotometry. Naphthalenes were detected in the coral mucus. In a similar experiment performed in Bermuda, significant quantities of chloroform-extractable ^{14}C radioactivity were present in mucus secreted by *M. annularis* after exposure of the animals to seawater containing ^{14}C-naphthalene. It is presumed that the ^{14}C activity in the coral mucus was in the form of unmetabolized ^{14}C-naphthalene since coelenterates are apparently unable to metabolize polycyclic aromatic hydrocarbons (Payne, 1977). These results suggest that the coral mucus (mostly mucopolysaccharides) can bind or adsorb aromatic hydrocarbons. Surface mucus may protect the underlying coral tissues from aqueous hydrocarbons and other pollutants and mucus secretion may be an avenue of hydrocarbon release from contaminated corals.

Butterfly fish *Chaetodon* sp. were observed to actively consume coral mucus. When corals which had been exposed for 24 h to the WSF of south Louisiana crude oil were placed in aquaria containing butterfly fish, the fish consumed the mucus produced by the corals and were allowed to do so for several hours. Low concentrations of naphthalenes were detected by UV spectrometry in several organs and tissues of corals (Table 31). Particularly noteworthy were the relatively high concentrations of naphthalenes in gall bladder, head, heart and brain. The distribution of naphthalenes in the fish tissues probably parallels that of tissue lipids. Replicate samples of whole fish were analyzed by GC and no petroleum hydrocarbons were detected at or above the 1 ppm level. This result is not surprising since the major mass of fish tissue (body musculature and tail) contained little or no naphthalenes. Thus, there appears to be only a limited potential for the contamination of mucus-consuming coral reef fish by ingestion of oil-contaminated mucus.

The coral *Madracis decatis* from Bermuda accumulated significantly more ($p < 0.025$) ^{14}C-naphthalene during exposure to 3·6 ppb ^{14}C-naphthalene in bright light during the day than at night in complete darkness (Table 32). Mean amounts of ^{14}C-naphthalene accumulated by the coral tissues were 0·45 ng/mg protein-N (ppm) and 0·21 ng/mg protein-N under light and dark conditions, respectively. The rate of

TABLE 31

CONCENTRATIONS OF TOTAL NAPHTHALENES, AS DETERMINED BY UV SPECTROPHOTOMETRY IN ORGANS AND TISSUES OF BUTTERFLY FISH Chaetodon sp. THAT HAD BEEN ALLOWED TO CONSUME MUCUS FROM OIL-CONTAMINATED CORALS. ALL VALUES ARE CORRECTED FOR BACKGROUND ABSORBANCE DETERMINED FOR THE CORRESPONDING ORGANS FROM UNEXPOSED FISH

Organ	Total naphthalenes mg/kg wet weight (ppm)
Brain	1·62
Eyes	0·24
Gut	N.D.[a]
Heart	2·47
Gall bladder	5·77
Liver	N.D.
Kidney	0·85
Gills	N.D.
Head (excluding brain)	4·51
Tail	N.D.
Body musculature	0·23

[a] N.D., none detected.

^{14}C-naphthalenes accumulation by different pieces of coral was highly variable, particularly in the light.

When the coral *Oculina diffusa* from Texas was exposed to a 10 per cent WSF of south Louisiana crude oil for up to 21 days, non-biogenic alkanes and aromatic hydrocarbons could not be detected by GC analysis at or above the 0·1 ppm level in the coral tissues.

In a second experiment at Texas A&M University, *O. diffusa* was exposed to a 10 per cent WSF of south Louisiana crude oil that had been spiked with ^{14}C-naphthalene. The initial concentration of hot and cold naphthalene combined was estimated to be 42 µg/liter (ppb). ^{14}C-naphthalene was lost rapidly from the exposure medium. Approximately 50 per cent of the ^{14}C-naphthalene originally present was lost from solution in 3 h. Only 36 per cent (15 ppb) remained after 7 h.

Radioactivity in the coral tissues increased rapidly and reached a level of 177 CPM/mg protein-N above background in 3 h (Fig. 39). This level of radioactivity was estimated to represent approximately 0·27 ng naphthalene/mg protein-N (ppm). Tissue ^{14}C activity then dropped to a mean of 104 CPM/mg protein-N at 7 h. When the corals were returned to

TABLE 32

THE AMOUNT OF ^{14}C-NAPHTHALENE ACCUMULATED BY THE BERMUDA REEF CORAL *Madracis decatis* DURING EXPOSURE FOR 12 h TO 3·6 ppb ^{14}C-NAPHTHALENE IN SEAWATER DURING THE DAY IN BRIGHT LIGHT OR AT NIGHT IN COMPLETE DARKNESS

Sample	Day			Night		
	Total protein-N (mg)	CPM/mg N[a]	ng naphthalene/mg N	Total protein-N (mg)	CPM/mg N	ng naphthalene/mg N
1	62·2	16·5	0·25	27·8	11·8	0·19
2	80·4	9·3	0·15	78·9	11·8	0·19
3	35·4	23·8	0·37	35·6	9·8	0·15
4	27·8	37·0	0·58	43·2	17·2	0·27
5	49·8	17·4	0·27	38·3	10·1	0·16
6	52·5	20·6	0·32	57·4	12·9	0·20
7	35·8	35·1	0·55	39·5	8·7	0·14
8	28·2	74·1	1·16	56·0	29·1	0·46
9	50·8	30·5	0·48	59·3	14·2	0·22
10	50·4	24·0	0·37	57·4	11·0	0·17
Mean	47·3	28·8	0·45	49·3	13·7	0·21

[a] CPM = counts per minute.

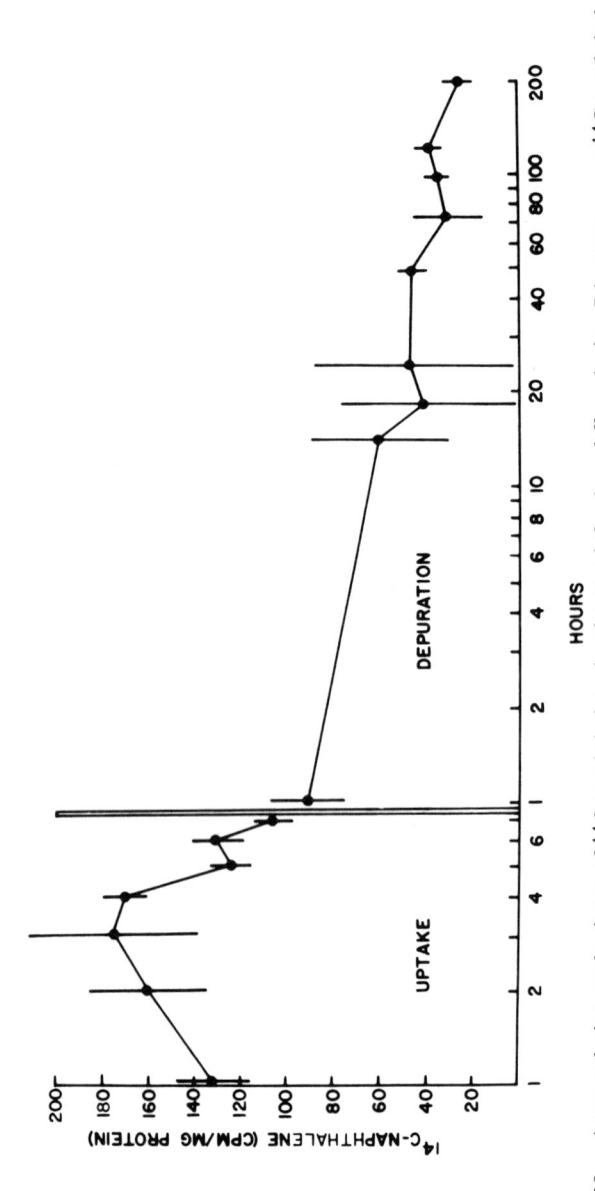

FIG. 39. Accumulation and release of ^{14}C-naphthalene by the coral *Oculina diffusa* during 7-h exposure to ^{14}C-naphthalene labeled WSF of southern Louisiana crude oil and following return of the corals to hydrocarbon-free seawater.

hydrocarbon-free seawater, they continued to release ^{14}C activity but at a slower rate. Tissue radioactivity dropped to 24 CPM/mg protein-N (0·04 ppm) after 8-days' depuration. By extrapolation of the depuration curve, it was estimated that depuration of accumulated naphthalene would be complete in approximately 14 days. The half-time ($t_{1/2}$) of naphthalene depuration from the coral tissues was approximately 24 h.

4.3.2 Uptake and Release of Petroleum Hydrocarbons: Field Studies

In three separate experiments, clams *Rangia cuneata* collected from relatively unpolluted San Antonio Bay, Texas were transplanted to stations under and near an oil separator platform in Trinity Bay, Texas for up to 100 days (Fucik *et al.*, 1977). After the exposure period, some of the clams were returned to the laboratory and placed in oil-free artificial seawater for depuration studies. In the first two experiments, clams were sampled at different times during exposure and depuration and analyzed spectrophotometrically for residues of naphthalenes. Sediment samples from exposure stations were also analyzed. In the third experiment, a detailed gas chromatographic analysis was performed on clams that had been exposed under the platform for three months. The depth of the water at the platform was approximately 3 m and the brine outfall was located 1 m above the bottom (Armstrong *et al.*, 1977, 1979).

The greatest uptake of naphthalenes was observed in those clams exposed beneath the platform (Fig. 40). At the end of the exposure period, tissue concentrations in these clams varied from 6·3 to 33·6 ppm naphthalenes. Little or no uptake of naphthalenes was observed in clams exposed at distances of 150–1800 m from the platform.

As expected, naphthalenes concentrations in sediments were highest under the platform and decreased with distance away from the platform. Sediment naphthalenes under the platform varied from 7·2 to 22 ppm during the exposure period. At other stations, maximum concentration observed was 10 ppm and, at most times, concentrations were substantially lower.

When allowed to depurate in the laboratory, clams rapidly released accumulated hydrocarbons. After 47 days, only a few clams retained detectable levels of naphthalenes.

Specimens analyzed by gas chromatographic methods contained high concentrations of aliphatic and aromatic hydrocarbons. Gas chromatograms of the aliphatic and aromatic fractions were characterized by a large unresolved hydrocarbon envelope typical of weathered oil. Measured concentrations in clams were 140 ppm total saturated hydrocarbons and

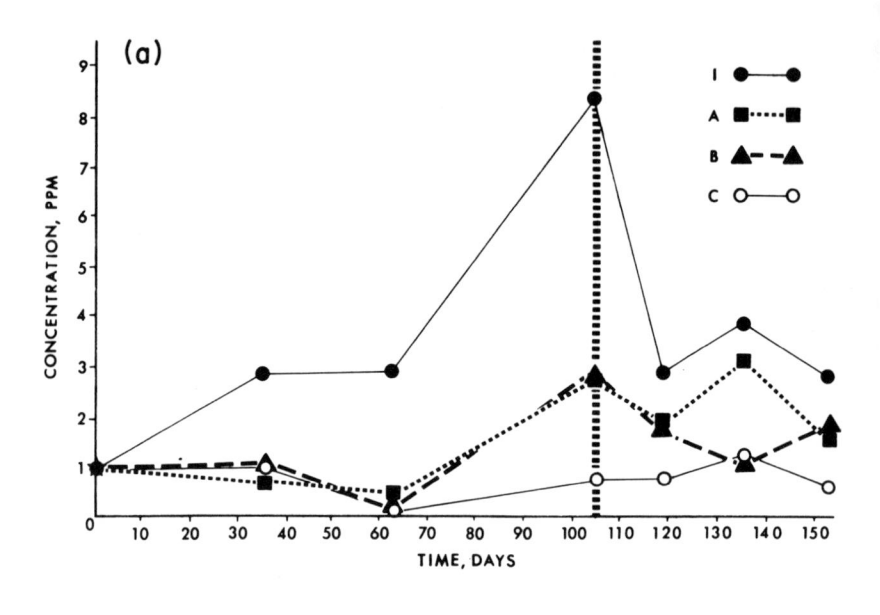

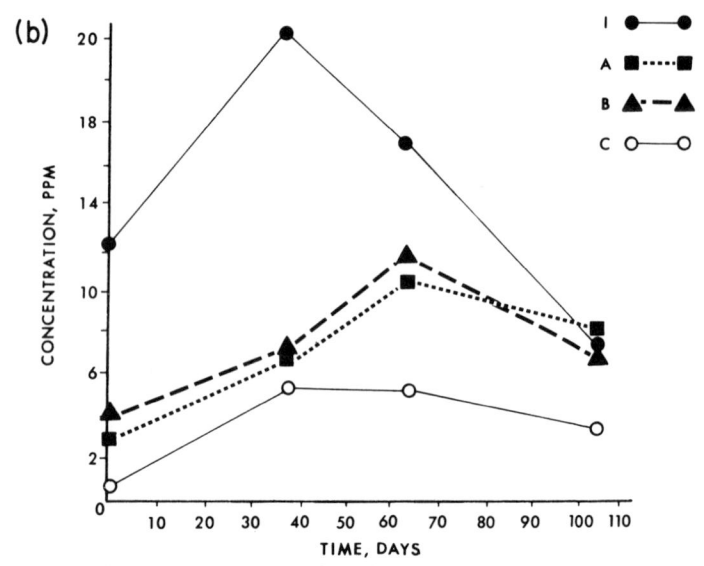

FIG. 40. a. Uptake and depuration of total naphthalenes by the marsh clam *Rangia cuneata* placed at different distances from an oil separator platform in Trinity Bay, Texas from August to December 1975. Depuration took place in the laboratory after 104-day exposure in the field and lasted for 47 days. b. Concentration of naphthalenes in sediments at stations 1, A, B, and C during the exposure period. The distances of stations from the outfall were: 1, 15 m; A, 150 m; B, 460 m; C, 915 m (from Fucik *et al.*, 1977, with permission of the American Petroleum Institute).

60 ppm total aromatic hydrocarbons. Individual peaks in the chromatograms were difficult to resolve but the saturated hydrocarbon envelope appeared to peak at C_{15} and the aromatic hydrocarbon envelope in the C_{18}–C_{20} range. Total petroleum hydrocarbon concentration in sediments under the platform (as measured by GC) was 96 ppm while that in the overlying water column was 10·5 ppb, all of the latter consisting of aromatic hydrocarbons.

These experiments demonstrate that the pattern of hydrocarbon uptake by molluscs exposed in the field is qualitatively similar to that observed in laboratory exposures. Significantly, clams contaminated by relatively chronic (three months or more) low-level exposure to oil under field conditions, were still able to depurate accumulated naphthalenes fairly rapidly when returned to an oil-free environment. Nevertheless, as stated earlier, the nutritional status of the animal may be an important factor influencing the rate of hydrocarbon release. In the laboratory, clams are not fed and are thus required to mobilize nutrient reserves (glycogen and lipid) for sustenance. This nutrient mobilization may facilitate hydrocarbon exchange with the ambient medium. The high concentration of hydrocarbons in sediments at the platform, their low concentration in the overlying water column, and the presence of 'weathered' oil in clam tissues all suggest that at least part of the oil accumulated by clams was obtained from sediments.

4.3.3 Effects of Salinity and Temperature on Hydrocarbon Uptake

Groups of grass shrimp *Palaemonetes pugio* were acclimated to salinities of 5, 15, and 35‰ and then exposed to 3 ppm aqueous naphthalene at their acclimation salinity for 12 h. At different times during the exposure period, groups of four animals from each salinity were sampled for naphthalenes analysis (Young, 1977).

Naphthalene uptake was rapid at all salinities (Fig. 41). Tissue naphthalene concentrations reached a maximum in 2 h. The highest level (57 ppm) was observed in animals acclimated and exposed at the intermediate salinity, 15‰. At 6 and 35‰ S, maximum naphthalene uptake was much less with 2-h values of 27·5 and 41 ppm, respectively. In all groups, tissue naphthalene concentrations dropped during the remainder of the exposure period.

In a similar experiment, groups of shrimp were acclimated to 2, 17, and 32‰ S seawater and exposed to 0·3 ppm phenanthrene. Again, uptake was maximal (14·7 ppm) at the intermediate salinity (Fig. 42) and substantially lower at 2 and 32‰ S (both approximately 9·8 ppm). When returned to

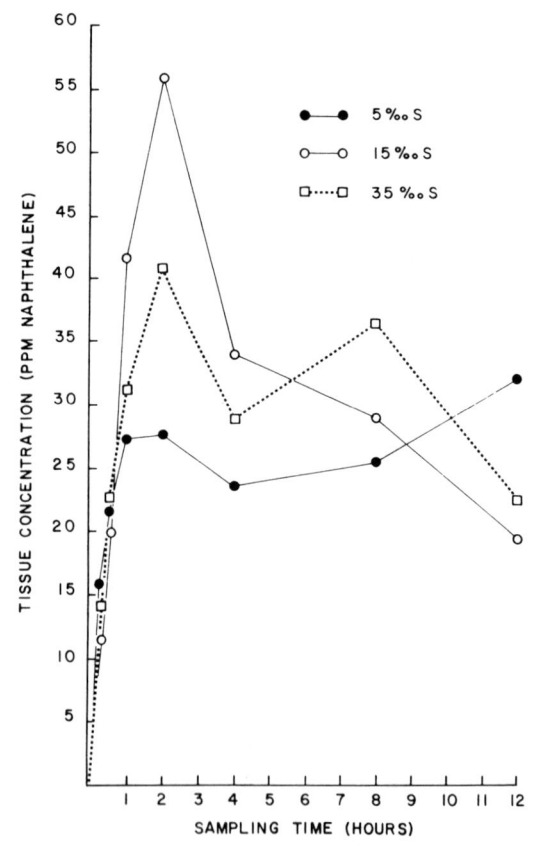

FIG. 41. Concentration of naphthalene in tissues of grass shrimp *Palaemonetes pugio* at different times during exposure to 3 ppm naphthalene in seawater. Grass shrimp were acclimated and exposed to naphthalene at salinities of 5, 15, and 35‰. Each datum represents the value for a pooled sample of four shrimp (from Young, 1977).

phenanthrene-free seawater at the acclimation salinity, all three groups released phenanthrene rapidly and at approximately the same rate. Tissue phenanthrene concentrations of less than 2 ppm were reached in 24 h.

Uptake of naphthalene and phenanthrene was greatest at salinities which are isosmotic with body fluids of the shrimp (15–17‰), and decreased at salinities in which shrimp actively osmoregulate. As indicated earlier, grass shrimp were less tolerant to naphthalene and phenanthrene at extreme salinities than they were at the isosmotic salinity. Analysis of water

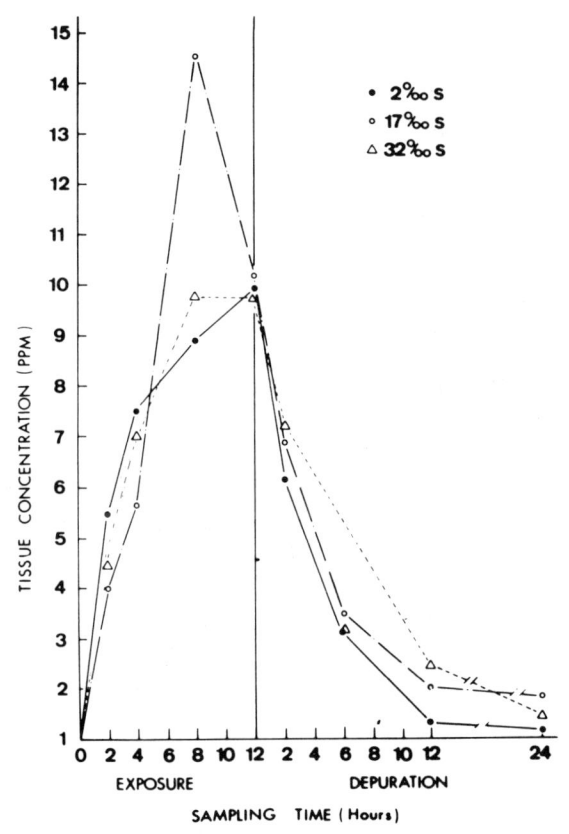

FIG. 42. Concentration of phenanthrene in tissues of grass shrimp at different times during exposure to 0·3 ppm phenanthrene and following return to hydrocarbon-free seawater. Shrimp were acclimated and exposed to phenanthrene at salinities of 2, 17, and 32‰. Each datum represents the value for a pooled sample of eight shrimp (from Young, 1977).

exchange kinetics with tritiated water revealed that water exchange rates between the animals and the medium were similar at all three salinities and in animals exposed to phenanthrene at each salinity. Thus, differences observed in hydrocarbon uptake rates at different salinities are as yet unexplained.

The temperate clam *Rangia cuneata* and the boreal clam *Protothaca staminea* were exposed to a 25 per cent WSF of south Louisiana crude oil for 3 days under various temperature–salinity regimes (Fucik and Neff, 1977). *Rangia cuneata* were exposed at temperature–salinity combinations

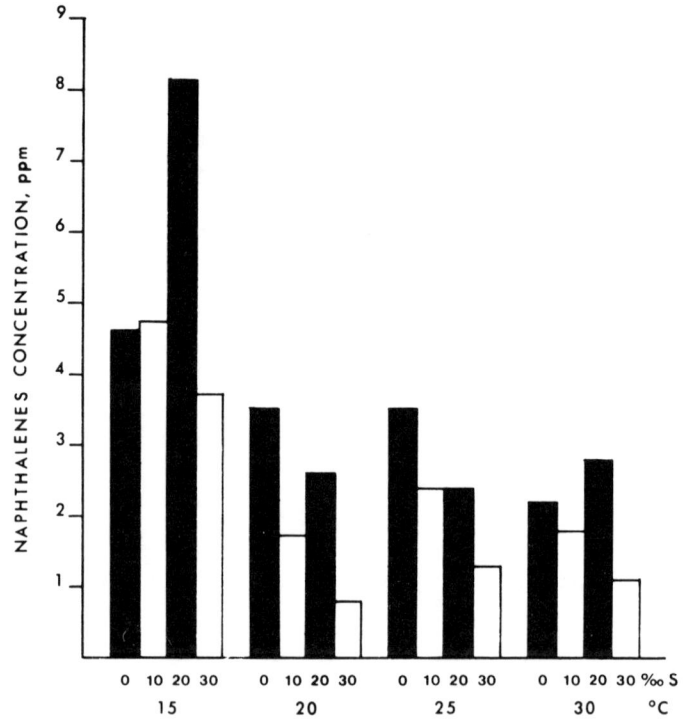

Fig. 43. The effect of salinity and temperature on the accumulation of naphthalenes during 3-day exposure to a 25 per cent WSF of south Louisiana crude oil by the marsh clam *Rangia cuneata*. Clams were acclimated to the different salinity–temperature combinations for 24 h before exposure (from Fucik and Neff, 1977, with permission of Pergamon Press).

of 15, 20, 25, and 30 °C and 0, 10, 20, and 30‰ S. Temperature–salinity combinations of 5, 10, and 15 °C and 25, 30, and 35‰ S were used in *P. staminea* exposures. Clams were then allowed to depurate for 3 days under these same temperature–salinity combinations.

In three uptake experiments, two using *R. cuneata* and one using *P. staminea*, the highest naphthalenes concentrations were measured in clams exposed at the lowest temperatures at all salinities (Figs. 43, 44, 45). When *Rangia* were acclimated to experimental temperature–salinity regimes for 24 h before exposure, maximal uptake took place at 15 °C and 20‰ S. With the pre-exposure acclimation period extended to 14 days, maximal naphthalenes uptake took place at 15 °C and 0‰ S. Maximal uptake by *P. staminea* occurred at 5 °C and 30‰ S. Statistical analysis

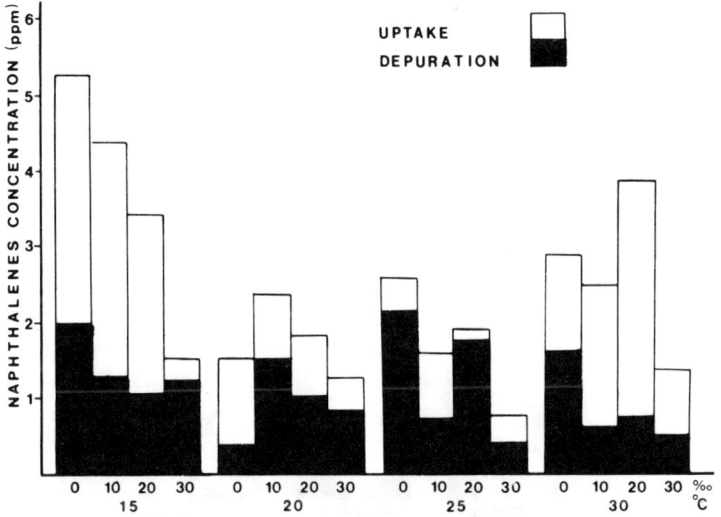

FIG. 44. The effect of salinity and temperature on the accumulation of naphthalenes during 3-day exposure to a 25 per cent WSF of south Louisiana crude oil by the marsh clam *Rangia cuneata*. Tissue naphthalenes concentrations following 3-day depuration in clean seawater are also given. Clams were acclimated to the different salinity–temperature combinations for 14 days before exposure (from Fucik and Neff, 1977, with permission of Pergamon Press).

confirmed that naphthalenes uptake in the different temperature–salinity groups in each experiment was significantly different. Temperature had the greatest effect on this difference; salinity had a slight effect in only one of the *R. cuneata* uptake experiments. Tissue naphthalenes concentrations in *P. staminea* were less than those measured in *R. cuneata* after the same exposure period. Temperature and salinity had no statistically significant effect on the rate of release of naphthalenes from either species when they were returned to hydrocarbon-free seawater.

Filtration rates of clams were measured at different experimental temperatures and were found to increase linearly with temperature in both species. Thus, tissues of clams at the highest temperatures were exposed to the largest volumes of oil-contaminated medium yet accumulated naphthalenes more slowly than did animals at lower temperatures.

The possibility that hydrocarbons disappeared from high temperature exposure media more rapidly than from low temperature media was investigated (Table 33). Though small differences in initial and final (3 day) concentrations of naphthalenes were observed, these did not bear a consistent relationship to temperature and were not statistically significant.

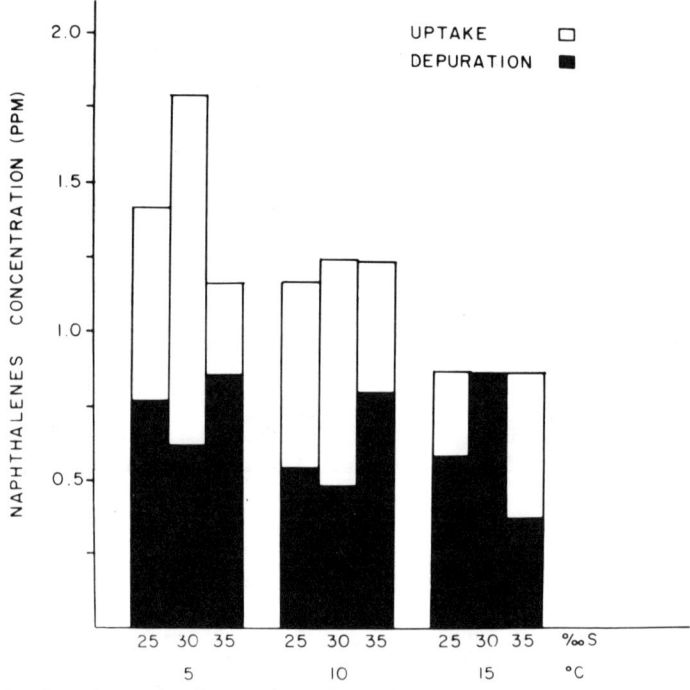

FIG. 45. Uptake and release of total naphthalenes by the clam *Protothaca staminea* exposed for 3 days to a 25 per cent WSF of south Louisiana crude oil and allowed to depurate for 3 days in clean seawater at different combinations of salinity and temperature (from Fucik and Neff, 1977, with permission of Pergamon Press).

In a subsequent experiment, *R. cuneata* were exposed to 225 ppb ^{14}C-naphthalene or ^{14}C-methylnaphthalene in seawater at 15 or 25 °C. Uptake of ^{14}C was greatest at 15 °C in both cases. Per cent efficiency of ^{14}C uptake in the naphthalene exposure was 27 per cent at 15 °C and 4·2 per cent at 25 °C. Per cent efficiency of ^{14}C uptake for methylnaphthalene was estimated at 26·7 and 7·2 per cent at 15 and 25 °C, respectively. Initial and final ^{14}C activity in the exposure media was similar at both temperatures.

Recently, Harris *et al.* (1977) demonstrated an inverse relationship between the temperature and the rate of ^{14}C-naphthalene accumulation by the copepod *Calanus helgolandicus*. Herbes (1977) provided a possible explanation for this phenomenon. He measured the effect of temperature on the adsorption of anthracene on to non-living yeast cells from solution (0·2 µg/liter). The fraction of anthracene adsorbed by the cells decreased significantly as temperature increased. The calculated heat of

TABLE 33
CONCENTRATIONS OF TOTAL NAPHTHALENES IN A 25 PER CENT WSF OF SOUTH LOUISIANA CRUDE OIL IMMEDIATELY AFTER PREPARATION AND AFTER 3 DAYS AT DIFFERENT COMBINATIONS OF TEMPERATURE AND SALINITY (FROM FUCIK AND NEFF, 1977, WITH PERMISSION OF PERGAMON PRESS)

Temperature (°C)	Rangia cuneata experiment[a]	
	Original concentration (ppm)	Final concentration (ppm)
15 (a)	0·103	0·018
(b)	0·101	0·016
20 (a)	0·098	0·012
(b)	0·101	0·011
25 (a)	0·099	0·011
(b)	0·098	0·009
30 (a)	0·096	0·013
(b)	0·098	0·012

Temperature (°C)	Protheca staminea experiment		
	Salinity (‰)	Original concentration (ppm)	Final concentration (ppm)
5 (a)	25	0·14	0·06
(b)	30	0·13	0·06
(c)	35	0·12	0·05
10 (a)	25	0·13	0·03
(b)	30	0·14	0·04
(c)	35	0·14	0·05
15 (a)	25	0·17	0·05
(b)	30	0·16	0·05
(c)	35	0·14	0·04

[a] In the *R. cuneata* experiment, naphthalenes concentrations were measured in only the 30‰ S water.

adsorption for this process was 5·2 kcal/mol which is characteristic of simple physical (Van der Waals) adsorption. As temperature rises, the strength of this weak chemical bond decreases, favoring desorption of PAH from the particles. Because partitioning of PAH between soluble and adsorbed phases is determined by relative rates of adsorptive and desorptive reactions, adsorption will be increasingly favored over desorption as temperature decreases. Therefore, temperature exerts its effect on PAH uptake primarily at the initial step of the process—adsorption of PAH from water on to the surface of a biological membrane.

4.3.4 Accumulation of Hydrocarbons from Sediment and Food

Bioavailability of sediment-adsorbed petroleum hydrocarbons to the clam *Rangia cuneata* was investigated in two experiments.

Sediments were stirred in 100 per cent WSF of south Louisiana crude oil and placed in small, rectangular aquaria to serve as substrate for *Rangia*. Mud samples were collected at the beginning and end of the experiment for analysis. Water samples were also taken for analysis at 6, 18, and 66 h during the experiment.

Clams were exposed in two ways in four aquaria. In order to determine uptake from the sediments, 10 clams in two aquaria were submerged in the oil-contaminated muds with only siphons exposed. *Palaemonetes pugio* shrimp were added to these aquaria to stir up and keep sediments in suspension. Ten clams were exposed in two other aquaria but these were placed on a perforated platform above the oil-contaminated mud. Minimal aeration prevented resuspension of the sediments. Therefore, clams were exposed only to those hydrocarbons which leached out of the mud into the water column. After 66 h of exposure in this system, three clams from each aquarium were analyzed for uptake of naphthalenes. The results of this analysis are presented in Table 34.

Only a small percentage of naphthalenes adsorbed on sediments actually leached into the water column. This was confirmed by the slight increase in naphthalenes concentrations in the water column after 66 h. Mean sediment naphthalenes concentrations were 1·79 and 1·51 ppm at 0 and 66 h, respectively. Corresponding values for the overlying water column were 0·008 ppm at 0 h and 0·022 ppm at 66 h.

Naphthalenes concentrations in the clams were low in each exposure group. Clams buried in sediment contained 0·3–0·83 ppm naphthalenes, while those not allowed to come into contact with sediment contained 0·28–2·45 ppm (Table 34). Since clams in direct contact with sediment contained no more naphthalenes than those suspended in the water

TABLE 34

CONCENTRATION OF TOTAL NAPHTHALENES IN CLAMS *Rangia cuneata* EXPOSED TO OIL-CONTAMINATED MUD AND WATER FOR 66 h. CLAMS IN AQUARIA 1 AND 2 WERE BURIED IN THE MUD; CLAMS IN 3 AND 4 WERE PLACED ON A PLATFORM ABOVE THE MUD

Aquarium number	Total naphthalenes (ppm)
1	0·30
	0·68
	0·46
	Average 0·48
2	0·38
	0·83
	0·76
	Average 0·66
3	2·45
	0·32
	0·28
	Average 1·02
4	0·33
	0·36
	0·38
	Average 0·36

column, any uptake observed can be attributed to uptake from water but not from sediment.

Clams in the second experiment were exposed to more heavily contaminated sediments. Eighteen liters of 10‰ S artificial seawater were added to 4 liters of San Antonio Bay sediment in a large, circular, all-glass mixing chamber. Two liters of No. 2 fuel oil were added and the mixture was agitated vigorously with a magnetic stirrer for 24 h. Sediments were allowed to settle out and the remaining oil and water layers removed with a siphon. The contaminated sediments were rinsed once for 24 h with clean seawater and placed in exposure aquaria. As in Experiment 1, a group of clams was allowed to burrow into the sediment while the other was suspended several centimeters above the sediment on a platform. During the exposure period,

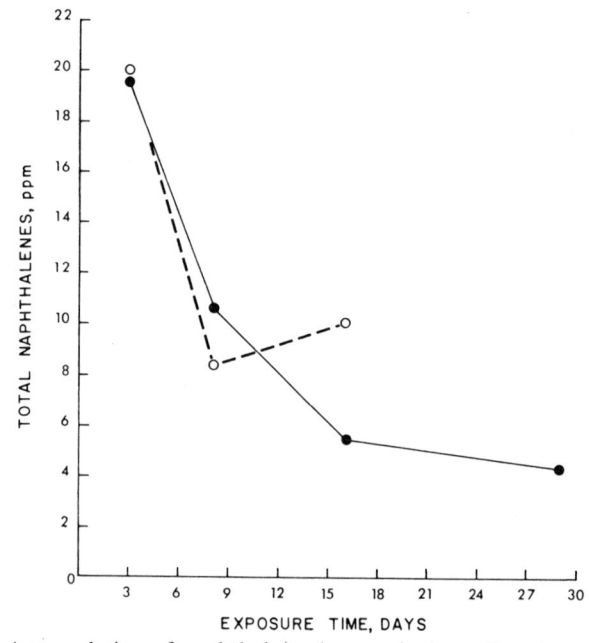

FIG. 46. Accumulation of naphthalenes by marsh clams *Rangia cuneata* during exposure to sediments heavily contaminated with No. 2 fuel oil. One group of clams was held several centimeters above the sediment (●——●) and another group was allowed to come into contact with the burrow into the sediment (○- - -○).

overlying water was recirculated continuously through charcoal filters to remove any desorbed hydrocarbons. Sediment, water, and clam samples were taken periodically for naphthalene analysis.

Sediments were heavily contaminated with oil by this procedure, initially containing 704 ppm total naphthalenes with dimethylnaphthalenes being the most abundant. Naphthalenes concentrations dropped during the course of the experiment with the largest drop occurring during the first 2–3 days. After 25 days, total naphthalenes concentration dropped to 241 ppm. Naphthalenes concentrations in water were near detection limits of the analytical technique with a maximum concentration of 0·093 ppm measured on day 4.

Highest concentrations of total naphthalenes were observed in both groups of clams on day 3 (Fig. 46). Mean naphthalenes concentration in the water column clams was 19·6 ppm and in the sediment clams 20·0 ppm, indicating that all uptake seen at this time was from the water column. Naphthalenes concentrations in clams suspended in the water column decreased progressively during the remainder of the exposure period. In clams immersed in sediment, naphthalenes concentration dropped to 8·35 ppm at day 8 but rose again to 10·13 ppm on day 16. Due to mortalities in the sediment exposed group, insufficient numbers of clams were available for analysis on day 29.

These data indicate that molluscs may have at least a limited ability to accumulate hydrocarbons directly from heavily contaminated sediments. Uptake efficiency from the water column is obviously much greater.

To investigate potential for hydrocarbon transfer from contaminated sediments to the sediment-living polychaete worm *Neanthes arenaceodentata*, worms were exposed to artificially contaminated sediments and at selected times during exposure tissues were analyzed for naphthalenes (Rossi, 1977). Substrates were contaminated by thoroughly mixing sediment with 100 per cent WSF of No. 2 fuel oil for 1 h. Once tainted, sediments were rinsed by mixing in hydrocarbon-free seawater for 1 h to remove those hydrocarbons not firmly adsorbed. Sediment used here was composed of two parts, approximately 5 per cent sand/gravel:95 per cent clay/silt/mud (mixture obtained from Trinity Bay) to one part powdered alfalfa. Polychaetes were exposed to this substrate in 1-liter culture dishes containing 600 ml hydrocarbon-free seawater which flowed through each dish unidirectionally (non-recirculating) at a rate of 15 ml/min. Twenty immature young adult worms were held in each of four dishes for 28 days. At selected times during exposure, worms and sediment were monitored for naphthalenes by UV spectrophotometry.

Naphthalenes were initially present in sediments at approximately 3 ppm and declined to concentrations of less than 1 ppm (μg/g dry weight sediment) during the experimental period. Observed decreases were presumably attributable to combinations of microbial degradation, photooxidation, and simple volatilization of compounds desorbed into the water from sediment particles. Periodic water sampling failed to detect measurable (>0·01 ppm) concentrations of naphthalenes in the overlying water column. Analyses of more than 20 replicate samples of worms living within contaminated sediments taken periodically throughout the 28-day exposure period showed that polychaetes contained <0·1 ppm total naphthalenes at all times during exposure. These data show that *Neanthes arenaceodentata* cannot be contaminated by exposure to fuel oil-derived naphthalenes adsorbed on sediments under the conditions described above.

The ability of *Neanthes arenaceodentata* to accumulate methylnaphthalene from its food was investigated also (Rossi, 1977). Powdered alfalfa, the normal food of this worm in laboratory culture, was mixed with seawater containing ^{14}C-methylnaphthalene for 1 h. The contaminated alfalfa was then rinsed for 5 min in isotope-free seawater to remove unadsorbed radioactivity and stored frozen until used. The contaminated food, which contained an estimated concentration of 10–15 ppm ^{14}C-methylnaphthalene, was fed to the worms daily for a maximum of 16 days. A total of 40 immature *N. arenaceodentata* were held individually in 100 ml culture dishes containing 50 ml unaerated seawater and each was offered 5 mg (dry weight) contaminated alfalfa each day. Worms were transferred to clean culture dishes each day and provided additional contaminated food. After 1, 2, 4, 8, or 16 days of feeding, worms were sacrificed for analysis of tissue radioactivity either immediately or following 24 h of feeding on isotope-free powdered Tetramin fish food to purge the gut of unassimilated radioactivity.

The polychaetes contained no radioactivity, following gut clearance, after daily ingestion of food contaminated with the equivalent of > 10 ppm ^{14}C-methylnaphthalene for up to 16 days in succession. After 8 or 16 days, radioactive material present in worms prior to gut clearance represented the amount still bound to alfalfa within the digestive tract of animals. Hexane-extractable radioactive material in fecal pellets gradually increased with duration of feeding time, indicating a corresponding decrease in efficiency of digestion within polychaete alimentary canals. This observation was expected since worms were in a starved condition at the onset of experimentation. Consequently, digestion early in the feeding period would be expected to be more efficient than later on.

The fate of radioactive alfalfa introduced to feeding chambers every 24 h throughout the experimental period was investigated. Approximately 10^4 total counts per minute were introduced at each feeding. A large majority of radioactive material (> 90 per cent) was found in chamber seawater after 24 h. As the nutritional condition of *N. arenaceodentata* improved with continued feeding, radioactivity increased in excreted feces, as well as in digestive tracts. Increased recovery of ^{14}C-methylnaphthalene in fecal pellet and tissue (gut) fractions was paralleled by decreased concentrations in seawater fractions. Subsequent analyses showed that greater than 85 per cent radioactive material recovered in seawater was in a hexane-extractable, or unmetabolized, form. These results offer further evidence that polychaetes cannot effectively accumulate diaromatic hydrocarbons from contaminated sediments or food.

When Gulf killifish *Fundulus similis* were force-fed ^{14}C-naphthalene in cod liver oil via stomach tube, 34 per cent of the radioactivity administered was associated with body tissues after 2 h (Dixit and Anderson, 1977).

TABLE 35
DISTRIBUTION OF ^{14}C-NAPHTHALENE IN THE TISSUES OF THE KILLIFISH *Fundulus similis* AT DIFFERENT TIMES AFTER ADMINISTRATION OF ^{14}C-NAPHTHALENE IN COD LIVER OIL INTO THE STOMACH (FROM DIXIT AND ANDERSON, 1977, WITH PERMISSION OF THE AMERICAN PETROLEUM INSTITUTE)

Tissue or body region	Per cent distribution of total radioactivity recovered at each sampling time		
	Sampling time (hours)		
	2	4	8
Gut	34	18	2·0
Liver	25	13	4·2
Gall bladder	21	44	79
Heart	6·5	8·7	9·3
Gills	2·0	4·0	0·9
Brain	3·1	4·6	3·2
Head	1·4	1·0	0·2
Mid-section	7·4	5·6	0·4
Tail-section	0·7	0·9	0·3
Per cent recovered of total administered	34	21	2·7

Thirty-four per cent of this was still associated with the gut (Table 35). Thus, approximately 12 per cent of the administered ^{14}C-naphthalene was assimilated by *Fundulus*. At 2 and 4 h, liver and gall bladder contained the highest proportions of the assimilated radioactivity. At 8 h, 79 per cent of the remaining radioactivity was in the gall bladder. Only 2·7 per cent of the radioactivity administered was recovered from the tissues after 8 h. These results strongly suggest that aromatic petroleum hydrocarbons absorbed from the gut are transported to the liver where they are metabolized rapidly and excreted in the bile.

4.3.5 Metabolism of ^{14}C-Naphthalene by *Neanthes arenaceodentata*

Several experiments were performed to determine whether polychaete worms *Neanthes arenaceodentata* are able to metabolize aromatic hydrocarbons (naphthalene) and, if so, the relative importance of metabolic transformation in the depuration of these compounds (Rossi, 1977). Immature adult specimens were exposed for 24 h to seawater containing 0·15 ppm ^{14}C-naphthalene. Two groups of worms were used, those not previously exposed to oil (controls) and worms which had been exposed continuously for three generations (F_3) to a 10 per cent WSF of

No. 2 fuel oil. After exposure, worms were transferred to |^{14}C-free seawater. Periodically water and tissue samples were taken and double extracted with n-hexane. Radioactive material in the hexane extracts and that remaining in the aqueous phase after extraction was measured by liquid scintillation spectrometry. All counts were corrected for background and quench. Previous studies have shown that hexane extraction removed 99 per cent of naphthalene from aqueous solution. Thus, radioactive material in the hexane extract was assumed to be unmetabolized ^{14}C-naphthalene and that in the aqueous phase to be polar metabolites (such as naphthol) and organic conjugates of naphthalene (Corner et al., 1973).

In order to ascertain the possible role of microbial flora associated with *N. arenaceodentata* in the metabolism of naphthalene, additional experiments were conducted in which worms were incubated in seawater containing 300 mg of penicillin/liter, 200 mg streptomycin sulfate/liter, and 50 mg chloramphenicol/liter before exposure to ^{14}C-naphthalene. This antibiotic mixture has been shown to inhibit bacterial and fungal activity associated with marine organisms (Anderson and Stephens, 1969).

Both previously unexposed (control) and chronically exposed (F_3) groups of worms accumulated and released ^{14}C-naphthalene in nearly identical fashion (Fig. 47). In addition, the quantitative role of ^{14}C-naphthalene metabolism in depuration appeared to be similar in the two groups. Amounts of hexane-extractable radioactive material accumulated by both groups were equivalent to 7-8 ppm naphthalene. Significant quantities of the ^{14}C-naphthalene had been converted to a non-hexane extractable form in both groups by the end of the exposure period. Some evidence of the differences in metabolic mechanisms between control and F_3 worms was indicated by the presence of significant quantities of converted radioactive material in F_3 animals after only 3 h of exposure. Control animals contained only unmetabolized radioactive material at this time. The exposure medium containing the F_3 worms contained significantly higher levels of metabolized radioactive material than that containing control worms. This observation supports the hypothesis that naphthalene metabolism by worms chronically exposed to petroleum differs somewhat from that of previously unexposed worms.

When worms were returned to isotope-free seawater, both groups released radioactive material in unmetabolized as well as metabolized form (Fig. 47). Approximately one-third of the radioactivity released by both groups during the first 24 h in clean seawater was in metabolized form.

Analyses of depuration water indicated possible metabolic differences between control and F_3 worms. However, these differences proved

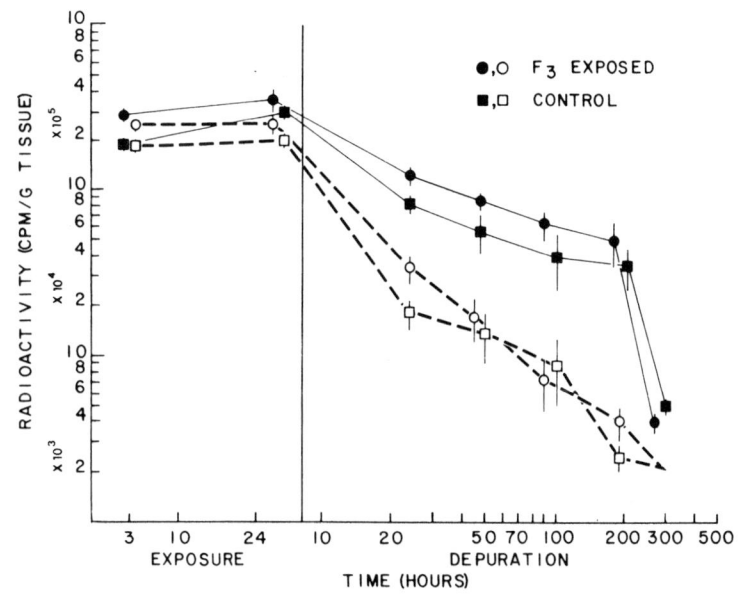

FIG. 47. Radioactivity (CPM/g tissue) in adult polychaete worms *Neanthes arenaceodentata* at different times during exposure to ^{14}C-naphthalene in seawater, and following return to isotope-free seawater. Open symbols indicate hexane-extractable (unmetabolized naphthalene) radioactivity. Dark symbols represent total extractable (naphthalene + naphthalene metabolites) radioactivity. Each datum represents the mean value for eight samples ± the standard deviation (vertical lines) (from Rossi, 1978b, with permission of Springer–Verlag).

quantitatively insignificant since tissue decreases of the two forms of radioactive material were essentially identical. Both groups retained significant quantities of radioactive compounds in the non-hexane extractable state for up to 192 h after exposure. Small amounts of radioactive metabolites were present even after 288 h in clean seawater. Both groups lacked detectable levels ($> 2 \times 10^3$ CPM/g or 0·05 ppm) of free ^{14}C-naphthalene in their tissues 288 h after exposure. These results indicate that *Neanthes arenaceodentata* is capable of metabolizing aromatic hydrocarbons. The ability of the closely related polychaete *Nereis* sp. to metabolize aliphatic and aromatic hydrocarbons was recently demonstrated (Lee, 1976; Lee *et al.*, 1977, 1979). The results presented above suggest that metabolism (chemical conversion or conjugation) does not necessarily result in an increase in mobility of aromatic hydrocarbons within host tissues. Metabolism does contribute significantly to the rate of aromatic hydrocarbon release.

Polychaetes treated with a mixture of antimicrobial agents for 24–48 h before uptake and release experiments accumulated, metabolized and depurated ^{14}C-naphthalene in much the same manner as did untreated worms. Thus, bacteria and fungi do not appear to play a part in uptake, metabolism, and discharge of naphthalene by *Neanthes arenaceodentata*.

The investigations reported here have demonstrated that marine molluscs, annelids, shrimp, and fish are all capable of rapidly accumulating in their tissues a wide spectrum of petroleum hydrocarbons from solution or dispersion in seawater. Aromatic hydrocarbons are considerably more water-soluble than alkanes of similar molecular weight (McAuliffe, 1966), rendering them more readily available for accumulation by marine organisms. This may explain the observation that in most cases aromatic hydrocarbons are accumulated to a considerably greater extent than are alkanes. When mixing conditions are sufficiently energetic to substantially increase the amount of dispersed oil in the aqueous phase, as was the case in the flow-through exposures reported above, very substantial quantities of alkanes may be accumulated, presumably by ingestion of dispersed oil droplets. Most n-alkanes are released rapidly when molluscs are returned to clean seawater, indicating that ingested oil droplets are not actually absorbed but are excreted with the feces and pseudofeces.

Blumer *et al.* (1970), Ehrhardt (1972), and Stegeman and Teal (1973) have suggested that marine molluscs accumulate different petroleum hydrocarbon types roughly in proportion to their relative concentrations in exposure water, but release n-alkanes more rapidly than aromatic hydrocarbons. It was hypothesized that n-alkanes, since they closely resemble biogenic hydrocarbons, are metabolized more readily than aromatic hydrocarbons by molluscs or their endogenous microflora.

We have shown that uptake and retention of aromatic hydrocarbons by molluscs is directly proportional to hydrocarbon molar volume in the series naphthalene, alkyl naphthalenes, phenanthrene. However, higher molecular weight aromatic hydrocarbons do not show this relationship. High molecular weight PAH, because of their extremely low aqueous solubilities, are probably not present in true solution in exposure media, rendering their uptake less efficient than that of lower molecular weight aromatics. Once accumulated, high molecular weight PAH are retained longer than low molecular weight hydrocarbons.

Boehm and Quinn (1976) showed that the presence of dissolved organic matter in seawater influenced the accumulation of hexadecane, phenanthrene, and a No. 2 fuel oil in different ways by the marine clam *Mercenaria mercenaria*. Dissolved organic matter decreased uptake of the n-alkane, had no effect on phenanthrene uptake, and decreased No. 2 fuel oil uptake. Effects of dissolved organic matter on hydrocarbon uptake were attributed to its ability to change the physical state of hydrocarbons in the exposure medium. All data so far published strongly support the hypothesis that patterns of hydrocarbon uptake and release by marine molluscs are dependent in large part on physical and chemical characteristics of the pollutant source.

The rate of aromatic hydrocarbon accumulation is highly species-dependent. Shrimp and fish accumulate aromatic hydrocarbons very rapidly. Tissue concentrations often reach maximum levels within the first hour of exposure. Longer exposure times often result in a drop in tissue hydrocarbon levels. Polychaete worms show a similar but less dramatic pattern of rapid hydrocarbon uptake. Clams and oysters, on the other hand, tend to accumulate petroleum hydrocarbons more slowly but continue to do so for as long as they are exposed to oil-contaminated seawater. Other investigators (Stegeman and Teal, 1973; DiSalvo et al., 1975; Boehm and Quinn, 1977) have shown that marine molluscs continue to accumulate oil hydrocarbons over exposure periods considerably longer than those used in the present investigations.

Of the petroleum-derived aromatic hydrocarbons investigated, naphthalenes are accumulated to the highest concentration by marine organisms. This is undoubtedly due in part to the fact that naphthalenes are somewhat more water-soluble than biphenyls, fluorenes, and phenanthrenes and are also present at higher concentrations in the No. 2 fuel oil used than the latter hydrocarbons.

Following oil exposure, all the test species released accumulated hydrocarbons when they were returned to hydrocarbon-free seawater. Fish and shrimp tended to release aromatic hydrocarbons more rapidly than clams and oysters. Release of hydrocarbons from polychaete worms is more complex. Juvenile and adult male worms released hydrocarbons rapidly, but gravid females retained hydrocarbons until spawning occurred. Developing oocytes in the coelom of gravid females are rich in lipid reserves not metabolically available to the parent. Thus, retention of hydrocarbons by these individuals would tend to support the hypothesis of Boehm and Quinn (1977) that hydrocarbons accumulated in stable hydrophobic tissue compartments are released only very slowly when the animal is returned to hydrocarbon-free seawater. The most stable hydrophobic compartment in non-gravid marine animals is depot fat. Hydrocarbons sequestered in these depots would tend to be retained or released very slowly until the animal mobilized its lipid reserves during starvation or oogenesis. Therefore, nutritional status or stage in the reproductive cycle may have a profound influence on hydrocarbon accumulation/release patterns in marine animals. This area merits further study.

Aromatic hydrocarbons are unevenly distributed and released in tissues of oil-contaminated fish and shrimp. It is interesting to note that edible portions of shrimp and fish, the abdominal muscle of shrimp and somatic musculature of fish, are the body regions which release naphthalenes most rapidly. Accumulation of much higher concentrations of naphthalenes in the digestive gland of shrimp and gall bladder of fish indicates that these organs may serve as sites of storage and excretion of lipid-soluble foreign materials. Fong (1976) reported that the highest concentrations of aromatic hydrocarbons accumulated from Kuwait crude oil by the clam *Mya arenaria* were in the intestine and hepatopancreas.

There is evidence that fish (Lee *et al.*, 1972*b*; Pedersen *et al.*, 1976; Corner, 1975; Payne and Penrose, 1975; Burns, 1976) and at least some species of crustaceans (Corner *et al.*, 1973, 1976*a*; Lee *et al.*, 1976, 1977) and polychaete worms (Lee, 1976; Lee *et al.*, 1976, 1977, 1979) are able to metabolize aromatic hydrocarbons and excrete them as water-soluble metabolites. Marine bivalve molluscs apparently have little or no ability to metabolize polycyclic aromatic hydrocarbons (Lee *et al.*, 1972*a*; Pohl *et al.*, 1974; Payne, 1977; Anderson, 1978; Vandermeulen and Penrose, 1978).

In the present investigation, the marine worm was shown to excrete up to one-third of accumulated ^{14}C-naphthalene as polar metabolites. Species variation in metabolic capabilities may account in part for the more rapid rate of release of aromatic hydrocarbons by shrimp, fish, and polychaetes than by clams and oysters. It should be pointed out that high concentrations of naphthalenes observed in this investigation in the digestive gland of shrimp and gall bladder of fish are present in native form and not as metabolites, since analytical techniques used would not detect the latter. Apparently, all species investigated are able to release large amounts of petroleum hydrocarbons from their tissues either by active excretion or passive outward diffusion of unmetabolized hydrocarbons. It should also be noted that polychaete worms retained naphthalene metabolites longer than unmetabolized naphthalene, indicating that a portion of the metabolite fraction may become covalently bound to some tissue fraction.

Although uptake and release of aromatic hydrocarbons by the species studied depends ultimately on active biological processes such as ventilation, blood circulation, excretion, and hydrocarbon metabolism, retention of hydrocarbons in animal tissues appears to be a passive process dependent on a partitioning of hydrocarbons between exposure water and tissue lipids (Neely et al., 1974; Hamelink and Spacie, 1977). Hydrocarbon release data indicate that binding of hydrocarbons to tissue lipids is probably through hydrophobic interaction (Stone, 1975). Thus, hydrocarbons remain exchangeable and when animals are returned to hydrocarbon-free seawater, the lipid/water partition coefficients for hydrocarbons favor their gradual release from tissues to the water. The rate of release depends in part on metabolic turnover rate of the tissue lipid pool. The observed increase in bioaccumulation factors with increased molecular weight and probable partition coefficient in the naphthalene–phenanthrene series tends to support this view.

Temperature and salinity affect many factors intrinsic to both physiological functions of organisms and behavior of hydrocarbons in seawater. It is therefore not surprising that these environmental variables also affect hydrocarbon uptake and release by marine organisms. However, patterns of interaction are complex and reflect the intricacy of biological processes involved. In the investigations discussed above, salinity had a marked effect on naphthalene and phenanthrene uptake by grass shrimp *Palaemonetes pugio* but not on naphthalenes uptake by clams *Rangia cuneata* and *Protothaca staminea*. The shrimp is an excellent osmoregulator whereas molluscs are osmoconformers. Since hydrocarbon uptake is depressed at salinities in which the shrimp osmoregulates, some component of the osmoregulatory machinery may govern the rate of hydrocarbon exchange between the animal and its environment, possibly by an alteration of integumentary permeability.

A wide variety of physiological and biochemical processes, physical state of tissue lipids, solubility of hydrocarbons in water and lipids, and other biological and physical factors are influenced by temperature. These in turn affect hydrocarbon uptake and retention. We showed that the filtration rate of *R. cuneata* (and thus perfusion rate at presumed hydrocarbon exchange sites) diminishes with decreasing temperature. A decrease in temperature can also be expected to reduce metabolic turnover rate of potential hydrocarbon sequestration sites, such as membrane lipids, facilitating hydrocarbon retention in these hydrophobic tissue compartments. On the other hand, some weak chemical bonds are strengthened at low temperatures (Somero and Hochachka, 1976), which would be expected to decrease exchange rates between the medium and lipid pools. Interactions are complex and the observed increase in naphthalenes uptake rate at lower temperatures undoubtedly represents the net effect of several temperature-mediated changes. Additional studies in this area may yield important information about molecular mechanisms of hydrocarbon uptake by marine animals.

The highest levels of petroleum hydrocarbons in oil-impacted marine environments are found in bottom sediments (McAuliffe, 1976). It is therefore of considerable importance to determine biological availability of these sediment-adsorbed hydrocarbons to benthic marine animals. Blumer *et al.* (1970) reported that benthic invertebrates from the West Falmouth, Massachusettes oil spill site remained heavily contaminated with oil for a considerable time after the spill. Since sediments from the spill site have retained petroleum hydrocarbons for more than six years, it is likely that animals were continually being recontaminated by oil desorbing from sediments. Lee (1976) showed that benthic meio- and macro-infauna are important in metabolic breakdown of sediment hydrocarbons, and suggested that hydrocarbons are biologically available to these species.

Prouse and Gordon (1976) reported that hydrocarbon concentrations in fecal casts of the deposit-feeding polychaete worm *Arenicola marina* are lower than those in surrounding oil contaminated sediments, implying uptake and metabolism of sediment-bound hydrocarbons by the worm. Lee *et al.* (1976) were able to induce the cytochrome P-450 mixed-function oxidase system in blue crabs *Callinectes sapidus* by feeding them hydrocarbon-contaminated food.

These reports provide only circumstantial evidence for direct uptake from sediments or food of petroleum hydrocarbons. Unequivocal evidence for direct uptake and adsorption of hydrocarbons from sediment and food is difficult to obtain. A distinction must be made between uptake directly from the sediment and from water containing hydrocarbons desorbed from sediments. In the natural environment, desorbed hydrocarbons can be expected to be rapidly diluted to low concentration in the overlying water column. A distinction must also be made between hydrocarbons present in gut contents and digestive diverticula and those which are actually absorbed into tissues. The former are likely to be rapidly lost from the animal by defecation. We were unable to demonstrate in our studies, accumulation in tissues of *Neanthes arenaceodentata* of sediment- or food-adsorbed ^{14}C-methylnaphthalene or ^{14}C-naphthalene. However, there was some evidence of direct uptake of hydrocarbons from heavily contaminated sediments by the clam *Rangia cuneata* in both laboratory and field studies. This important area definitely deserves further investigation.

CHAPTER 5

Effects of Petroleum on the Behavior of Marine Animals

5.1 INTRODUCTION

The majority of marine animals possess highly complex, stereotyped patterns of behaviour. These behavioral patterns are highly species-specific and are an essential component of the functional biology and ecology of a population. Many components of individual, intra- and interspecific interactive behaviors in marine animals are triggered by chemical stimuli. This response to environmental chemicals is called 'chemotaxis'. Specific chemotactic responses of an animal may be elicited by slight changes in concentration of certain naturally occurring inorganic or organic substances in the ambient medium, by specific organic chemicals (pheromones) released by other members of the same species, or by specific organic chemicals (allomones and kairomones) released by another species (Kittredge et al., 1973).

Marine organisms are extremely sensitive to the presence of these chemical stimuli. Because of this sensitivity and the extremely low concentrations at which natural chemical cues occur in the ocean, chemical pollution may interfere with natural chemotaxis of marine organisms in several ecologically maladaptive ways (Blumer et al., 1973; Sutterlin, 1974).

Pollutant chemicals may either mimic or mask natural chemical signals, eliciting inappropriate responses in exposed organisms. Contaminants may damage chemosensory apparatus (Gardner, 1972; DiMichele and Taylor, 1978) or produce varying degrees of narcosis (Crisp et al., 1967) to render the animals refractory to chemical stimuli. In addition, chemical pollutants which are sensed as being noxious may elicit alarm or defensive behavioral responses in marine animals. Because of the importance of chemical communication to marine animals and the receptiveness of sensory apparatus involved, marine animals can be expected to respond behaviorally to pollutant concentrations orders of magnitude lower than those inducing acute toxicity.

Several problems emerge in using behavior as an index of sublethal pollutant stress. Normal behavior of most marine organisms is extremely complex and variable and is influenced by a wide variety of natural biotic and abiotic factors. It is often, therefore, difficult to distinguish pollutant-induced behavioral modification from natural behavioral variation. Also, any observed pollutant-mediated behavioral modification should be interpretable in terms of predicted impact on long term survival of the individual or population.

Several recent studies have dealt with behavioral responses of marine animals to oil pollution. These types of behavioral responses—locomotion/ orientation, feeding/predation, and alarm reactions—have received particular attention.

Hargrave and Newcombe (1973) reported that the crawling rate of the intertidal snail *Littorina littorea* was increased in the presence of dispersed bunker C residual oil but decreased by brief exposure to a low toxicity oil dispersant (Corexit 8666) or to oil/dispersant mixtures. Dicks (1973, 1976) showed that the intertidal limpet *Patella vulgata* readily became detached from its rock substrate when exposed to Kuwait crude oil; detachment was probably due to narcosis. Locomotory activity of two arctic marine invertebrates, the amphipod *Onisimus affinis* and the medusa *Halitholus cirratus*, was significantly impaired by 15–20 ppm concentrations of four crude oil-in-water dispersions (Percy and Mullin, 1977). Pembina and Normal Wells crude oils had greatest effect on locomotor activity of both species. Recovery of normal activity after 24 h in clean seawater was more complete in the medusa than in the amphipod. Lee and Nicol (1977) reported that positive phototaxis of coastal zooplankton was partially impaired by exposure to 20 per cent WSF of a No. 2 fuel oil (about 3·8 ppm total hydrocarbons). Bigford (1977) studied geotactic, phototactic, and pressure responses of larval rock crabs *Cancer irroratus* exposed to water-accommodated fractions of a No. 2 fuel oil. Geonegative (upward) movements in the water column were significantly depressed in early stage larvae and enhanced in late stage larvae by aqueous hydrocarbon concentrations of 0·1 and 1·0 ppm. Oil exposure tended to impair positive phototactic response in those larval stages normally positively phototactic. Control larvae responded to an increase in hydrostatic pressure by upward movement in the water column. This response was partially abolished by exposure to the WSF. The tactic responses described allow larvae to maintain a favorable position in the water column; impairment of tactic responses could decrease larval survival.

Jacobson and Boylan (1973) reported that kerosene at concentrations as low as 4 µg/liter (ppb) inhibited feeding chemotaxis in the mud snail *Nassarius obsoletus*. In other experiments, the snails were attracted by 1 ppm and repelled by 10 ppm No. 2 fuel oil (API reference oil No. III) added to water (Atema, 1976).

Exposure for 30 min to seawater extract of crude oil or to solutions of benzene, naphthalene or 2,2'-binaphthal completely inhibited feeding behavior in the crab *Pachygrapsus crassipes* (Kittredge, 1971; Takahaski and Kittredge, 1973). Dispersed La Rosa crude oil at a nominal concentration of 10 ppm was repellant to lobsters *Homarus americanus* and inhibited feeding activity (Atema and Stein, 1974). Water-soluble fractions of the oil were without effect. However, low concentrations of kerosene (nominal concentration 10 µg/liter) attracted lobsters and stimulated feeding activity (Atema *et al.*, 1973; Atema, 1976). The branched cyclic fraction of the kerosene seemed to cause attraction and feeding, while the polar aromatic fraction seemed to elicit repulsion and feeding inhibition responses. No histological changes in odor receptors on the antennae were observed. A wide variety of other behavioral responses of lobsters to oil have been reported (Blumer *et al.*, 1973). Blue crabs *Callinectes sapidus* could detect 0·12 ppm naphthalene in seawater but showed no food searching or feeding response (Pearson and Olla, 1979). Five species of marine crustaceans responded with increased motor activity to low concentrations of n-hexane in seawater (Idoniboye-Obu, 1977). Thompson *et al.* (1977) reported that scarlet prawns *Plesiopenaeus edwardsianus* were attracted in large numbers to a benthic oil deposit off Aruba, Dutch West Indies.

Predation rate of the gastropod oyster drill *Drupa granulata* on the mussel *Mytilus variabilis* was measured following 7-day exposure to high sublethal concentrations of Iranian crude oil (Eisler, 1973). Predation rate was three times higher in controls than in the group in which both predator and prey had been exposed to oil. When only one species was exposed to oil, intermediate predation rates were observed.

An easily quantified behavioral response of fish to noxious chemicals is opercular rate and coughing rate. These responses were measured in pink salmon fry *Oncorhynchus gorbuscha* exposed to the WSF of Prudhoe Bay crude oil (Thomas and Rice, 1975; Rice *et al.*, 1976). Opercular rates increased significantly for as long as 9–12 h after exposure to WSF concentrations of 2·8–3·5 ppm. The pattern of the cough response was the same as that of opercular rate changes but was detected in relatively fewer fish.

In all but a few of these studies, the concentrations of petroleum hydrocarbons in the water column required to elicit significant behavioral impairment were not actually measured or were higher (1–20 ppm) than would be expected in all but the most severely oil-contaminated marine environments. In many cases, the concentrations of hydrocarbons causing significant behavioral change were not greatly different from those which are acutely toxic or cause physiological impairment. Thus, with few exceptions, the behavioral responses measured are not significantly more sensitive than other biological responses as indices of sublethal hydrocarbon stress in marine animals. The extreme complexity and variability of normal behaviors militates against the general use of behavior as an index of sublethal pollutant stress.

In preliminary experiments, we investigated the effects of oil WSF on locomotary behavior of marine molluscs. Results of these studies are described below.

5.2 MATERIALS AND METHODS

An egg string belonging to a marine gastropod mollusc, the lightning whelk *Busycon contrarium* (Conrad), was collected on the front beach at Galveston, Texas and maintained in the laboratory. Many snails began hatching from capsules and within 2 days after the capsule began hatching, enough snails were gathered to begin experiments.

Four-inch finger bowls were used as test containers. Each bowl had a circle, 3 cm in diameter, marked in the center on the outside bottom. Test solutions consisted of clean seawater (control), 5, 10, 25, 50, 75 and 100 per cent dilutions of south Louisiana crude oil WSF (1–19·8 ppm total hydrocarbons). Twenty snails, all less than 48-h old, were used at each WSF concentration. For each dilution, all 20 snails were put into a single finger bowl, 200 ml of the test solution added, and the snails then placed inside the 3-cm circle.

A circular piece of nitex mesh, slightly larger than the inside diameter, was wedged in the bowl at the waterline to prevent snails from crawling out. Solutions were changed daily and observations were made 1, 4, and 24 h after dosing to determine:

1. The position of snails in the bowl, i.e., how many were in the 3-cm circle, outside of it (the distance from the edge of the circle to the

side was 3 cm), or on the side (about 2·5 cm from the bottom of the bowl to the waterline)
2. The length of the siphon of each snail, i.e., no siphon visible, siphon length equal to width, twice width, or three times width

All snails were measured periodically after the experiment began. Because the snails were approximately the same age and size, only 20 snails were measured before the experiment began and these were taken as a representative sample. The size of these 20 snails ranged from 0·43–0·51 cm with a mean of 0·47 ± 0·02 cm.

Adult specimens of the surf or coquina clam *Donax variabilis* (Philippi) were collected from the front beach at Galveston. These clams are common on sandy Texas beaches during the summer months and are usually found in large numbers along the surf line. Coquina clams burrow just below the sand surface in a band along the surf line with siphons extended to the surface. Clams emerge or are washed from the sand with incoming waves and rapidly burrow back down into the sand as waves recede. The spade-like foot is used for burrowing and can be extended for a length equal to or greater than the length of the shell. Coquinas feed on microscopic organisms in the water or adhering to sand grains. Siphons are used for feeding and respiration by drawing water across the gills.

The activity of the foot and siphons was measured on clams maintained in clean seawater (control) and seawater containing 5, 10, and 25 per cent WSF (1·0–4·9 ppm total hydrocarbons) of south Louisiana crude oil (all at 25‰ S, 27°C). Ten clams per group were placed in 4-in finger bowls to which 200 ml of clean seawater, or seawater plus WSF were added, one finger bowl per exposure concentration. All clams were 1·7–1·8 cm in length. Solutions were changed daily and the activity of the clams recorded 1 and 24 h after the solutions were changed. Activity was determined in two ways. First, undisturbed resting clams were observed in order to determine the amount of extension of the foot and siphons; each clam was given a rating from 0–10 according to these lengths (Fig. 48). A clam which had both foot and siphon extended greater than one-half the shell length was given a '10' whereas a closed clam was given '0'. Other designations were as follows:

9. Long siphon and medium or short foot
8. Medium or short siphon and long foot
7. Medium or short siphon and foot
6. Medium siphon and short foot
5. Short siphon and medium foot

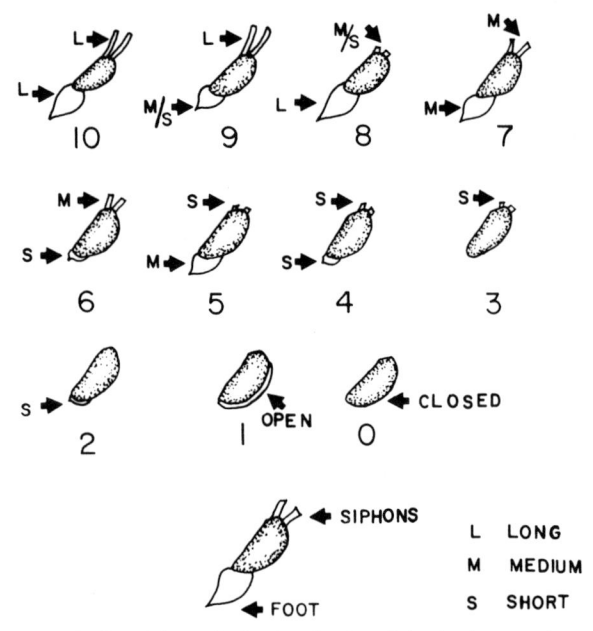

Fig. 48. Method of ranking relative levels of activity in the coquina clam *Donax variabilis*. Activity is ranked from 0 to 10 according to the degree of extension of the siphons and foot from the shell and whether the shell is open or closed.

4. Short siphon and short foot
3. Short siphon and no foot
2. No siphon and short foot
1. Shell open but no siphon or foot

where long = greater than one-half shell length; medium = less than one-half shell length; short = just extending from between valves.

Second, each test bowl was shaken and the number of clams which showed a burrowing response with the foot was recorded. Other responses noted were whether burrowing started immediately after the shaking stopped (within 5 s) or later, and if the foot remained extended or was withdrawn into and remained in the shell.

5.3 RESULTS AND DISCUSSION

In the experiments with *Busycon contrarium*, all snails in control, 5, 10, and 25 per cent WSF solutions of south Louisiana crude oil were actively crawling within 24 h after initial dosing. All snails in the 50, 75, and 100 per cent WSF were either tightly withdrawn into the shell or the foot was barely extended and pointed upward with no movement. These died within 6 days after no movement had taken place. The results reported below, therefore, are concerned with the control, 5, 10, and 25 per cent WSF groups.

Figures 49 and 50 show the position of snails in the bowls one and 24 h after renewal of exposure solutions, respectively, during weeks one, three and eight of exposure. Figure 49 shows that after 1 h of exposure in the control solution (clean seawater) almost all of the snails remained in the 3-cm circle and this pattern did not change with time. In the 5, 10, and 25 per cent WSF solutions, there appeared to be increased crawling activity with an increase in WSF concentration. This crawling activity increased with time in the 5 per cent WSF, remained somewhat constant in the 10 per cent WSF and decreased with time in the 25 per cent WSF.

After 24 h in the exposure media (Fig. 50), none or less than 5 per cent of the snails in the control and 5 per cent in the WSF solutions remained in the 3-cm circle, except in the 5 per cent WSF, during week one. In the 10 and 25 per cent WSF, there was still a movement of snails outside the circle during the 24 h period but some (from 5 to 43 per cent) remained in the circle, apparently having made no effort to crawl.

Data concerning siphon length of the snails is variable but the pattern was for most snails in the control and 5 per cent WSF to have the siphon extended two times its width with some extended one and some three times its width. In the 10 per cent WSF, the length of the siphon was almost always equal to its width. Snails in the 25 per cent WSF rarely had a siphon showing.

Measurements on growth after eight weeks of exposure show there was no significant difference in growth between snails in the control and 5 per cent WSF, but there was a very significant difference in growth between snails in the control and those in the 10 and 25 per cent WSF. Oil-exposed snails were significantly smaller than the controls.

These experiments were repeated on snails exposed to WSF of No. 2 fuel oil. The results were very similar to those obtained for south Louisiana crude oil.

Behavior of *Donax variabilis* is summarized in Figs. 51 and 52. Figure 51 is based on observations 1 h after renewal of exposure solutions and Fig. 52 for observations 24 h after renewal of exposure solutions. Figure 51 illustrates that in the control group almost all clams showed a 4–7 rating during the entire 14-day exposure period. In the 5, 10, and 25 per cent WSF exposure solutions the majority of clams showed a 0–4 rating during the

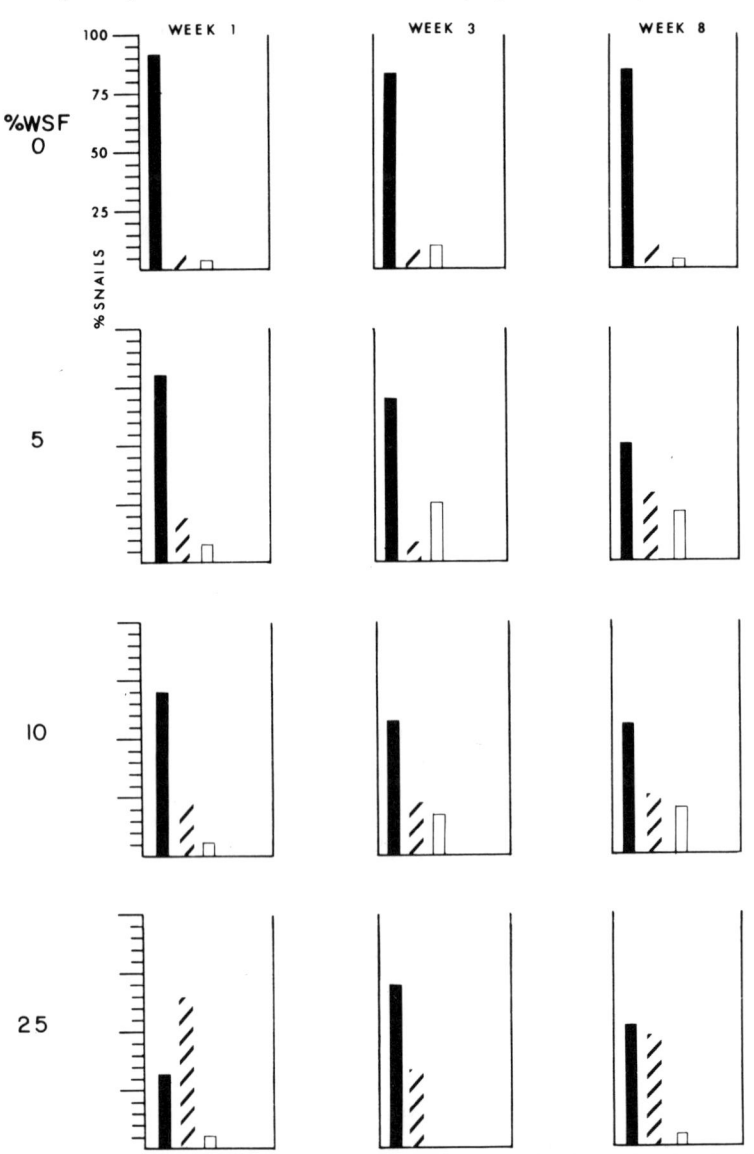

FIG. 49. The positions of snails *Busycon contrarium* in finger bowls 1 h after changing the south Louisiana crude oil WSF exposure solutions and after one, three, or eight weeks continuous exposure to the WSF which was renewed daily. Black bar, snails within circle; striped bar, snails outside circle; white bar, snails on side of bowl.

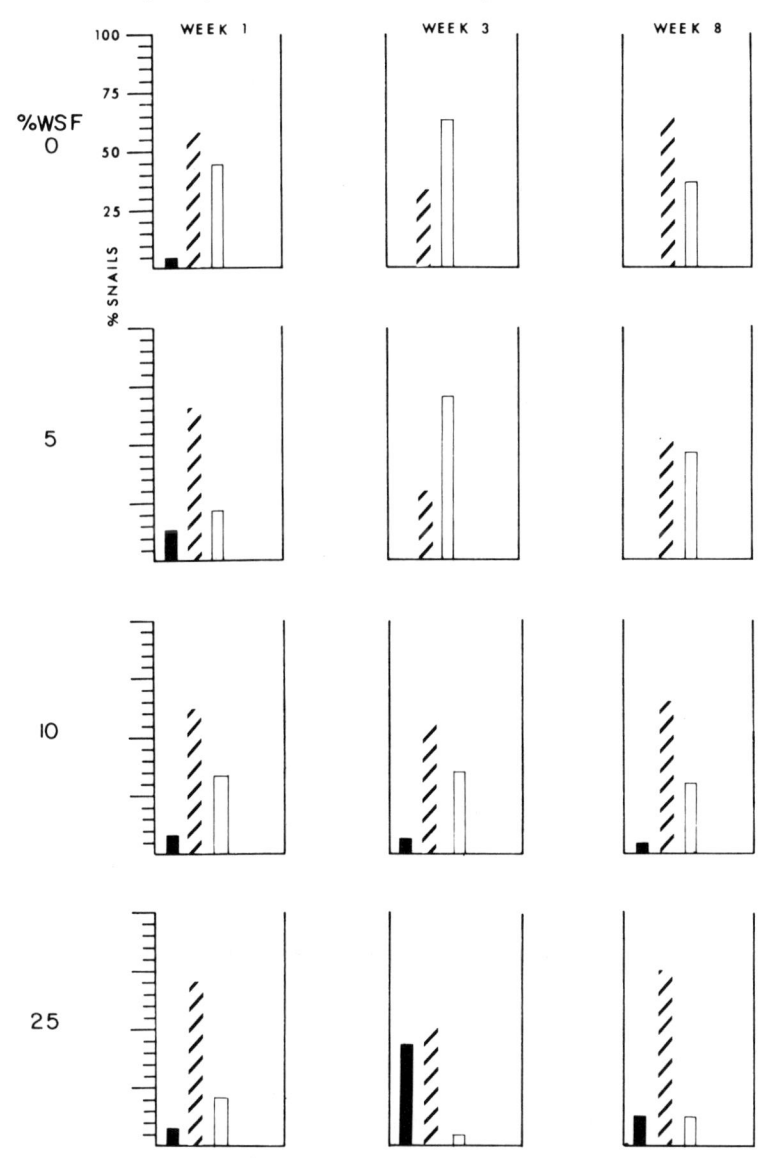

FIG. 50. The positions of snails *Busycon contrarium* in finger bowls 24 h after changing the south Louisiana crude oil WSF exposure solutions and after one, three, or eight weeks continuous exposure to the WSF which was renewed daily. Black bar, snails within circle; striped bar, snails outside circle; white bar, snails on side of bowl.

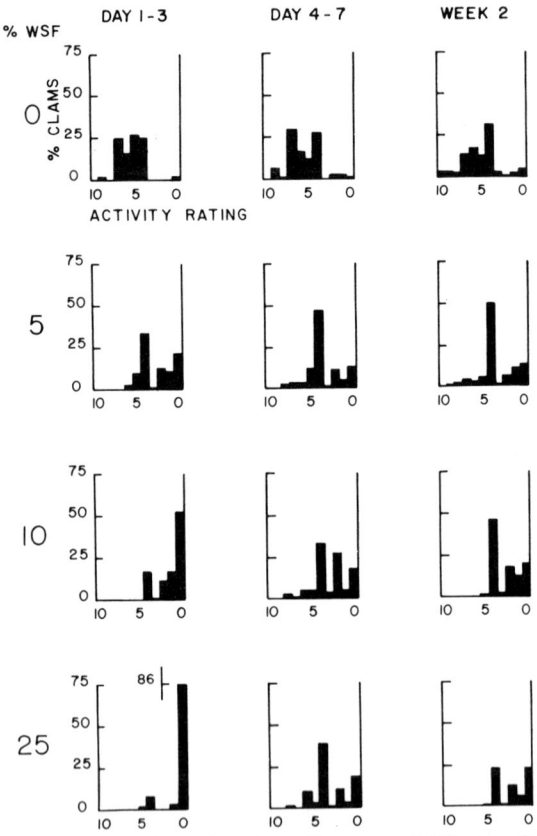

FIG. 51. Relative activity of coquina clams *Donax variabilis* 1 h after daily dosing with the WSF of south Louisiana crude oil and at different time intervals during chronic exposure to the WSF.

first 3-day exposure. About the third day, however, some clams in each exposure concentration appeared to become somewhat acclimated to the solutions and showed ratings up to 8 with a sharp increase in rating 4. A few clams remained in the 0–3 rating. This trend lasted throughout the second week. Observations after 24 h (Fig. 52) showed the same trends as 1 h observations except for a shift to somewhat higher ratings, probably due to a decrease in exposure concentrations with time.

No mortalities were observed in control, 5, and 10 per cent WSF during the two week observation but 50 per cent mortality occurred in the 25 per cent test solution at the end of one week. Ten clams each were exposed to 50, 75, and 100 per cent WSF. Most remained closed and all died within 5 days.

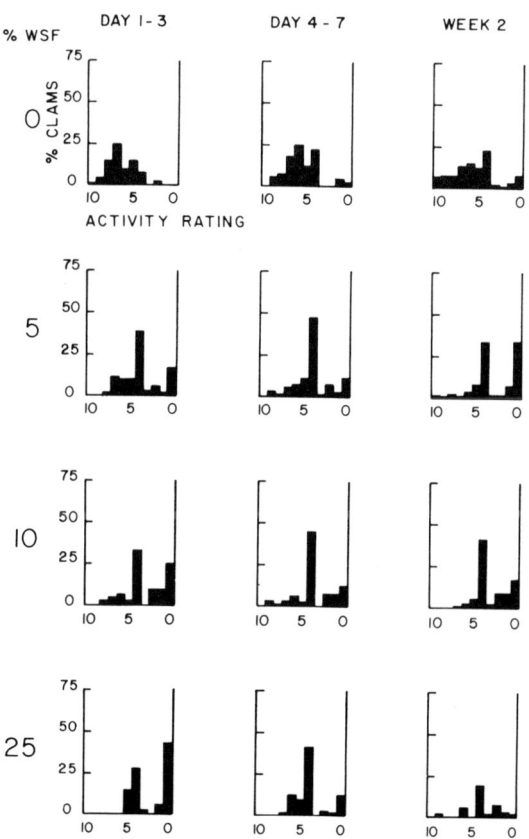

FIG. 52. Relative activity of coquina clams *Donax variabilis* 24 h after daily dosing with the WSF of south Louisiana crude oil and at different time intervals during chronic exposure to the WSF.

Results of the burrowing response study after shaking indicated a decrease in burrowing activity with an increase in WSF concentration during the first few days, then an increase to or near control levels with acclimation.

The above experiments were run twice, with fresh clams each time. Clams in the second test group were slightly smaller, but results were the same as reported above. These experiments were also conducted using No. 2 fuel oil with essentially identical results.

Results of experiments with the snail *Busycon contrarium* are similar to those reported by Hargrave and Newcombe (1973) for the snail *Littorina littorea*. In both cases, sublethal exposure to oil stimulated locomotory activity. However, normal activity of *Donax variabilis* was depressed by oil exposure. Increased locomotor activity of snails and decreased activity with partial or complete shell closure of clams can be considered to be avoidance reactions to irritant chemicals in the ambient medium.

It is important to note that oil-induced behavioral changes in the two species studied decreased during chronic exposure, approaching control behaviour by the end of the experiment. Apparently, these molluscs are able to acclimate to sublethal oil exposure. Behavioral acclimation to chronic exposure to low level hydrocarbon exposure is an important area of research that has been largely neglected.

References

Acheson, M. A., R. M. Harrison, R. Perry and R. A. Wellings (1976) Factors affecting the extraction and analysis of polynuclear hydrocarbons in water. *Water Res.* **10**: 207–12.
Allen, H. (1971) Effects of petroleum fractions on the early development of a sea urchin. *Mar. Pollut. Bull.* **2**: 137–40.
American Petroleum Institute (1958) Determination of volatile and non volatile oily material. Infrared spectrometric method, No. 733–58. A.P.I., Washington DC.
American Public Health Association (1971) Standard methods for the examination of water and wastewater. A.P.H.A., Washington, DC, 874 pp.
Anderson, J. W. (ed.) (1975) Laboratory studies on the effects of oil on marine organisms: an overview. A.P.I. Publ. No. 4249. American Petroleum Institute, Washington, DC, 70 pp.
Anderson, J. W., D. B. Dixit, G. S. Ward and R. S. Foster (1977a) Effects of petroleum hydrocarbons on the rate of heart beat and hatching success of estuarine fish embryos. In: F. J. Vernberg, A. Calabrese, F. P. Thurberg and W. B. Vernberg (eds.) *Physiological Responses of Marine Biota to Pollutants.* Academic Press, New York, pp. 241–58.
Anderson, J. W., L. J. Moore, J. W. Blaylock, D. L. Woodruff and S. L. Kiesser (1977b) Bioavailability of sediment-sorbed naphthalenes to the sipunculid worm, *Phascolosoma agassizii.* In: D. A. Wolfe (ed.) *Fate and Effects of Petroleum Hydrocarbons in Marine Organisms and Ecosystems.* Pergamon Press, New York, pp. 276–85.
Anderson, J. W., J. M. Neff, B. A. Cox, H. E. Tatem and G. M. Hightower (1974a) Characteristics of dispersions and water-soluble extracts of crude and refined oils and their toxicity to estuarine crustaceans and fish. *Mar. Biol.* **27**: 75–88.
Anderson, J. W., J. M. Neff, B. A. Cox, H. E. Tatem and G. M. Hightower (1974b) The effects of oil on estuarine animals: toxicity, uptake and depuration, respiration. In: F. J. Vernberg and W. B. Vernberg (eds.) *Pollution and the Physiology of Marine Organisms.* Academic Press, New York, pp. 285–310.
Anderson, J. W. and G. C. Stephens (1969) Uptake of organic material by aquatic invertebrates. VI. Role of epiflora in apparent uptake of glycine by marine crustaceans. *Mar. Biol.* **4**: 243–9.

Anderson, R. S. (1978) Benzo[a]pyrene metabolism in the American oyster *Crassostrea virginica*. Environmental Protection Agency. Ecological Research Series. EPA-600/3-78-009. 18 pp.
Armstrong, H. W., K. Fucik, J. W. Anderson and J. M. Neff (1977) Effects of oilfield brine effluent on benthic organisms in Trinity Bay, Texas. A.P.I. Publ. No. 4291. American Petroleum Institute, Washington, DC, 82 pp.
Armstrong, H. W., K. Fucik, J. W. Anderson and J. M. Neff (1979) Effects of oilfield brine effluent on sediments and benthic organisms in Trinity Bay, Texas. *Mar. Environ. Res.* **2**: 55–69.
Armstrong, P. B. and J. S. Child (1965) Stages in the normal development of *Fundulus heteroclitus*. *Biol. Bull.* **128**: 143–66.
Atema, J. (1976) Sublethal effects of petroleum fractions on the behavior of the lobster, *Homarus americanus*, and the mud snail, *Nassarius obsoletus*. In: M. Wiley (ed.) *Estuarine Processes. Vol. 1, Uses, Stresses and Adaptation to the Estuary*. Academic Press, New York, pp. 302–12.
Atema, J., S. M. Jacobson, J. H. Todd and D. B. Boylan (1973) The importance of chemical signals in stimulating behavior of marine organisms: effects of altered environmental chemistry on animal communication. Pages 177–97 In: G. Glass (ed.). *Bioassay Techniques in Environmental Chemistry*. Ann Arbor Science Publ., Inc., Ann Arbor, Michigan.
Atema, J. and L. S. Stein (1974) Effects of crude oil on the feeding behavior of the lobster *Homarus americanus*. *Environ. Pollut.* **6**: 77–86.
Balon, E. K. (1975) Terminology of intervals in fish development. *J. Fish. Res. Bd Canada* **32**: 1663–70.
Barnes, D. J. and C. J. Crossland (1977) Coral calcification: sources of error in radioisotope techniques. *Mar. Biol.* **42**: 119–29.
Barr, A. J., J. H. Goodnight, J. P. Sall and J. T. Helwig (1976) *A Users Guide to SAS76*. SAS Institute, Inc., Raleigh, North Carolina, 329 pp.
Bellan, G., D. J. Reish and J. P. Foret (1971) Action toxique d'un detergent sur le cycle de development de la polychaete *Capitella capitata* (Fab.). *C.R. Acad. Sci., Paris. Ser. D* **272**: 2476–9.
Bellan, G., D. J. Reish and J. P. Foret (1972) The sublethal effects of a detergent on the reproduction, development, and settlement in the polychaetous annelid *Capitella capitata*. *Mar. Biol.* **14**: 183–8.
Bend, J. R. and M. O. James (1978) Xenobiotic metabolism in marine and freshwater species. In: D. C. Malins and J. R. Sargent (eds.) *Biochemical and Biophysical Perspectives in Marine Biology*, Vol. 4. Academic Press, New York, pp. 126–88.
Benson, A. A. and L. Muscatine (1974) Wax in coral mucus: energy transfer from corals to reef fishes. *Limnol. Oceanog.* **19**: 810–14.
Berdugo, V., R. P. Harris and S. C. O'Hara (1977) The effect of petroleum hydrocarbons on reproduction of an estuarine planktonic copepod in laboratory cultures. *Mar. Pollut. Bull.* **8**: 138–43.
Bieri, R. H. and V. C. Stamoudis (1977) The fate of petroleum hydrocarbons from a No. 2 fuel oil spill in a seminatural estuarine environment. In: D. A. Wolfe (ed.) *Fate and Effects of Petroleum Hydrocarbons in Marine Organisms and Ecosystems*. Pergamon Press, New York, pp. 332–44.

Bigford, T. S. (1977) Effects of oil on behavioral responses to light, pressure and gravity in larvae of the rock crab, *Cancer irroratus* Say. *Mar. Biol.* **43**: 137–48.

Birkeland, C., A. A. Reimer and J. R. Young (1976) Survey of marine communities in Panama and experiments with oil. Environmental Protection Agency, Office of Research and Development, Environmental Research Laboratory, Narragansett, Rhode Island. EPA-600/3-76-028. 177 pp.

Blumer, M. (1976) Polycyclic aromatic hydrocarbons in nature. *Sci. Amer.* **234**: 34–45.

Blumer, M., J. M. Hunt, J. Atema and L. Stein (1973) Interaction between marine organisms and oil pollution. Environmental Protection Agency, Ecological Research Series-R 3-73-042.

Blumer, M., G. Souza and J. Sass (1970) Hydrocarbon pollution of edible shellfish by an oil spill. *Mar. Biol.* **5**: 195–202.

Blumer, M. and W. W. Youngblood (1975) Polycyclic aromatic hydrocarbons in soils and recent sediments. *Science* **188**: 53–5.

Boehm, P. D. and J. G. Quinn (1976) The effects of dissolved organic matter in sea water on the uptake of mixed individual hydrocarbons and number 2 fuel oil by a marine filter-feeding bivalve (*Mercenaria mercenaria*). *Estuar. Cstl. Mar. Sci.* **4**: 93–105.

Boehm, P. D. and J. G. Quinn (1977) The persistence of chronically accumulated hydrocarbons in the hard shell clam, *Mercenaria mercenaria*, *Mar. Biol.* **44**: 227–33.

Bohon, R. and W. F. Claussen (1951) The solubility of aromatic hydrocarbons in water. *J. Amer. Chem. Soc.* **73**: 1571–8.

Boyd, J. F. and R. C. Simmonds (1974) Continuous laboratory production of fertile *Fundulus heteroclitus* (Walbaum) eggs lacking chorionic fibrils. *J. Fish Biol.* **6**: 389–94.

Boylan, D. B. and B. W. Tripp (1971) Determination of hydrocarbons in seawater extracts of crude oil and crude oil fractions. *Nature, London* **230**: 44–7.

Burns, D. A. (1976) Microsomal mixed function oxidases in an estuarine fish, *Fundulus heteroclitus*, and their induction as a result of environmental contamination. *Comp. Biochem. Physiol.* **53B**: 443–6.

Burwood, R. and G. C. Speers (1974) Photo-oxidation as a factor in the environmental dispersal of crude oil. *Estuar. Cstl. Mar. Sci.* **2**: 117–35.

Byrne, C. J. and J. A. Calder (1977) Effect of the water-soluble fractions of crude, refined and waste oils on embryonic and larval stages of the quahog clam *Mercenaria* sp. *Mar. Biol.* **40**: 225–31.

Cahnmann, H. J. and M. Kuratsune (1957) Determination of polycyclic aromatic hydrocarbons in oysters collected in polluted water. *Anal. Chem.* **29**: 1312–17.

Caldwell, R. S., E. M. Calderone and M. H. Mallon (1977) Effects of seawater-soluble fraction of Cook Inlet crude oil and its major aromatic components on larval stages of the Dungeness crab, *Cancer magister* Dana. In: D. A. Wolfe (ed.). *Fate and Effects of Petroleum Hydrocarbons in Marine Organisms and Ecosystems.* Pergamon Press, New York, pp. 210–20.

Carlberg, S. R. and C. B. Skarstedt (1972) Determination of small amounts of nonpolar hydrocarbons (oil) in sea water. *J. Cons. int. Explor. Mer.* **34**: 506–15.

Chalker, B. E. and D. L. Taylor (1976) Light-enhanced calcification and the role of oxidative phosphorylation in calcification of the coral *Acropora cervicornis*. *Proc. Roy. Soc. London, Ser. B.* **190**: 323–31.

Chalker, B. E. and D. L. Taylor (1978) Rhythmic variations in calcification and photosynthesis associated with the coral *Acropora cervicornis* (Lamarck). *Proc. Roy. Soc. London, Ser. B.* **201**: 179–89.

Chambers, J. E. (1979) Induction of microsomal mixed-function oxidase system components in striped mullet by short term exposure to crude oil. *Toxicol. Let.* **4**: 227–30.

Christiansen, M. E. and F. C. Stormer (1978) Effects of the water-soluble fraction of Ekofisk crude oil on zoeal larvae of the crab *Hyas araneus*. *Ambio* **7**: 23–5.

Clausen, C. D. and A. A. Roth (1975) Estimation of coral growth rates from laboratory ^{45}Ca-incorporation rates. *Mar. Biol.* **33**: 85–91.

Corner, E. D. S. (1975) The fate of fossil fuel hydrocarbons in marine animals. *Proc. Roy. Soc. London, Ser. B.* **189**: 391–413.

Corner, E. D. S., R. P. Harris, C. C. Kilvington and S. C. M. O'Hara (1976a) Petroleum compounds in the marine food web: Short-term experiments on the fate of naphthalene in *Calanus*. *J. Mar. Biol. Assn. U.K.* **56**: 121–33.

Corner, E. D. S., R. P. Harris, K. J. Whittle and P. R. Mackie (1976b) Hydrocarbons in marine zooplankton and fish. In: A. P. M. Lockwood (ed.) *Effects of Pollutants on Aquatic Organisms*. Cambridge University Press, Cambridge, pp. 71–106.

Corner, E. D. S., C. C. Kilvington and S. C. M. O'Hara (1973) Qualitative studies on the metabolism of naphthalene in *Maia squinado* (Herbst). *J. Mar. Biol. Assn. U.K.* **53**: 819–32.

Crisp, D. J., A. O. Christie and A. F. A. Ghobashy (1967) Narcotic and toxic action of organic compounds on barnacle larvae. *Comp. Biochem. Physiol.* **22**: 629–49.

Cucci, T. L. and C. E. Epifanio (1979) Long-term effects of water-soluble fractions of Kuwait crude oil on the larval and juvenile development of the mud crab *Eurypanopeus depressus*. *Mar. Biol.* **55**: 215–20.

Davies, I. W., R. W. Harrison, R. Perry, D. Ratnayaka and R. A. Wellings (1976) Municipal incinerator as source of polynuclear aromatic hydrocarbons in environment. *Environ. Sci. Technol.* **10**: 451–3.

Davies, T. T., M. A. Crenshaw and B. M. Heatfield (1972) The effect of temperature on the chemistry and structure of echinoid spine regeneration. *J. Paleontol.* **46**: 874–83.

Davis, W. R. and D. J. Reish (1975) The effect of reduced dissolved oxygen concentrations on the growth and production of oocytes in the polychaetous annelid *Neanthes arenaceodentata*. *Rev. Intern. Oceanog. Medit.* **37**: 3–16.

DeAngelis, E. and G. G. Giordano (1974) Sea urchin egg development under the action of benzo[a]pyrene and 7,12-dimethylbenz[a]anthracene. *Canc. Res.* **34**: 1275–80.

Dicks, B. (1973) Some effects of Kuwait crude-oil on the limpet, *Patella vulgata*. *Environ. Pollut.* **5**: 219–29.

Dicks, B. (1976) The importance of behavioral patterns in toxicity testing and ecological prediction. In: J. M. Baker (ed.) *Marine Ecology and Oil Pollution*. John Wiley and Sons, New York, pp. 303–20.

Dillon, T. M., J. M. Neff and J. S. Warner (1978) Toxicity and sublethal effects of No. 2 fuel oil on the supralittoral isopod *Lygia exotica*. *Bull. Environ. Contam. Toxicol.* **20**: 320-7.

DiMichele, L. and M. H. Taylor (1978) Histopathological and physiological responses of *Fundulus heteroclitus* L. to naphthalene exposure. *J. Fish. Res. Bd Canada* **35**: 1060-6.

DiSalvo, L. H., H. E. Guard and L. Hunter (1975) Tissue hydrocarbon burden of mussels as potential monitor of environmental hydrocarbon insult. *Environ. Sci. Technol.* **9**: 247-51.

Dixit, D. and J. W. Anderson (1977) Distribution of naphthalenes within exposed *Fundulus similis* and correlations with stress behaviour. In: *Proceedings 1977 Oil Spill Conference (Prevention, Behavior, Control, Cleanup)*. American Petroleum Institute, Washington, DC, pp. 633-6.

Drew, E. A. (1973) The biology and physiology of alga-invertebrate symbiosis. III. *In situ* measurements of photosynthesis and calcification in some hermatypic corals. *J. Exp. Mar. Biol. Ecol.* **13**: 165-79.

Dunn, B. P. and J. Fee (1979) Polycyclic aromatic hydrocarbon carcinogens in commercial seafoods. *J. Fish Res. Bd Canada* **36**: 1469-76.

Dunn, B. P. and H. F. Stich (1975) The use of mussels in estimating benzo[*a*]pyrene contamination of the marine environment. *Proc. Soc. Exp. Biol. Med.* **150**: 49-51.

Dunn, B. P. and H. F. Stich (1976*a*) Monitoring procedures for chemical carcinogens in coastal waters. *J. Fish. Res. Bd Canada* **33**: 2040-6.

Dunn, B. P. and H. F. Stich (1976*b*) Release of the carcinogen benzo[*a*]pyrene from environmentally contaminated mussels. *Bull. Environ. Contam. Toxicol.* **15**: 398-401.

Dunn, B. P. and D. R. Young (1976) Baseline levels of benzo[*a*]pyrene in southern California mussels. *Mar. Pollut. Bull.* **7**: 231-4.

Edwards, R. R. C. (1978) Effects of water-soluble oil fractions on metabolism, growth and carbon budget of the shrimp *Crangon crangon*. *Mar. Biol.* **46**: 259-65.

Eganhouse, R. P. and J. A. Calder (1976) The solubility of medium molecular weight aromatic hydrocarbons and the effects of hydrocarbon co-solutes and salinity. *Geochim. Cosmochim. Acta.* **40**: 555-61.

Ehrhardt, M. (1972) Petroleum hydrocarbons in oysters from Galveston Bay. *Environ. Pollut.* **3**: 257-71.

Eisler, R. (1973) Latent effects of Iranian crude oil and a chemical oil dispersant on Red Sea molluscs. *Israel J. Zool.* **22**: 97-105.

Epifanio, C. E. (1971) Effects of dieldrin in sea water on the development of two species of crab larvae, *Leptodius floridanus* and *Panopeus herbstii*. *Mar. Biol.* **11**: 356-62.

Ernst, V. V., J. M. Neff and J. W. Anderson (1977) Effects of the water-soluble fraction of No. 2 fuel oil on the development of the estuarine fish, *Fundulus grandis* Baird and Girard. *Environ. Pollut.* **14**: 25-35.

Fabacher, D. L. and H. Chambers (1971) A possible mechanism of insecticide resistance in mosquito fish. *Bull. Environ. Contam. Toxicol.* **6**: 372-6.

Falk-Petersen, I.-B. (1979) Toxic effects of aqueous extracts of Ekofisk crude oil, crude oil fractions, and commercial oil products on the development of sea urchin eggs. *Sarsia* **64**: 161-9.

Farrington, J. W., J. M. Teal and P. L. Parker (1976) Petroleum hydrocarbons. In: Goldberg, E. D. (ed.) *Strategies for Marine Pollution Monitoring.* John Wiley and Sons, New York, pp. 3–34.

Farrington, J. W. and B. W. Tripp (1977) Hydrocarbons in western North Atlantic surface sediments. *Geochim. Cosmochim. Acta.* **41**: 1627–41.

Fazio, T. (1971) Analysis of oyster samples for polycyclic hydrocarbons. In: *Proceedings of the 7th National Shellfish Sanitation Workshop.* FDA, Washington, DC, pp. 238–43.

Fong, W. C. (1976) Uptake and retention of Kuwait crude oil and its effects on oxygen uptake by the soft-shell clam, *Mya arenaria. J. Fish. Res. Bd Canada* **33**: 2774–80.

Fossato, V. U. and E. Siviero (1974) Oil pollution monitoring in the lagoon of Venice using the mussel *Mytilus galloprovincialis. Mar. Biol.* **25**: 1–6.

Friedel, R. A. and M. Orchin (1951) *Ultraviolet Spectra of Aromatic Compounds.* John Wiley and Sons, New York, pp. 1–52.

Fucik, K. W., H. W. Armstrong and J. M. Neff (1977) The uptake of naphthalenes by the clam, *Rangia cuneata*, in the vicinity of an oil-separator platform in Trinity Bay, Texas. In: *Proceeding 1977 Oil Spill Conference (Prevention, Behavior, Control, Cleanup).* American Petroleum Institute, Washington, DC, pp. 637–40.

Fucik, K. W. and J. M. Neff (1977) Effects of temperature and salinity on naphthalenes uptake in the temperate clam *Rangia cuneata* and the boreal clam *Protothaca staminea*. In: D. A. Wolfe (ed.) *Fate and Effects of Petroleum Hydrocarbons in Marine Organisms and Ecosystems.* Pergamon Press, New York, pp. 305–12.

Gardner, G. R. (1972) Chemically induced lesions in estuarine or marine teleosts. Symp. Fish. Pathol., Armed Forces Institute Path., Washington, DC, NMWQL Contribution No. 47.

Glaser, O. (1929) Temperature and heart rate in *Fundulus* embryos. *Brit. J. Exp. Biol.* **6**: 325–39.

Goldberg, E. D. (1975) The mussel watch—a first step in global marine monitoring. *Mar. Pollut. Bull.* **6**: 111.

Gordon, D. C., Jr., P. D. Keizer and J. Dale (1974) Estimates using fluorescence spectroscopy of the present state of petroleum hydrocarbon contamination in the water column of the Northwest Atlantic Ocean. *Mar. Chem.* **2**: 251–61.

Goreau, T. F. (1959) The physiology of skeleton formation in corals. I. A method for measuring the rate of calcium deposition by corals under different conditions. *Biol. Bull.* **116**: 59–75.

Goreau, T. F. and N. I. Goreau (1959) The physiology of skeleton formation in corals. IV. On isotopic equilibrium exchanges of calcium between corallum and environment in living and dead reef-building corals. *Biol. Bull.* **119**: 416–27.

Gornall, A. G., C. J. Bardawill and M. M. David (1949) Determination of serum proteins by means of the biuret reaction. *J. Biol. Chem.* **177**: 751–66.

Grossling, B. F. (1976) An estimate of the amounts of oil entering the oceans. In: *Sources, Effects and Sinks of Hydrocarbons in the Aquatic Environment.* American Institute of Biological Sciences, Washington, DC, pp. 5–36.

Gruger, E. H., Jr., M. M. Wekell, P. T. Numoto and D. R. Craddock (1977) Induction of hepatic aryl hydrocarbon hydroxylase in salmon exposed to petroleum dissolved in seawater and to petroleum and polychlorinated biphenyls, separate and together, in food. *Bull. Environ. Contam. Toxicol.* **17**: 512–20.

Hamelink, J. L. and A. Spacie (1977) Fish and chemicals: the process of accumulation. *Ann. Rev. Pharmacol. Toxicol.* **17**: 167–77.

Hansen, N., V. B. Jensen, H. Appelquist and E. Morch (1978) The uptake and release of petroleum hydrocarbons by the marine mussel *Mytilus edulis*. *Prog. Wat. Technol.* **10**: 351–9.

Hardy, R., P. R. Mackie, K. J. Whittle and A. D. McIntyre (1974) Discrimination in the assimilation of n-alkanes in fish. *Nature, London* **252**: 577–8.

Hargrave, B. T. and C. P. Newcombe (1973) Crawling and respiration as indices of sublethal effects of oil and a dispersant on an intertidal snail *Littorina littorea*. *J. Fish. Res. Bd Canada* **30**: 1789–92.

Harris, R. P., V. Berdugo, E. D. S. Corner, C. C. Kilvington and S. C. M. O'Hara (1977) Factors affecting the retention of a petroleum hydrocarbon by marine planktonic copepods. In: D. A. Wolfe (ed.) *Fate and Effects of Petroleum Hydrocarbons in Marine Organisms and Ecosystems*. Pergamon Press, New York, pp. 286–304.

Hase, A. and R. A. Hites (1976) On the origin of polycyclic aromatic hydrocarbons in recent sediments: biosynthesis by anaerobic bacteria. *Geochim. Cosmochim. Acta.* **40**: 1141–3.

Heatfield, B. M. (1970) Calcification in echinoderms: effects of temperature and Diamox on incorporation of calcium-45 *in vitro* by regenerating spines of *Strongylocentrotus purpuratus*. *Biol. Bull.* **139**: 151–63.

Heatfield, B. M. (1971a) Origin of calcified tissues in regenerating spines of the sea urchin *Strongylocentrotus purpuratus* (Stimpson): a quantitative radioautographic study with tritiated thymidine. *J. Exp. Zool.* **178**: 233–46.

Heatfield, B. M. (1971b) Growth of the calcareous skeleton during regeneration of spines of the sea urchin *Strongylocentrotus purpuratus* (Stimpson): a light and electron microscopic study. *J. Morphol.* **134**: 57–89.

Heatfield, B. M. and D. F. Travis (1975a) Ultrastructural studies of regenerating spines of the sea urchin *Strongylocentrotus purpuratus*. I. Cell types without spherules. *J. Morphol.* **145**: 13–49.

Heatfield, B. M. and D. F. Travis (1975b) Ultrastructural studies of regenerating spines of the sea urchin *Strongylocentrotus purpuratus*. II. Cell types with spherules. *J. Morphol.* **145**: 51–71.

Herbes, S. E. (1977) Partitioning of polycyclic aromatic hydrocarbons between dissolved and particulate phases in natural waters. *Water Res.* **11**: 493–6.

Hites, R. A. (1976) Sources of polycyclic aromatic hydrocarbons in the aquatic environment. In: *Sources, Effects and Sinks of Hydrocarbons in the Aquatic Environment*. American Institute of Biological Sciences, Washington, DC, pp. 325–32.

Hoffman, E. J. and J. G. Quinn (1978) A comparison of Argo Merchant oil and sediment hydrocarbons from Nantucket Shoals. In: *In the Wake of the Argo Merchant*. Center for Ocean Management Studies, University of Rhode Island, Kingston, Rhode Island, pp. 80–8.

Hoffman, E. J. and J. G. Quinn (1979) Gas chromatographic analyses of Argo Merchant oil and sediment hydrocarbons at the wreck site. *Mar. Pollut. Bull.* **10**: 20–4.

Holden, A. V. (1962) A study of the absorption of ^{14}C-labelled DDT from water by fish. *Ann. Appl. Biol.* **50**: 467–77.

Hyland, J. L. and E. D. Schneider (1976) Petroleum hydrocarbons and their effects on marine organisms, populations, communities and ecosystems. In: *Sources, Effects and Sinks of Hydrocarbons in the Aquatic Environment*. American Institute of Biological Sciences, Washington, DC, pp. 463–506.

Idoniboye-Obu, B. (1977) Recording bioelectric action potentials of marine decapod crustacea by remote electrodes: a bioassay procedure for monitoring hydrocarbon pollution. *Environ. Pollut.* **12**: 159–66.

International Agency for Research on Cancer (1973) Evaluation of the carcinogenic risk of chemicals to man. III. Certain polycyclic aromatics and heterocyclic compounds. I.A.R.C. Monogr., World Health Organization, Lyon, France.

Jacobson, S. M. and D. B. Boylan (1973) Effect of seawater soluble fraction of kerosene on chemotaxis in a marine snail, *Nassarius obsoletus*. *Nature, London*. **241**: 213–15.

Kallio, R. E. (1976) The variety of petroleums and their degradations. In: *Sources, Effects and Sinks of Hydrocarbons in the Aquatic Environment*. American Institute of Biological Sciences. Washington, DC, pp. 214–23.

Katz, L. M. (1973) The effects of water soluble fractions of crude oil on larvae of the decapod crustacean *Neopanope texana* (Sayi). *Environ. Pollut.* **5**: 199–204.

Keck, R. T., R. C. Heess, J. Wehmiller and D. Maurer (1978) Sublethal effects of the water-soluble fraction of Nigerian crude oil on the juvenile hard clams, *Mercenaria mercenaria*. *Environ. Pollut.* **15**: 109–19.

Kittredge, J. S. (1971) Effects of the water-soluble component of oil pollution on chemoreception by crabs. US Dept. of Commerce; NTIS. AD-738-505. 5 pp.

Kittredge, J. S., F. T. Takahashi, J. Lindsey and R. Lasker (1973) Chemical signals in the sea: marine allelochemics and evolution. *Fish. Bull.* **72**: 1–11.

Koons, C. B. and D. E. Brandon (1975) Hydrocarbons in water and sediment samples from Coal Oil Point area, offshore California. *Proc. 1975 Offshore Tech. Conf.* III. pp. 513–21.

Koons, C. B. and P. H. Monaghan (1976) Input of hydrocarbons from seeps and recent biogenic sources. In: *Sources, Effects and Sinks of Hydrocarbons in the Aquatic Environment*. American Institute of Biological Sciences, Washington, DC, pp. 84–107.

Korn, S., N. Hirsch and J. W. Struhsaker (1976) Uptake, distribution, and depuration of ^{14}C-benzene in northern anchovy, *Engraulis mordax*, and striped bass, *Marone saxatilis*. *Fish. Bull.* **74**: 545–51.

Kuhnhold, W. W.(1972) The influence of crude oils on fish fry. In: M. Ruivo (ed.) *Marine Pollution and Sea Life*. Fishery News (Books) Ltd, London, pp. 315–18.

Kuhnhold, W. W. (1974) Investigations on the toxicity of seawater-extracts of three crude oils on eggs of cod (*Gadus morhua* L.). *Berr. dt. wiss. Komm. Meeresforsch.* **23**: 165–80.

Kurelec, B., S. Britvic, M. Rijavec, W. E. G. Muller and R. K. Zahn (1977)

Benzo[a]pyrene monooxygenase induction in marine fish—molecular response to oil pollution. *Mar. Biol.* **44**: 211–16.

LaRoche, G., R. Eisler and C. M. Tarzwell (1970) Bioassay procedures for oil and oil dispersant toxicity evaluation. *J. Wat. Pollut. Contr. Fed.* **42**: 1982–9.

Laughlin, R. B., Jr., O. Linden and J. M. Neff (1979) A study on the effects of salinity and temperature on the disappearance of aromatic hydrocarbons from the water-soluble fraction of No. 2 fuel oil. *Chemosphere*, **10**: 741–9.

Laughlin, R. B., Jr. and J. M. Neff (1979) Interactive effects of salinity, temperature and polycyclic aromatic hydrocarbons on the survival and development rate of larvae of the mud crab *Rhithropanopeus harrisii*. *Mar. Biol.* **53**: 281–91.

Laughlin, R. B., Jr., L. G. L. Young and J. M. Neff (1978) A long-term study of the effects of water-soluble fractions of No. 2 fuel oil on the survival, developmental rate and growth of the mud crab *Rhithropanopeus harrisii* (Gould). *Mar. Biol.* **47**: 87–95.

Lee, C. C., W. K. Craig and P. J. Smith (1974) Water-soluble hydrocarbons from crude oil. *Bull. Environ. Contam. Toxicol.* **12**: 212–17.

Lee, R. F. (1976) Metabolism of petroleum hydrocarbons in marine sediments. In: *Sources, Effects and Sinks of Hydrocarbons in the Aquatic Environment.* American Institute of Biological Sciences, Washington, DC, pp. 334–44.

Lee, R. F., E. Furlong and S. Singer (1977) Metabolism of hydrocarbons in marine invertebrates. Aryl hydrocarbon hydroxylase from the tissues of the blue crab, *Callinectes sapidus*, and the polychaete worm, *Nereis* sp. In: C. S. Giam (ed.) *Pollutant Effects on Marine Organisms.* D. C. Heath Co., Lexington, Mass., pp. 111–24.

Lee, R. F., W. S. Gardner, J. W. Anderson, J. W. Blylock and J. Barwell-Clarke (1978) Fate of polycyclic aromatic hydrocarbons in controlled ecosystem enclosures. *Environ. Sci. Technol.* **12**: 832–8.

Lee, R. F., C. Ryan and M. L. Neuhauser (1976) Fate of petroleum hydrocarbons taken up from food and water by the blue crab. *Callinectes sapidus. Mar. Biol.* **37**: 363–70.

Lee, R. F., R. Sauerheber and A. A. Benson (1972a) Petroleum hydrocarbons: uptake and discharge by a marine mussel *Mytilus edulis. Science* **177**: 344–6.

Lee, R. F., R. Sauerheber and J. H. Dobbs (1972b) Uptake, metabolism and discharge of polycyclic aromatic hydrocarbons by marine fish. *Mar. Biol.* **17**: 201–8.

Lee, R. F., S. C. Singer, K. R. Tenore, W. S. Gardner and R. M. Philpot (1979) Detoxification system in polychaete worms: Importance in the degradation of sediment hydrocarbons. In: W. B. Vernberg, A. Calabrese, F. P. Thurberg and F. J. Vernberg (eds.) *Marine Pollution: Functional Responses.* Academic Press, New York, pp. 23–37.

Lee, W. Y. and J. A. C. Nicol (1977) The effects of the water soluble fractions of No. 2 fuel oil on the survival and behavior of coastal and oceanic zooplankton. *Environ. Pollut.* **12**: 279–92.

Linden, O. (1974) Effects of oil spill dispersants on the early development of Baltic herring. *Ann. Zool. Fennici.* **11**: 141–8.

Linden, O. (1975) Acute effects of oil and oil/dispersant mixtures on larvae of Baltic herring. *Ambio.* **4**: 130–3.

Linden, O. (1976a) Effects of oil on the reproduction of the amphipod *Gammarus oceanicus*. *Ambio.* **5**: 36–7.
Linden, O. (1976b) Effects of oil on the amphipod *Gammarus oceanicus*. *Environ. Pollut.* **10**: 239–45.
Linden, O. (1978) Biological effects of oil on early development of the Baltic herring *Clupea harengus membras*. *Mar. Biol.* **45**: 273–83.
Linden, O., R. B. Laughlin, Jr., J. R. Sharp and J. M. Neff (1980) The combined effect of salinity, temperature and oil on the growth pattern of embryos of the killifish, *Fundulus heteroclitus* Walbaum. *Mar. Environ. Res.* **3**: 129–44.
Linden, O., J. R. Sharp, R. Laughlin, Jr. and J. M. Neff (1979) Interactive effects of salinity, temperature and chronic exposure to oil on the survival and development rate of embryos of the estuarine killifish *Fundulus heteroclitus*. *Mar. Biol.* **51**: 101–9.
Litchfield, J. T. (1949) A method for rapid graphic solution of time per cent effect curves. *J. Pharmacol. Exp. Therap.* **96**: 399–408.
Litchfield, J. T. and F. Wilcoxon (1949) A simplified method for evaluating dose effects experiments. *J. Pharmacol. Exp. Therap.* **96**: 99–113.
Lonning, S. and B. E. Hagstrom (1975) The effects of crude oils and the dispersant Corexit 8666 on sea urchin gametes and embryos. *Nor. J. Zool.* **23**: 121–9.
Lowry, O. H., N. J. Rosenbrough, A. L. Farr and R. J. Randall (1951) Protein measurement with the Folin phenol reagent. *J. Biol. Chem.* **193**: 265–75.
Loya, Y. and B. Rinkevich (1979) Abortion effect in corals induced by oil pollution. *Mar. Ecol. Prog. Ser.* **1**: 77–80.
Lucas, A. and S. LeRoux (1975) Mise en evidence de la toxicite de devers petroles bruts vis-a-vis des larves de moule. *C.R. Acad. Sci., Paris.* **280**: 2381–4.
Lyes, M. C. (1979) Bioavailability of a hydrocarbon from water and sediment to the marine worm *Arenicola marina*. *Mar. Biol.* **55**: 121–7.
Lysyj, I. and E. C. Russell (1974) Dissolution of petroleum-derived products in water. *Water Res.* **8**: 863–8.
Mackay, D. and W. Y. Shiu (1976) Aqueous solubilities of weathered northern crude oils. *Bull. Environ. Contam. Toxicol.* **15**: 101–9.
Mangum, C. P. (1970) Respiratory physiology in annelids. *Amer. Scient.* **58**: 641–7.
Marty, J. C. and A. Saliot (1976) Hydrocarbons (normal alkanes) in the surface microlayer of seawater. *Deep-Sea Res.* **23**: 863–73.
May, W. E., S. P. Wasik and D. H. Freeman (1978) Determination of the solubility behavior of some polycyclic aromatic hydrocarbons in water. *Anal. Chem.* **50**: 997–1000.
McAuliffe, C. D. (1966) Solubility in water of paraffin, cycloparaffin, olefin, acetylene, cycloolefin and aromatic hydrocarbons. *J. Phys. Chem., Wash.* **70**: 1267–75.
McAuliffe, C. D. (1976) Surveillance of the marine environment for hydrocarbons. *Mar. Sci. Commun.* **2**: 13–42.
McAuliffe, C. D., A. E. Smalley, R. D. Groover, W. M. Welsh, W. S. Pickle and G. E. Jones (1975) Chevron main pass block 41 oil spill: chemical and biological investigations. In: *Proceedings of the 1975 Conference on Prevention and Control of Oil Pollution*. American Petroleum Institute, Washington, DC, pp. 555–66.
McCain, B. B., H. O. Hodgins, W. D. Gronlund, J. W. Hawkes, D. W. Brown,

M. S. Myers and J. H. Vandermeulen (1978) Bioavailability of crude oil from experimentally oiled sediments to English sole (*Parophrys vetulus*) and pathological consequences. *J. Fish. Res. Bd Canada* **35**: 657–64.

Mileikovsky, S. A. (1970) The influence of pollution on pelagic larvae of bottom invertebrates in marine nearshore and estuarine waters. *Mar. Biol.* **6**: 350–6.

Mix, M. C., R. T. Riley, K. I. King, S. R. Trenholm and R. L. Schaffer (1977) Chemical carcinogens in the marine environment. Benzo[*a*]pyrene in economically-important bivalve molluscs from Oregon estuaries. In: D. A. Wolfe (ed.) *Fate and Effects of Petroleum Hydrocarbons in Marine Organisms and Ecosystems*. Pergamon Press, New York, pp. 421–31.

Mix, M. C. and R. L. Schaffer (1979) Benzo[*a*]pyrene concentrations in mussels (*Mytilus edulis*) from Yaquina Bay, Oregon during June 1976–May 1978. *Bull. Environ. Contam. Toxicol.* **23**: 677–84.

Moore, S. F. and R. L. Dwyer (1974) Effects of oil on marine organisms: a critical assessment of published data. *Water Res.* **8**: 819–27.

National Academy of Sciences (1975) Petroleum in the marine environment. Ocean Affairs Board, N.A.S., Washington, DC. 107 pp.

Neely, W. B., D. R. Branson and G. E. Blau (1974) Partition coefficient to measure bioconcentration potential of organic chemicals in fish. *Environ. Sci. Technol.* **8**: 1113–15.

Neff, J. M. (1979) *Polycyclic Aromatic Hydrocarbons in the Aquatic Environment. Sources, Fates and Biological Effects*. Applied Science Publishers Ltd, London. 262 pp.

Neff, J. M. and J. W. Anderson (1975) An ultraviolet spectrophotometric method for the determination of naphthalene and alkyl naphthalenes in the tissues of oil-contaminated marine animals. *Bull. Environ. Contam. Toxicol.* **14**: 122–8.

Neff, J. M., J. W. Anderson, B. A. Cox, R. B. Laughlin, Jr., S. S. Rossi and H. E. Tatem (1976*a*) Effects of petroleum on survival, respiration and growth of marine animals. In: *Sources, Effects, and Sinks of Hydrocarbons in the Aquatic Environment*. American Institute of Biological Sciences, Washington, DC, pp. 515–40.

Neff, J. M., B. A. Cox, D. Dixit and J. W. Anderson (1976*b*) Accumulation and release of petroleum-derived aromatic hydrocarbons by four species of marine animals. *Mar. Biol.* **38**: 279–89.

Neff, J. M. and C. S. Giam (1977) Effects of Arochlor 1016 and Halowax 1099 on juvenile horseshoe crabs *Limulus polyphemus*. In: F. J. Vernberg, A. Calabrese, R. P. Thurberg and W. B. Vernberg (eds.) *Physiological Responses of Marine Biota to Pollutants*. Academic Press, New York, pp. 21–35.

Nicol, J. A. C., W. H. Donahue, R. T. Wang and K. Winters (1977) Chemical composition and effects of water extracts of petroleum on eggs of the sand dollar *Melitta quinquiesperforata*. *Mar. Biol.* **40**: 309–16.

Ott, F. S., R. P. Harris and S. C. M. O'Hara (1978) Acute and sublethal toxicity of naphthalene and three methylated derivatives to the estuarine copepod, *Eurytemora affinis*. *Mar. Environ. Res.* **1**: 49–58.

Pancirov, R. J. (1974) Compositional data on API reference oils used in biological studies: a No. 2 Fuel Oil, a Bunker C, Kuwait crude oil and south Louisiana crude oil. Rept. No. AID. 1BA. 74, American Petroleum Institute, Washington, DC. 16 pp.

Pancirov, R. J. and R. A. Brown (1975) Analytical methods for polynuclear aromatic hydrocarbons in crude oils, heating oils, and marine tissues. In: *Proceedings 1975 Conference on Prevention and Control of Oil Pollution*. American Petroleum Institute, Washington, DC, pp. 103–13.

Pancirov, R. J. and R. A. Brown (1977) Polynuclear aromatic hydrocarbons in marine tissues. *Environ. Sci. Technol.* **11**: 989–92.

Payne, J. F. (1976) Field evaluations of benzopyrene hydroxylase induction as a monitor for marine petroleum pollution. *Science* **191**: 945–6.

Payne, J. F. (1977) Mixed function oxidases in marine organisms in relation to petroleum hydrocarbon metabolism and detection. *Mar. Pollut. Bull.* **8**: 112–14.

Payne, J. F. and N. May (1979) Further studies on the effect of petroleum hydrocarbons on mixed-function oxidases in marine organisms. In: M. A. Q. Khan, J. J. Lech and J. J. Menn (eds.) *Pesticides and Xenobiotic Metabolism in Aquatic Organisms*. A.C.S. Symposium Series. No. 99. American Chemical Society, Washington, DC, pp. 339–47.

Payne, J. F. and W. R. Penrose (1975) Induction of aryl hydrocarbon (benzo[a]pyrene) hydroxylase in fish by petroleum. *Bull. Environ. Contam. Toxicol.* **14**: 112–16.

Peakall, D. B. (1975) PCBs and their environmental effects. *C.R.C. Crit. Revs. Environ. Contr.* **5**: 469–508.

Pearse, V. B. and L. Muscatine (1971) Role of symbiotic algae (zooxanthellae) in coral calcification. *Biol. Bull.* **141**: 350–63.

Pearson, W. H. and B. L. Olla (1979) Detection of naphthalene by the blue crab, *Callinectes sapidus*. *Estuaries* **2**: 64–5.

Pedersen, M. G., W. K. Hershberger, P. K. Zachariah and M. R. Juchau (1976) Hepatic biotransformation of environmental xenobiotics in six strains of rainbow trout (*Salmo gairdneri*). *J. Fish. Res. Bd Canada*. **33**: 666–75.

Percy, J. A. and T. C. Mullin (1977) Effects of crude oil on the locomotory activity of arctic marine invertebrates. *Mar. Pollut. Bull.* **8**: 35–40.

Perkins, E. J. (1979) The need for sublethal studies. In: H. A. Cole (ed.) *The Assessment of Sublethal Effects of Pollutants in the Sea*. The Royal Society, London pp. 27–41.

Pocock, D. M. E., J. R. Marsden and J. G. Hamilton (1971) Lipids in an intertidal polychaete and their relation to maturation of the worm. *Comp. Biochem. Physiol.* **39A**: 683–97.

Pohl, R. J., J. R. Bend, A. M. Guarino and J. R. Fouts (1974) Hepatic microsomal mixed-function oxidase activity of several marine species from coastal Maine. *Drug. Metabol. Dispos.* **2**: 545–55.

Prouse, N. J. and D. C. Gordon, Jr. (1976) Interactions between the deposit-feeding polychaete *Arenicola marina* and oiled sediment. In: *Sources, Effects and Sinks of Hydrocarbons in the Aquatic Environment*. American Institute of Biological Sciences, Washington, DC pp. 407–22.

Reish, D. J. and T. L. Richards (1966) A culture method for maintaining large populations of polychaetous annelids in the laboratory. *Turtox News* **44**: 16–17.

Rice, S. D., D. A. Moles and J. W. Short (1975) The effect of Prudhoe Bay crude oil on survival and growth of eggs, alevins and fry of pink salmon *Oncorhynchus gorbuscha.* In: *Proceedings, 1975 Conference on Prevention and Control of Oil Pollution.* American Petroleum Institute, Washington, DC. pp. 503–7.

Rice, S. D., J. W. Short and J. F. Karinen (1976) Toxicity of Cook Inlet crude oil and No. 2 fuel oil to several Alaskan marine fishes and invertebrates. In: *Sources, Effects and Sinks of Hydrocarbons in the Aquatic Environment.* American Institute of Biological Sciences, Washington, DC, pp. 394–406.

Rice, S. D., R. E. Thomas and J. W. Short (1977) Effects of petroleum hydrocarbons on breathing and coughing rates, and hydrocarbon uptake-depuration in the pink salmon fry. In: F. J. Vernberg, A. Calabrese, F. P. Thurberg and W. B. Vernberg (eds.) *Physiological Responses of Marine Biota to Pollutants.* Academic Press, New York, pp. 259–77.

Roesijadi, G., J. W. Anderson and J. W. Blaylock (1978a) Uptake of hydrocarbons from marine sediments contaminated with Prudhoe Bay crude oil: influence of feeding type of test species and availability of polycyclic aromatic hydrocarbons. *J. Fish. Res. Bd Canada* **35**: 608–14.

Roesijadi, G., D. L. Woodruff and J. W. Anderson (1978b) Bioavailability of naphthalenes from marine sediments artificially contaminated with Prudhoe Bay crude oil. *Environ. Pollut.* **15**: 223–9.

Rossi, S. S. (1977) Bioavailability of petroleum hydrocarbons from water, sediments and detritus to the marine annelid, *Neanthes arenaceodentata.* In: *Proceedings 1977 Oil Spill Conference (Prevention, Behavior, Control, Cleanup).* American Petroleum Institute, Washington, DC, pp. 621–6.

Rossi, S. S. and J. W. Anderson (1976) Toxicity of water-soluble fractions of No. 2 fuel oil and south Louisiana crude oil to selected stages in the life history of the polychaete, *Neanthes arenaceodentata. Bull. Environ. Contam. Toxicol.* **16**: 18–24.

Rossi, S. S. and J. W. Anderson (1977) Accumulation and release of fuel-oil-derived diaromatic hydrocarbons by the polychaete *Neanthes arenaceodentata. Mar. Biol.* **39**: 51–5.

Rossi, S. S. and J. W. Anderson (1978a) Effects of No. 2 fuel oil water-soluble-fractions on growth and reproduction in *Neanthes arenaceodentata* (Polychaeta: Annelida). *Water, Air, Soil Pollut.* **9**: 155–70.

Rossi, S. S. and J. W. Anderson (1978b) Petroleum hydrocarbon resistance in the marine worm *Neanthes arenaceodentata* (Polychaeta: Annelida) induced by chronic exposure to No. 2 fuel oil. *Bull. Environ. Contam. Toxicol.* **20**: 513–21.

Rossi, S. S., J. W. Anderson and G. S. Ward (1976) Toxicity of water-soluble fractions of four test oils for the polychaetous annelids, *Neanthes arenaceodentata* and *Capitella capitata. Environ. Pollut.* **10**: 9–18.

Rossi, S. S. and J. M. Neff (1978) Toxicity of polynuclear aromatic hydrocarbons to the marine polychaete *Neanthes arenaceodentata. Mar. Pollut. Bull.* **9**: 220–3.

Roubal, W. T., T. K. Collier and D. C. Malins (1977) Accumulation and metabolism of carbon-14 labeled benzene, naphthalene, and anthracene by young coho salmon (*Oncorhynchus kisutch*). *Arch. Environ. Contam. Toxicol.* **5**: 513–29.

Sauer, T. C., Jr., W. M. Sackett and L. M. Jeffrey (1978) Volatile liquid hydrocarbons in the surface coastal waters of the Gulf of Mexico. *Mar. Chem.* **7**: 1–16.

Schwarz, F. P. and S. P. Wasik (1976) Fluorescence measurements of benzene, naphthalene, anthracene, pyrene, fluoranthene and benzo[a]pyrene in water. *Anal. Chem.* **48**: 524–8.

Sharp, J. R., K. W. Fucik and J. M. Neff (1979) Physiological basis of differential sensitivity of fish embryonic stages to oil pollution. In: F. J. Vernberg, W. B. Vernberg and A. Calabrese (eds.) *Marine Pollution: Functional Responses.* Academic Press, New York, pp. 85–108.

Singer, S. C. and R. F. Lee (1977) Mixed function oxygenase activity in blue crab, *Callinectes sapidus:* tissue distribution and correlation with changes during molting and development. *Biol. Bull.* **153**: 377–86.

Smith, R. L. and J. A. Cameron (1979) Effect of water soluble fraction of Prudhoe Bay crude oil on embryonic development of Pacific herring. *Trans. Amer. Fish. Soc.* **108**: 70–5.

Smith, S. (1957) Early development and hatching. In: M. E. Brown (ed.) *The Physiology of Fishes.* Vol. I. Academic Press, New York, pp. 323–59.

Somero, G. N. and P. W. Hochachka (1976) Biochemical adaptations to temperature. In: R. C. Newell (ed.) *Adaptation to Environment. Essays on the Physiology of Marine Animals.* Butterworths, London, pp. 125–90.

Speers, G. C. and E. V. Whitehead (1969) Crude petroleum. Chapter 27. In: G. Eglinton and M. R. J. Murphy (eds.) *Organic Geochemistry: Methods and Results.* Springer-Verlag, Berlin, pp. 638–75.

Sprague, J. B. (1969) Measurement of pollutant toxicity to fish—I. Bioassay methods for acute toxicity. *Water Res.* **3**: 793–821.

Sprague, J. B. and D. W. McLeese (1968) Toxicity of kraft pulp mill effluent for larval and adult lobsters, and juvenile salmon. *Water Res.* **2**: 753–60.

Stegeman, J. J. (1978) Influence of environmental contamination on cytochrome P-450 mixed function oxygenases in fish: implications for recovery in the Wild Harbor Marsh. *J. Fish. Res. Bd Canada* **35**: 668–74.

Stegeman, J. J. and J. M. Teal (1973) Accumulation, release and retention of petroleum hydrocarbons by the oyster, *Crassostrea virginica. Mar. Biol.* **22**: 37–44.

Stone, W. J. (1975) Hydrophobic interaction of alkanes with liposomes and lipoproteins. *J. Biol. Chem.* **250**: 5368–70.

Stromgren, T. (1976) Skeleton growth of the hydrocoral *Millipora complanata* Lamarck in relation to light. *Limnol. Oceanog.* **21**: 156–60.

Struhsaker, J. W. (1977) Effects of benzene (a toxic component of petroleum) on spawning Pacific herring, *Clupea harengus* Pallasi. *Fish. Bull.* **75**: 43–9.

Suess, M. J. (1976) The environmental load and cycle of polycyclic aromatic hydrocarbons. *Sci. Total Environ.* **6**: 239–50.

Sutterlin, A. M. (1974) Pollution and the chemical senses of aquatic animals—perspective and review. *Chem. Senses and Flavor* **1**: 167–78.

Sweeney, B. M. (1976) Circadian rhythms in corals, particularly Fungiidae. *Biol. Bull.* **151**: 236–46.

Takahashi, F. T. and J. S. Kittredge (1973) Sublethal effects of the water soluble component of oil: Chemical communication in the marine environment. Louisiana State University, Sea Grant, LSU-56-73-01.

Tatem, H. E. (1977) Accumulation of naphthalenes by grass shrimp: effects on respiration, hatching, and larval growth. In: D. A. Wolfe (ed.) *Fate and Effects of Petroleum Hydrocarbons in Marine Organisms and Ecosystems*. Pergamon Press, New York, pp. 201–9.

Teal, J. M., K. Burns and J. Farrington (1978) Analyses of aromatic hydrocarbons in intertidal sediments resulting from two spills of No. 2 fuel oil in Buzzards Bay, Massachusetts. *J. Fish. Res. Bd Canada* **35**: 510–20.

Thomas, R. E. and S. D. Rice (1975) Increased opercular rates of pink salmon (*Oncorhynchus gorbuscha*) fry after exposure to water-soluble fraction of Prudhoe Bay crude oil. *J. Fish. Res. Bd Canada* **32**: 2221–4.

Thompson, H. C., Jr., R. N. Farragut and M. H. Thompson (1977) Relationship of scarlet prawns (*Pleiopenaeus edwardsianus*) to a benthic oil deposit off the north-west coast of Aruba, Dutch West Indies. *Environ. Pollut.* **13**: 239–53.

Thorson, G. (1966) Some factors influencing the recruitment and establishment of marine benthic communities. *Neth. J. Sea Res.* **3**: 267–93.

Trinkaus, J. P. (1967) *Fundulus*. In: F. H. Wilt and N. K. Wessels (eds.) *Methods in Developmental Biology*. T. Y. Gowell Co., New York, pp. 113–22.

Ustach, J. R. (1979) Effects of sublethal oil concentration on the copepod, *Nitocra affinis*. *Estuaries* **2**: 273–6.

Vandermeulen, J. H. and T. P. Ahern (1976) Effect of petroleum hydrocarbons on algal physiology: review and progress report. In: A. P. M. Lockwood (ed.) *Effects of Pollutants on Aquatic Organisms*. Cambridge University Press, Cambridge, pp. 107–25.

Vandermeulen, J. H., N. D. Davis and L. Muscatine (1972) The effects of inhibitors of photosynthesis on zooxanthellae in corals and other marine invertebrates. *Mar. Biol.* **16**: 185–91.

Vandermeulen, J. H. and W. R. Penrose (1978) Absence of aryl hydrocarbon hydroxylase (AHH) in three marine bivalves. *J. Fish. Res. Bd Canada* **35**: 643–7.

Varanasi, U. and D. C. Malins (1977) Metabolism of petroleum hydrocarbons: accumulation and biotransformation in marine organisms. In: D. C. Malins (ed.) *Effects of Petroleum on Arctic and Subarctic Marine Environments and Organisms. Vol. II. Biological Effects*. Academic Press, New York, pp. 175–270.

Vernberg, W. B., H. McKellar, Jr. and F. J. Vernberg (1978) Toxicity studies and environmental impact assessment. *Environ. Manag.* **2**: 23–243.

Walters, J. M., R. B. Cain, I. J. Higgins and E. D. S. Corner (1979) Cell-free benzo[a]pyrene hydroxylase activity in marine zooplankton. *J. Mar. Biol. Assn. U.K.* **59**: 553–63.

Walton, D. G., W. R. Penrose and J. M. Green (1978) The petroleum-inducible mixed-function oxidase of cunner (*Tautogolabrus adspersus* Walbaum 1792): some characteristics relevant to hydrocarbon monitoring. *J. Fish. Res. Bd Canada* **35**: 1547–52.

Warner, J. S. (1975) Determination of sulfur-containing petroleum components in marine samples. In: *Proceedings, 1975 Conference on Prevention and Control of Oil Pollution*. American Petroleum Institute, Washington, DC. pp. 97–101.

Warner, J. S. (1976) Determination of aliphatic and aromatic hydrocarbons in marine organisms. *Anal. Chem.* **48**: 578–83.

Wells, P. G. and J. B. Sprague (1976) Effects of crude oil on lobster larvae in the laboratory. *J. Fish. Res. Bd Canada* **33**: 1604–14.

Whittle, K. J., J. Murray, P. R. Mackie, R. Hardy and J. Farmer (1977) Fate of hydrocarbons in fish. In: A. D. McIntyre and K. J. Whittle (eds.) *Petroleum Hydrocarbons in the Marine Environment.* Cons. Intern. Explor. Mer. Vol. 171. Charlottenlund Slot, Denmark, pp. 139–42.

Wilson, K. W. (1975) The laboratory estimation of the biological effects of organic pollutants. *Proc. Roy. Soc. London, Ser. B* **189**: 459–77.

Wilson, K. W. (1976) Effects of oil dispersants on the developing embryos of marine fish. *Mar. Biol.* **36**: 259–68.

Wilson, R. D., P. H. Monaghan, A. Osanik, L. C. Price and M. A. Rogers (1974) Natural marine oil seepage. *Science* **184**: 857–65.

Young, G. P. (1977) Effects of naphthalene and phenanthrene on the grass shrimp *Palaemonetes pugio* (Holthuis). Master's Thesis. The Graduate College, Texas A&M University, College Station, Texas. 67 pp.

Young, R. H. and A. J. Sethi (1975) Compositional changes of a fuel oil from an oil spill due to natural exposure. *Water, Air, Soil Pollut.* **5**: 195–205.

Young, S. D., J. D. O'Connor and L. Muscatine (1971) Organic material from scleractinian coral skeletons. II. Incorporation of ^{14}C into protein, chitin, and lipid. *Comp. Biochem. Physiol.* **40B**: 945–58.

Youngblood, W. W. and M. Blumer (1975) Polycyclic aromatic hydrocarbons in the environment: homologous series in soils and recent marine sediments. *Geochim. Cosmochim. Acta.* **39**: 1303–14.

ZoBell, C. E. (1971) Sources and biodegradation of carcinogenic hydrocarbons. In: *Proceedings Joint Conference on Prevention and Control of Oil Spills.* American Petroleum Institute, Washington, DC. pp. 441–51.

Glossary of Selected Technical Terms

API gravity: a measure of the relative weight of a crude oil, estimated by the formula, $(141 \cdot 5/\text{s.g.}) - 131 \cdot 5$, where s.g. equals the specific gravity of the oil in grams per milliliter at a particular temperature (usually 60°F). Crude oils are classified as follows: °API > 30, light crude; °API = 18–30, medium crude; °API < 18, heavy crude.

Blastula: an early stage of embryonic development in animals in which the cells form a hollow ball.

Brood tube: a tube constructed of mucus or other material in which the adults of some polychaete worms protect their embryos and early larvae.

Coughing response: forceful momentary reversal of the direction of water flow over the gills of teleost fish; used to clean clogged gill lamellae.

Chorion: the external sac surrounding the embryo in an egg.

Coelom: a body cavity of triploblastic animals lying within the mesoderm and completely lined by it.

EC_{50}: median effective concentration; the concentration of a drug, pollutant, or other substance causing a particular sublethal response in 50 per cent of test organisms.

Embryo: the early stages of development of an organism.

ET_{50}: median effective time; the time required for a given stimulus to elicit a particular response in 50 per cent of test organisms.

Fecundity: The rate of production of offspring by an organism, the number of offspring per reproductive cycle, or the number offspring produced during the life of an individual.

Gastrulation: The process in animal development, in which the blastula is transformed into a gastrula.

Hypoxia: condition of low oxygen tension in the respiratory environment.

Integument: skin or other outer covering of the body of an animal.

Larva: an early developmental stage of some organisms that follows the embryo and does not resemble the adult form of the organism.

LC_{50}: median lethal concentration; the concentration of a toxin that causes 50 per cent mortality among the test organisms.

Lipophylic: having a high affinity for or solubility in lipids; characteristic of many non-polar organic compounds.

Melanin: A dark brown or black pigment common in the integument of many animals and sometimes found in other organs; usually occurs within special pigment cells.

Mesenchyme: a meshwork of loosely associated often star-shaped cells; found in the embryos of vertebrates and adults of some invertebrates.

Metamorphosis: an abrupt transition from one developmental stage to another, e.g., from a larva to an adult.

Metatrochophore: a larval stage of some polychaete worms and certain other invertebrates.

Metric ton: a unit of weight equal to 1000 kg or 2204·62 avoirdupois pounds.

Molting: the periodic shedding of an old exoskeleton.

Morphogenesis: the development of form, size, and other features of a particular organ or part of the body.

Nauplius: a larva with three pairs of appendages—future head limbs—characteristic of some crustacean species.

Necrosis: the death of cells of a circumscribed piece of tissue or of an organ.

Oocyte: a cell that eventually gives rise by meiosis to an ovum or egg.

Oogenesis: the development of a mature egg from germinal cells in the female gonad.

Opercular rate: the rate at which the opercula of teleost fish are expanded and compressed to pump water over the gills.

Organogenesis: the formation and development of organs in a developing embryo.

Osmoregulation: the regulation of water content and the concentration of total dissolved solutes in the body fluids of an organism.

Ovarian egg: mature egg still present in the ovary of the female.

Ovigerous: carrying, internally or externally, mature unfertilized or fertilized eggs.

PAH: polycyclic aromatic hydrocarbon, sometimes also called polynuclear aromatic hydrocarbon; a carbon–hydrogen compound containing two or more fused benzene rings.

Phyletic: pertaining to race, species or major organismic subdivision (phylum).

Porphyrin: organic pigment containing a tetrapyrrole structure and usually derived from heme, hematin or chlorophyll.

ppb: parts per billion; micrograms per kilogram or per liter.

ppm: parts per million; milligrams per kilogram or per liter.

ppt: parts per trillion; nanograms per kilogram or per liter.

Salinity: concentration of dissolved solutes in seawater; expressed as parts per thousand and the symbol ‰ S.

Serous fluid: any watery or serum-like fluid.

Supernumerary zoea: one or more extra larval stages in crustacean development sometimes produced in response to stress or suboptimal environmental conditions.

Teratogenesis: the production of anomalous or deformed offspring by an animal due to exposure of the parent to exogenous stressors or toxins.

Trochophore: a top-shaped larva of marine molluscs and polychaete worms that bears a girdle of cilia and is usually planktonic.

Veliger: a larval stage of many marine bivalve and gastropod molluscs. It often follows the trochophore stage and possesses a larval shell.

Velum: an organ developing from the prototroch in a molluscan veliger larva. It is ciliated and is used for locomotion and feeding.

Visceral mass: a central body region of molluscs containing the digestive tract, digestive gland, excretory organs and sometimes the gonads.

Zoea: larval stage or stages in most decapod crustaceans characterized by a large cephalothoracic shield, two compound eyes, median nauplius eye, and biramous thoracic appendages.

Zygote: a diploid cell formed by the fusion of two gametes; a fertilized egg.

Index

Acetone, 7
Acropora cervicornis, 114
Acropora palmata, 114
Aequipecten irradians, 95
Analysis techniques, 5
Anthracene, 128
Arenicola marina, 142
Artemia salina, 64
Artemia salina nauplii, 65, 66

Benzene, 145
Benzo[a]pyrene, 12, 30, 31, 107
2,2′-Binaphthal, 145
Bioaccumulation factors, 105, 107
Bioassay procedures, 9–10
Biochemical synthesis, 2
Busycon contrarium, 146, 148, 150, 151, 153

Calanus helgolandicus, 128
Calcium deposition
  reef corals, by, 80–9
  sea urchin spines, in, 89–92
Callinectes sapidus, 96, 98, 142, 145
Cancer irroratus, 144
Cancer magister, 37
Capitella capitata, 9, 17, 22, 75
Carcinogens, 2, 94
Chaetodon sp., 117, 118
Chrysene, 12, 31, 107
Clams, 105

Clupea harengus membras, 38
Clupea harengus pallasi, 36
Colanus helgolaneicus, 98
Coral zooxanthellae, 77–80
Coregonus, 59
Crangon crangon, 38
Crassostrea virginica, 93, 95, 96
Crassostrea virginica (Gmelin), 100–2
Cyprinodon variegatus, 45–50, 61, 62
Cyprinodon variegatus (Lacépède), 22

Diadema antillarum, 89–92
Diadema antillarum (Phillipii), 39–41
Dibenzanthracene, 31
Dibenzothiophene, 89, 91, 100
Dimethylnaphthalenes, 47
Donax variabilis, 148, 149, 152–4
Drupa granulata, 145

Embryonic development, 42–62
Engraulis mordax, 97
Enteromorpha, 9
ET_{50}, 32, 34
Eurypanopeus depressus, 70
Eurytemora affinis, 36

Favia fragum, 79, 82
Favia fragum (Esper), 40
Fish, 45–62
Fluoranthene, 12, 30

Index

Fluorene, 29
Food, 130-5
Fundulus grandis, 58, 60
Fundulus heteroclitus, 39, 45-52, 61, 62
Fundulus similis, 45, 47, 48, 62, 110, 111, 134, 135
Fundulus similis (Baird and Girard), 22

Gadus morhua, 38, 97
Gammarus oceanicus, 35, 37

Halitholus cirratus, 144
Homarus americanus, 37, 94, 145
Hyas araneus, 37
Hydrocarbons, 1-4, 6-7
  accumulation and release, 93-142
  aliphatic, 115, 121
  analysis, 8
  aromatic, 2, 32-4, 95, 114, 121, 123, 139, 140
  biogenic, 2, 3
  chlorinated, 25
  compositions, 11, 14
  concentrations of, 4, 14-16, 40
  high molecular weight, 3
  levels of, 3
  see also Polycyclic aromatic hydrocarbons

Kerosene, 145

Larval development and growth, 42-5, 62-72
LC_{50} values, 9-10
Leander tenuicornis (Say), 22
Ligia exotica (Roux), 20-2
Limulus polyphemus, 77
Littorina littorea, 144, 153

Macoma inquinata, 98
Madracis decatis, 77, 78, 82-5, 88, 89, 117, 119

Madracis decatis (Lyman), 40
Marine animals, behaviour, 143-54
Marone saxatilis, 97
Melitta quinquiesperforata, 37
Menidia beryllina (Cope), 22
Mercenaria sp., 36
Mercenaria mercenaria, 38, 96, 138
Methylnaphthalenes, 97
Methylphenanthrenes, 12, 29, 100
Millepora sp., 40, 77-9, 81, 85-9
Molluscs, 42-5, 94, 123
Montastrea annularis, 79, 84-6, 114, 117
Montastrea annularis (Ellis and Solander), 40
Mucus, 117
Mussels, 94, 95
Mya arenaria, 140
Mysidopsis almyra (Bowman), 22, 23
Mytilus californianus, 94, 95
Mytilus edulis, 36, 94-6
Mytilus variabilis, 145

Naphthalenes, 8, 16, 19, 20, 30, 31, 34, 45, 57, 73, 89, 99, 105-14, 117, 121, 123, 124, 127, 129, 130, 132-5, 145
Nassarius obsoletus, 145
Neanthes arenaceodentata, 17, 22-31, 39, 62, 72-7, 111, 113, 115, 133-42
Neanthes arenaceodentata (Moore), 9
Neopanope texana, 37, 70
Nereis sp., 137
Nitocra affinis, 36

Oculina diffusa, 80, 83, 85, 86, 89, 118, 120
Oculina diffusa (Lamarck), 40, 41
Oil-in-water dispersions (OWD), 7, 8, 12-17, 20, 21, 39, 99, 107
Oil seepage estimation, 1
Oncorhynchus gorbuscha, 61, 71, 97, 145
Onisimus affinis, 144
Oyster tissues, 105

Pachygrapsus crassipes, 145
Palaemonetes pugio, 31, 32, 63–6, 71, 123, 124, 130, 141
Palaemonetes pugio (Holthuis), 22–4
Paracentrotus lividus, 37, 38
n-Paraffins, 100, 105
Parophrys vetulus, 98
Patella vulgata, 144
Penaeus aztecus, 31, 109
Penaeus aztecus (Ives), 22, 23
Penaeus setiferus (Linn), 23
Penaeus setiferus (Young), 32–4
Perylene, 12
Petroleum
  influx to marine environment, 1
  toxicity, 5–6
Phascolosoma agassizii, 98
Phenanthrene, 29, 34, 65, 80, 85, 88, 89, 106, 123–5
Photosynthetic carbon fixation, 77–80
Platynereis dumerili (Andouin and Milne Edwards), 22
Plesiopenaeus edwardsianus, 145
Polychaete worms, 9, 17, 22, 23, 25, 72, 111, 115
Polycyclic aromatic hydrocarbons (PAH), 2, 28, 29–31, 93, 94, 96, 106, 107, 130, 138
Protothaca staminea, 98, 125–7, 141
Psammechinus miliaris, 37

Rangia cuneata, 121, 122, 125–32, 141, 142
Rangia cuneata (Gray), 100, 103–6, 108

Reef corals, 80–9, 114
Reference oils, 6, 10, 39, 99
Reproduction, 35, 72–7
Rhithropanopeus harrisii, 39, 65, 66, 69, 70

Salinity effects, 32–4, 123–30
Salmo, 59
Sea urchin spines, 89–92
Sediment, 130–5
Strongylocentrotus droebachiensis, 37
Strongylocentrotus pallidus, 37
Strongylocentrotus purpuratus, 41
Stylophora pistillata, 36

Temperature effects, 123–30
Thais haemostoma, 61
Thais haemostoma (Conrad), 42–5
Toxicity
  comparisons
    different species, to, 22
    marine larvae, to, 23–8
    petroleum-derived aromatic hydrocarbons, 28
    reference oils, 17–20
    WSF and OWD, 20–2
  petroleum, of, 5–6
  salinity effects on, 32–4

Water-soluble fractions (WSF), 7, 8, 14, 16, 17, 20–5, 39, 41, 99
Weathering effects, 5

Language Contact in the History
of English

STUDIES IN ENGLISH MEDIEVAL LANGUAGE AND LITERATURE

Edited by Jacek Fisiak

Advisory Board:
John Anderson (Edinburgh), Norman Blake (Sheffield),
Ulrich Busse (Osnabrück), Olga Fischer (Amsterdam),
Richard Hogg (Manchester), Dieter Kastovsky (Vienna),
Marcin Krygier (Poznań), Roger Lass (Cape Town),
Peter Lucas (Dublin), Donka Minkova (Los Angeles),
Ruta Nagucka (Cracow), Akio Oizumi (Kyoto),
Katherine O'Brian O'Keeffe (Notre Dame, USA),
Matti Rissanen (Helsinki), Hans Sauer (Munich),
Liliana Sikorska (Poznan), Jeremy Smith (Glasgow)

Band 1

PETER LANG
Frankfurt am Main · Berlin · Bern · Bruxelles · New York · Oxford · Wien

Dieter Kastovsky/Arthur Mettinger
(eds.)

Language Contact
in the History of English

PETER LANG
Europäischer Verlag der Wissenschaften

Die Deutsche Bibliothek - CIP-Einheitsaufnahme

Language contact in the history of English / Dieter Kastovsky ;
Arthur Mettinger (eds.). - Frankfurt am Main ; Berlin ; Bern ;
Bruxelles ; New York ; Oxford ; Wien : Lang, 2001
  (Studies in English Medieval Language and Literature ;
  Vol. 1)
  ISBN 3-631-36243-9

ISSN 1436-7521
ISBN 3-631-36243-9
US-ISBN 0-8204-4745-5

© Peter Lang GmbH
Europäischer Verlag der Wissenschaften
Frankfurt am Main 2001
All rights reserved.

All parts of this publication are protected by copyright. Any
utilisation outside the strict limits of the copyright law, without
the permission of the publisher, is forbidden and liable to
prosecution. This applies in particular to reproductions,
translations, microfilming, and storage and processing in
electronic retrieval systems.

Printed in Germany 1 2 3 4 6 7

To the Memory of Andrei Danchev and Herbert Penzl

Contents

Dieter Kastovsky & Arthur Mettinger:
Introduction 9

David Burnley:
French and Frenches in fourteenth-century London 17

†Andrei Danchev & Merja Kytö:
The Middle English "*for to* + infinitive" construction:
A twofold contact phenomenon? 35

Hans-Jürgen Diller:
Verbs of verbal communication in the English Renaissance:
A lexical field under language contact 57

Richard Dury:
The history of the English language in the context
of the history of the European languages 69

Andreas Fischer:
Lexical borrowing and the history of English:
A typology of typologies 97

Udo Fries:
Foreign place names in the ZEN-Corpus 117

Raymond Hickey:
Language contact and typological difference:
Transfer between Irish and Irish English 131

Thomas Kohnen:
The influence of "Latinate" constructions in Early Mondern English:
Orality and literacy as complementary forces 171

Lucia Kornexl:
"Unnatural Words"? Loan formations in Old English glosses 195

Manfred Markus:
Duplications of vowels in Middle English spelling 217

Gabriella Mazzon:
Language contact in the history of Englishes,
or the genesis of extraterritorial varieties 233

Contents

Ruta Nagucka:
    Latin prepositional phrases and their Old English equivalents 251

Gabriele Rinelli:
    Scandinavian and native social terms in Middle English:
    The case of *cherl/carl* 267

Nikolaus Ritt:
    The spread of Scandinavian third person plural pronouns
    in English: Optimisation, adaptation and evolutionary stability 279

Herbert Schendl:
    Code-switching in medieval English poetry 305

Robert P. Stockwell & Donka Minkova:
    The partial-contact origins of English parameter verse:
    The Anglicization of an Italian model 337

Laura Charlotte Wright:
    Models of language mixing:
    Code-switching versus semicommunication
    in medieval Latin and Middle English accounts 363

Index of subjects 377

Index of names 401

Dieter Kastovsky & Arthur Mettinger (Vienna)
Introduction

Every European language – and certainly many if not most non-European languages – has been subject to language contact in various ways, be it with neighbouring living languages, be it by dialect mixture, i.e. contact between variants of the same language, or be it by the contact with some prestigious language no longer spoken but of historico-cultural importance, as, e.g., Latin and Greek (Neo-Latin) in the European context. But it would seem that no European language has been subject to this phenomenon as profoundly as English, so much so that some linguists would even deny this language its place within the family of Germanic languages and rather treat it as a Franco-Germanic creole (with more Romance than Germanic features), cf., e.g., Bailey – Maroldt (1977). This is why, when thinking about a new topic for our series of conferences on various aspects of the history of English (cf. Kastovsky – Bauer (eds.) 1988; Kastovsky (ed.) 1991, 1994), we decided to devote the next meeting to Language Contact in the History of English (LACHE). The conference took place in Tulln, Lower Austria, in July 1994. Unfortunately, there were major problems with the publication of the papers due to some unforeseeable circumstances such as changes in the editorial policy of the envisaged publisher, which caused a substantial delay. The editors are therefore very grateful to Jacek Fisiak for offering his new series "Studies in English Medieval Language and Literature" with Peter Lang as a host for this collection, and to the contributors for their goodwill and patience. All papers selected from the original conference contributions have been substantially revised and brought up to date (i.e. up to 1999) and provide, we think, interesting spotlights on the various contact situations and their level-specific aspects in the history of English. It goes without saying, of course, that no overall coverage of the topic was possible, nor was it intended, but on the whole both from the point of view of the contact languages involved and the theoretical approaches used, as well as the linguistic levels affected by the contact situation, the papers included in this volume should provide some new insights into the phenomenon of language contact as such and into its role in the history of English. Moreover, the obvious gaps might also spark off new research in this area.

Apart from language-internal contact between variants of English, four languages played a major role in the shaping of this language: Latin,

Celtic, Scandinavian and French. Of these, Latin perhaps had the most extended influence, temporally speaking, especially as regards the vocabulary. This influence already began on the Continent before the migration of the Germanic tribes to the British Isles, continued in the course of the Christianisation of the Germanic immigrants, then was intensified during the Benedictine Reform in the late tenth and the eleventh centuries, and reached its climax during the Renaissance period in the sixteenth and early seventeenth centuries, which is reflected by notions such as "hard words" or "inkhorn terms" referring to the over-extensive borrowing from Latin. It is therefore not surprising that six out of the seventeen papers in this volume deal with the Latin influence on English in one way or another: two papers (Schendl and Wright) discuss code-switching between English and Latin in medieval poetry and business records (macaronic style), i.e. strictly speaking not the influence of one language on another, but rather the interaction of these languages, which, however, may well have been a precondition for the influence scenario, where one language borrowed something from the other; Kornexl and Diller deal with lexical matters in Old and Middle English, and Kohnen and Nagucka investigate the impact of Latin on English syntax, coming to the conclusion that the direct influence of Latin may not have been as great as it is sometimes assumed.

As was noticed repeatedly during discussions at the conference, it is amazing how little attention has so far been paid to the possible influence of Celtic during the earlier stages of the history of English, in particular as a potential substratum at the phonological, especially prosodic level. Thus, it might well have been the case that the language-inherent prosodic system with stress-timing was strengthened by Celtic influence, producing a stronger initial expiratory stress than in the Continental Germanic dialects, which may in turn have triggered certain phonological and ultimately morphophonemic and morphological developments such as the weakening and loss of unstressed (final) syllables at a much earlier period and to a much greater degree as was the case with the Continental Germanic dialects/languages. What we usually find in histories of the English language are just remarks as to how few loan-words made it into English in the Old English period, and that, maybe, the rise of the progressive is due to Celtic influence, but, generally, it is assumed that the Celtic influence, at least in the Old English and Middle English periods, was negligible – a somewhat doubtful assumption. This situation is reflected by the absence of papers dealing with this topic in the present volume. But at least one paper (Hickey's "Language contact and typological difference: transfer between Irish and Irish English") makes up for this gap and is devoted to the contact between Celtic and English, although only in connection with certain features of Irish English which can be attributed to the contact with Irish in Ireland since the seventeenth century.

Scandinavian fares somewhat better in this respect, cf. the papers by Danchev – Kytö, Rinelli, and Ritt, which focus on syntactic, lexical and morphological influences. The same is true of French, which is usually regarded as the most important foreign factor influencing the history of English during the Middle English period. Various aspects of this contact situation, e.g. sociolinguistic factors, spelling, code-switching, syntax, are dealt with by Burnley, Danchev – Kytö, Markus, Schendl and Wright.

Finally, one language usually not that prominent in discussions of language contact except in connection with loan-words, viz. Italian, proves to have played an extremely important role in the development of the Middle English pentameter verse as instigated, e.g., by Chaucer. Thus Stockwell – Minkova convincingly argue in their paper "The partial-contact origins of English pentameter verse: the Anglicization of an Italian model" that the English pentameter is a combination of an inherent lexical and phrasal tendency toward iambic rhythm and the decasyllabic Romance meter based on syllable counting, where the former is superimposed upon the latter, but where the latter also sometimes surfaces.

The papers included in this volume not only address language- or level-specific problems, quite a few also deal with general questions of language contact from a theoretical point of view, highlighting sociolinguistic, typological, or methodological aspects.

Thus, Dury in "The history of the English language in the context of the history of the European languages" argues that a history of the English language should also take into account parallels between English and the histories of neighbouring languages. Such parallels may of course be due to language contact, but they may also be the result of independent parallel developments triggered by a shared general cultural background (investigated under the term "areal linguistics" or "*Sprachbund*"), or by some inherent structural tendency (Sapir's "drift"), neither of which have so far been given enough consideration in existing histories.

Ritt's "The spread of Scandinavian third person plural pronouns in English: optimisation, adaptation and evolutionary stability" is much more than a simple attempt at accounting for the replacement of the Old English *h*-pronouns *hie, hiera, him* by the Scandinavian *þ*-pronouns *they, their, them*. Rather, he uses this puzzling development as a test case for evaluating two different approaches to language change, the structural, model-theoretic and the functional-genetic, psycholinguistic-physiological one. The former, more traditional approach would approach this problem from the point of view of abstract linguistic structures and their potential deficiencies, which allow innovations to succeed, because they improve the structural contrasts of the system. But Ritt argues that this does not really solve the problem of how such innovations manage to change the speech behaviour of the individual speaker and therefore opts for a model constructed on the analogy with the "survival of the fittest" postulate and the genetic selection theory based on it, coming from biology and evolution theory.

Burnley's "French and Frenches in fourteenth-century London" is as much an exercise in illustrating the linguistic levels at which French influenced English and the way it did this, as it is a study of the sociolinguistic situation in London at the time in question and the role of French and Latin in the speech community, which determined the degree and manner of the influence of French on English. He argues that the active and passive knowledge of French varied considerably in the speech community, which in turn affected the way in which French patterns were integrated into English.

A rather different, but very important theoretical problem is addressed by Mazzon in "Language contact in the history of Englishes, or The genesis of extraterritorial varieties", which reviews the question to what extent intralinguistic contact (i.e. contact between variants of the same language with concomitant dialect-levelling) and interlinguistic contact (i.e. contact between different languages with potential creolisation) have played a role in the development of primary extraterritorial Englishes (e.g. American, Australian, South African English) and secondary extraterritorial Englishes, i.e. English spoken as a non-native *lingua franca*.

A borderline case between general theoretical and level-specific papers is Fischer's "Lexical borrowing and the history of English: a typology of typologies". On the one hand, the papers deals with lexical borrowing exclusively, but, on the other hand, it does so from a general point of view, looking at the various criteria that have been used to classify lexical borrowings. Criteria used and typologies of borrowings based on them include aspects from morphology, semantics and sociolinguistics, which, Fischer argues, should be integrated into a comprehensive approach.

This brings us to the papers dealing with lexical borrowings, which again address both descriptive and theoretical aspects. Thus, Kornexl investigates the status and function of Old English glosses in her paper

"'Unnatural words"? Loan-formations in Old English glosses". She comes to the conclusion that glosses may not necessarily reflect normal linguistic usage, but treating glossing as a method in its own right, adhering to an over-literal rendering of the original, might also be too simplistic. Glossators were probably better versed in the morphology and semantics of the source language (Latin) than we are inclined to believe, and their goal of literal renderings of the originals may well have been an attempt at a morphosemantic explication of the Latin word.

Rinelli's "Scandinavian and native social terms in Middle English: the case of *cherl/carl*" is an exercise in sociolinguistic lexicology, showing how terms indicating social status underwent a semantic change in Middle English (certainly also in connection with a change in the social status of the original referents of the respective terms). In the instances discussed, the semantic change involves pejoration, and, at the same time, a shift from a definite referential sphere (i.e. a particular social position) to a connotative and less specific one (i.e. a derogative term).

Diller's "Verbs of verbal communication in the English Renaissance: a lexical field under language contact" is an investigation into the restructuring of a particular lexical field due to language contact, in this case Latin. This immediately raises the question of the reason for borrowing and its impact on the general structure of a semantic domain, involving both functional as well as esthetical aspects, but also the impact of extralinguistic, cultural factors triggering lexical borrowings. It is a pity that there are only very few studies so far which combine this lexical-semantic approach with the indispensable extralinguistic, cultural background.

Finally, Fries's paper, "Foreign place names in the ZEN-Corpus", highlights the general question as to how place-names are rendered in different languages, e.g., when are they adapted, when are they left unchanged, and whether there is any historical development in these practices. At the same time he illustrates how one can make use of one of the most important tools for present-day historical studies, viz. machine-readable corpora, for such an investigation.

Syntactic change as a result of language contact is much less obvious than changes at other linguistic levels. It is therefore not surprising that the three papers dealing with syntactic phenomena are much less assertive with regard to the role of contact in the respective change than the papers dealing with changes at other linguistic levels.

Thus, Danchev – Kytö in their "The Middle English *'for to* + infinitive' construction: a twofold contact phenomenon?" argue that the introduction of the *for to* construction is probably due to a combination of Scandinavian and French influence, enhanced by an internal need for emphasising the idea of purpose, which is behind this construction, a tendency corroborated by typological and second language acquisition data. On the other hand, the subsequent progressive loss of *for* in this construction seems to be an internal development and might well be attributed to a decreolisation process.

Kohnen in "The influence of "Latinate" constructions in Early Modern English: orality and literacy as complementary forces" demonstrates that certain so-called Latinate constructions such as participial constructions are indeed due to language contact, but that their spread in the language is governed by factors such as text type, and the degree of orality or literacy of the text type in question, which is why the spread of a construction is always gradual and results in considerable synchronic variability.

Nagucka is even more hesitant to recognise Latin influence in her paper "Latin prepositional phrases and their Old English equivalents", arguing that the translation of Latin prepositional phrases by prepositional phrases in Old English reflects a native structural pattern rather than a calque of Latin.

Finally, Markus in his paper "Duplication of vowels in Middle English spelling" corroborates the assumption that double spelling of vowels in Middle English was influenced by contact with French, but also shows that the considerable variation of the spelling is due to language-internal developments such as the gradual neutralisation of the feature [length] in connection with various lengthening and shortening processes in Late Old English and Middle English (cf. Ritt 1994).

As has already been mentioned above, two papers, viz. Schendl's "Code-switching in medieval English poetry", and Wright's "Models of language mixing: code-switching versus semicommunication in Medieval Latin and Middle English accounts", deal with macaronic writing, i.e. the switch between two or more languages within a sentence. These papers thus do not actually investigate language contact in connection with linguistic change, but rather the synchronic co-existence of different languages in a linguistic community and the way this linguistic community used these languages side by side. But it is exactly this synchronic co-existence of several languages which is a precondition for contact-induced change, and studies like these are therefore of great importance for the understanding of language change induced by language contact.

Since the papers contained in this volume go back to papers presented at the LACHE conference in 1994, it is high time to finally acknowledge our debt of gratitude to the benefactors who made this conference a most

Introduction

enjoyable occasion – at least for the conference organisers, but we also immodestly think, for the conference participants, too. First and foremost, our thanks go to the Bundesminister für Wissenschaft und Forschung, Herrn Dr. Scholten, and the Landeshauptmann of Niederösterreich, Herrn Dr. Pröll, both for accepting the patronage of the conference and for subsidising it substantially. We also gratefully acknowledge the assistance of the British Council, who again helped with travel grants. We also thank the Mayor and the Town Councillor for Cultural Affairs of Tulln for the hospitality extended to the conference in the form of a memorable reception. Thanks also again go to the staff of the Hotel Rossmühle, our venue, and in particular to its chef for the lovely meals we were served. And, as always, the conference secretary, Frau Christine Klein, performed miracles behind the scenes and did everything for the smooth running of the conference. Those who have attended the other conferences before know that it is she who does the work, while the conference organisers bask in the sun. Last, but by no means least, we have to thank a newcomer to the scene, Ms. Anna-Maria Adaktylos, linguist and computer freak, for doing a great job with the editing of the manuscript. Without her, the contributors would have had to wait even longer for their product to finally reach the general public.

References

Bailey, Charles-James N. — Karl Maroldt
1977 "The French lineage of English", in: Jürgen Michael Meisel (ed.), Langues en contact – pidgins – creoles – languages in contact. (Tübinger Beiträge zur Linguistik 75.) Tübingen: Narr, 21-53.

Kastovsky, Dieter (ed.)
1991 *Historical English syntax.* (Topics in English Linguistics 2.) Berlin: Mouton de Gruyter.
1994 *Studies in Early Modern English.* (Topics in English Linguistics 13.) Berlin: Mouton de Gruyter.

Kastovsky, Dieter — Gero Bauer (eds.)
1988 *Luick revisited: papers read at the Luick Symposium at Schloß Liechtenstein, 15-18 September 1985.* (Tübinger Beiträge zur Linguistik 288.) Tübingen: Narr.

Ritt, Nikolaus
  1994   *Quantity adjustment: vowel lengthening and shortening in Early Middle English.* (Cambridge Studies in Linguistics. Supplementary volume.) Cambridge: Cambridge University Press.

John David Burnley (Sheffield)
French and Frenches in fourteenth-century London

1. Complexity and variation

The contact between English and French in London during the latter half of the fourteenth century is a far from simple matter. The idea that we are concerned with straightforward contact between two synchronic languages, or even with creolisation based on spoken sources, is too simplistic. The situation is complicated by the quantity of apparently conflicting evidence, and by the number of distinct factors of mode and style which must be taken into consideration. As a problem of linguistic description, its difficulty lies in reconciling diversity. Recent outlines seek to record more details of the variety in use and context within the contact situation, and have been enlightening (Richter 1979), but even some of these (Bailey − Maroldt 1977; Kibbee 1991) have been subject to criticism by specialists impatient with the historical linguist's unfamiliarity with the detail of primary sources (Rothwell 1996, 1998). My purpose in this paper is to illustrate some of the pragmatic complexity of the circumstances of the use of French in texts from fourteenth-century London and to question the validity of over-simple dichotomies between English and French languages.

London English itself is complicated, even leaving aside the contribution of French. A distinct local language may be identifiable in a few documents, but most written records from the capital are dialectally diverse *mischsprache*. Indeed, in linguistic-descriptive terms, it is in fact quite difficult to say what London English is beyond the combination of a few diagnostic features identified by scholars such as Brunner, Flasdieck, Heuser, Morsbach or Samuels. The particular diagnostic features themselves change through the centuries. Yet, at least when studying English, we routinely recognise this diversity in the object of our study. But when anglicists refer to "French" in an English context, it is by no means certain that we are so scrupulous about the complexity of the object of our reference. Traditionally, in discussing the history of English, a distinction is made between the French of Paris and that of "Stratford atte Bow", but this simple division between Continental and Anglo-French is really only the beginning (Rothwell 1985; Short 1996). The discernible varieties within Anglo-French may be far more complicated. Use in distinct contexts creates distinct styles, and some words adopt senses not recorded on the continent. This last point has been used, indeed, to argue that independent productivity in the lexicon of Anglo-French indicates the existence of a vernacular distinct from Continental French (Legge 1980; Rothwell 1993a, 1993b). As a vernacular, it might be expected that, like English, Anglo-French too might show diatopic and diaphasic variation. A fully functional vernacular would be widely used both in different regions and for different purposes; and since it has a written mode, we should expect a spoken one also, perhaps with its own distinct variation. Yet, although French was quite widely written in England at the end of the fourteenth century, it is not at all clear that it can be regarded as a spoken vernacular. The extent to which it was spoken outside a narrow social group is very questionable. It soon had to be deliberately learned, and, indeed, its speakers had lost the perception of themselves as French before the end of the twelfth century (Short 1996: 173-174). The impact of French upon English was in fact largely as a written language used for a variety of special purposes. Although relatively homogeneous synchronically, the stylistic complexity of written French as a linguistic presence in England was ensured by the depth of its history, where it had been used alongside English for ten generations. French was modified to some degree by this contact, but was constantly renewable from Continental sources, and when competently written, it was relatively free of English interference. By contrast, its influence on English was very great; and it is this influence, rather than the status of French itself, which concerns us next.

There are, perhaps, good functional reasons for the adoption of huge quantities of French words into the Middle English lexicon, but often it is more revealing when considering French influence to attend to the social circumstances in which they were used. To take the example of the outstanding literary figure of the age, the vocabulary of Chaucer notoriously contains slightly more words of Romance origin than those from Germanic (Kaplan 1932; Mersand [1939]); and it is equally well known that the rate of lexical borrowing from French peaked in the last quarter of the fourteenth century (Baugh 1935; Dekeyser 1986; Jespersen 1962: 86-87). Chaucer's home was in London; his career was connected with the royal court, his avocation was literature. All three circumstances brought him into close contact with the French, the language of the aristocratic communication, of literary and courtly refinement, and of administration. London, with its great concentration of scribes, often recruited from the provinces, its shifting and mixed forms of native English, and its functional, literate bilingualism, was the most dynamic centre in medieval

England, both in terms of its varied and shifting population and its mixture of language varieties.

French in medieval London was one possible linguistic choice. It is encountered in documents in many forms, any of which we are accustomed to designate, from a modern English point of view, simply as "French" or sometimes as "French influenced". I should like to make a few distinctions in the ways in which this "French-ness" is encountered. Firstly – and I mention this only to set it aside for treatment by more competent scholars – are the various kinds of continuously written French prose and verse: passages in official records of the *French Chronicle of London* (1259-1342), the extensive verse of John Gower (a near Londoner), or the lost French lyrics of William Montagu, Earl of Salisbury. Many Englishmen sought to follow the aristocratic practice of writing letters in French, and there were those who could write fluently and effectively in pure French (the poet Hoccleve composed a formulary in French shortly before his retirement from the Privy Seal Office) (Bentley – Young 1965). But many fell short of competence in this art and more or less conventional apologies for their errors are the norm (Ellis (ed.) 1827, 1: 18-19).

French was written for various purposes and with various degrees of skill, but if we define French as discourse which uses French lexical items within the grammatical and syntactical patterns proper to the French language, French in England is not restricted to lengthy and continuous texts. Laura Wright has been studying texts in which indisputably French phrases and clauses occur, characteristically as the terms of technical discourse alongside English, in what she calls a "macaronic" mixture (Wright 1992, 1996). In formal respects comparable, although stylistically quite distinct and less densely packed, are the many French phrases to be found in popular narrative poetry. All fourteenth-century narrative poems from the London area make extensive use of French adverbial phrases, often at the end of the line and rhyming on another French word; for example *saunz fable – stable; par ma fey – contrey; tut entoure – honoure*. They certainly fulfil a literary purpose in supplying a rhyme, but this can not be the primary reason for their use. Brief French phrases also occur within the line and are by no means always the reflection of technical language, but can serve as indexes of style (cf. appendix A).

The lexical contact between two languages is by no means limited either to their independent co-existence or their mutual adoption of words and phrases. Far more interesting, and often more difficult to interpret, is that form of interaction which results in "calques". These are defined as the linguistic forms resulting from a process by which the structures proper to one language are filled by the forms proper to another, as, for example, when the English phrase *make mone* seems to be a loan translation of the French phrase *faire complainte*. In this paper, I shall extend the usual application of the word to consider the possibility that in addition to lexical, grammatical, and syntactic calques, other patterns of language use – call them stylistic or pragmatic calques – may also be found.

2. Grammatical contact

Firstly, let us consider briefly what appears to be a frequent grammatical calque. This is the use of adjectives with plural inflection in some English texts. As a formal feature, plural marking of adjectives is by no means necessarily French. In English at the end of the fourteenth century, adjectival plural inflexion is quite common over an area south-east of a line from the Wash to the Solent. A final -e is used to mark both weakness and plurality of adjectives in the best manuscripts of the *Canterbury Tales*, by Gower, in Hoccleve's autograph poetry, and by the anonymous author of the prose *Equatorie of the Planetis*. Certain fourteenth-century authors – and not simply for purposes of verse – maintained an inflexional distinction in English adjectives which had originated in Old English, and which had been regularised and extended to some early French borrowings by the beginning of the century. The distribution of this feature among London and other texts is hard to account for in purely dialectal terms, since its appearance is geographically and chronologically discontinuous: it is a feature of neither the Auchinleck manuscript (c. 1330), of City documents, nor of Chancery records. Nor is it used by Edward, Duke of York, in his *Master of Game* (1406), or (as far as can be judged) by Usk in his *Testament of Love*. Yet it is represented in the language of Kent by the *Ayenbite of Inwit* (1340) and in that of Norfolk by *Havelok the Dane* (Samuels 1972; Smithers (ed.) 1987: lxxxi; Morris – Gradon (eds.) 1979: 63-69). Social as much as dialectal differentiation might explain its use by Gower, Chaucer and Hoccleve. But whatever the explanation, the origins of this grammatical distinction are clearly unconnected with French influence. It could be argued on purely formal grounds that awareness of a native singular/plural contrast in adjectives might have encouraged the adoption of a similar contrast from French. In fact, the evidence seems to be against it. French adjectival plurals were widely imitated, often accompanied by post-position of the adjective, but their distribution between texts seems quite distinct from that of the

English adjectival plurals, and they are almost restricted to phrases consisting of words and derivational affixes of French origin. When not collocated with words of French etymology, they are combined in phrases of a technical nature or implying high social status:

(1) *I knowlech wel, that for certain that I among other communed and asked of certeins Clercs, whethir that we myght give up our homage for drede of our lyves, or non (Rolls of Parliament III.379 (1397))*

(2) *and the Answers of certeins Lordes, that is atte saye, of the forsayd Ducs of Aumarle (Rolls of Parliament III.451 (1399))*

These French plural adjectives show all the indications of being part of a literary language, which in the case of rhymes in the Auchinleck manuscript extends also to the use of French masculine singular nouns ending in consonant + *s*, and similar rhymes intended to attract the eye rather than be heard: *graunt mercys – twiis (7 Sages of Rome*, 938); *Frere Menours and Jacobins / Frere Carmes and frere Austines (St Patrick's Purgatory*, 823-824); *In whom he biʒat Ywayns / Hende and noble and kniʒt certeyns (Arthour & Merlin*, 2614-2615). It is notable that, just as there is no connection between the English grammatical marking of plural adjectives and French adjective plurals, the grammar here is often neither English nor French. Rather, significant features of French grammar are used metalinguistically, and function purely symbolically to create an association with French without properly observing the grammatical structures of the language (cf. appendix B).

Syntactical calques are less easy to identify with certainty, since it is often impossible to rule out a native origin for the structures. Parallels like *cynk semeygnes et plus (Fch Chronicle*, 1) and *an hundred kniʒtes and mo (Sir Orfeo*, 143) are hard to adjudicate. Nevertheless, syntactic parallels, when occurring in contexts subject to other kinds of French influence, are persuasive of calques in whichever direction (Orr 1962).

(3) a. *al be that he was (Canterbury Tales*, I.297)
vs. *Et, tout soit le feble chastel pris (Livre de Seyntz Medicines*, p. 20)
b. *Embrouded was he (Canterbury Tales*, I.89)
vs. *Laide estoit e pale e folee (Roman de la Rose*, 198)
c. *he remembreth hym of (Legend of Good Women*, 1105)

      vs. *li suvenge ... del beivre* (Thomas, *Tristran*, 2493)
d.   *ryng of gold* (*Canterbury Tales*, X.155)
      vs. *anel d'or esmeré* (*Tristran*, 953-954)
e.   *have I in remembrance* (*Canterbury Tales*, V 714)
      vs. *Qu' ad tant eü en remembrance* (*Tristran*, 178)
f.   *better man drank neuer win* (*Guy of Warwick*, 6830)
      vs. *Meliur n'ad en crestienté* (*Tristran*, 2652)

3. Stylistic influences

The influence of French is also evident in late fourteenth-century English prose style. Some years ago Diane Bornstein identified in the writings of Chaucer and Caxton a characteristic style, which she called "the clergial style" (Bornstein 1977). Although there were numerous misunderstandings inherent in this description, she was justified in recognising a style which was modelled on that developed in the Latin Chancelleries, transformed into French, and then imitated in English. This style, better called the "curial style", is distinguished by the use of great lexical elaboration, consisting of extensive borrowing from French and Franco-Latin, by elaborated phrasal forms of indirect address, by very long sentences which exploit lexical repetition, synonymous doublets, and anaphoric referential devices (such as *the same* + noun, *the saide* + noun, *the whiche* + noun) (Burnley 1986a, 1986b; Rasmussen 1958: 32-53).

    A major source for the curial style is the records of parliamentary business which were written in English by Chancery scribes with increasing frequency after 1386; but it is by no means restricted to this milieu. Very often scribes were reflecting the style of petitions which had been received, and these were in epistolary form. Accordingly, formal letters at this period – such as those sent by the Mayor and Citizenry of London to Henry V in France – also exhibit this style (Chambers — Daunt 1931: 63-89; Legge (ed.) 1941: 63-67). It is a style characteristic of official business at the close of the fourteenth and throughout the fifteenth century. But it is not restricted to legal and administrative documents, since it is soon imitated in the dedicatory prologues and epilogues of literary works. It is especially noticeable in the works of Caxton, where it used to be ascribed to imitation through translation of the Burgundian court style. But its ubiquity in Anglo-Norman letters and in records written in French from the thirteenth century onwards indicates its true origin.

The curial style in the fifteenth century is frequently found in works translated into English from French, but it is important to recognise that its occurrence is not restricted to those contexts where it owes its existence simply to the substitution of English lexical items into a word-order determined by the French original. Such stereotyped examples are to be found, but it is equally the case that English translators can find their own syntactical patterns, as well as their own words, thus, in effect creating their own version of the curial style (Bornstein 1978). This stylistic calquing was a conscious and deliberate process. It represents not the syntactical modification of English at a deep level, but the calquing of a style. English writers internalised the principles of writing this way, driven certainly by extra-linguistic motives. Rather than translate into English structures, they sought to imitate recognisable surface patterns in French. The process is one of emulation rather than translation. *Aemulatio* was the name given by Quintilian (X v, 2-6) to a process of paraphrasing one mode of writing by another in a spirit of competition. Cicero explains a similar technique behind his rendering of Greek orations:

> nec converti ut interpres, sed ut orator, sententiis isdem et earum formis tanquam figuris, verbis ad nostram consuetudinem aptis. In quibus non verbum pro verbo necessi habui reddere, sed genus omne verborum vimque servavi. Non enim ea me adnumerare lectori putavi oportere, sed tanquam appendere.

> 'And I did not translate them as an interpreter, but as an orator, keeping the same ideas and the forms, or as we might say, the "figures" of thought, but in language which conforms to our usage. And in so doing I did not hold it necessary to render word for word, but I preserved the general style and force of the language. For I did not think I ought to count them out to the reader like coins, but pay by weight, as it were.' (*De optime genere oratorum* IV, 14)

His method involved finding techniques in Latin to render an impression of the Attic style of the Greek original. Although it is difficult to prove that fourteenth-century writers of English perceived any analogy between the relationship of French to English and the historical one of Greek to Latin, the concept of the *translatio studii* was familiar to them, and such views of the emulating relationship between Latin and English became explicit in the sixteenth century, as in the *Civil Conversations* of George Pettie.

4. Pragmatic influences

As a pragmatic calque, I should like to cite the conscious and discriminating use of *ye* and *thou* in English, which emerged in the thirteenth century, persisted in widespread use until the seventeenth century and which survives today in dialect between the Tees and the Trent. The use of *ye* as a marked form of address to a single individual is accompanied by a kind of lexical concord in Middle English, by which address forms indicating positive politeness in Brown — Levinson 's terms (*dame, lord, sire*) attract the use of *ye*, whilst those indicating negative politeness (*brother, sone, doughter*) – as well as derogatory forms of address – attract the use of *thou* (Brown — Levinson 1987: 101-210). Medieval French shows a similar pattern of usage. The major influence on English in this respect is undoubtedly French, but the usage of both languages was ultimately indebted to formal epistolary practice in Latin. French eventually incorporated this usage fully into its grammatical system, so that the plural form of address to a single individual was accompanied by plural verb concord. In English this subject-verb concord was maintained in the indicative, but, although distinct singular and plural forms of the imperative existed in Chaucer's language, they were not fully incorporated into this system. The selection of *ye* will be accompanied by indirect modes of address, but it is only sporadically extended to a polite plural form of the imperative. This pragmatic calque is once again behavioural and expressed in lexis, a stylistic marker without real impact on the grammar.

5. Speaker attitudes

The preceding examples suggest a degree of deliberate emulation of French models of linguistic behaviour in terms of admired styles and the patterns of linguistic politeness. Recognisably French forms are used as the emblem of a behavioural ideal. Let us now return to that scatter of French phrases traceable in most literary sources. Although attested only in literary sources, it seems likely that some may reflect actual colloquial exchange. We can divide them broadly into two types: those reflecting the technical vocabularies of hunting, battle-cries and court life on the one hand, and those which are simply adverbial tags on the other. The latter type is not restricted to any particular discourse; neither does this class of phrase communicate significant cognitive information. What then was the motivation for their use? They are normally attached to an assertion, and their function seems to be to express the attitude of the speaker to what has been said. They therefore have an interpersonal role, directing the focus of attention within a direct speech exchange towards the attitudes of the speaker. English expressions could have been found to fulfil this purpose, but the choice of French ones obviously carried some extra and desirable connotation. It is perhaps significant that, in fiction, they tend to be used by heroes.

There is a general truth in the suggestion that non-cognitive use of French has a special role in Middle English, since it is notorious too in letters, where it is not unusual to find salutations and signatures in French, when the body of the narrative is in English. French phrases, or alternatively calques upon French linguistic usage, had a sociolinguistic rôle as an indication of status. What kind of status varies with circumstances. Those styles identifiable with the life of the gentry create a link with the well known assertion in *Arthour and Merlin* (23) that "*Freynssh vse þis gentil man*", but other styles were associated more generally with worldly experience or refinement. Caxton's adoption of the "curial style" for his prologues placed his work at the summit of the aspirations of his "audience", just as the selection of the same style in official letters from the citizenry of London declares their willing participation in the business of the court.

The social status of French rather than functional need motivates much of its influence on English, but we should beware of over-simplification of the kind just quoted from *Arthour and Merlin*. It is worth recording that attitudes to the status of French-derived styles were not monolithic. This is nowhere better illustrated than in the use of the pronouns *ye* and *thou*. The distinction in their use was not made throughout every level in society, nor was it driven simply by the social status of the language user. Practice may vary in different works of the same author: in Chaucer's *The Franklin's Tale*, *Tale of Melibee* or *Troilus and Criseyde*, the distinction is very carefully made, following the pattern familiar in French. But in *The Parson's Tale*, *The Miller's Tale*, or *The Shipman's Tale*, no careful discrimination is discernible. Clearly this distinctive use of *ye* and *thou* is not determined by Chaucer's own knowledge or by his expectation of different audiences. Langland makes little use of the distinction between *ye* and *thou*, except when representing the King's court and the flattery of beggars. Indeed his use of *thou* to the King in a speech which is described as "clergial", and which is glossed by imitation in the *Crowned King* (42-50), is evidence in English of a broader European rejection among the learned of the conscious use of a style of address adopted by certain individuals as a mark of their social position (Burnley 1990). The point is that *ye/thou* discrimination in Middle English was never properly assimilated into the language: it remained a conscious lexical choice. This choice reflected social divisions. It was unknown to the unsophisticated, enthusiastically adopted by the polite and cultivated, and deliberately rejected by some social critics. These distinctions made it suited to particular styles of speech and writing, but difficult to handle if treated as a general linguistic feature of Middle English.

Literate people in Chaucer's time were heavily influenced by French; and this is evident not only in the forms of their language, but in features of linguistic behaviour, and in their techniques of writing. But, to what extent were they aware of this fact? There is a tendency for philologists, when discussing the Middle English lexicon, to consider as French any words which can be shown to have French etymology. But in the synchronic investigation of fourteenth-century London English this is a fallacy, for it confuses distinct modes of analysis, the diachronic with the synchronic. To Chaucer, the words *fruyt* and *peple*, or to Edward, Duke of York, the word *search*, were no more French than they are to us. Although all three are indeed etymologically French, they are specifically mentioned by these authors as the *English* translations of other foreign equivalents. The point is that much that is identifiable to a modern linguist as French, was not apparent as such to medieval people. We must make a distinction between the French which had been thoroughly assimilated into English and that which was synchronically recognisable as foreign. Medieval Englishmen spoke a language which had been greatly influenced by French at the lexical level, but were not constantly aware of that fact. Linguistic awareness is a matter of contextual salience or foregrounding: words, phrases, calques, and styles could no doubt have been identified as French had the need arisen, but in everyday use would often pass for something else. Their ultimate Frenchness was not at issue.

Edward, Duke of York, who wrote his hunting manual, the *Master of Game*, in 1406, is an excellent example of this blindness to linguistic origins co-existing with an awareness of stylistic variety. He is anxious to impart to his readers a detailed knowledge of the correct terms to use on the hunting field; and this includes some knowledge of those that might be encountered, as he says, "*biyonde the see*" (is it a calque on *outremer*?). But he goes on to say that he does not intend his pupils to adopt the terminology of France and he would prefer that they use the English terms. The terms he actually gives and regards as so important are a special restricted language, but one which in fact largely descends from

Anglo-Norman (cf. appendix C). He is oblivious to this etymology and regards them simply as the terms of hunting necessary to any English gentleman.

The case of Edward suggests that we might benefit from viewing the French influence we can detect in the language from the perspective of the users of the language. To reconstruct the role of French in the language contact situation of fourteenth-century London simply from formal analysis and a modern historical perspective might well prove misleading.

Contemporary London English literature contains many references to French, and even some uses of French encapsulated in an English context. We can categorise them roughly as follows:

French is the language of a politically powerful and culturally intimidating neighbour, who is by turns envied, hated and admired.

The French language is exotic and foreign.

French in England is the language of privilege. It is not known to ordinary people and is imperfectly known even to the educated.

The message comes through loud and clear that French is regarded in the fourteenth century as a foreign language to the vast majority of the English population, who characteristically fail to recognise the close affinity between French and their own language. How are we, then, to reconcile the fact that so many records remain in the form of French texts written in England; that Englishmen wrote in French; that the English language itself is so obviously profoundly affected by more than three hundred years' contact with French?

6. The ambiguity of the term "French"

We must accept that in reference to later medieval England the word "French" is highly ambiguous (Short 1996). It could refer to the language and people of the kingdom of France (however defined), and then, to most Englishmen, would be felt to be foreign. But it could also refer to an English sociolect: a variable collection of linguistic usages employed by a privileged group within English society. Generally speaking, the perception of French influence upon English was in terms of this latter conception of French. In this view, French-influenced discourse was a stylistic choice which had connotations with a world which had for many generations been occupied by French, spoken or written. It was considered to be a style which indicated social origin or aspirations. What is difficult to grasp from the modern perspective is that this attitude extended not merely to the use of French lexis or syntactic patterns, or the emulation of style characteristic of the world of Anglo-French, but in fact also to writings entirely and purely in French. Medieval conceptions of writing focused narrowly on content and import. The language chosen should match the inherited norm for the content, but the shift between English, French or Latin of appropriate kinds was not considered to be comparable to translation between foreign languages in the modern period. It was seen rather as a code shift. As Edward of York used Anglo-French hunting terms whilst offering those from "beyond the sea" as exotic alternatives, the Chancery scribe who recorded the first opening of parliament in English thought nothing of stating that the words were spoken "en la forme qui s'ensuit", and recording the English address in French. French was the inherited style proper to the business of making records (Clanchy 1979).

Thus, French to the medieval Londoner was as ambiguous a concept as it seems to the scholar looking back with a modern perspective. It was used alongside English in certain specified circumstances. It was fairly freely mixed with English. It was incorporated into English in ways where the connection with the contemporary French of France might be perceived, but it was more often overlooked.

If we are to understand the role and status of French in later medieval England, we must be aware of this complexity. Both the kind of French and its contact with English differ from century to century. It belongs to specific modes and styles, which also differ according to date. Within these stylistic domains different aspects of French may be unconsciously adopted or – especially in the later period – more deliberately emulated. Within the synchrony of the late fourteenth century were many inherited diachronies. In the fourteenth century, French influence upon English, whether in the forms of French grammar, characteristic word-order, adopted phrases, emulated styles, or patterns of linguistic behaviour, were no longer much to do with any familiar conception of creolisation, nor are they best understood in terms of a purely linguistic analysis. The relationship between English and French at this time was a matter of linguistic competence in a multicultural society (Short 1992), and it was an aspect rather of Hyme's communicative competence than competence in the Chomskyan sense (Hymes 1971; Gumperz 1969). Mastery of French and overtly French-influenced modes of discourse formed part of the stylistic architecture, part of the eloquence, of later medieval England.

References

Aungier, George James (ed.)
    1844   *Chroniques de London depuis l'an 44 Hen. III à l'an 17 Edw. III*. (Camden Society 28.) London: Bowyer Nichols.

Bailey, Charles-James N. — Karl Maroldt
    1977   "The French lineage of English", in: Jürgen Michael Meisel (ed.), *Langues en contact – pidgins – creoles – languages in contact.* (Tübinger Beiträge zur Linguistik 75.) Tübingen: Narr, 21-53.

Baugh, Albert Croll
    1935   "The chronology of French loan-words in English", *Modern Language Review* 50: 90-93.

Bentley, Elna — Jean Young
    1965   *The formulary of Thomas Hoccleve.* [Ph.D. dissertation, Emory University, University Microfilms Order No. 65-11, 503].

Bornstein, Diane
    1977   "French influence on fifteenth-century English prose as exemplified by the translation of Christine de Pisan's *Livre du corps de policie*", *Mediaeval Studies* 39: 369-386.
    1978   "Chaucer's *Tale of Melibee* as an example of "Style clergial"", *Chaucer Review* 12: 236-564.

Brown, Penelope — Stephen C. Levinson
    1987   *Politeness: some universals in language usage.* (Studies in Interactional Linguistics 4.) Cambridge: Cambridge University Press.

Brunner, Karl
    1920   "Die Reimsprache der sogenannten kentischen Fassung der "Sieben Weise Meister"", *Archiv* 140: 199-205.

Burnley, John David
    1986a   "Christine de Pizan and the so-called *Style Clergial*", *Modern Language Review* 81: 1-6.
    1986b   "Curial prose in England", *Speculum* 61: 593-614.
    1990   "Langland's clergial lunatic", in: Helen Phillips (ed.), *Langland, the mystics and the medieval English religious tradition: essays in honour of S. S. Hussey.* Cambridge: Brewer, 31-38.

Chambers, Raymond Wilson — Marjorie Daunt
    1931   *A book of London English.* Oxford: Clarendon.

Clanchy, Michael Thomas
1979　　　　*From memory to written record: England 1066-1307.* London: Arnold.
Dekeyser, Xavier
1986　　　　"Romance loans in Middle English", in: Dieter Kastovsky — Aleksander Szwedek (eds.), *Linguistics across historical and geographical boundaries: in honour of Jacek Fisiak on the occasion of his fiftieth birthday,* vol. 1. (Trends in Linguistics: Studies and Monographs 32.) Berlin: Mouton de Gruyter, 253-265.
Ellis, Henry (ed.)
1827　　　　*Original letters illustrative of English history.* 2nd series, 4 vols. London: Harding and Lepard.
Flasdieck, Hermann Martin
1931　　　　"Review of B. A. Mackenzie, *The early London dialect* (Oxford 1928)", *Beiblatt zur Anglia* 42: 33-47.
Gumperz, John J.
1969　　　　"Communication in multilingual societies", in: S. Tyler (ed.), *Cognitive anthropology.* New York: Holt Rinehart, 435-449.
Heuser, Wilhelm
1914　　　　*Altlondon mit besonderer Berücksichtigung des Dialekts.* Osnabrück: Liesecke.
Hymes, Dell
1971　　　　*On communicative competence.* Philadelphia: University of Pennsylvania Press.
Jespersen, Otto
1962　　　　*Growth and structure of the English language.* (9th edition.) Oxford: Blackwell.
Kaplan, Theodore
1932　　　　"Gower's vocabulary", *Journal of English and Germanic Philology* 31: 395-402.
Kibbee, Douglas A.
1991　　　　*For to speke Frenche trewely: the French language in England, 1000-1600: its status, description and instruction.* (Amsterdam Studies in the Theory and History of Linguistic Science 3, 60.) Amsterdam — Philadelphia: Benjamins.

Legge, Dominica
 1980 "Anglo-Norman as a spoken language", in: *Proceedings of the Battle conference on Anglo-Norman studies 2 (1979).* Woodbridge: Boydell and Brewer, 108-117.
Legge, Dominica (ed.)
 1941 *Anglo-Norman letters and petitions from All Souls MS. 182.* (Anglo-Norman Text Society 3.) Oxford: Blackwell.
Mersand, Joseph
 1937 *Chaucer's romance vocabulary.* New York: Comet.
 [1939] [Reprinted Port Washington, N. Y.: Kennikat.]
Morris, Richard — Pamela Gradon (eds.)
 1979 *Ayenbite of Inwit,* vol II: *Introduction, notes and glossary.* (EETS OS 278.) London: Oxford University Press.
Morsbach, Lorenz
 1888 *Ueber den Ursprung der neuenglischen Schriftsprache.* Heilbronn: Henninger.
Orr, John
 1962 *Old French and modern English idiom.* Oxford: Blackwell.
Rasmussen, Jens
 1958 *La prose narrative française du XVe siècle.* Copenhagen: Munksgaard.
Richter, Michael
 1979 *Sprache und Gesellschaft im Mittelalter: Untersuchungen zur mündlichen Kommunikation in England von der Mitte des elften bis zum Beginn des vierzehnten Jahrhunderts.* (Monographien zur Geschichte des Mittelalters 18.) Stuttgart: Hiersemann.
Rothwell, William
 1985 "Stratford atte Bowe and Paris", *Modern Language Review* 80: 39-54.
 1993a "The "Faus Franceis d'Angleterre": later Anglo-Norman", in: Ian Short (ed.), *Anglo Norman anniversary essays.* London, 309-326.
 1993b "The legacy of Anglo-French: *faux amis* in French and English", *Zeitschrift für romanische Philologie* 109: 16-46.
 1996 "Playing "follow my leader" in Anglo-Norman studies", *French Language Studies* 6: 177-210.
 1998 "Arrivals and departures: the adoption of French terminology into Middle English", *English Studies* 79: 144-165.

Samuels, Michael Louis
- 1963 "Some applications of Middle English dialectology", *English Studies* 44: 81-94.
- 1972 "Chaucerian final *-e*", *Notes and Queries* 217: 445-448.

Short, Ian
- 1992 "Patrons and polygots: French literature in twelfth-century England", *Anglo-Norman studies: proceedings of the Battle conference* 14 (1991): 229-249.
- 1996 "*Tam Angli quam Franci*: self-definition in Anglo-Norman England", *Anglo-Norman studies: proceedings of the Battle conference* 18 (1995): 153-175.

Smithers, Geoffrey Victor (ed.)
- 1987 *Havelok*. Oxford: Clarendon.

Wright, Laura Charlotte
- 1992 "Macaronic writing in a London archive, 1380-1480", in: Matti Rissanen — Ossi Ihalainen — Terttu Nevalainen — Irma Taavitsainen (eds.), *History of Englishes: new methods and interpretations in historical linguistics.* (Topics in English Linguistics 10.) Berlin — New York: Mouton de Gruyter, 762-770.
- 1996 *Sources of London English: medieval Thames vocabulary.* Oxford: Clarendon.

Appendix A

French phrases (and their rhymes) used in the Auchinleck manuscript

courtly life	chivalry	general
Deu vous doint boniour	*fiz a puteyn : lain* (*7 Sages* 1629)	*saun fable : stable* (*7 Sages*, 652)
	fiz a puteyn : main (*GW* 682)	*saun faille : entaille*
		saunz doutance : ignorance
a choger	*Ore tost a ly a ly*	*saunz demorrance : countenance*
a lauer	*Ore sa tost*	*Jeo crey [ceo] ben : yseen*
as armes	*so ho* (ironically – from hunting)	*ieo vus dy : enemy*
Or sus		*saunz retours : traitours*
		par amure : tenure
		par force
		par maistre þrie
		par ma fey : contrey
		par charite : see
		par aventure
		tut entoure : honoure
		bele amy
		a debles

Appendix B

Note the stylistic provenience and the near-restriction to Romance formative affixes. Examples are rare before the last two decades of the fourteenth century. The frequency with the adjective *certein* suggests a written language mannerism.

John David Burnley

In Chaucer's works:

goodes pardurables; tables tolletanes; goodes temporels; places delitables; vniversels speces; goodes espirituels; nowmbyres porcionables; propocionels conuenientz; agreables shadewes; thingis certeins

Examples from early fifteenth-century documentary sources:

lordes spirituels and temporels
certeins Clercs
certeins Lordes
certeins matiers
certeins recordes
thyng mevables
diverses notables causes
resonables causes

defectif & noyantʒ
chymeneys defectifs
lettres testimonyalx sufficeantz
lettres patentʒ
lord appelantz
vitailles convenables
his lord temporals
insuffycyantz paraches

Appendix C

The "English" terms of hunting from The Master of Game

> If eny hounde fyndeþ of hur þer as she haþ be and is goo froom þens in to anoþer place he shal say þus to his houndes as lowde as he may, Ha cy douce cy et venuz, arere, sohowe. And he se þat she be goon in to þe playn or in þe feld or in areable lond or in to þe woode if his houndis fynde wel of hire þan he shal say, la douuce amy it ad est illeoqs, and þerwiþ he shal say, Sohow illeoqs cy douce cy vaillaunt, and twies Sohowe, and whan he is commen þer as he supposeth þe hare wil dwell þan he shal say þus, la douce la est il venuz. And þerwiþ thries, Sohow, and no more. (*Master of Game*, 104)

†Andrei Danchev (Sofia) & Merja Kytö (Uppsala)
The Middle English "*for to* + infinitive" construction: a twofold contact phenomenon?

1. Purpose

The purpose of this paper is:

Firstly, to examine once again the possibility, suggested by some authors (e.g. Danchev 1989: 38-39; Mätzner 1885: 57-59; Shearin 1903: 18) of foreign – Scandinavian and/or French – influence in the emergence and spread of the *for to* + infinitive construction in Middle English. This will be done against the background of additional empirical data that do not seem to have been taken into account so far.

Secondly, we will examine the equally, if not even more puzzling, later retreat of the *for to* infinitive. Do we have to speak of *for*-insertion, or of *for*-deletion? Why did the so called *for to* filter rule (Chomsky – Lasnik 1977) appear in Standard English? We will try to offer tentative answers to this and some other questions.

2. Methodological note

This Middle English development will be viewed within the broader framework of the language change typology in terms of general, group and idiosyncratic changes (as outlined in Danchev 1990, 1991 and elsewhere), which takes into account cross-language (including pidgins and creoles, as well as second[1] language learner interlanguages). As has been pointed out by the present and other authors, such an approach makes it possible to weigh more realistically the arguments for and against the externally and internally motivated language change explanations.

In this particular case we claim that only within the languages involved, namely English and possibly Scandinavian and/or French, no truly convincing conclusion can be reached as to whether *for*-insertion was an internally or externally motivated change. Additional evidence is needed to break out of such methodological impasses.

---

[1] "Second" and "foreign" language acquisition are treated as synonymous terms here.

3. Chronological parameters of the *for to* infinitives

In addition to the extensive available data in the existing literature we have also checked on the *Helsinki Corpus (HC)*. The earliest example, mentioned by a number of authors (e.g. Mustanoja 1960: 514; Shearin 1903: 18; Svartvik — Quirk 1970: 399[2]), seems to be:

(1) and ic bidde eou alle daet ge bien him on fultume at dis cristendome Godes gerichtten for to [emphasis here and elsewhere ours] setten and to driuen.
(rogamus etiam uos quatenus eidem si necesse fuerit auxiliari velitis ad christianitatem sustinendam).
(*Codex Diplomaticus Aevi Saxonici* 1066: IV.306, 3).

The last attested example is much more difficult to pinpoint. It is generally agreed that after the fifteenth century the frequency of the *for to* forms began to decline, but authors disagree concerning the last attested instance. Thus, for instance, following Visser (1963) and Lightfoot (1979), Fischer (1988: 74) accepts that "the last instances of the *for to* infinitive are found in the early part of the seventeenth century". In the *Helsinki Corpus* (in which the Early Modern English section extends from 1500 to 1710), we came across

(2) ..., and I entertained him with those things that I had provided for to entertain my Friends at the Funeral.
(*The Trial of Titus Oates* IV, 83).

dated 1685, that is, the end of the seventeenth century.
Even more recent instances are found in Dickens, e.g.:

(3) Your husband is here for to take you home.
(*David Copperfield* 1849-1850, quoted in Davis 1984: 69).

Altogether there are quite a few such instances in this particular novel.[3]

---

[2] In the literature this publication is usually referred to as Quirk — Svartvik (1970). However, at the end of their paper (the only place where the authors' names are indicated) Svartvik comes first, followed by Quirk.
[3] In *David Copperfield* we counted twenty-one instances of *for to* infinitives with the following verbs: *ask, bring, carry, change, come, complain, do* (2), *drink, eat, marry* (3), *see* (2), *seek, take* (4), *tell, tend*. Most of the verbs are transitive and in the majority of cases

The Middle English "for to + infinitive" construction

More precise dating is complicated by the fact that although the *for to +* infinitive construction went out of use in the standard language, it has been preserved in some dialects (Orton — Sanderson — Widdowson (eds.) 1978), especially in northern British English dialects, in Belfast English (see Corrigan 1993; Davis 1984 and the references therein), e.g.:

(4)   For to *do that would be foolish* (Davis 1984: 56).

(5)   ... *up at the roof was this basket thing made of sally rods and plastered with cow-dung* for to *draw off the smoke.*
        (Bliss 1979, quoted in Corrigan 1993),

as well as in some North American dialects – e.g. in Ozark and Ottawa Valley English (Irish-based in some of the rural areas):

(6) a. *Mary wants* for to *leave.*
    b. *Mary would be happy* for to *get some bagels.*
    c. *This is a machine* for to *cut hair with.*
        (For details and comments, see Carroll 1993 and the references therein).

It can therefore be claimed that in English as a whole (or as a composite notion) the *for to* infinitive has been preserved, albeit its frequency and distribution have become considerably restricted.

---

the overall meaning of the construction is clearly purposive. Here are some typical examples (excerpted from the Everyman Library edition, 1907, London — Toronto: Dent).

(a) ... *and here's Ham come* fur to *take you home* (p. 413).
(b) *He got me them papers as I wanted* for to *carry me through* ... (p. 547).
(c) *They would set me down at their cottage doors, and give me what-not* for to *eat and drink, and show me wheer to sleep* (p. 547).
(d) *He's never been heerd* fur to *complain* (p. 550).
(e) *..., you know as I have been – wheer not –* for to *seek my dear niece* (p. 640).
(f) *..., and yet was hoping* fur to *see her without blame* (p. 692).

Interestingly, all such examples occur exclusively in the speech of Mr. Pegotty, who lived in the area of Yarmouth in East Norfolk. It can therefore be assumed that towards the middle of the nineteenth century the *for to*-infinitive was still fairly current in parts of East Anglia.

4. External influence hypotheses

4.1. The Scandinavian influence hypothesis

There exists the following evidence for Scandinavian influence:

First of all, the relatively early date of the first attested instance in English (1066 – see example (1) above) would point to Scandinavian, rather than to French influence (that is, if we accept the foreign influence hypotheses at all).

Secondly, practically the same construction – *for at* – occurs in Danish and in Swedish, as well as in northern and north-eastern Middle English dialects. *For till* occurred too and is still preserved in Scottish dialects (see e.g. Brunner 1962: 342). The Scandinavian form could evidently have been both borrowed and/or translated (see also Farkas 1994).

Thirdly, the *for to* + infinitive construction is preserved today mostly in the dialects that were subjected to Scandinavian influence during the late Old English and Early Middle English periods.

On the other hand, the fact that although spreading and even prevailing in some Middle English texts, the *for to* + infinitive construction after all did not become permanently established in Standard English would seem to support Thomason — Kaufman's (1988: 298) assertion that "in spite of the relatively large number of grammatical elements of Norse origin in Norsified ME, their effect on English structure was almost trivial".[4]

4.2. The French influence hypothesis

The following two main variants of this hypothesis could be envisaged.

Variant 1. According to some authors (e.g. Einenkel 1916: 14-17), only the ME *for* + bare infinitive, e.g. in:

(7) *þat he were mid heom ilome* For *teche heom of his wisdome* (ON 1766),

should be regarded as due to French influence.

Variant 2. Having once entered the language, from use with the bare infinitive only, *for* could later have spread to the *to*-infinitive thus providing the needed new emphasis of the purposive meaning. In such a case

---

[4] This is a highly controversial point which requires very thorough consideration.

this would be an instance of the receptor language borrowing an element needed by its own trend of development (this in terms of Vachek 1962 and some other linguists of the Prague Circle: for some details see also the survey in Danchev 1988).

Mätzner (1885: 58) points to parallel prepositional infinitives in some Romance languages – French *por, pour,* Italian *per,* Spanish *por* and especially to the Old Provencal and Old French *por a* forms. Further on Mätzner states that "Die Einwirkung des Afr. scheint hier unleugbar", at the same time referring also to the impact of Scandinavian (Mätzner 1885: 59).

The probability of such a scenario is supported by quite a few other cases of French prepositional usage transfer to Middle English, e.g. in phrases such as *by sea and by land* (from *par mer et par terre*) replacing the Old English *on* in *ge on sae ge on land,* numerous phrases of "the city of London" type (from French *de* – quoted in Thomason – Kaufman 1988: 126), *son to the king* (Fr. *fils au roi*) and a number of others (for more examples see Orr 1962; Prins 1948; Workman 1940).

At this point it may be recalled that in the various scales of accessibility to borrowing (compared in Lefebvre 1984: 12) prepositions come after nouns; adjectives come after nouns, verbs, adjectives and adverbs. Contact-induced (including translation) change involving prepositions must therefore be taken as an indication of relatively strong foreign influence (cf. also Danchev 1988).

On the other hand, however, the early attestation date of *for to* (see example (1) above) would seem to militate against the primacy of strong French influence.

4.3. The bilateral influence hypothesis

As a third hypothesis one could evidently envisage the possibility of combined Scandinavian and French influence. Scandinavian influence would account for the appearance of *for to* infinitives in Early Old English (example (1) above), whereas the chronologically later French influence (partly also through translations of numerous texts from French into English) would explain the marked spread of *for to* forms to the southern dialects of Middle English, reflected in the writings of Chaucer, Caxton and other authors. Although not stated explicitly, this seems to be the position of Mätzner (1885: 57-59).

5. Internal development hypotheses

5.1. Brief survey of existing opinions

As is usually the case in mainstream historical linguistics, the prevailing opinion in the literature is in terms of an internally motivated change (for general surveys of this tendency see Thomason — Kaufman 1988; Danchev 1988). In the case of *for to* this view has been summed up very clearly by Fischer (1988: 73):

> To account for this introduction of *for*, the traditional notion still provides the best explanation: namely that it was introduced in ME in order to emphasise the idea of purpose, earlier (in OE) expressed by the *to*-infinitive (in contrast to the plain infinitive). All the earliest instances of the *for to* infinitive clearly express purpose (cf. Mustanoja, 1960: 514). In the course of the ME period the meaning of *for* weakened (just as *to* had weakened and needed strengthening by *for*) and the result was that the *to*- and the *for to* infinitives came to be used almost indiscriminately and finally became interchangeable.

According to Fischer (1988: 85, note 11) this is seen in the following two examples from Chaucer:

(8)   *This prison caused me nat for to crye* (28).

(9)   *That caused hym to sette hymself afyre* (83).

Further on (1988: 74) Fischer points out that after the weakening of *for* the strengthening of the purposive meaning was taken over by *(so) as to* + infinitive and *in order to* + infinitive constructions.

Interestingly, the possibility of external influence is admitted by earlier authors with a basically philological orientation, whereas practically all more recent authors work on the assumption of an internal development. The same division has also been noted in connection with other developments.

5.2. Paradigmatic and syntactic parameters of *for to* infinitives

A search for internally motivated change factors would obviously involve an investigation of all the paradigmatic and syntagmatic aspects of the ME *for to* + infinitive construction. As it happens, these aspects have already been researched by a number of authors.

Thus, for example, Svartvik – Quirk (1970: 399) offer a classification in terms of functions: noun modification, adjective modification etc. They indicate also that "there appear to be two major areas where the *for to* infinitive is preferred, that is as adjunct and, with a subject, as complement of transitive verbs". Further on they disagree with Kerkhof's (1966: 49) claim that the two marked (*to* and *for to*) infinitives are used "promiscuously" (Svartvik – Quirk 1970: 410).

As has been pointed out (see e.g. Kaartinen – Mustanoja 1958: 179), the choice of a bare, a *to-*, or a *for to*-infinitive would often also depend on purely metrical factors, especially in poetry.

According to Svartvik – Quirk (1970: 398) "the *for to*-infinitive takes all the functions of the *to*-infinitive except complement of the equative verb *be*."

The fact that intervening elements could separate *for* and *to* suggests that Chaucer still perceived those two components of the doubly marked infinitive as independent language units.

Svartvik – Quirk (1970: 403-404) found that in Chaucer's English separation from the governing verb promoted the use of *(for) to* with infinitival complements.

On the other hand, however, numerous instances of fused *forto, vorto* etc. spellings, even in relatively early texts, e.g. in

(10)   for her nabbe ich man iseid bute forte *munegin mon or wummion* (*Helsinki Corpus, AW*, 164),

suggest that *for* had become weakened and was not perceived (at least by some sections of the speech community) as consisting of two independent language units any more.

Jack (1991) offers statistical data showing that the *for to* infinitives were much less frequent in Middle English than the *to* infinitives (with some exceptions such as AW). Further on Jack points out that most often the element intervening between *for* and *to* is either an object or an adverb, e.g. in

(11) a.   for *hereword* to *hauen* (TH 93/10)
     b.   mare for *worl[d]es scome þenne* for *heore sunne* to *beten* (LHA 27/ 33-34).

Therefore in examples of this type the original structure was *for* + [*to* + infinitive], not [*for* + *to*] + infinitive. In the course of time, however, *for to* evidently came to function as a single element (Jack 1991: 316).

While Svartvik — Quirk (1970: 403-404) thus found that in Chaucer's English separation from the governing verb promoted the use of *(for) to* with infinitival complements, Jack (1991: 325) found no evidence of this in Early Middle English prose. He also considers that

> whether there was any significant difference in force between adverbial infinitives preceded by *to* and those preceded by *for to* is difficult to judge. In principle it seems reasonable to assume that when *for to* first came to be used as an infinitival marker it was more emphatic than *to*; but the increasing use of *for to*, and its transition from use primarily in the expression of purpose to being an alternative marker found in most situations in which *to* was employed, must have eroded any distinction between the two. (Jack 1991: 326)

Like some other authors, Jack also notes that the infinitive with *for to* is common only as an adverbial adjunct, where it is found in 44% of instances (Jack 1991: 334). Of interest is Jack's statement (1991: 336) that "infinitival complements with *(for) to* may appear where there is some element of futurity, e.g. following such verbs as *hopen, munten,* and *sechen*". Finally, Jack (1991: 336) sees "the development of *for to* + infinitive as a re-enactment of the earlier development of *to* + infinitive", thus echoing Fischer (1988: 72) above.

Checking on the combinability of *for to* with various verbs we have not found any significant difference from the *to* infinitives (beside the ones already mentioned in the literature).

In her recent survey Fanego (1994: 191) sums up the state of research as follows:

> the choice of infinitive marking in ME seems to have been controlled by factors such as:
> a) the grammatical function of the clause, that is, whether this is a subject, an object, or other;
> b) metre;
> c) the separation between the infinitive and its governing verb;
> d) the fronting of an element within the infinitive clause; and
> e) semantic factors of various kinds.

It should be noted that (like the other authors mentioned in this section) Fanego does not envisage the possibility of foreign influence. Besides, Fanego's survey tends to gloss over certain differences between the authors considered by us (this, and the rather limited number of publications referred to in her study, may be accounted for by restrictions of space in the volume).

The interplay of various paradigmatic and syntagmatic factors underlies also generativist attempts to explain the change of rules. The weakness of such approaches lies in their failure to offer satisfactory explanations for the factors triggering the so called "rule reanalysis" at a given point of time and in a given variety of the language (for details see, e.g., Carroll 1983; Davis 1984; Lightfoot 1979 and some more recent endeavours[5]). On the other hand, however, some of the generativist studies have made explicit a number of relevant points.

Thus, e.g., worth noting are Davis's (1984) observations that the position of traditional grammarians, e.g. of Jespersen (1940: 212) who states that *"for to* ... originally expressed purpose, but was used extensively in ME as a simple equivalent of *to* with the infinitive", does not apply to Belfast English. In Belfast English there are a number of places where only *to* occurs and *for to* is impossible. For example, *for to* may not be used as an alternative to *to* after a raising predicate such as

(12)   **It seems* for to *be raining*.

Middle English has fewer constraints in the use of *for to*, but further research is needed along these lines.

Although the consideration of the various paradigmatic and syntagmatic parameters of Middle English *for to* is relevant above all for the internalist analyses, these findings must evidently be taken into account when examining the likelihood of external influence.

What both more traditional and generativist accounts have in common is the failure to answer the questions posed by the well-known actuation question: why did *for*-insertion occur at a given time and in a given dialect?

---

[5] We are grateful to Olga Fischer for drawing our attention to some recent work by Jarad. Unfortunately, however, we have been unable to obtain any of his publications.

6. Cross-language evidence

6.1. Areally connected languages

The arguments in favour of externally triggered causation will gain plausibility if the emergence of the *for to* infinitive in Middle English can be shown to be part of a global change pattern, observable in other contact situations involving a variety of languages.

In this particular case, the validity of the French-English transfer hypothesis would be strengthened if similar instances in other French-Germanic language contact situations could be found.

Thus, for instance, it could be predicted that similar forms should exist in the German dialects along the French-German border. This assumption was followed up and it turned out indeed that according to Bruch (1953: 145)

> Eine ähnliche Beeinflussung [from French – addition ours] der Syntax über die Sprachgrenze hinweg liegt vor im Wenkersatz 20: «zum Dreschen bestellt». Das Frz. gebraucht in diesem Fall «*pour*» (für) mit Infinitiv: das gesamte linksrheinische Gebiet von Kleve bis Basel meldet in mehr oder minder starker Streuung, besonders dicht an der Grenze:

(13)   für zu *dreschen*   'for to thresh'.

On the basis of these data one could also expect the occurrence of *for to* infinitives in other French-Germanic contact areas, for instance in Flemish, although we have not had access to any such data yet. Our assumption of the possible existence of such forms there is also bolstered by the fact that, like Luxembourg German, Flemish too has *go*-future forms borrowed from French (see Danchev – Kytö 1991 and the references therein).

A similar prediction could be made concerning German-Italian (the latter language having practically the same preposition – *per*) contact areas. A search through the literature revealed that such instances exist indeed in the German dialect spoken by Walser communities in North-West Italy (in the southern valleys of the Monte Rosa massif – Aosta Valley and Piedmont). The example reported in Ramat (1992: 323, in another connection) is:

(14)   *géeld en der buschò heiber nit khät* fer z *goa em wertschhus*
'money in the pocket have-we not had *for to* go to-the inn'.

The above-mentioned data would seem to warrant the tentative conclusion that when the Germanic infinitive expresses purpose it tends to use two prepositions.

Thus, some authors (e.g. Brunner 1962: 342; Einenkel 1916: 16) equate the German *um zu* with *for to* in examples like

(15)   *se kyng hit dide* for to *hauene sibbe* (ASC, 1127)
       'der König tat es *um* eine Familie *zu* haben'.

As a matter of fact, Shearin (1903: 17) notes that such an intensification of the preposition "is general in English and cognate languages", e.g. *um zu* in the later development of German and *om te* in later Netherlandish.

Going outside the Germanic languages, one cannot leave unmentioned the existence of the practically isomorphous structure in Irish Celtic, which according to Hickey (this volume) underlies Irish English sentences like

(16)   *He went to Cork* for to *buy a car.*
       Chuaigh go dti Corcaigh chun gluaistean a cheannaigh.

Hickey regards such examples as instances of "supportive transfer" from (Celtic) Irish.

All these data, to the best of our knowledge presented together like this for the first time in extensive *for to* literature, give rise to various inferences, but before trying to draw any conclusions let us also take a look at some learners' interlanguage data. At this stage we will just try to show that there seems to exist a global Romance-Germanic contact-induced change pattern resulting in the adoption by some Germanic languages and dialects of *for* as an equivalent of French *pour* and Italian *per* as an additional purposive marker of the infinitive. The Celtic influence evidence is clearly significant, too.

6.2. Second language acquisition (interlanguage) data

In this section we will turn first of all to instances of English language acquisition by learners who are native speakers of French.

We do not have any data yet about the interlanguages of French learners of English in France itself, but very relevant evidence was found in the interlanguages of French Canadian learners of English.

Zobl (1984: 205 – also referred to in Danchev 1990) reports examples such as:

(17) a. *He needed money for to get a clean shirt.*
     b. *He go for to meet the girl.*

In addition to such instances, Zobl also adduces cases of *for* plus the bare infinitive, which are even more straightforward instances of negative L1 transfer. Referring to the parallel development in Middle English (although he states erroneously (Zobl 1984: 206) that *for* occurred with infinitives "from the end of the 12th Century to the end of the 16th Century (approximately)"; see, however, section 3.), Zobl asserts that "the description of developmental continua in isolation from historical continua misses significant invariant constraints which hold for change processes. Moreover, in failing to take these constraints into account, explanations of developmental phenomena become biased in favour of specifics of the target language" (Zobl 1984: 204). In discussing Andersen's nativisation hypothesis further on Zobl claims that

> nativization optimalizes linguistic resources for performability, rendering them more regular and transparent in response to processing and storage requirements. The essence of optimalization is demarking, the operation of which involves
> 1. the selection and reinterpretation of semantically transparent, morphologically independent forms to serve functions for which the TL [target language] input provides grammaticized and semantically opaque forms;
> 2. the regularization of form-meaning relationships by exploiting language-universal semantic and functional affinities. (Zobl 1984: 204)

Concerning the spread of *for to* Zobl (1984: 206) considers that "as in Middle English, the spread of FOR is an instance of rule generalisation with system-internal motivation." In other words, he does not identify this change explicitly with language contact.

At any rate, although in some cases Zobl seems prone to overgeneralisations given the limited data (see Silva-Corvalán's comments in the same volume), his article contains some highly interesting and relevant points.

Worth noting above all is the quest for what could be described as a global pattern. Zobl's claim that *for* insertion amounts to "demarking" should be noted too. We will come back to those points further on.

In her comments on Zobl's paper Silva-Corvalán (1984: 220) notes that he "draws interesting parallels between L2 acquisition processes and historical language change", but further on, though admitting that "Zobl's proposals are thought-provoking and attractive," she finds that "some of the evidence presented ... seems inconclusive" (1984: 221-222). Silva-Corvalán winds up her comments by stating that "more research is needed that will focus on the acquisition of an L2 by speakers of typologically quite different languages" (1984: 222).

This is what we will proceed to do now.

Zobl mentions research on *for* insertion in the English speech production of Spanish learners. Clahsen *et al.* (1983: 371) report the following example in the speech production of a Spanish *Gastarbeiter* in Germany:

(18)   **um man musse was arbeite fur was zu essen.*[6]

Similar cases, e.g.

(19)   **I study English for to go abroad.*

as well as collocations with the bare infinitive, e.g.

(20)   **I go to school for study English.*

occur frequently in the interlanguages of Italian learners of English (oral communications of Dr. Adriana Mattei from the University of Naples and of Dr. Rinelli from the University of Brindisi).

For-insertion has been noted also in the interlanguages of Greek learners of English. In his book *Common mistakes in English* Fitikides (1963: 78) explicitly warns Greek learners of English against saying

(21)   **I came here for to learn English (Greek).*

Examples of this type are also frequent in the interlanguages of Bulgarian learners of English, especially during the beginner stages. The practically isomorphous Bulgarian construction is *za da*[7] + the Bulgarian equivalent of the infinitive (for a contrastive analysis of English *to* and Bulgarian *za da*, see Molhova 1968).

---

[6] We are grateful to those German colleagues at the Tulln Conference who pointed out to us that this example could also be given an alternative syntactic parsing.

[7] The *da* particle in Bulgarian is usually described as a conjunction, not a preposition. The Bulgarian functional equivalent of the ME *for to* infinitive is usually described as consisting of the preposition *za* (= English *for*), the conjunction *da* + a subordinate clause. This construction is quite well known to Balkan *Sprachbund* researchers.

There seems no doubt that *for* insertion occurs in a wide variety of learner interlanguages based on a number of different native languages. As a whole interlanguages are known to favour morphological transparency and less marked forms. This ties in with Zobl's claim that *for* insertion is an unmarking process and goes counter the isomorphism postulate of natural[8] syntax. We will follow up the implications of this in the following final section of this paper.

7. Discussion

The evidence for borrowing from Scandinavian and/or French is fairly strong. On the other hand, however, the need for additional emphasis is an internal development. Emphasis itself can be considered as a kind of expressivity, and, as has been pointed out before, the striving after expressivity could be regarded as one of the basic factors of internal development. This idea has been formulated by a variety of authors and in variety of ways.

To sum up, the following causation hypotheses could be advanced.
1. Scandinavian influence alone. This hypothesis is easy to motivate and support. The arguments in its favour have already been adduced (see subsection 4.1.).
2. French influence alone. At first sight this hypothesis seems less plausible than the previous one. On second thoughts, however, bearing in mind the strong linguistic impact of numerous translations from French, especially in the area of prepositional usage, the assumption of French influence cannot be dismissed lightly. Moreover, the cross-language and second language acquisition evidence strongly supports the French influence version (see subsection 4.2.).
3. Combined Scandinavian and French influence. This possibility has been envisaged too (see subsection 4.3.). It carries the advantage of being a compromise combining the two possibilities (1. and 2.) considered above and actually emerges as the most plausible hypothesis.
4. A "need" in the receptor language. The "emphasis need" referred to above was filled by a native element, but the model for the syntagmatic pattern it entered was provided by Scandinavian and/or French. This would then be a "classical", so to speak, example of the "weak constraints" borrowing hypothesis promoted by Jakobson, Vachek and other Prague

---

[8] As we have pointed out before, the very term "natural" is unsuitable, because it can be claimed that everything in the real world is "natural".

school linguists. The existence of a global change pattern of a general nature lends plausibility to such a proposition.
5. A purely internal development in terms of demarking, interlanguage optimalisation and universal grammar (see Zobl 1984). On the other hand, however, the very mention of "interlanguage" presupposes contact of some kind, so that Zobl's position on this particular aspect of *for*-insertion remains somewhat unclear.

Two interesting topics remain. The one is whether this development can be regarded as part of the overall "drift" towards analytic structure, and the second concerns the reasons for its retreat from Standard English.

It is easy to demonstrate that *for*-insertion, which can also be viewed as a kind of prepositional serialisation (on the analogy of the verbal serialisation of some creole languages[9] – cf. Seuren 1991), is a further step in the movement towards analyticity. As has been pointed out before (Danchev 1992), one can distinguish between formal and semantic analyticity. The former distinction is in terms of bound versus free morphemes and obviously applies to *for*-insertion. As regards semantic analyticity, this is revealed in the "meanings : forms" relationship. In Old English the purposive meaning was conveyed by just one form – the preposition *to*. Due to the semantic bleaching of *to* it became necessary to insert *for*, which changed the "meaning : form" ratio from 1 : 1 (the isomorphism of natural syntax) to 1 : 2. The latter state could be described as "analytic redundancy", various analogues of which have also been observed in learner interlanguages, e.g. in

(22)   *I will be go to London.

(The respective Bulgarian learners had not been exposed to the future progressive forms yet, so that example (22) cannot be interpreted as an attempt to say I *will be* going...).

This is the opposite strategy to the elimination of form words and auxiliaries as described in so called "foreigner talk" (Ferguson 1975). It seems to be prompted by the learners' realisation at a later L2 acquisition stage that form words are important after all and that their redundant use is a safer strategy than their omission.

---

[9] It is a tempting possibility to envisage prepositional serialisation alongside verbal serialisation. The latter has been referred to as "an expressive category" (Seuren 1991: 201) and if this is also true of prepositional serialisation it would confirm the arguments of all those scholars who have spoken of *for* as means of *to* reinforcement.

There remains the problem of accounting for the loss of the *for*-insertion (or the acquisition of the "*for-to* filter") rule. As a matter of fact, it is easier to explain the appearance of *for* than its subsequent disappearance from Standard English. If the twofold Scandinavian and French influence hypothesis is correct and there is no cogent reason for rejecting it, then one might well wonder why the combination of Scandinavian and French influence did not result in the permanent establishment of the *for*-insertion rule, which is more transparent from a morphological and semantic point of view.

The explanation advanced by some authors (e.g. Fischer 1988: 74; Lightfoot 1979), namely the emergence of the *for* + NP + *to* construction, first attested in 1385 (Fischer 1988: 72, confirming Lightfoot 1979), seems fairly plausible. It should also be noted, of course, as Fischer (1988: 85, note 7) does, that in dialects in which *for to* + infinitive is preserved, it coexists with *for* + N + *to* + infinitive, which weakens the above-mentioned assumption. Be that as it may, the fact remains that with very few exceptions throughout the Middle English period the *for to* infinitives were much less frequent than the *to* infinitives. Thus, for example, in the *Book of London English 1384-1425*, researched by Kaartinen — Mustanoja (1958), the ratio of *to* and *for to* is 5 : 1 (see also the data in Jack 1991).

Another possibility to explain the nearly general loss of *for* in the *for to* + infinitive construction is to regard it as a decreolisation feature, if we accept the prepositional serialisation version, of course. There is hardly any doubt that a number of creolisation-like features occurred in Middle English (for details see Danchev 1997), but whether we could speak of prepositional serialisation in the same way in which typical creole verbal serialisations are treated in the literature, remains a somewhat open question. In any case, it is a possibility to bear in mind.

In our survey we hope to have shown the heuristic usefulness of typological and second language acquisition data which reveal a global pattern of *for* (or its equivalents) insertion with an enhancing purposive function in various languages. From a cross-language point of view the double marking of purpose seems to be the preferred means, especially in the Germanic languages. The doubly marked structure evidently contributes to transparency and could thus also be regarded as an interlanguage feature. Its later loss, or, what we could also describe as the "*for to* loss puzzle", could perhaps be interpreted as a subsequent interlanguage complication (or decreolisation).

References

Bliss, Alan J.
1979 *Spoken English in Ireland, 1600-1740*. Dublin: Dolmen (quoted in Corrigan 1993).

Bruch, Robert
1953 *Grundlegung einer Geschichte des Luxemburgischen*. Luxemburg: Publications littéraires et scientifiques du Ministère de l'éducation nationale.

Brunner, Karl
1962 *Die englische Sprache: ihre geschichtliche Entwicklung*, vol. 2: *Die Flexionsformen und ihre Verwendung*. (2nd edition.) Tübingen: Niemeyer.

Carroll, Susanne
1983 "Remarks on FOR-TO infinitives", *Linguistic Analysis* 12: 415-451.

Chomsky, Noam — Howard Lasnik
1977 "Filters and control", *Linguistic Inquiry* 8: 425-504.

Clahsen, Harald — Jürgen Michael Meisel — Manfred Pienemann
1983 *Deutsch als Zweitsprache. Der Spracherwerb ausländischer Arbeiter*. (Tübinger Beiträge zur Linguistik A 3.) Tübingen: Narr.

Corrigan, Karen P.
1993 "Hyberno-English syntax: nature versus nurture in a creole context", *Newcastle and Durham Working Papers in Linguistics* 1: 95-131.

Danchev, Andrei
1988 "Language contact and language change", *Folia Linguistica* 22: 37-53.

1989 "On global patterns of interlingual influence", in: *Second symposium on English and Greek: description and/or comparison of the two languages*. Thessaloniki: Aristotle University, 34-46.

1990 "Some aspects of a language change typology", in: Werner Bahner — Joachim Schildt — Dieter Viehweger (eds.), *Proceedings of the fourteenth international congress of linguistics, Berlin/GDR, 10-15 August 1987*. Berlin: Akademie-Verlag, 1340-1342.

1991 "Language change typology and some aspects of the SVO development in English", in: Dieter Kastovsky (ed.), *Historical English syntax*. (Topics in English Linguistics 2.) Berlin — New York: Mouton de Gruyter, 103-124.

1992 "The evidence for analytic and synthetic developments in English", in: Matti Rissanen — Ossi Ihalainen — Terttu Nevalainen — Irma Taavitsainen (eds.), *History of Englishes: new methods and interpretations in historical linguistics*. (Topics in English Linguistics 10.) Berlin — New York: Mouton de Gruyter, 25-41.

1997 "The Middle English creolization hypothesis revisited", in: Jacek Fisiak (ed.), *Studies in Middle English linguistics*. (Trends in Linguistics: Studies and Monographs 103.) Berlin — New York: Mouton de Gruyter, 79-108.

Danchev, Andrei — Merja Kytö

1991 "The '*go*'-futures in English and French viewed as an areal feature", paper presented at the workshop on verbal periphrases at the tenth international conference on historical linguistics, 12-18 August 1991, Amsterdam.

Davis, Alison

1984 "Behind the *for-to* filter: *for-to* infinitives in Belfast English and the theory of government", *Sheffield Working Papers in Language and Linguistics* 1: 56-71.

Einenkel, Eugen

1916 *Geschichte der englischen Sprache. II. Grundriß der germanischen Philologie 6. Historische Syntax*. (3rd edition.) Strassburg: Trübner.

Fanego, Teresa

1994 "Infinitive marking in Early Modern English", in: Francisco Fernández — Miguel Fuster — Juan José Calvo (eds.), *English historical linguistics 1992: papers from the seventh international conference on English historical linguistics, Valencia, 22-26 September 1992*. (Amsterdam Studies in the Theory and History of Linguistic Science 4, 113.) Amsterdam — Philadelphia: Benjamins, 191-203.

Farkas, Judit

1994 "Prepositions in the four versions of *Cursor Mundi*", paper presented at the Conference on Language Contact in the History of English, Tulln, 17-21 July 1994.

Ferguson, Charles A.
1975   "Toward a characterization of English foreigner talk", *Anthropological Linguistics* 17: 1-14.
Fischer, Olga
1988   "The rise of the *for* NP *to* V construction: an explanation", in: Graham Nixon — John Honey (eds.), *An historic tongue: studies in English linguistics in memory of Barbara Strang.* London — New York: Routledge, 67-88.
Fitikides, T. J.
1963   *Common mistakes in English.* (5th edition.) London: Longman.
Jack, George
1991   "The infinitive in Early Middle English prose", *Neuphilologische Mitteilungen* 92: 311-341.
Jespersen, Otto
1940   *A Modern English grammar on historical principles, part 5: Syntax*, vol. 4. London: Allen and Unwin.
Kaartinen, Anja — Tauno F. Mustanoja
1958   "The use of the infinitive in *A book of London English* 1384-1425", *Neuphilologische Mitteilungen* 59: 179-192.
Kerkhof, Jelle
1966   *Studies in the language of Geoffrey Chaucer.* Leiden: Universitaire Pers Leiden.
Lefebvre, Claire
1984   "Grammaires en contact: définition et perspectives de recherche", *Revue québécoise de linguistique* 14: 11-47.
Lightfoot, David W.
1979   *Principles of diachronic syntax.* (Cambridge Studies in Linguistics 23.) Cambridge: Cambridge University Press.
Mätzner, Eduard
1880-1885   *Englische Grammatik.* 3 vols. (3rd edition.) Berlin.
Molhova, Žana
1968   "Anglijsko-bǎlgarski sintaktični paraleli" [English-Bulgarian syntactic parallels], in: Ivan Lekov — Emil Georgiev — Simeon Rusakiev — Nikolai Dilevski — Konstantiv Popov (eds.), *Slavistični izsledvanija* [Slavonic studies]. Sofia: Nauka i izkustvo, 319-340.
Mustanoja, Tauno F.
1960   *A Middle English syntax, I: Parts of speech.* Helsinki: Société néophilologique.

Orr, John
1962   *Old French and Modern English idiom.* Oxford: Blackwell.
Orton, Harold — S. F. Sanderson — John D. A. Widdowson (eds.)
1978   *The linguistic atlas of England.* London: Helm.
Prins, Anton Adriaan
1948   "French phrases in English", *Neophilologus* 32: 28-39, 73-82.
Ramat, Anna Giacalone
1992   "The pairing of structure and function in syntactic development", in: Marinel Gerritsen — Dieter Stein (eds.), *Internal and external factors in syntactic change.* (Trends in Linguistics: Studies and Monographs 61.) Berlin — New York: Mouton de Gruyter, 317-339.
Rutherford, William E. (ed.)
1984   *Language universals and second language acquisition.* (Studies in Language: Typological Studies in Language 5.) Amsterdam — Philadelphia: Benjamins.
Seuren, Pieter
1991   "The definition of serial verbs", in: Francis Byrne — Thom Huebner (eds.), *Development and structures of creole languages: essays in honor of Derek Bickerton.* (Creole Language Library 9.) Amsterdam — Philadelphia: Benjamins, 193-205.
Shearin, Hubert Gibson
1903   "The expression of purpose in Old English prose", *Yale Studies in English* 18: 1-149.
1909   "The expression of purpose in Old English poetry", *Anglia* 32: 235-252.
Silva-Corvalán, Carmen
1984   "Comments on the paper by Zobl", in: William E. Rutherford (ed.), 219-222.
Svartvik, Jan — Randolph Quirk
1970   "Types and uses of non-finite clause in Chaucer", *English Studies* 51: 393-411.
Thomason, Sarah Grey — Terrence Kaufman
1988   *Language contact, creolization, and genetic linguistics.* Berkeley: University of California Press.
Vachek, Josef
1962   "On the interplay of external and internal factors in the development of language", *Lingua* 11: 433-448.

Visser, Fredericus Theodorus
  1963-1973   *An historical syntax of the English language.* 3 vols. Leiden: Brill.
Workman, Samuel K.1940 *Fifteenth century translation as an influence on English prose.* New York: Octagon Books.
Zobl, Helmut
  1984   "Uniformity and source-language variation across developmental continua", in: William E. Rutherford (ed.), 185-218.

Text abbreviations

ASC - Anglo-Saxon Chronicle.
AW - Ancrene Wisse (this is quoted in Svartvik — Quirk (1970), but there is no indication of the edition).
HC - The Helsinki Corpus of English Texts, diachronic part. 1991. Helsinki: The Department of English, The University of Helsinki.
LHA - Lambeth Homilies, Group A, in: Richard Morris (ed.), 1867-1868. Old English homilies ... of the twelfth and thirteenth centuries, First Series, texts I-V and X-XIII. (EETS OS 29 and 34.) London (quoted in Jack 1991).
ON - The Owl and the Nightingale (quoted in Einenkel 1916, no indication of edition).
TH - Trinity Homilies, in: Richard Morris (ed.), 1873. *Old English homilies of the twelfth century,* Second Series, texts I-XXXIV. (EETS OS 53.) London (quoted in Jack 1991).

Hans-Jürgen Diller (Bochum)
Verbs of verbal communication in the English Renaissance: a lexical field under language contact

1.1. The vocabulary of a language is also the depository of the cultural and intellectual history of the language community concerned. Language contact is also a contact between cultures. The culinary, military and legal terminologies which the Normans imported into the English language are well known. But what they show is the linguistic and cultural heritage of English rather than the contacts between two language communities. Our knowledge of present-day English (PDE) vocabulary tells us that the two most extensive contacts in the history of English have been with French and Latin. But our knowledge of present-day English vocabulary alone would not tell us that at the peak time of contact French was very much a living language while Latin was a comparatively dead one. "Comparatively" is not meant facetiously, for there certainly are different degrees of deadness in languages. Medieval Latin was clearly less dead than the classical Latin of the humanists, which in turn was decidedly less dead than Latin is today.

If we want to study language contact in the context of cultural contact, then these differences are important. And the differences between degrees of deadness, while certainly more elusive, may be just as important as the difference between dead and living. These differences will become clearer when we consider the means by which a donor language community can export its language material to a recipient language community. Three factors which work for linguistic import and export are the power, the prestige and the material culture or technology of the donor language community. In contact with a living donor language all three factors can be brought to bear. A dead donor language can exert influence only through the prestige of its culture – of its intellectual culture to be precise.

1.2. Dead and living donor languages will also differ with respect to the class which acts as gatekeeper – gate opener rather – within the recipient language community. In contact with a living donor language all classes of people with some degree of physical and intellectual mobility can act as gatekeepers, such as sailors, soldiers, administrators, traders. All of these groups are capable of, and will often prefer, oral communication. In contact with a dead language, there is only one class of possible gatekeepers: the scholars. And the only channel through which linguistic material can be imported is the written one. This is a severe restriction, but it makes the entry route easier to trace: the time elapsing between first entry into the language and first attestation in a written record is likely to be shorter than in contact with a living donor language. And since our dictionaries usually represent the writings of (humanist) scholars rather faithfully, we can hope to reconstruct the trajectory of loans from Latin fairly accurately.

1.3. A great help in this enterprise is the *Chronological English dictionary* (*CED*: Finkenstaedt — Leisi — Wolff (eds.) 1970), which lists the lemmata of the *Shorter Oxford dictionary* in the order of their first citation. The *Chronological English dictionary* is now almost a quarter of a century old, but it still has advantages which even the CD-ROM version of the *OED* cannot match: it contains a code for the language of origin, and it indicates for each item whether it is listed in the *Advanced learner's dictionary* or in the *General service list* by Michael West (1953). It thus provides information on the currency and/or frequency of a word in twentieth-century English. The *CED* enables us therefore to make a distinction the importance of which has been emphasised by Coseriu (1974: 67ff.): that between innovation (*Neuerung*) and reception (*Übernahme*). Innovations are to Coseriu mere facts of speech, which happen all the time without receiving much notice. Receptions, on the other hand, are facts of the linguistic norm or even the linguistic system and are therefore of greater linguistic interest. The problem of the dictionary, and especially the historical dictionary, is that it does not, and in fact cannot and should not, systematically distinguish between facts of speech and facts of language. A nonce-use in Shakespeare is certainly worth recording, but it remains a nonce-use and thus a speech fact nevertheless. The combination of information from the *ShOD* and the *ALD* allows us to be more confident that we are dealing with language rather than with speech facts.

In a way, the *CED* is more useful than the *OED* on CD-ROM because it is old-fashioned: it is a book, and you can browse through it. Doing that, you will discover "facts" about the history of the English vocabulary which you had not thought of before. It can serve the historical linguist pretty much as an aerial photograph serves the archaeologist: it indicates promising ground and suggests the best place for the first dig.

2.1. Browsing through the *CED* I felt that Latinate words which had entered the language in the first half of the sixteenth century might repay further study. Counting only those lexical items which are also listed in the *ALD*, I obtained a list of 806. I therefore decided to limit my study to verbs of the first thirty years of the sixteenth century. The reasons for this decision are mainly practical: the first half of the sixteenth century, though not the peak period for imports from Latin, is a particularly interesting period for verbal borrowings. While in the fifteenth century verbs are underrepresented in the vocabulary of Latin origin, this ratio begins to be reversed with the sixteenth century, as a spot check of a number of selected years revealed. Two trends seem to be at work here, the second reinforcing the first: (1) owing to morphological changes, especially the loss of the infinitive ending, the language had become more hospitable to verbs of foreign origin; (2) this tendency encouraged writers whose preferred text types required a high proportion of verbs to draw on the Latin stock. This hypothesis would find support in the striking fact that most of the verbs end in -*ate*, originally a participial ending. With the facilitation of conversion which the loss of the infinitive ending brought about, this ending was available for the formation of Latinate verbs, too. Other Latinate verbs are derived from past participles as well, e.g. *evict* (1503), *obsess* (1503), *post* (1520), *flex* (1521), *deduct* (1524), etc. Verbs to be derived from Latin infinitives or finite verb forms soon become a distinct minority (ca. 20 % for 1500-1530, ca. 10 % for 1520-1530). Moreover, such verbs usually belong to a small family of simplexes, such as -*suade*, -*ride*, -*fer*. Taken together, these facts would suggest that the rate of borrowing is not only determined by extra-linguistic, cultural factors, but that the structure of the recipient language has to be taken into account, too.

Table 1. Share of verbs and Latinisms in neologisms

year	new words total	new Latinate words abs.	% of total	new verbs abs.	% of total	Latinate verbs abs.	% of verbs	% of Latin words
1450	439	109	24,8	85	19,4	14	16,5	12,8
1500	243	43	17,7	39	16	7	17,9	16,3
1548[a]	384	125	32,5	81	21,1	29	35,8	23,2
1550	183	51	27,9	24	13,1	8	33,3	15,7
1582[b]	135	32	23,7	28	20,7	8	28,6	25

[a] The year of Grafton's edition of Hall's *Chronicle*
[b] The year of the (Roman Catholic) Rheims translation of the New Testament

2.2. Browsing through the early sixteenth century in the *Chronological English Dictionary* revealed yet another striking fact which promised to be rewarding for the study of Latin-English language contact: a large number of the Latinate verbs appeared to belong to what could broadly be described as "verbs of verbal communication": a table of the Latinate verbs

Hans-Jürgen Diller

first attested in the first three decades of the sixteenth century bears this out:

Table 2. Latinate verbs entering English 1500-1530

1500	cable	2 vb		1528	alleviate	2 vb		
1500	implore	vb	S[a]	1528	appropriate	2 vb		
1500	postpone	vb		1528	confer	vb	S?	
1502	afforest	vb		1528	deduce	vb	S	
1502	inflate	2 vb		1528	precipitate	3 vb		
1502	unify	vb		1529	accumulate	2 vb		
1503	evict	vb	S	1529	circumscribe	vb		
1503	obsess	vb		1529	confute	vb	S	
1509	aggregate	2 vb		1529	correspond	vb		
1509	generate	2 vb		1529	exact	2 vb	S	
1510	investigate	vb		1529	extenuate	vb	S?	
1511	consolidate	2 vb		1529	extort	vb		
1513	alienate	2 vb		1529	insert	2 vb		
1513	denude	vb		1529	insinuate	vb		
1513	derogate	2 vb	S	1529	intoxicate	2 vb		
1513	describe	vb	S	1529	neglect	2 vb	M	
1513	dissuade	vb	S	1529	prognosticate	vb	S	
1513	explain	vb	S	1530	accelerate	2 vb		
1513	occur	vb		1530	aggravate	2 vb		
1513	persuade	vb	S	1530	arrogate	vb	S	
1513	sequestrate	vb		1530	attenuate	1 vb		
1514	contend	vb	S	1530	compact	5 vb		
1516	refute	vb	S	1530	complete	2 vb		
1520	post	8 vb		1530	conjugate	1 vb	S	
1521	flex	2 vb		1530	consummate	2 vb		
1523	attribute	3 vb	S(M)	1530	contract	3 vb	S	
1523	remunerate	vb		1530	contribute	vb		
1524	deduct	2 vb		1530	convince	vb	S	
1524	exonerate	vb	(S)	1530	copper	3 vb		
1525	annihilate	2 vb		1530	corroborate	2 vb	(S)	
1526	abrogate	2 vb	S	1530	decorate	2 vb		
1526	captivate	vb		1530	dedicate	2 vb	S	
1526	communicate	vb	S	1530	delegate	3 vb	S	
1526	contaminate	2 vb		1530	deride	vb		
1526	denigrate	vb	S	1530	digress	vb	S	
1526	diffuse	2 vb		1530	excogitate	vb	M	
1526	evacuate	vb		1530	fumigate	vb		
1526	excommunicate	1 vb	S	1530	habituate	2 vb		
1526	illustrate	vb		1530	lament	2 vb	S	
1526	infer	vb	M	1530	occasion	2 vb		
1526	protect	vb		1530	penetrate	vb		
1526	reiterate	2 vb		1530	perpetuate	2 vb		

Verbs of verbal communication in the English Renaissance

1526	suffocate	2 vb		1530	promulgate	vb	S
1526	suggest	vb	S	1530	prostitute	2 vb	
1527	consult	1 vb	S	1530	relate	2 vb	S
1527	ventilate	vb	S	1530	repine	2 vb	

[a] S = verb of speaking; (S) "speaking" meaning developed after first occurrence;
M = mental verb; (M) "mental" meaning developed after first occurrence.

Checking the meanings of these verbs in the *OED*, one is struck by the fact that many of them entered English with more than one meaning, often with a more "abstract" and a more "concrete" one. In most cases, the more concrete meaning did not survive. The following list, drawn from the previous one, illustrates this fact:

Table 3. Polysemous verbs of verbal communication[a]

derogate	1513:	to repeal in part (e.g. a law);
	1560:	to cause to seem inferior, to take away from a right, excellency, privilege.
dissuade	1513:	with abstract object: to d. a course of action;
	1534:	with pers. obj.: to d. sb. from ...
explain	1513:	to "unfold", make clear (e.g. Scripture)
	1549:	to smoothe out (e.g. e. their foreheads)
	1607:	to spread out flat (a material object)
	1736:	to make clear the cause of ...
persuade	~	dissuade.
refute	1513:	to refuse, reject;
	1545:	to prove (a person) to be in error;
	1597:	to disprove [an opinion] by argument, prove to be false.
contend	1514:	to strive earnestly;
	1529:	to strive in opposition, to engage in conflict or strife, to fight
	1530:	to strive in argument or debate, to argue.
exonerate	1524:	to relieve of a burden;
	1575:	to free from blame.
communicate	1529:	to impart information to sb.;
	1582:	to give a share of;
	1731:	of vessels etc.: to open into each other by a common channel;
	1959:	to convey one's thoughts, feelings successfully; Collins: to allow to be sensed.
denigrate	1526:	to sully, blacken, stain (the character of);
	1583:	to darken mentally;
	1623:	to make black (literally).

61

ventilate	1527:	to examine or investigate freely;
	1609:	to fan or winnow.
confer	1528:	to contribute;
	1545:	to hold conference.
deduce	1528:	to conduct (a process);
	1529:	to infer.
extenuate	1529:	to estimate or state at a low figure;
	1533:	to make slender or weak;
	1599:	to beat (metal) into thin plates;
	1741:	to plead partial excuses for.
prognosticate	1529:	to foretell;
	1533:	(of things) to betoken, indicate beforehand.
relate	1530:	to narrate;
	1590:	to bring back, restore (Spenser);
	1697:	to bring (a thing or person) in relation to another.
arrogate	1530:	to adopt (that which is proper to another) ("the bishops arrogate unto themselves something of the Pharisees' pride");
	1537:	to claim and assume as a right that to which one is not entitled.
convince	1530:	to overcome in argument, to confute;
	1632:	to bring [a person] to acknowledge the truth of.
corroborate	1530:	to confirm (a law);
	1533:	to invigorate;
	1706:	to strengthen (an opinion etc.) by arguments or agreeing statements.
digress	1530:	to deviate from the subject in discourse;
	1541:	to diverge from the right path.

[a] Verbs in bold print show "pre-borrowing polysemy", on which term see the following paragraph.

Most of these polysemies do not seem to be the result of internal development within the borrowing language, i.e. English, but of repeated borrowing from Latin. Such a claim is extremely hard to substantiate, but to understand the language contact situation it is useful to assume a simplified theoretical case. Language user A is the first to use a Latin lexeme L, in sense S_1. Language user B is the second to use the same lexeme, but in sense S_2. Usually it will not even be easy to know whether B knew A's use of L. And even if we could prove that he did know it, we can hardly prove that he was thinking of it when he used L in sense S_2. What we can know is whether or not the two senses of L appear already in Latin. If they do not, the problem's solution is easy: the polysemy has developed within the borrowing language, in our case in English. If they do appear in Latin, chances are that we are dealing with renewed borrowing, although there is no proof. There is however another important clue: senses usually develop in certain directions, for instance from concrete to abstract, or simply by an extension of possible semantically defined contexts. "Concrete to abstract" can perhaps be regarded as a subdivision of "extension of semantic contexts". Whenever the two (or more) senses in which the loan-word is used exist already in the donor language, I shall speak of "pre-borrowing polysemy", when one of the (at least) two senses does not exist in the donor language, I will speak of "post-borrowing polysemy". I should add that in all cases of post-borrowing polysemy I observed a sense development which was in agreement with the directions which I have sketched. Also, the sense that does not exist in Latin has always been the later one in English. When the two senses existed in Latin, the order of attestation in English could be either way: in the postulated direction or against it.

2.3. Lest this become too abstract, I illustrate what I have said so far with one example. It so happens that both kinds of polysemy can be illustrated with the lexeme *EXPLAIN*. In table 3 I have distinguished four different senses (which are of course only a selection from the *OED*). Three of these exist also in Latin: 1513, 1549, 1607. The fourth, 1736, apparently does not (Georges [1962]: *s.v. explanare*). The second and third senses can be called concrete: tangible, physical objects are "explained", i.e. made plain. The objects of the third sense, lumps of metal or clay for instance, are also extended, spread out, in the process. Both lexemes with their formatives *ex-* and *out* show that the enlarging of a surface is seen under a metaphor of spacial movement. From the concrete senses of *EXPLANARE* it is easy to reach *EXPLANARE* 'to explain' in the current sense: a text is written on a sheet of paper or vellum which may be folded or scrolled; by unfolding or unscrolling it we enlarge its surface, and we make it 'plain', flat. To this extent we are still in the realm of the concrete. By extension from the sheet of paper to the text written on it we can say that the text is now 'plain', there for everyone to see. Seeing is a common metaphor for understanding (Sweetser 1990: 38). If someone explains the text to us, makes it plain to us, we will understand it. We are now in the realm of the abstract. It is important to remember that this step was taken already in Latin antiquity. But in the sixteenth to seventeenth century we are in an age when man tries to understand not only texts but also the general course of events. The systematic, repeatable relations between events become an object of study no less than the systematic, repeatable relations between the elements of a text or sentence. To identify these systematic, repeatable relations is in both cases called *explain*. Thus, the new meaning is created by the extension of possible semantic contexts. The first user of the new sense need not even have been aware of his innovation. Thus, while between senses S_3 (1607) and S_1 (1513) the chronological order is against the "natural" direction, the development from S_1 to S_4 (1736) is in this direction. We can assume, therefore, that senses S_1 to S_3 were borrowed directly from Latin, without any development within English, whereas S_4 was an internal development from S_1 within English. The first three senses can thus be called pre-borrowing polysemy, whereas S_4 creates post-borrowing polysemy.

2.4. On my count, 14 out of the 19 verbs in Table 3 show cases of pre-borrowing polysemy: *DEROGATE, EXPLAIN, REFUTE, EXONERATE, COMMUNICATE* (1529/1582; 1731 and 1959 are more likely to be post-borrowing polysemies), *DENIGRATE, VENTILATE, CONFER, DEDUCE, EXTENUATE, RELATE, ARROGATE, CORROBORATE, DIGRESS*. In the case of *CONTEND* we should also assume pre-borrowing polysemy on the ground of our criteria, but the order in which the senses are attested in English would also permit sense development within English. It is striking that nearly all the verbs showing pre-borrowing polysemy belong with at least one of their senses to fields other than that of verbal communication.[1] The frequency of non-VC senses suggests that the contact with Latin was too broad to be confined to one conceptual field. It should be noted, however, that a search of the Early Modern English part of the *Helsinki Corpus* showed instances of non-VC senses only in the case of one verb, viz. *DIGRESS*, which occurred at least twice in the sense of 'deviating' (from duty or justice). The infrequency of non-VC senses may be simply due to the fact that the texts which do contain them were not excerpted in Helsinki, but it also suggests that such texts cannot have been very numerous.

3.1. From the point of view of language contact pre-borrowing polysemy is of course the much more interesting phenomenon. When, as in our case, the pre-borrowing polysemy extends beyond the boundaries of a lexical field, it seems reasonable to assume that the main factor responsible for borrowing was not the need of the recipient language for new terms and concepts, but the desire to adorn the language. And no matter how great our respect for the *OED*, we do not really believe that an early Tudor gold-

---

[1] Since the expressions "verbal communication sense" and "non-verbal communication" are are rather unwieldy and will have to be repeated a number of times, I will henceforth use the expressions "VC sense" and "non-VC sense".

smith would say to his apprentice "Explain that lump of gold", meaning 'flatten it out into a dish or whatever reasonably flat shape'. The discussion on the virtues of the English language which we find in the rhetoricians of the second half of the century is clear proof of a desire to adorn the language. The fact that, prior to the rhetorical debate, we find the described practice, suggests that the desire preceded the writing about it. This is an interesting cultural fact in itself. It is, however, a different thing to assess the extent of the polysemy. The evidence of the *Helsinki Corpus* suggests that the non-VC senses never penetrated very deep into the language. But the often very extensive documentation of the *OED*, frequently extending over many centuries, does not suggest this. Modern frequency-based dictionaries often do not even mention the non-VC senses. In the parlance of the *COBUILD* dictionary: they are not "real English". Whether they were "real English" in the sixteenth century is difficult to decide. It would be interesting to know whether pre-borrowing polysemy is confined to purely literate contacts as between Latin and a vernacular, or is also to be found in contact situations which involve two living languages.

3.2. While pre-borrowing polysemy is a strong indicator of the donor language's prestige, post-borrowing polysemy does not say anything about the nature of the contact. It suggests, however, that the lexeme which after borrowing developed new senses has become firmly entrenched in the recipient language, for only frequently used words are likely to develop new senses (Manczak 1985). If post-borrowing polysemy turns out to be frequent in an entire lexical field, we may assume that the lexical field in question and the cultural sphere represented by it are developing vigorously and that this development owes something to the culture of the donor language. This is no doubt true of the field of verbal communication, more particularly of certain specialised sub-fields, such as rhetoric and the art of reasoning. But these sub-fields, once started with the help of the donor language, take their independent developments.

4.1. A particularly interesting sub-class of verbs of verbal communication are those verbs which may be called "verbs of verbal influencing". Without claiming that I have isolated a well-defined semantic field, all these verbs have in common that they designate a process in which a person changes, by verbal means, the behaviour of another over whom he has no authority. In Late Middle English we find *AVISEN, MOVEN, CONSEILEN* and probably others. *MOVEN* and *CONSEILEN* in particular are used to render *PERSUADERE* and *SUGGERERE* in the Wycliffite translations of the Vulgate. The early sixteenth century adds

PERSUADE, CONVINCE, CONCILIATE and *SUGGEST*, with no clear or consistent distinctions of meaning. The new verbs share one semantic feature: the "goal" of the persuading is in the persuader's interest. This is a feature which they have in common with *MOVE*, ME *MOVEN*. While ME *CONSEILEN* can also be used to describe such situations, it is not limited to them. In Cruse's terminology we could say that the feature "goal of persuasion in Agent's interest" is "possible" but not "criterial" in the verb *CONSEILEN* (cf. Cruse 1986: 16), whereas it is criterial for the other verbs mentioned. The fact that (at least) four new verbs were introduced in a comparatively short time is thus not explicable in functionalist terms. The innovators were presumably driven by the wish to embellish the language.

4.2. If innovation (Coseriu's *Neuerung*) is motivated aesthetically rather than functionally, reception (*Aufnahme*) is more functional. This can best be shown by the various uses of *CONVINCE*. The verb appears in the senses of PDE 'to convict', 'to win over', 'to conquer' and 'to convince'. In sixteenth-century English you could convince a crime, a criminal, and also a debater. For the speaker of present-day English this is rather bewildering. We can see unity behind this variety only if we remember the meaning of Latin *CONVINCERE* 'to defeat utterly'. You can be utterly defeated in a law court and in a philosophical, theological or juridical debate. The leading metaphor is clearly the debate as a battle. Since the decisions of a law court may also be reached in a debate, the law court sense is not so different from the 'debate' senses. To convince a crime is of course not utterly to defeat a crime, but there are analogies like *PERSUADE HARMONY, DISSUADE REBELLION, CONCILIATE AMITY*. In all these instances the human object is suppressed, leaving that syntactic slot free for the "goal" or "purpose" of the action. Sixteenth-century English knew a freedom of syntactic contexts, as opposed to semantic contexts, which later English has lost.

To return to the modern sense of *CONVINCE*, we can define it as follows (largely on the basis of the *OED*): A has convinced B when B has acknowledged the validity of A's arguments. This sense, according to the *OED*, is first attested in 1632, 102 years after the word's first attestation in the language. Before 1632 a person was regarded as "convinced" when a panel of judges or *doctores* thought so; what he thought himself did not matter. When you said of a man in the sixteenth century that he could not be convinced, you did not mean that he was stubborn but that he was a skilful debater. Similar sense developments, always sense specialisations, could be shown for other verbs, such as *REFUTE* and *EXTENUATE*. The case of *EXPLAIN* has been discussed already.

4.3. The new sense which these verbs develop after borrowing reflect refinements in the art of logical reasoning. Perhaps we can also describe the development as one from rhetoric to logic. This is not to say that those who first introduced the verbs in a VC sense were not interested in logic, but it does mean that for the earlier senses of these verbs rhetorical effect was more criterial than logical accuracy. If in that sense the development was from rhetoric to logic, then the sense development of these verbs is closely connected with cultural developments, something we might call the culture of arguing or reasoning.

5. Until now I have written about the development of this culture as if it took place within one language community, and there are arguments, such as "natural sense development", which support such a view. But clearly the development which I sketched is not peculiar to Early Modern English. The features which appear to have become criterial for English *CONVINCE* and *EXPLAIN* are criterial for German *ÜBERZEUGEN* and *ERKLÄREN* as well, and presumably for French *CONVAINCRE* and *EXPLIQUER*, too. Thus there seems to be parallel development here. But is it really "parallel", if you press the metaphor? Parallels, after all, meet only in the infinite. And of course there were contacts between the intellectuals of those language communities in which we discover "parallel developments". Whether the sense developments in question were indeed parallel is impossible to decide at the moment. A careful classification of the contexts of verbs of similar meaning in some of the major languages of early modern Europe, including Latin, would be the first necessary step towards an answer. A second step would be the tracing of international intellectual contacts.

References

Coseriu, Eugenio
    1974   *Synchronie, Diachronie und Geschichte: das Problem des Sprachwandels.* (Internationale Bibliothek für allgemeine Linguistik 3.) Tübingen: Fink.
Cruse, David Alan
    1986   *Lexical semantics.* (Cambridge Textbooks in Linguistics.) Cambridge: Cambridge University Press.

DuCange, Charles
    1883-1887   *Latin dictionary.* Cambridge: Cambridge University Press.
Finkenstaedt, Thomas — Ernst Leisi — Dieter Wolff (eds.)
    1970   *A chronological English dictionary.* Heidelberg: Winter.
Forcellini, Egidio
    1858-1860   *Totius Latinitatis lexicon.* Padua.
Georges, Karl Ernst
    1913   *Ausführliches lateinisch-deutsches Handwörterbuch.* 2 vols.
    [1962]   [Reprinted Hannover: Hahn.]
Hornby, Albert Sydney — Edward Vivian Gatenby — H. Wakefield
    1963   *The advanced learner's dictionary of current English.* London: Oxford University Press.
Manczak, Witold
    1985   "Semantic development of borrowings", in: Jacek Fisiak (ed.), *Historical semantics: historical word-formation.* (Trends in Linguistics: Studies and Monographs 29.) Berlin: de Gruyter, 367-375.
Onions, Charles T. (ed.)
    1959   *The shorter Oxford dictionary on historical principles.* (3rd edition.) Oxford: Clarendon.
Sweetser, Eve E.
    1990   *From etymology to pragmatics: metaphorical and cultural aspects of semantic structure.* (Cambridge Studies in Linguistics 54.) Cambridge: Cambridge University Press.
Traugott, Elizabeth Closs
    1985   "On regularity in semantic change", *Journal of Literary Semantics* 14: 155-173.
West, Michael
    1953   *A general service list of English words.* London: Longman.

Richard Dury (Bergamo)
The history of the English language in the context of the history of the European languages

1. Introduction

The following contribution takes the form of a proposal for a type of English language history that will be sensitive not only to distinctive differences but also to continuing and interesting similarities with related and neighbouring languages, due in part to contact between languages and cultures, and in part to shared developments that have occurred independently.

2. Parallel developments in the European languages

Similarities between languages have traditionally been studied mainly as a means to historical reconstruction of unattested proto-forms of parent languages. History written in the normal direction of "time's arrow", on the other hand, has traditionally presumed regular divergence of related languages: "changes in the same linguistic structure when they occur independently, as through geographical isolation, always lead to different total end results" (Scott – Erickson 1968: 16-17). We may recognise a natural tendency of historical writing to emphasise divergence (dialectalisation, split, fragmentation, etc.), which has the advantage of creating a clear – even dramatic – narrative sequence and, in a narrative of a single language history, contributes to a sense of distinctive linguistic and national identity.

Languages have been traditionally seen as individual, homogenous and isolated: the *Stammbaum* encourages this view and most pure linguistics starts from this abstraction. There has been little in the way of a complementary study of persistent similarities and parallel losses and innovations as part of a historical narrative. We may, however, see some tendencies in this direction.

Perhaps the first real sign of an interest in common developments in different languages comes from Antoine Meillet in a short essay (1918 [1965]) entitled "Convergence des développements linguistiques", in which he talks of "le parallélisme des développements indépendants" (Meillet 1918 [1965]: 63) and "innovations ... orientées dans une même direction" (Meillet 1918 [1965]: 65). He sees a general tendency of all Indo-European languages to evolve from inflected to uninflected words, and it is "l'action constante de cette tendence" that explains the parallel development of these languages (Meillet 1918 [1965]: 68).

A few years later, Edward Sapir introduces a fascinating metaphor to this area of study when, at the beginning of Chapter 8 of *Language* (Sapir 1921), he uses the term "drift" in the sense of parallel developments in separated but related languages or dialects.[1] He remarks that "[t]he momentum of the more fundamental, the pre-dialectal drift is often such that languages long disconnected will pass through the same or strikingly similar phases" (Sapir 1921: 170). He gives the example of Proto-Germanic *musiz*, and its parallel development in English and German, noting that the Old English *i*-umlaut change of the vowel in the plural form (*mys*) precedes the similar form of Middle High German (*müsi*) by three or more centuries. He concludes that there was clearly some general tendency or group of tendencies operational in Proto-Germanic that then "pushed" later English and German along parallel lines of development.

Examining all the changes in the two sets of words *foot, feet* and *mouse, mice* he summarises the shared changes of the two languages: (1) weakening of non-stressed syllables, (2) change of stem-vowel under the influence of the following vowel (*i*-umlaut), (3) more close articulation of half-close long vowels (Proto-Germanic *fot* vs. Early Modern English *fut* and Early Modern High German *fus*), (4) diphthongisation of close vowels (OE *mus* > ModE *mouse*; OHG *mus* > ModHG *Maus*), adding that these parallel developments cannot be accidental: they are rooted in a "pre-dialectal drift" (i.e. a change that began to affect the common ancestor language before English and German became two separate dialects).

This meaning of "drift" was then taken up by later linguists. Zellig S. Harris (1951, quoted in Malkiel 1981: 549) suggests that such changes not due to diffusion in genetically related languages "along parallel lines of development" may derive from the structure of the parent language which "may have contained certain imbalances and irregularities, or may have otherwise favored the occurrence of certain changes rather than others".

---

[1] Although in his main chapter devoted to "drift" (chapter 7) Sapir uses the term in the sense of "divergent drift", it is the use of the term in chapter 8 which has been taken up by other linguists. Malkiel's attempt to discourage this usage, in his admirable essay tracing the history of the term (1981: 566), has been of no avail and typologists, discussing equivalent changes in several languages, talk freely of "typological drift" and "syntactical drift".

Here is an attempt to explain this mysterious phenomenon in a way that will be taken up by typologists.

3. Linguistic typology

All these thoughts around Sapir's idea of "drift" take a new direction when Greenberg in 1963 suggests interesting correlations between the ordering of various morpho-syntactical elements found in even unrelated languages, which may thereby be seen as being of the same linguistic "type".

Later contributions to modern typological linguistics have added a diachronic dimension – partly, perhaps, as a way of explaining apparent exceptions: the language is seen in the process of transition from one type to another and elements that conflict with the dominant typological patterns are seen as open to pressure to conform. This introduces the idea of "typological drift" for a single language and "parallel typological drift" for groups of languages.

The possibility of typological convergence caused by language contact has had to be faced by typologists. Markey (1978) and Ramat (1986) refer to the "problem" of whether identical changes in related subgroups in a contact situation result from contact phenomena, or from parallel typological/genealogical drift. To some extent it may be possible to distinguish between linguistic features that tend to spread by contact and those that are resistant to such diffusion (and so may be more genetically stable within the same system). Nichols (1995, quoted in Ramat 1998: 229-230), for example, concludes that word-order patterns are liable to diffuse geographically while head/dependent marking is resistant to borrowing. It is certainly also possible that both factors may act together (Bechert 1990: 139; Hoenigswald 1990: 444-445; Ureland 1986: 59).

4. Areal linguistics and areal typology

Areal linguistics typically seeks for explanations of parallel developments in the same geographic area not in shared (genetic/typological) characteristics, but in contact and diffusion: "all areal linguistic phenomena involve diffusion" (Campbell 1985: 26; cf. also Simpson 1994: 212). Note that the linguistic features "spread" in a process of "diffusion", while the language systems are said to "converge" – two different perspectives on the same evolution.

Studies aimed at a synchronic typological description of languages in the same geographical area belong to a new subdivision of language study, "areal typology". This branch of studies is similar to "areal linguistics" (discussed below), except that it has a more exclusive interest in synchronic structures and the identification of perhaps deeper levels of coherence binding together bundles of shared linguistic features. It is concerned with "the diffusion of ... structural schemes which are relevant from a typological point of view" (Ramat 1998: 228).[2] Hence it too is necessarily interested in contact-induced diffusion.

The study of "loan-words" has a long history and has an established place in historical linguistics and histories of the various European languages. Also based on a supposition of contact and interference (when two languages are used by the same persons) is Schmidt's Wave model of linguistic change (1872), traditionally applied to phonological and morphological diffusions among related dialects of the same language, or to explaining problems in the construction of a simple Indo-European family tree (as in Bloomfield (1933: 316, fig. 3): "Some overlapping features of special resemblance among the Indo-European languages, conflicting with the family-tree diagram").

Sets of non-inherited similarities of quite separate languages (rather than related dialects) spoken in geographically contiguous areas have occupied more recent linguists. Where geographically adjacent languages share a "significant" (but undefined) number of non-inherited features ("a bundle of common features" (Ramat 1998: 228), "a trait complex" (van der Auwera 1998: 259)), they are said to belong to a *Sprachbund* (Trubetzkoy 1930)[3], "convergence area" (Weinreich 1954), or "linguistic area" (Emeneau 1980).

Although such a linguistic area was soon defined for the Balkan languages, little work has been done until recently on the inter-relations of the languages of Western Europe, partly because "the leading scholars in Germanic, Romance, and English studies did not consider the problem of establishing such a typology of linguistic areas to be central to their

---

[2] Examples of work in this branch of studies are the "EUROTYP Project", which has now begun publishing the first of its nine monographic volumes (Siewierska (ed.) 1997; Feuillet (ed.) 1997; and van der Auwera — Ó Baoil (eds.) 1998), and a recent special issue of *Language Sciences* (1998, vol. 20) edited by Paolo Ramat.
[3] Trubetzkoy's formulation was made at a conference in 1928; in the following year Franz Boas (1929, quoted in Ramat 1998: 227) refers to the "striking similarities" of "neighboring languages" probably due to "the diffusion of grammatical processes over contiguous areas".

research" (Ureland 1990: 475), and because "most of the comparative linguistic work on European languages" has been reconstruction work "focused on the historical and genetic relationships between these languages" (Kuteva 1998: 289). We should not forget, however, Benjamin Lee Whorf's concept (1941 [1956]) of "Standard Average European" (SAE), the pervasive basic similarity of all the European languages deriving from similar mental categories expressed in language, in the similarity of their basic morphosyntactic categories.

Lewey (1942) talks of an "Atlantic linguistic area" of morphologically isolating languages (Spanish, French, Italian, English, and North Germanic languages, but not German, which he classifies in a group with Hungarian as an inflected language). Becker (1948) sees the whole European language area as a *Sprachbund* because of common external factors: similar developments of standards and movements of language awareness, the common influence of prestige languages, Latin and French, loans of lexical items and rules of word building. Décsy (1973) groups together German, French, English, Italian and Russian (like Becker, mainly on historical rather than linguistic grounds).

Other linguists who have accepted the reality of some areal grouping of European languages include Bechert – Bernini – Buridant ((eds.) 1990: ix), who talk of Western European languages as representing "a particular group – a sort of "Sprachbund""; Simpson (1994: 212) denies the name "*Sprachbund*" to English and contiguous languages but sees them as a grouping with "areal similarities"; B. J. Blake, however, in the same publication (1994: 1954) says without qualification that Western Europe is a language area. Now Haspelmath (1998: 271) concludes that "a sizeable body of new comparative research on syntactic features of European languages supports the thesis that the core European languages show a fairly high degree of structural similarity" and refers to "the SAE *Sprachbund*". In the same publication Kuteva (1998: 289) refers to "a European Sprachbund" formed of languages where a long period of interaction has led to "a relatively high degree of linguistic uniformity with regard to a number of important aspects of language structure".

5. The linguistic area of Europe and language history

While synchronic areal-typological studies of European languages are now quite numerous, there have been very few detailed diachronically focused studies of the European language area: little development of "a history of the European languages" or a "historical areal linguistics of Europe".

Bechert (1988) admits to this lack of research when presenting his maps of the distribution of key morphosyntactical features in present-day European languages, noting "a continuity of transitions" and "large-scale [East-West] directionality in distribution" which he sees as "due to historical processes that are yet to be analysed" (Bechert 1988: 24).

Hock, in his handbook of historical linguistics (1986: § 16.3.6. "Europe – sprachbund as dialect continuum") puts down a foundation stone for historical areal linguistics when he says that "prior to the development of the notion of the monolingual nation-state, much of medieval and early modern Europe was a convergence area" (Hock 1986: 505). Taking as his explanatory model "[diffusion] through a dialect-like chain of bilingualism" (Hock 1986: 510), he mentions as "one development which swept virtually the whole area" the cliticisation of the auxiliary, its movement into clause-second position and then the change to SVO as the unmarked word-order of most of the Indo-European languages of Europe – a change that also spread to the Uralic languages of Europe and the languages of North Africa. Like Bechert he sees an interesting geographical distribution of European SVO word-order combined with prepositions (vs. postpositions) and N + Gen (vs. Gen + N).

A second European innovation noted and mapped by Hock is the system of contrasting definite and indefinite articles in Greek, Romance and Germanic languages; and a third is the spread of discontinuous and then postverbal sentence negation, apparently starting in Old Norse and spreading through German, French and English. Two other developments do not affect English: the innovation of a verb-based system of pronominal clitics in Romance and Balkan languages; and the replacement of the simple past by the present perfect in colloquial French, Romantsch, southern German and northern Italian. Cumulatively, these developments lead him to talk of a "European sprachbund" (if only in inverted commas, Hock 1986: 507).

The above-mentioned development of discontinuous and then postverbal sentence negation in the European languages has been studied extensively by Bernini – Ramat (1992) and Ramat – Bernini (1990). They present the evolution of European sentence negation in historical times from early preverbal negative morphemes (present in the majority of world languages), to (first) discontinuous morphemes and (then) postverbal morphemes. They see this development as a probable example of areal diffusion of a structural pattern over time (Bernini — Ramat 1992: 67).

Hock's idea of much of medieval and early modern Europe as a convergence area is taken up by Emanuele Banfi, general editor of a series of volumes called "Lingue d'Europa", published from 1993, which hopes to emphasise (at least in the programme of intentions) the development of the European languages "in a substantially unitary direction, guided ... by a constant interaction ... with Greco-Roman civilization, and its immediate successor, Christianity" (Banfi (ed.) 1993: 2). In the first volume of the series Marcello Meli (Banfi (ed.) 1993: 122) also emphasises the importance of contact between European languages: for the Germanic languages in the Middle Ages, he says, "it is difficult to identify features that distinguish one language from the other without taking account of the more-or-less marked influence of one language on another".

The parallel developments that these linguists have noted in different European languages may take the form of (1) shared retentions,[4] or (2) shared innovations resulting in (2.a) common abandonment of archaic features or (2.b) common "novel" features.

These shared innovations may be the result of (1) similar (genetic or typological) structure leading to similar evolution; (2) contact influence of one language on the other (especially during the period of migration of the Germanic peoples in late Antiquity and the early Middle Ages); (3) shared contact influence of a third language (a pre-Indo-European language; Latin, French[5]); (4) other shared cultural influences (a common European culture from the Middle Ages onwards); or combinations of any of the above.[6] Where a structural similarity is found in many languages, then, we

---

[4] An example of a shared retention is the persistence of the correlative/equitive "diptych structure" with explicit markers of parameter and standard (as in Lat. *tam ... quam*) in the light of the rarity of this structure in non-European languages (Haspelmath — Buchholz 1998, quoted in Ramat 1998: 231-232). Ramat sees this as an important feature of the Standard Average European *Sprachbund* (though, as it is a shared feature not spread by contact, he is clearly using the term *"Sprachbund"* in a new way, as the product not only of contact but also of genetic and typological factors).

[5] The common European written culture and the common use and imitation of Latin is Whorf's main explanation for Standard Average European. French has also played a significant superlect role: it was a learnt cultural language all over Europe c.1200-1400, reflected in the large number of loans: c.3000 French words being first attested in English in the fourteenth century (Baugh — Cable 1951 [1990]: § 133, Jespersen 1905 [1990]: § 95), and c.2000 French loans into German in the same century (von Polenz 1978: 53).

[6] Haspelmath's five possibilities for the origin of the significant features shared by the Western European languages (1998: 272) are basically the same as the above, except that

may have a "banal" or "natural" innovation, which will still, however, be of interest where there is a concentration of such changes or a coincidence in time. Nevertheless, features found rarely in world languages will have greater weight as defining traits of a *Sprachbund*.

The importance of language contact seems to be borne out by the way that the maps of Greenbergian features presented by Hock and Bechert are remarkably similar to dialect maps, suggesting similar processes of diffusion, as shown in Table 1.

Table 1. Stylised map of European language areas showing a basic west-east ordering of Greenbergian word-order features (based on Hock 1986 and Bechert 1988)

Celtic languages	"central area", "SAE"	semi-relic area: Dutch and German	North Germanic languages	Finnish, Estonian	Uralic (except Balto-Finn. and Hungarian), Turkish
VSO prepositions	SVO prepositions	SVO/SOV prepositions	SVO prepositions	SVO post-positions	SOV post-positions
N + Gen	N + Gen	N + Gen	Gen + N	Gen + N	Gen + N

Recent historical work (Clanchy 1987) has emphasised the multilingual nature of medieval English (and European) society and the frequency and apparent naturalness with which speakers switched from one language to another. It is now easier to imagine a population of bilinguals in the post-migratory centuries, concentrated in certain areas of general cultural innovation, imitated because of their prestige and power (partly derived from their command of language), themselves imitators of translation-influenced written language and of the shared Latin superlect. These speakers could have functioned as the agents of many of the parallel linguistic developments that have been noted. I suggest, however, that the interesting parallel developments of the Western European languages[7] probably involve typological drift, language contact and the influence of a certain unity of cultural history.

---

he separates my third point: (1) retention of proto-Indo-European structures, (2) influence of a pre-Indo-European language, (3) contacts during the migratory and post-migratory period, (4) Latin and shared Latin culture, (5) the common European culture of modern times.
[7] The same languages, of course, have also simultaneously diverged in other areas: "linguistic change is characterized ... by a constant and subtle interplay of divergence and convergence" (Birnbaum 1985: 3).

Haspelmath, it is true, excludes superlect influence of Latin "because most SAE features were absent from Latin" (1998: 285). This would seem to refer to Classical Latin, however, while the very different Vulgar Latin and Medieval Latin in many of their varieties were much closer to the structures of the evolving vernaculars and so may have acted as a vector for further encouraging and spreading of such change. It is clear that more work on Medieval Latin is needed from this point of view. Joseph (1987: 87, note 3), for example, speculates that "At some point Spoken Latin must have developed a definite article, since all of the modern Romance languages have one. Except for a few dialects ... the source was ... [the] demonstrative *ille*".

Haspelmath also excludes as significant contributors to Standard Average European (1) retention of Indo-European features (since many common features of Standard Average European evolve late, in the first millennium "before our eyes"), (2) pre-Indo-European substratum influence (for the same reason), and (3) the common European culture of modern times (since the time depth is not sufficient). His explanatory framework for Standard Average European convergence is therefore "the great migrations at the transition between antiquity and the Middle Ages" when "language contact must have been particularly intensive and effective". This is the time frame for the evolution of the *have*-perfect, the analytic participial passive, "anti-causative" word derivation, negation patterns after negative indefinite pronouns, "nominal experiencers", and verb fronting in yes/no questions. My suggestion is that the fluid linguistic situation of partial multilingualism maintained by an elite for some centuries after the end of the migratory period may also have contributed to convergence. Kuteva (1998: 309) also refers to a clearly wider time frame than that allowed by Haspelmath when he talks of the "geographical closeness of the discourse communities in Europe and the massive linguistic and cultural communication between these communities".

6. English, a European linguistic area and histories of the English language

Despite the frequent noting of parallel developments in the European languages, histories of the English language have generally ignored them. Despite the possibility that "Old English, Old Frisian, Old Saxon and (probably) Old Franconian were so similar that they were probably to a large extent mutually comprehensible", "traditional scholarship ... has commonly treated these languages in terms of the detailed (largely phonological) differences between them, taking the similarities for granted" (Milroy 1996: 175-176).

The consideration of similarities and parallel developments in other European languages is surely relevant to the search for causes of change. The disappearance from normal use of Dutch *du* in the same century as the disappearance of English *thou*, for example, could be used to test hypotheses concerning association of the latter case with "the rise of the middle class and everyone wanting the "better" pronoun *you*" (D. Stein 1992: 142). In cases of imperfect convergence, it may be supposed that the factors present in the system and context of one language were not present in the other (D. Stein 1992: 132).

The value of such an approach is acknowledged by Görlach (1993: 157-159) in his review of Leuvensteijn – Berns (1992) (a comparativistic collection of papers on dialect and standard language in the English, Dutch and German language areas). Those papers on Dutch and German, he says, are interesting for "the light they throw on developments and conditions identical or contrasting with English".

The noting of parallel developments can also prevent "insular" explanations of change. Knowles (1997: 42), observing that the loss of inflections in English is often explained by the contact situation in the Danelaw, adds that "[t]his does not explain why Danish itself lost most of these endings too. Nor does it explain why nouns, adjectives and articles lost their endings in the Latin of different parts of the Roman empire after the Germanic invasions. This is a widespread phenomenon to be examined on a European scale, and not a local problem of the north of England".

The first recognition of parallel developments in a "history of the English language" that I have found[8] is by Lindelöf, who in *Elements of the history of the English language* (1895) sees similarities of the Germanic languages as the result of the Wave Theory and remarks that "English and German offer many analogies in the course of their development" (§ 15-17). Huchon (1923: xii) makes a typical passing mention that English formed periphrastic tenses using the verbs *have, shall* and *will* "par curieux accord avec les langues scandinaves".

---

[8] Based on a study of all the (mainly one-volume) comprehensive accounts in the British Library.

The first example of a history of the English language that takes extended notice of the parallel changes in other European languages is an interesting "Übersicht über die Gemeinsamkeit im Leben der deutschen und der englischen Sprache" which concludes the 1949 *Geschichte der englischen Sprache* by Wolfgang Jungandreas. He begins by remarking that "[i]nternal development shows (and the same goes for the Scandinavian languages) how the common Proto- and West-Germanic base has often led to the same changes and forms", and goes on to list some of these shared changes, together with external "cultural and linguistic influences ... of general European significance" that have shaped both languages (Jungandreas 1949: § 482). A selection of the table translated into English is given in table 2.

Table 2. Jungandreas's list of changes shared by English and German (excluding external cultural and linguistic influences)

ENGLISH, century		GERMAN, century
8	Addition of *t* to second person singular present: OE *thu bindes(t)*, OHG *du bindist*	9/10
9-11	Weakening of unstressed vowels	10/11
OE	Weakening of unstressed -*m* > -*n*: OE *dagum* > *dagun*, OHG *tagum* > *tagun*	OHG
10	Loss of *h* before *l, n, r* in word-initial position	9/10
OE	Use of demonstrative pronoun as article: OE *se cyning*, OHG *der kuning*	OHG
lOE	Use of verbs *sollen* and *wollen* for analytic future	OHG
OE	Completion of *i*-umlaut	lOHG
c.1100	W-Germanic *sk-* > *sh-*	11
12	Lengthening of vowels in open syllables	12/13
ME	Further weakening of unstressed vowels. Consequence: loss of inflectional morphology Levelling of stem vowel of second person singular preterite of strong verbs with first/third person: ME *thu bunde* > *thu bandest*, MHG *du næme* > *du namest*	MHG
14	Vocalic levelling of stem vowels of singular and plural pret. of strong verbs	13/14
15	Diphthongisation of W-Germanic long *i:* and *u:* *min hus* > ModE *my house*, ModHGerman *mein Haus*	12-14
15	Emergence of standard literary language	16

A later example of this kind of contextualisation is Roger Lass's chapter on "English and Germanic revisited" (1987); two recent one-volume histories by Knowles (1997) and Smith (1996) also make interesting occasional observations that place English language history in a wider context.[9]

7. Suggestions for further research

As an aid to further research and in an attempt to set out the essential features of the linguistic evolution of English that can be profitably studied in a wider European context, I would like to supplement the above table with some further notes concerning historical changes in English that are somehow paralleled in the histories of other European languages. To my knowledge, it is the first attempt to list such similarities from a diachronic rather than a typological point of view. I have excluded most cases of changes resulting from simple bilateral contact and also those features that are shared by English with other European languages throughout historical times (i.e. what we may see as common retentions from Indo-European) though these are still interesting for the characterisation of English language history within in a wider context.

7.1. Morphology and syntax

7.1.1. General evolution from synthetic to analytic forms in Romance and Germanic languages. We could see this as the result of complementary evolution reinforced by areal diffusion.
It is manifested in a remarkable series of parallel developments:
(1) the simplification, levelling and loss of inflectional morphology for nouns, verbs, adjectives, pronouns, and numbers (especially advanced in English, mainland Scandinavian and Dutch, cf. Thomason — Kaufman 1988: 315-321);
(2) the increased use of prepositions;
(3) the evolution of the definite article (derived from the demonstrative)[10] and of the indefinite article;

---

[9] Knowles (1997), for example, mentions parallels of morphological simplification in English and Danish (Knowles 1997: 42) and notes similarities in long vowel evolution in English, Dutch and German (Knowles 1997: 59), in open-syllable lengthening in English and French (Knowles 1997: 59) and in unstable raising and lowering of /er/ and /ar/ in English and French (Knowles 1997: 59).
[10] This is one of Haspelmath's eleven features of Standard Average European: "In virtually all SAE languages with sufficient documentation we can observe the grammaticalization of definite and indefinite articles in historical texts" (1998: 281). It is a

(4) the obligatory use of unstressed subject pronouns;
(5) auxiliary verbs (especially *be* and *have* for analytic past tenses; *be* for analytic passives, not found in proto-Indo-European; *shall* and *will* for an analytic future in North Germanic, English, Dutch and Middle High German).[11] The *have*-perfect in particular (one of Haspelmath's eleven features of Standard Average European) is a strong marker of a European *Sprachbund*, since it is "limited to Western Europe" (Dahl 1995: 19f., quoted in Ramat 1998: 232; cf. also Kuteva 1998: 274-275). Kuteva (1998) demonstrates a "European uniformity in auxiliation" (the grammaticalisation of lexical verbs to auxiliary verbs and then to affixes), and, in addition to the *have*-perfect, points out that another typical feature of European auxiliation is the use of verbs with general lexical meanings for the auxiliary (the use of lexically specific verbs being found on the periphery of Europe and elsewhere in the world).
(6) the loss of comparison indicated by case (Lat. *te major*, OE *migtigra þu*) and the generalisation of the analytic comparison of adjectives with the standard marked by a particle (OE *migtigra þonne þu*).[12]

These related changes were probably caused by an interaction of
(1) the affirmation of SVO word-order (making some syntactic morphology redundant, making articles useful for the marking of theme and rheme; cf. Lehmann 1978: § 8.3; Ramat 1986: 114);
(2) the loss of unstressed vowels and final nasals (hence of word-final inflectional morphology);
(3) the very success of the evolving analytic morphology (making inflections redundant, cf. Morani 1992: 224);

---

feature that is still spreading: a definite article seems to be developing today in Slavic languages in contact with Germanic and Romance languages, Czech *ten* being used as an article in modern times in expressions such as "the last request", where the referent is active in the context. A similar pseudo-article has also developed in Slovenian, also in contact with German; in Bulgarian/ Macedonian a postposed article has developed under influence from Romanian (Giuliano Bernini, personal communication). In Finnish too a definite article "is currently arising from a demonstrative pronoun (Haspelmath 1998: 281)."

[11] The use of modal and incoative verbs to form future tenses is, it is true, attested in many world languages (Comrie 1993: 108), but the parallelism of development and lack of subsequent divergence in the basic pattern suggests possible mutual influence between languages.

[12] One of Haspelmath's eleven features of Standard Average European (Haspelmath 1998: 278-279).

(4) archilect influence of medieval Latin (which evolved some of the analytic forms first, cf. Décsy 1988: 117; Meli 1993: 97);
(5) loans and calquing between the European languages;
(6) near-universal analytical and simplificational strategies of users of a second language, particularly marked perhaps in the fluid linguistic situation of medieval Europe.

7.1.2. Development of progressive constructions using auxiliary + gerund/ present participle in Celtic languages, English, Middle French, Italian, Spanish and Portuguese (possibly due to influence of Celtic reinforced by mutual contact; or by contact via French or Latin).

7.1.3. Development of the *going to* "andative future" (first signs in the later fifteenth century, increasing grammaticalisation from the mid-seventeenth century) and similar developments involving verbs of motion in French, Spanish, Catalan, Portuguese, Southern Dutch. Since there is a similar development in Hebrew, Arabic dialects and some African languages we may see this as a near universal development (Danchev — Kytö 1994: 70), but its concentration in European languages suggests typological/genetic similarities, or language contact, in particular of French with English, Southern Dutch and Western German dialects (cf. Danchev – Kytö 1994: 71).

7.1.4. Evolution of periphrastic tense-carrying uses of *do* and *tun* from the twelfth (German) and thirteenth (English) centuries, presumably an independent development (Lightfoot 1979: 117; D. Stein 1992: 134).

7.1.5. Development of "impersonal" subjects like *it* in *it's raining*.

7.1.6. The change from "patient (or: dative) experiencer" (*it pleases me*) to "agent (or: nominative) experiencer" (*I like it*). This is one of Haspelmath's eleven Standard Average European traits. West European languages now use predominantly nominative experiencers with the most common experiential predicates (Bossong 1997, quoted in Haspelmath 1998: 276-277).

7.1.7. Development of discontinuous and subsequently postverbal sentence negation in both Germanic and Romance languages. Probably the result of a combination of (1) internal factors (parallel typological evolution towards SVO and consequent tendency towards placing determiners to the right of head-words: V-O, N-Adj, N-Gen, V-Neg; cf. Vennemann 1974, 1989) and (2) external factors (language contact and areal diffusion from Germanic languages to French, North Italian dialects and Romantsch, cf. Bernini (1992); Bernini — Ramat (1992); Ramat — Bernini (1990); Schwegler (1983)).[13]

7.1.8. Development and spread of post-posed genitives.

7.1.9. Spread of "verb + particle" constructions (especially English and Icelandic); probably connected with word-order changes (cf. change from preverbal to postverbal negation) and contact with North Germanic (Hiltunen 1983: 104; Samuels 1972: 163-164).

7.1.10. The development of complex adverbial subordinators (*as if, d'autant plus que, nihilominus* etc.) in English, French and other European languages (in contrast with Celtic languages and languages from the easternmost parts of Europe). Here the contact seems to have been via the prestige superlect of Latin or French and through the diffusion of rhetorically sophisticated text types (Ramat 1998: 233-234, discussing Kortmann 1998).

7.1.11. The use of object pronouns in subject position (*them's our chairs*). This may be a universal tendency (yet to be demonstrated), but it is certainly common not only in English but in a range of European languages. Kjellmer (1986) refers to a similar phenomenon in Swedish, where *de* 'they' has been almost completely replaced in the spoken language by *dem/dom* 'them', and a similar case in Dutch is noticed by Stroop (1992: 166): "The moment is not far off, I believe, when the pronoun *hun* ('them') in subject function will become the normal form, and *ze* deviant". There is also notable object-for-subject syncretism of personal pronouns in Italian in the plural and third person, carried further in dialects (e.g. Bergamasco where the original object forms *me* and *te* function as both subject and object).

7.2. Phonology

Here, too, we may see a set of interconnected phenomena, associated in this case with the change in Germanic from a free to a fixed word accent – a change that has also affected Romance languages, either through contact with Germanic languages, or because of a common influence, perhaps a non-Indo-European substrate (Ramat 1986: 243, note 1). Linked changes

---

[13] One cannot exclude a contribution from a universal development found in many languages: emphasis that then becomes normal, which leads to a new means of showing emphasis, cf. Jespersen (1917 [1961]) and Schwegler (1983). However, Jespersen's idea of a universal cycle needs to be modified by taking into account the rarity of discontinuous negation in world languages (only 7%, according to Hagège, quoted in Ramat 1998: 236).

may include (1) *i*-umlaut; (2) the redistribution of long vs. short vowels according to syllable structure; and (3) the indistinct nature of vowels in unaccented syllables, hence too the weakening of inflections.

7.2.1. *i*-umlaut: affects all Germanic languages at different periods in their evolution, apparently independently, though "possibly all the typological and structural pre-conditions existed in common Germanic" (Ramat 1986: 44-45).

7.2.2. Lengthening of vowels in open syllables, shortening of vowels in closed syllables in Germanic languages; apart from OE *bacan* > ModE *bake*, "principles similar to MEOSL [Middle English open-syllable lengthening], with similar outputs, are recorded in Scandinavian and Low German" (Haugen 1976: 258-259); we may also see a similar development in OHG *faran* > ModHG *fahren* and in the diphthongisation of the vowel in an open syllable in French (Lat. *pedem* > Fr. *pied*).

7.2.3. Simplification of geminate consonants in Germanic languages, French and North Italian dialects.

7.2.4. Palatisation and affrication of velar plosives before close front vowels. The period of the onset of this change is "nearly identical in OE, OFris and OSax" (Markey 1978: xlii); similar changes also took place in Romance languages.

7.2.5. "West Germanic vowel shifts": raising and diphthongisation of long vowels in West Germanic languages, and lowering of short vowels, including the first elements of diphthongised long close vowels (Bynon 1977: 192; Labov — Yaeger — Steiner 1972: 99-110; Samuels 1972: 145). Hence the evolution from earlier /i:/ of the diphthongs in present E *time*, German *Zeit*, Dutch *tijd*, and from earlier /u:/ of the diphthongs in *house*, *Haus*, and *huis*. This change possibly results from a shared tendency to make more prominent by diphthongisation an information-bearing long stressed vowel of the root.

7.2.6. Loss of *h* in non-standard English dialects, Romance languages (first attested in Burgundy and Berry for French) and southern Dutch.

7.2.7. Middle English lowering of the vowel in *er* sequences (e.g. *ferme* > *farm*, and many other examples, e.g. *desart, sarvant*), also found in fourteenth-century French (Lat. *perfectum* > Fr. *parfait*; OFr. *lerme* > *larme*, and in many other examples, later re-formed because *er* was considered etymologically "correct" and because *ar* became associated with stigmatised Parisian speech). Even if the original change to *ar* in English was independent of French, the "correction" to *er* could well be due to cultural contact, with English writers (as in the case of other etymological respellings) following the lead given by the French.

7.2.8. Voicing of initial fricatives in southern English dialects and in Dutch (*fork* > *vork, sand* > *zand*); "the result of diffusion from Continental Franconian dialects into English in early OE times" (Lass 1987: 316). German has undergone a similar change only for initial *-s* (though the spelling remains <s->, not <z->), cf. *Sand* /zant/.

7.3. Orthography[14]

7.3.1. West European languages share the same alphabet and many spelling and textual conventions.
7.3.2. The sixteenth-century spread of spelling innovations from French: the apostrophe (1521), <i> vs. <j> distinction (Meigret, 1542); <u> vs. <v> distinction (La Ramée, 1562).
7.3.3. The capitalisation of most/all nouns in English and German in the seventeenth and eighteenth centuries is undoubtedly connected (the common prestige of highly nominal plain styles; the mutual influence of spelling reformers and printing houses).
7.3.4. Double consonants kept as a spelling convention in Germanic languages to indicate a preceding short vowel.
7.3.5. The common medieval and Renaissance tendency to suggest etymology in spelling. French usage here may have influenced English. (e.g. the <h> in *hom* already in the *Chanson de Roland* (c.1000)); fifteenth-sixteenth-century French forms like *author, doubt* probably precede English respellings.

7.4. Lexis

There are many close similarities in the inventory of words and concepts of the various European languages as a result of borrowing and calquing; we also find a rapid spread of neologisms; common loans from Latin; loans of word-formation elements; loans of isomorphous prepositional phrases.

---

[14] Much of the evolution in this and the following sections belongs to the modern period, in contrast to morphosyntactic convergence which has an older and longer period of evolution.

7.5. Discourse and pragmatic features

7.5.1. Shared expansion in the use of written records and archives from the twelfth and thirteenth centuries (Clanchy 1987). Gradual change from Latin to the vernacular in legal and official documents, thirteenth to fourteenth centuries; similar great expansion in the use of the vernaculars in the sixteenth century.

7.5.2. European diffusion of language styles, especially in written language: imitation of Latin rhetorical schemes[15]; exuberant language foregrounding Renaissance and Baroque styles found all over Europe; the plain styles in the seventeenth and eighteenth centuries.

The technical registers of science and learning evolved in all the European languages together, by a combination of mutual influence (learning at a certain level being international) and by parallel development. "Both in the Italian of Galileo and in the English of Isaac Newton we find more or less identical syndromes of grammatical features" (Halliday 1992: 75).

7.5.3. European-wide evolution of politeness and associated forms of second person address: use of second person plural pronoun for single interlocutor in Romance and Germanic languages in the later medieval period. E *thou* and Dutch *du* both virtually disappear in the sixteenth century (van den Toorn 1977: 526).

7.6. Standardisation, language consciousness and linguistics

7.6.1. Parallel development of standard languages from the twelfth and thirteenth centuries, which reflects the parallel development of the modern state; chancery scribes in England, France and Germany (using a common handwriting style which spread from Italy in the thirteenth century) were influential in regularising spelling (N. Blake (ed.) 1992: 13). van Coetsem (1992: 56) thinks that the early social stigmatisation of French regional language varieties may have influenced other European cultures. Fisher (1996: 65) comments on the strange lack of comparativist studies of the decline of dialects and the emergence of standard languages in

---

[15] "Methods of constructing chains of subordinate and coordinate statement were part of the heritage acquired by the Middle French avant-gardistes from the writings of their Roman masters. Influential in this regard were Titus Livy and, above all, Cicero" (Joseph 1987: 143). Later, the Senecan style of brevity and balance became more influential (Joseph 1987: 146-147).

Europe: "I do not know of a discussion that points out how similar this process was in various countries and discusses the implications of this similarity for our general understanding of the nature of standard languages".

7.6.2. The rapid spread of printing 1440-1500; the international movement of printers and international trade in books; the adoption by printers of common spelling and textual practices.

7.6.3. A common move towards codification of European languages in the seventeenth and eighteenth centuries; a shared ideal of explicitness, fixity and exclusiveness. "Double negative" constructions were classed as incorrect in English, Dutch (Mioni 1986: 220) and German (von Polenz 1978: 98) from the sixteenth to the eighteenth century. Language academies were founded in most European countries on the model of academies in Italy and France. English dictionaries, grammars and works on spelling reform come a little later than Continental examples and were undoubtedly influenced by them.

7.6.4. Similar Renaissance debates, descriptions, analyses and rule-making for vernacular languages. Vernacular grammars and works on spelling reform are first written in Italy, Spain and France (the first treatise of a modern European language is Nebrija's study of Castillian, 1492). The sixteenth-century debate over the national languages involves similar positions about purism, Latin loans, new formations from the language's internal resources and the use of dialect terms.

7.6.5. Interacting evolution of dictionary-making from the late fifteenth century. Material and schemes of organisation were borrowed between dictionary makers in different countries.

Latin-French dictionaries were used as models for Latin-English dictionaries by John Vernon (1552), Thomas Cooper (1565), and Thomas Thomas (1587) (G. Stein 1985: 167, 208-209, 314). The Latin-French lexicographical tradition was also drawn on by Petrus Cholevius for his German-Latin dictionary (1541). The idea of dictionaries of underworld slang seems to have spread quickly across Europe in the sixteenth century.

Schäfer (1989: 1-10) points out that the monolingual vernacular dictionary may derive from monolingual glossary appendixes compiled by translators, and a number of early examples of these were translated from one vernacular into another: John Day's "gatheryng of certayne harde wordes in the newe Testamente", for example, is a translation of Jean Girard's appendix to the Geneva Bible, "Receuil d'aucuns mots difficiles" (1546).

8. Conclusion

The aims of this contribution have been threefold. First, methodological: to recommend the inclusion of European areal references in surveys of the history of the English language. Second, as an aid to further research: to make a listing of as many areas of evolution (many due to contact) within the European *Sprachbund* in which English has taken part. And third: to identify a missing or largely ignored linguistic discipline, historical areal linguistics, to which it is hoped that *Anglistik* scholars might make some contribution, so that the history of the English language may be seen in a powerful explanatory context, the history of the European languages.

References

Asher, Robert E. (ed.)
   1994   *The encyclopedia of language and linguistics.* Oxford: Pergamon.
Auwera, Johan van der
   1998   "Revisiting the Balkan and Meso-American linguistic areas", *Language Sciences* 20: 259-270.
Auwera, Johan van der — Dónall P. Ó Baoil (eds.)
   1998   *Adverbial constructions in the languages of Europe.* (Empirical Approaches to Language Typology/EUROTYP 20, 3.) Berlin — New York: Mouton de Gruyter.
Banfi, Emanuele (ed.)
   1993   *La formazione dell'Europa linguistica: le lingue d'Europa tra la fine del I e del II millenio.* (Biblioteca di cultura 187, Lingue d'Europa 1.) Firenze: La nuova Italia.
Baugh Albert Croll — Thomas Cable
   1951 [1990] *A history of the English language.* (3rd edition.) London — New York: Routledge and Kegan Paul.
Bechert, Johannes
   1988   "Convergence and individuality of languages", in: Vicenzo Orioles (ed.), *Tipologie della convergenza linguistica: atti del convegno della Società Italiana di Glottologia, Bergamo, 17-19 dic. 1987.* (Biblioteca della Società Italiana di Glottologia 12.) Pisa: Giardini, 11-26.
   1990   "The structure of the noun in European languages", in: Johannes Bechert — Giuliano Bernini — Claude Buridant (eds.), 115-140.

Bechert, Johannes — Giuliano Bernini — Claude Buridant (eds.)
1990 *Toward a typology of European languages.* (Empirical Approaches to Language Typology 8.) Berlin — New York: Mouton de Gruyter.
Becker, H.
1948 *Der Sprachbund.* Leipzig: Humboldt-Bücherei Gerhard Mindt.
Bernini, Giuliano
1992 "Per una tipologia areale delle lingue europee", in: Antonio G. Mocciaro — Giulio Soravia (eds.), 37-48.
Bernini, Giuliano — Paolo Ramat
1992 *La frase negativa nelle lingue d'Europa.* Bologna: Il Mulino.
Birnbaum, Henrik
1985 "Divergence and convergence in linguistic evolution", in: Jacek Fisiak (ed.), 1-24.
Blake{ XE "Blake, B. J." }, B. J.
1994 "Language classification", in: Robert E. Asher (ed.), 1952-1957.
Blake, Norman (ed.)
1992 *The Cambridge history of the English language,* vol. II: *1066-1476.* Cambridge: Cambridge University Press.
Bloomfield, Leonard
1933 *Language.* New York: Holt, Rinehart and Winston.
Boas, Franz
1929 "Classification of American Indian languages", *Language* 5: 1-6 (quoted in Ramat 1998).
Bossong, Georg
1997 "Le marquage de l'expériant dans les langues d'Europe", in: Jack Feuillet (ed.), 259-294 (quoted in Haspelmath 1998).
Bynon, Theodora
1977 *Historical linguistics.* (Cambridge Textbooks in Linguistics.) Cambridge: Cambridge University Press.
Campbell, Lyle
1985 "Areal linguistics and its implications for historical linguistics", in: Jacek Fisiak (ed.), 25-56.
Clanchy, Michael Thomas
1987 *From memory to written record: England 1066-1307.* London: Arnold.

Coetsem, Frans van
 1992 in: J. van Leuwensteijn — J. Berns (eds.).
Comrie, Bernard
 1993 "La famiglia linguistica indoeuropea: prospettive genetiche e tipologiche", in: Anna Giacalone Ramat — Paolo Ramat (eds.), 95-121.
Dahl, Östen
 1995 "Areal tendencies in tense-aspect systems", in: Pier Marco Bertinetto *et al.* (eds.), *Temporal reference, aspect and actionality*, vol. 2: *Typological perspectives.* (Linguistica 13.) Torino: Rosenberg and Sellier, 11-27.
Danchev, Andrei — Merja Kytö
 1994 "The construction *be going to infinitive* in Early Modern English", in: Dieter Kastovsky (ed.), *Studies in Early Modern English.* (Topics in English Linguistics 13.) Berlin: Mouton de Gruyter, 59-77.
Décsy, Gyula
 1973 *Die linguistische Struktur Europas: Vergangenheit, Gegenwart, Zukunft.* Wiesbaden: Harrassowitz.
 1988 *A select catalog of language universals.* Bloomington: Eurolingua.
Emeneau, Murray Barnson
 1980 *Language and linguistic area.* Stanford: Stanford University Press.
Feuillet, Jack (ed.)
 1997 *Actance et valence dans les langues de l'Europe.* (Empirical Approaches to Language Typology/EUROTYP 20, 2.) Berlin — New York: Mouton de Gruyter.
Fisher, John H.
 1996 *The emergence of Standard English.* Lexington: Kentucky University Press.
Fisiak, Jacek (ed.)
 1985 *Papers from the sixth international conference on historical linguistics.* (Amsterdam Studies in the Theory and History of Linguistic Science 4, 34.) Amsterdam: Benjamins.
Görlach, Manfred
 1993 Review of Leuvensteijn — Berns (eds.) 1992. *English World Wide* 14: 157-159.

Greenberg, Joseph Harold
   1963   "Some universals of grammar with particular reference to the order of meaningful elements", in: Joseph Harold Greenberg (ed.), *Universals of language: report of a conference held at Dobbs Ferry, N. Y., 13-15 April 1961.* Cambridge, Mass.: MIT Press, 58-90.

Hagège, Claude
   1982   *La structure des langues.* (Que sais-je? 2006.) Paris: Presses Universitaires de France. (quoted in Ramat 1998)

Halliday, Michael
   1992   "New ways of meaning: the challenge to applied linguistics", in: Martin Pütz (ed.), *Thirty years of linguistic evolution: studies in honour of René Dirven on the occasion of his sixtieth birthday.* Philadelphia — Amsterdam: Benjamins, 59-96.

Harris, Zellig S.
   1951   *Structural linguistics.* Chicago — London: The University of Chicago Press (quoted in Malkiel 1981).

Haspelmath, Martin
   1998   "How young is Standard Average European?", *Language Sciences* 20: 271-287.

Haspelmath, Martin — Oda Buchholz
   1998   "Equative and similative constructions in the languages of Europe", in: Johan van der Auwera — Dónall P. Ó Baoil (eds.), 277-334 (quoted in Ramat 1998).

Haugen, Einar
   1976   *The Scandinavian languages: an introduction to their history.* London: Faber and Faber.

Hiltunen, Risto
   1983   *The decline of the prefixes and the beginnings of the English phrasal verb: the evidence from some Old and Middle English texts.* (Turun Yliopiston julkaisuja B, Humaniora 160.) Turku: Turun Yliopisto.

Hock, Hans Henrich
   1986   *Principles of historical linguistics.* Berlin — New York — Amsterdam: de Gruyter.

Hoenigswald, Henry M.
   1990   "Language families and subgroupings", in: Edgar C. Polomé (ed.), 441-454.

Huchon, René Louis
- 1923   *Histoire de la langue anglaise*, vol. 1. Paris: Colin.

Jespersen, Otto
- 1905   *Growth and structure of the English language*. Oxford: Blackwell.
- [1990]   [Reprinted Oxford: Blackwell.]
- 1917   "Negation in English and other languages".
- [1961]   [Reprinted in: *Selected writings of Otto Jespersen*. London: Allen and Unwin.]

Joseph, John Earl
- 1987   *Eloquence and power: the rise of language standards and standard languages*. (Open Linguistics Series.) London: Pinter.

Jungandreas, Wolfgang
- 1949   *Geschichte der englischen Sprache*. Göttingen: Vandenhoek und Ruprecht.

Kjellmer, Göran
- 1986   " 'Us Anglos are a cut above the field": on objective pronouns in nominative contexts", *English Studies* 5: 445-449.

Knowles, Gerry
- 1997   *A cultural history of the English language*. London: Arnold.

Kortmann, Bernd
- 1998   "Adverbial subordinators in the languages of Europe", in: Johan van der Auwera — Dónall P. Ó Baoil (eds.), 457-561.

Kuteva, Tania
- 1998   "Large linguistic areas in grammaticalisation: auxiliation in Europe", *Language Sciences* 20: 289-311.

Labov, William — Malcah Yaeger — Richard Steiner
- 1972   *A quantitative study of sound change in progress*, vol. 1. Philadelphia: The U.S. Regional Survey.

Lass, Roger
- 1987   *The shape of English: structure and history*. London: Dent.

Lehmann, Winfred Philip
- 1978   "Toward an understanding of the profound unity underlying languages", in: Winfred Philip Lehmann (ed.), *Syntactic typology: studies in the phenomenology of language*. Austin — London: University of Texas Press, 395-432.

Leuvensteijn, J. van — J. Berns (eds.)
1992   Dialect and standard language in the English, Dutch, German and Norwegian language areas: proceedings of the colloquium "Dialect and standard language", Amsterdam, 15-18 October 1990. (Verhandelingen, Afd. Letterkunde/Koninklijke Nederlandse Akademie van Wetenschappen, N. R. 150.) Amsterdam: North Holland.

Lewey, Ernst
1942   "Der Bau der europäischen Sprachen", *Proceedings of the Royal Irish Academy* 48 (cii).

Lightfoot, David W.
1979   *Principles of diachronic syntax.* (Cambridge Studies in Linguistics 23.) Cambridge: Cambridge University Press.

Lindelöf, Uno Lorenz
1895   *Grunddragen af Engelska Språkets Historiska Lhudoch Formlära.*
[1911   English translation: *Elements of the history of the English language.* Seattle: University of Washington Press.]

Malkiel, Yakov
1981   "Drift, slope, and slant: background of, and variations upon, a Sapirean theme", *Language* 59: 535-576.

Markey, Thomas L.
1978   *A North Sea Germanic reader.* München: Fink.

Meillet, Antoine
1918   "Convergence des développements linguistiques", *Revue philosophique de la France et de l'étranger* 85: 97-110.
[1921]   [Reprinted in: Antoine Meillet, *Linguistique historique et linguistique générale.* Paris: Champion, 61-75.]
[1965]   [Reprinted Paris: Champion.]

Meli, Marcello
1993   "Le lingue germaniche", in: Emanuele Banfi (ed.), 91-130.

Milroy, James
1996   "Linguistic ideology and the Anglo-Saxon lineage of English", in: Juhani Klemola *et al.* (eds.). *Speech past and present: studies in English dialectology in memory of Ossi Ihalainen.* (Bamberger Beiträge zur englischen Sprachwissenschaft 38.) Frankfurt/Main: Lang, 169-186.

Mioni, Alberto M.
1986   *Lingue germaniche moderne: strutture, diffusione, storia.* Padova: CLESP.

Mocciaro, Antonia G. — Giulio Soravia (eds.)
    1992   *L'Europa linguistica: contatti, contrasti, affinità di lingue: Catania, 10-12 settembre 1987.* (Publicazioni della Società di Linguistica Italiana 30.) Roma: Bulzoni.

Morani, Moreno
    1992   "La sostituzione della desinenza casuale col singtagma preposizionale nominale: una "modificazione conservativa"?" in: Antonia G. Mocciaro — Giulio Soravia (eds.), 219-226.

Nichols, Johanna
    1995   "Diachronically stable structural features", in: Henning Andersen (eds.), *Historical linguistics 1993: selected papers from the eleventh international conference on historical linguistics, Los Angeles, 16-20 August 1993.* (Amsterdam Studies in the Theory and History of Linguistic Science 4, 124.) Amsterdam — Philadelphia: Benjamins, 337-355 (quoted in Ramat 1998).

Polenz, Peter von
    1978   *Geschichte der deutschen Sprache.* (9th edition.) Berlin — New York: de Gruyter.
    1984   "Die Geschichtlichkeit der Sprache und der Geschichtsbegriff der Sprachwissenschaft", in: Werner Besch — Oskar Reichmann — Stefan Sonderegger (eds.), *Sprachgeschichte: ein Handbuch zur Geschichte der deutschen Sprache und ihrer Erforschung.* (Handbücher zur Sprach- und Kommunikationswissenschaft 2.) Berlin — New York: de Gruyter, 1-8.

Polomé, Edgar C. (ed.)
    1990   *Research guide on language change.* (Trends in Linguistics: Studies and Monographs 48.) Berlin: Mouton de Gruyter.

Ramat, Anna Giacalone — Paolo Ramat (eds.)
    1993   *Le lingue indoeuropee.* Bologna: Il Mulino.

Ramat, Paolo
    1986   *Introduzione alla linguistica germanica.* Bologna: Il Mulino.
    1998   "Typological comparison and linguistic areas: some introductory remarks", *Language Sciences* 20: 227-240.

Ramat, Paolo — Giuliano Bernini
    1990   "Area influence versus typological drift in Western Europe: the case of negation", in: Johannes Bechert — Giuliano Bernini — Claude Buridant (eds.), 25-46.

Samuels, Michael Louis
    1972   *Linguistic evolution, with special reference to English.* Cambridge: Cambridge University Press.

Sapir, Edward
    1921   *Language: an introduction to the study of speech.* New York: Harcourt, Brace and World.

Schäfer, Jürgen
    1989   *Early Modern English lexicography.* Oxford: Clarendon.

Schwegler, Armin
    1983   "Predicative negation and word-order change: a problem of multiple causation", *Lingua* 61: 297-334.

Scott, Charles T. — Jon L. Erickson
    1968   *Readings for the history of the English language.* Boston: Allyn and Bacon.

Siewierska, Anna (ed.)
    1997   *Constituent order in the languages of Europe.* (Empirical approaches to language typology/EUROTYP 20, 1.) Berlin - New York: Mouton de Gruyter.

Simpson, J. M. Y.
    1994   "Areal linguistics", in: Robert E. Asher (ed.), 206-212.

Smith, Jeremy J.
    1996   *An historical study of English: function, form and change.* London — New York: Routledge.

Stein, Dieter
    1992   "*Do* and *tun*: a semantics and varieties based approach to syntactic change", in: Marinel Gerritsen — Dieter Stein (eds.), *Internal and external factors in syntactic change.* (Trends in Linguistics: Studies and Monographs 61.) Berlin — New York: Mouton de Gruyter, 131-155.

Stein, Gabriele
    1985   *The English dictionary before Cawdrey.* (Lexicographica: Series maior 9.) Tübingen: Niemeyer.

Stroop
    1992   in: J. van Leuvensteijn — J. Berns (eds.).

Thomason, Sarah Grey — Terrence Kaufman
1988 *Language contact, creolisation and genetic linguistics.* Berkeley: University of California Press.

Toorn, M. C. van den
1977 "De problematiek van de Nederlandse aanspreekvormen", *Niewe Taalgids* 70 (vi).

Trubetzkoy, Nikolaj
1930 "Proposition 16", in: *Actes du premier congrès international de linguistes à la Haye.* Leiden: Brill.

Ureland, Per Sture
1986 "Some contact-linguistic structures in Scandinavian languages", in: Peter H. Nelde — Per Sture Ureland — Iain Clarkson (eds.), *Language contact in Europe: proceedings of the working groups 12 and 13 at the thirteenth international congress of linguistics, Tokyo, 29 August - 4 September 1982.* (Linguistische Arbeiten 168.) Tübingen: Niemeyer, 31-80.
1990 "Contact linguistics", in: Edgar C. Polomé (ed.), 471-506.

Vennemann, Theo
1974 "Topics, subjects and word order: from SVX to SVX via TXV", in: John M. Anderson — Charles Jones (eds.), *Historical linguistics.* Amsterdam: North Holland, 339-371.
1989 "Language change as language improvement", in: Vincenzo Orioles (ed.), *Modelli esplicativi della diachronia linguistica.* Pisa: Giardini, 11-35.

Wagner, Heinrich
1959 *Das Verbum in den Sprachen der Britischen Inseln.* Tübingen: Niemeyer.

Weinreich, Uriel
1954 "Is structural dialectology possible?", *Word* 10: 388-400.

Whorf, Benjamin Lee
1941 "The relation of habitual thought and behaviour to language".
[1956] [Reprinted in: Benjamin Whorf, *Language, thought, and reality.* New York: MIT Press — Wiley.]

Andreas Fischer (Zürich)
Lexical borrowing and the history of English: a typology of typologies

1. Introduction

It is a truism that the history of English is characterised by a series of unusually intensive situations of language contact[1] and that Modern English, as a consequence, is a lexically mixed language. For this reason, most "traditional" histories of the language and all studies of the English lexicon in particular treat lexical borrowings in the history of English, usually by discussing them according to the historical sequence of contact situations.[2] Within this primary classification further distinctions are often made, for example according to the semantic domains the loan-words belong to or according to their internal chronology. Not surprisingly, more "structurally" oriented histories find the lexicon more difficult to deal with and treat it briefly, if at all.[3] In this paper I will not discuss this historical classification any further, but will look instead at attempts to classify lexical borrowings in different ways. Such classifications are basically independent of English and its history, and in principle they can be applied to all situations of language contact. It is for this reason that they are called typologies here, but no attempt will be made to link this modest use of the term with the larger claims of language typology. In the following I will look at altogether three "types of typologies" and will investigate what they can or could contribute to a deeper understanding of contact situations. They are:
a. Typologies that focus on the morphological structure and the etymology of borrowings. They will be called morpho-etymological or simply morphological typologies (section 2.).

---

[1] Note that this is true only from a European point of view. In the context of languages worldwide the situation of English is probably not so unusual; see Thomason – Kaufman (1988).
[2] This can be done either by discussing the lexicon including loan-words in connection with the various historical periods of English or by devoting a whole, historically organised section to the vocabulary. The former principle is used by Baugh – Cable (1993) or Bolton (1989), the latter by Berndt (1984) or Koziol (1984), to give just a few examples.
[3] See, for example, Bourcier (1981) or Lass (1987).

b. Typologies that look at the lexical and semantic consequences of borrowing, that is at the effects of borrowing on the vocabulary of the receiver language. They will be called lexicosemantic or semantic typologies (section 3.).
c. Typologies that classify the contact situations and their effects on processes of borrowing. They will be called socio-historical or sociolinguistic typologies (section 4.).

2. Morphological typologies

2.1. Morphological typologies focus on the formal aspects of borrowings and take into account both their ultimate etymology and their morphology as lexemes of the borrowing language.[4] They were developed, independently, as it seems, by Werner Betz and by Einar Haugen. Betz developed his typology on the basis of Latin borrowings in Old High German (first 1936, latest version 1974). Like all his successors he makes a basic distinction between what he calls *"Lehnwörter"* 'imported lexical items' and *"Lehnprägungen"* 'substituted lexical items', but puts particular emphasis on fine distinctions within the latter category (*"Lehnbildungen"* versus *"Lehnbedeutungen"*, and further subdivisions). The contact situation of Old High German and Latin, of course, closely resembles that of Old English and Latin, and Helmut Gneuss's monograph *Lehnbildungen und Lehnbedeutungen im Altenglischen* (1955) is essentially an application of Betz's typology to Old English: as indicated in the title, it concentrates on *"Lehnprägungen"*.[5] Einar Haugen's typology (1950), by contrast, goes back to his study of Norwegian borrowings in American English (see also Haugen 1953). Haugen (1950: 214-215) distinguishes the three categories of (1) loan-words (Betz's *"Lehnwörter"*), which "show morphemic importation without substitution", (2) the intermediate loanblends, which "show morphemic substitution as well as importation", and (3) loanshifts, which "show morphemic substitution without importation" (Betz's *"Lehnprägungen"*). Although Haugen uses a different, more structuralist terminology ("importation" and "substitution"), his basic

---

[4] In this paper the terms "source language" and "borrowing language" will be used throughout, but there are, of course, alternative terminologies such as "donor language" and "receiver language".
[5] See also Gneuss (1992). In this state-of-the-art survey Gneuss mainly discusses lexical borrowing in Old English, from Latin, Greek, Hebrew, the Celtic, Scandinavian and West Germanic languages, and French, but he also touches on issues raised in this paper.

analytic categories are the same as Betz's, but he adds the hybrid category of loanblends. All subsequent classifications are indebted to Betz and/or Haugen in one way or another, but the two typologies are perhaps most happily combined by Carstensen (1968), who fuses not only their substance, but also their terminologies, thus providing German and English terms for each category. Recent years have brought further refinements, but no essentially new insights. Jorgensen – Ferré's typology (1986) with sixteen "slots" altogether is easily the most detailed, but despite its apparent complexity it does not really add any new basic categories: the terminology and, above all, the four basic categories are Betz's, while the distinction between "*nicht-hybride*" and "*hybride Entlehnungsarten*" ultimately comes from Haugen. A typology such as Jorgensen – Ferré's, while extremely fine from an analytical point of view, is obviously too detailed to be practical. It comes as no surprise, therefore, that the new *Anglizismen-Wörterbuch* by Carstensen – Busse (1993-1996) uses a modified and greatly simplified system. Doing away with most of the fine differentiations and technical terms, it uses only three labels (Carstensen – Busse 1993: 53*-66*, especially 59*). Note that the first two correspond exactly to Betz's old distinction between "*Lehnwörter*" and "*Lehnprägungen*":

> aus engl. x, wobei x ein ins Deutsche übernommenes englisches Sprachzeichen ist, z.B. dt. *Job* aus engl. *job*

> nach engl. x, wobei das englische Sprachzeichen nicht mehr erkennbar ist, sondern das englische Vorbild mit deutschem Sprachmaterial nachgebildet wird, z.B. *Erste Dame* nach engl. *first lady*

> zu engl. x, wobei ein englisches Sprachzeichen als Ausgangspunkt der Entlehnung erkennbar ist, das aber im Deutschen morphologisch verändert worden ist, z.B. dt. *Twen* zu engl. *twenty.*

2.2. How have morphological typologies been used in describing language contact in the history of English, and how useful have they been? At this point it is important to remember that the various studies pursue different aims. The focus of many of them is systematic, their primary intention being to provide a classificatory scheme for the accommodation of all types of borrowings, irrespective of source language(s) and borrowing language.⁶ As shown above, enough such typologies are in existence, and there is no urgent need for new ones. Established typologies, on the other hand, may also be used to describe and explain particular situations of language contact, and it is here that more work needs to be done in at least two areas. First, a look at any history of the language or at a book such as Mary Serjeantson's *A history of foreign words in English* (1935) will show that when discussing borrowings they frequently deal with importations only, while substitutions are neglected. One simple reason for this may be that substitutions are, of course, much harder to detect than importations: while a term like *kindergarten* is immediately identifiable as an importation from German, an expression like *airship* is only identifiable as a substitution if one knows that it is actually a translation of German *Luftschiff*.⁷ A second reason for the apparent neglect of substitutions may be that they are, in fact, less frequent than importations, but this is exactly one of the questions that needs to be answered for any contact situation. Thus (this being the second area in need of further work) surprisingly little effort has been devoted to the question of why certain types of borrowing seem to be more frequent than others, depending on period, source language, text type, speech community or language policy, to name only the most obvious factors. A clear chronological difference, for example, is apparent in the Latin influence on English: Anglo-Saxon translators of Latin texts favoured substitutions, while the tendency after the Norman Conquest was to use importations. This difference between the pre- and post-Conquest vocabularies of English is particularly obvious in grammatical and biblical terminology: one may compare Ælfric's *stæfgefeg* and *dælnimend* with later *syllable* and *participle*, or pre-conquest *þrines* and *sundorhālga* with later *trinity* and *pharisee*. When source languages are compared with each other, it seems that Latin has led to more substitutions in English (especially in Old English, as just noted) than either Scandinavian or French, but this problem needs further investigation. Text type may have an influence as well: the practice of glossing (current especially in the Anglo-Saxon period) favours substitution since one of its main aims is to make opaque foreign words transparent by providing maximally transparent native equivalents. As shown by Lucia Kornexl (this volume) such substitutions may be nonce-formations that owe their existence solely to the process of glossing. An

---

⁶ This holds true, even though, as Betz's and Haugen's studies show, most such schemes have been established on the basis of one contact situation involving a single pair of languages.
⁷ A comprehensive list of German borrowings in English is given in Pfeffer (1987).

example of differences between speech communities is provided by Standard (i.e. German) High German versus Swiss High German: the former has traditionally favoured substitutions (or, at any rate, transparent, purely German words) while the latter has always been more open to importations: *(Rund-) funk* 'radio' versus *Radio, Gehsteig* 'pavement' versus *Trottoir, (Fahr-/Eintritts-) karte* 'ticket' versus *Billett*. I am not aware of any comparable systematic differences between different English-speaking speech communities, past or present, but it is quite conceivable that they did or do exist, and there is an urgent need for comparative work in this area.[8] (Predominantly) British English *pavement* or *autumn* (both from French) versus American English *sidewalk* or *fall* (both native English) are two isolated examples, but they may well be representative of a systematic difference between the two varieties. John Algeo, for example, writes (private communication): "British English is far more susceptible to French influence than is American English for the obvious reason of geographical proximity and the concomitant opportunities for cultural exchange." A speech community's receptivity to borrowings, finally, may be influenced by official or unofficial language policies. The policy, in present-day France, against *Franglais* words, or the Nazi campaign to "purify" German are prime examples of official policies supporting substitutions over importations, while the attempts of English language purists such as Sir John Cheke or William Barnes to replace importations by newly coined substitutions illustrate unofficial policies: "Saxonisms" such as *foreword* for *preface* and *folklore* for *tradition* established themselves (but did not replace the older borrowings), while *folkwain* for *omnibus* and *sky-sill* for *horizon* remained unsuccessful.[9]

2.3. In conclusion, therefore, it can be stated that several, partly overlapping morphological typologies are in existence and that there is no urgent need for further refinements. The typologies that are available, however, have been underused for describing and analysing language contact in the history of English, and a great deal of comparative work remains to be done.

---

[8] The following typologies could serve as useful starting points, although they are not primarily concerned with etymology: Benson – Benson – Ilson (1986: 14-174) provide a classification of differences between British and American English including a comprehensive word-list, Algeo (1989) sets up a typology of "interdialectal variation", and Görlach (1990) attempts to classify "heteronymy in international English".
[9] For Cheke see Bolton (1982: 215-216) and Görlach (1978: 150-152); for Barnes see Leisi (1985: 82-83) and McArthur (1992: 106, *s.v.* Barnes, William).

3. Semantic typologies

3.1. Lexical intake through borrowing will inevitably have consequences for the total vocabulary of the borrowing language, and it is these that are captured in what I have chosen to call lexico-semantic or simply lexical typologies. I am not aware of an "onlie begetter" here: possibly the earliest classification of this kind was proposed by Otto Jespersen in his book on *Growth and structure of the English language* (1905 [1982]: §§ 62-72), but similar ones are found in numerous histories of the language and studies of the English lexicon. The following typology synthesises these, but uses a strict terminology (the term *lexical* referring to word-form or lexeme, the term *semantic* to word-sense or sememe). Like the morphological typologies just discussed it is purely descriptive and does not attempt to describe reasons for borrowing. Four basic categories can be distinguished, namely (a) lexical and semantic intake, (b) lexical intake followed by loss of the older word, also: lexical replacement, (c) lexical intake followed by semantic change, also: semantic differentiation, and (d) semantic change due to language contact, also: semantic borrowing.

3.2. Lexical and semantic intake. In this case, borrowings enter a language as "names" of new, that is, hitherto unknown or at least unlexicalised objects, institutions or ideas. The net result is lexical augmentation, in that the borrowing language gains new lexemes together with the concepts they represent. Examples can be cited from the concrete as well as the abstract spheres. Foodstuffs entering the Germanic world from the Roman one came together with their names (OE *cīese* 'cheese', *pipor* 'pepper', *senep* 'mustard', or *wīn* 'wine') and the same process can be observed with culinary items taken to the United States by German immigrants *(hamburger, pretzel, sauerkraut, or wiener)*. As examples from the sphere of abstract terms one may cite Christian Latin loan-words in Old English such as *discipul* 'disciple' and *sealm* 'psalm' or recent German loan-words such as *angst* and *weltanschauung*.

It is in this category that the semantic classification intersects with the morphological one, in that a new concept may be expressed either by a word taken from the source language (that is, an importation) or with a newly coined or a semantically extended native lexeme (that is, a substitution): in Old English, for example, Latin *discipulus* appears both as *discipul* and as *leorningcniht*. I have already suggested that there is room here for further, methodologically stringent diachronic and contrastive studies. It is also in this category that the cultural influence of particular source-languages can be assessed best.

3.3. Lexical intake followed by loss of the older word, also: lexical replacement. Here a borrowed word first comes into contact and conflict with a synonymous older word and then eventually replaces it. The net result is zero, since one and the same concept is expressed first by one word (usually a native one), which is then replaced by a borrowed item. As examples one could mention earlier (Old English) *æ* and *niman* that were replaced by the Scandinavian words *law* and *take* with the same sense (Fischer 1989; Rynell 1948), or earlier (Old English) *sinscipe* and *æ*, that were replaced by French *marriage* and *matrimony* (Fischer 1986).

Two special cases may be mentioned here. One consists of ultimately unsuccessful replacements, a case that can be illustrated with the English words for the seasons of the year (Fischer 1994). The Romance borrowing *hiems* for 'winter' is attested three times in the OED (1450, 1568 and 1605), but it was never a serious rival for *winter*. *Ver(e)* and *primetemps* 'spring' were also borrowed in the Middle English period and seem to have established themselves for a while, but disappeared again (in the seventeenth and late fifteenth century respectively). Another fourteenth-century borrowing from French, *autumn*, eventually replaced the native *harvest* in Standard British English, while American English mainly settled for the native English *fall* (newly coined in the sixteenth century, probably as a counterpart of the slightly earlier *spring*). Replacement, obsolescence and loss, as these examples show, are far from straightforward processes and need careful study.

The second special case is constituted by newly coined native words (Betz's "*Lehnbildungen*"), which are intended to replace well-established borrowings, usually as a deliberate attempt to purify English. The best-known examples are coinages by Saxonists such as the Dorset poet William Barnes, mentioned above as examples of unofficial language policies.

Both these special cases show that when there is competition between synonymous words the "newer" one, be it borrowed or native (newly coined) does not always win. The chances for the ultimate success of a lexical item thus cannot be predicted on the basis of age or etymology alone.

3.4. Lexical intake followed by semantic change, also: semantic differentiation.[10] This case begins like 3.3. above, namely with the borrowing of a lexeme which has to compete with an existing native

---

[10] Differentiation may also happen through syntax, collocational restrictions, dialect, register and so on; see Fischer (1997). See also Ullmann (1962: chapter 6).

Andreas Fischer

synonym or near-synonym. In case 3.3. one of the two words disappears, while here the co-existence of the two eventually leads to semantically differentiated doublets. This is an extremely frequent process, the net results of which is lexical augmentation as in 3.2. plus semantic differentiation. Among the very many sets of examples one might cite English-Scandinavian doublets such as *rind* and *bark*, *hide* and *skin* or *heaven* and *sky*, English-French doublets such as *calf* and *veal* or *freedom* and *liberty*, or English-Latin ones such as *brotherly* and *fraternal* or *learned* and *erudite*.

3.5. Semantic change due to language contact, also: semantic borrowing. These are cases where an existing native word changes its meaning under the influence of language contact, usually because of phonological and/or semantic similarity with a word in the source language. Change due to (above all) phonological similarity is evidenced by *earl* (OE *eorl* 'warrior', semantic change due to Scandinavian *jarl* 'nobleman'), while Old English words of Germanic origin that took on Christian meanings after the conversion show change due to semantic similarity (*hālig* for *sanctus* 'holy', *synn* for *peccatum* 'sin'). Since older senses are usually lost in the process (or taken over by other lexemes), the net result here, as in 3.3., is often zero.

3.6. As stated at the beginning of this section, a semantic typology like the one just sketched is purely descriptive, but attempts have been made to classify the reasons for lexico-semantic changes. One such example is to be found in Görlach (1974), adapted from Hans Käsmann's *Studien zum kirchlichen Wortschatz des Mittelenglischen* (1961). Görlach (1974: 140-141) lists the following reasons for borrowing:

- (A) Lücken des heimischen Wortschatzes
- (1) Das Wort wird mit dem neuen Inhalt (und der Sache) übernommen
- (2) Für einen bekannten Inhalt ist keine Bezeichnung vorhanden
- (3) Für eine Inhaltsnuance ist der vorhandene Ausdruck unzureichend differenziert

- (B) Vorausgehende Schwächung des einheimischen Wortschatzes
- (4) Der Inhalt wurde (versuchsweise) durch eine Vielzahl von Ausdrücken wiedergegeben
- (5) Der Inhalt wird durch einen Ausdruck wiedergegeben, der infolge von Homonymie, Polysemie, als Element einer unproduktiven Wortbildungsklasse gefährdet ist
- (6) Für einen gefühlsmäßig stark belasteten Ausdruck (einschließlich der Tabus) wird ein neutrales Wort benötigt

(C) Assoziative Beziehungen
(7) Entlehnung nach vorausgehender Aufnahme eines Wortes derselben Wortfamilie (Konsoziation)
(8) Die Entlehnung wird begünstigt durch ein vorhandenes einheimisches Wort, das in Ausdruck/Inhalt ähnlich ist
(9) Angleichungen: ein schon entlehntes Wort wird durch eine Neuentlehnung ersetzt [beziehungsweise] in seiner Lautform und/oder Bedeutung angeglichen

(D) Besondere außersprachliche Gründe
(10) Entlehnung für den Reim benötigter Wörter
(11) Nicht durch Sachzwang (sondern durch Mode, Prestige der Fremdsprache, Stilwert) motivierte Übernahmen
(12) Infolge von Ratlosigkeit oder Bequemlichkeit des Übersetzers stehengebliebene Wörter der Quelle (*parole*-Phänomen?)

(E) Ursachen nicht (mehr?) festzustellen
(13) Der große Rest unklarer Fälle

Like all attempts to "explain" language change, this one suffers from the fact that explanations can only be guessed at and that actual, verifiable proof is hard to come by. Any such typology, therefore, will remain tentative. As Görlach's category E shows, moreover, there will always be a large residual category of cases for which no explanation is readily available. Finally, while this is a classification of mainly internal, linguistic reasons (see, however, category D), extra-linguistic, socio-historical ones may be even more important, and it is doubtful whether these can be satisfactorily classified at all.

3.7. Less work has gone into setting up descriptive semantic typologies than morphological ones, but I hope to have shown that semantic typologies are viable instruments, although there will be many practical problems when one tries to place each lexical borrowing into one of the categories. Semantic typologies of one sort or another, in fact, have been used quite frequently, but - I would claim - not systematically enough. It would be useful, for example, to establish how frequent categories 3.2. to 3.5. are relative to one another for each of the major source languages. Thus questions like "How many of all Scandinavian borrowings have replaced native terms, how many have led to semantic differentiation?" or "How does French compare with Scandinavian in this respect?" could be multiplied, and they could also be usefully combined with morphological ones such as "How many borrowings from language x are importations, how many are substitutions, and what semantic domains do they belong to?". The main problems here are the size of the vocabulary on the one hand, and a general insecurity about the setting up of semantic domains and the handling of senses on the other. Discussing all borrowings from one of the major source languages in sufficient detail will often be too big a task. Case studies, however, are a useful beginning, and can be extended step by step as one's resources increase. Setting up semantic domains as well as defining and delimiting senses presents further problems for which no theoretically neat solutions are available. Here the *Historical thesaurus of English* promises to be a very useful tool, but unfortunately it will not be ready for some time. The *Historical thesaurus of English* will present the complete vocabulary of English from the earliest records to the present, arranged according to semantic domains modelled on those of *Roget's thesaurus*.[11] With information from this thesaurus it will be much easier to study the lexical and semantic dynamics within a particular semantic domain throughout the history of English.

4. Sociolinguistic typologies

4.1. Sociolinguistic typologies primarily focus on the socio-historical aspects of contact situations, but it should be possible, at least in theory, to combine their categories with kinds of lexical borrowing. The earliest typology of this kind that I am aware of is to be found in Leonard Bloomfield's *Language* (1935) and appears to be his innovation.[12] Taking into account mainly the geographical "distance" involved in language contact, Bloomfield distinguishes between cultural, intimate and dialect borrowing. His definitions, it must be said at this point, are minimal and often not as precise as one would wish; in the following summary they are expanded and clarified where necessary. In cultural borrowing "the borrowed features come from a different language [not usually or not necessarily adjacent to the receiver language]" (Bloomfield 1935: 444); cultural borrowing, in other words, does not result from an immediate contact of the speech communities of source language and borrowing language. Intimate borrowing, by contrast, occurs when "two [different] languages are spoken in what is topographically and politically a single community" (Bloomfield 1935: 461). Dialect borrowing, finally, takes place,

---

[11] The *Historical thesaurus of English* will combine the contents of the *OED* with the Old English vocabulary not listed in the *OED*. For Old English see *A thesaurus of Old English* (Roberts – Kay – Grundy 1995). On the two projects see Kay (1994) and Roberts (1994).
[12] It is not found in Bloomfield's much earlier book *An introduction to the study of language* (1914 [1983]).

when "the borrowed features come from [another variety of the same language] within the same speech area" (Bloomfield 1935: 444). Bloomfield's distinction between cultural and intimate borrowing is useful to characterise language contact situations especially in the past, but it has become blurred in this century because of the way air travel, the media and mass communication shorten and sometimes obliterate distances: the influence of English on German in the present probably has to be characterised as intimate, although the two languages are not "spoken in what is topographically and politically a single community".[13] Bloomfield's tripartite scheme can be usefully supplemented by the division of contact-situations into superstrate, substrate and adstrate situations, which was originally developed in and for Romance linguistics.[14] It may be seen as a differentiation of Bloomfield's intimate borrowing which takes into account status or power relations between the two speech communities. In a typical adstrate situation the two speech communities are about equal in status, while inequality of status leads to substrate or superstrate influences, that is, influences "from below" (substrate) or "from above" (superstrate). In the history of English, to take the most obvious examples, Scandinavian loan-words in late Old English and Middle English exemplify adstrate borrowing, Celtic loan-words in Old English substrate borrowing, and French loan-words in Middle English superstrate borrowing.[15]

It is relatively easy to link these contact situations with generalisations about numbers and semantic domains of lexical borrowings: in adstrate situations the borrowed vocabulary may be substantial and tends to be of an everyday nature; in substrate situations there will be relatively few loan-words and the majority of the borrowed vocabulary will concern indigenous topography, fauna, flora and culture (customs and artefacts); in superstrate situations, finally, borrowing may be intense and will include learned, technical and scientific vocabulary. It is difficult, however, to be more specific than that and it soon becomes clear that such a

---

[13] However, intimate borrowing in Bloomfield's sense of the word happened and possibly still happens in German immigrant communities in the United States.
[14] The three terms were not coined as part of a grand scheme, but evolved gradually. "Substrate", the oldest, probably goes back to the Italian linguist Gradazio Isaia Ascoli (1881/1882). "Adstrate" was first used by the Dutch linguist Marius Volkoff in 1932, and "superstrate" goes back to the Italian Matteo Bartoli (1931/1933) and the Swiss Walther von Wartburg (1936). I owe this information to my colleague Gerold Hilty, emeritus Professor of Romance linguistics at the University of Zurich.
[15] These labels are appropriate in a general way, but they grossly simplify the individual contact situations. This is especially true of the contact of English with French (and Latin).

typology does not account for a number of important factors. Cultural and superstrate borrowing, in particular, may happen through the written as well as the spoken medium, and the source languages in this case may be "dead" ones such as Latin. Moreover, in situations of direct contact (Bloomfield's "intimate borrowing"), loan-words may either be "borrowed" from the source language by speakers of the borrowing language in a situation of language maintenance, or else they may be "taken along" by bilingual speakers who give up one of their languages in favour of the other (that is, in language shift).

4.2. The distinction between language maintenance and language shift is at the root of the latest and possibly most comprehensive attempt at setting up a sociolinguistic typology of language contact, namely the one by Sarah Grey Thomason – Terrence Kaufman in their book *Language contact, creolization and genetic linguistics* (1988).[16] In the section in question (chapter 4, "Language maintenance"), Thomason – Kaufman look at situations where a language is maintained despite contact with other languages, and their main criterion is intensity of contact. In their own words, greater intensity of contact, or greater cultural pressure on borrowing-language speakers is constituted by: "[1] length of time – enough time for bilingualism to develop and for interference features to make their way into the borrowing language; [2] many more source-language speakers than borrowing-language speakers; and [3] either sociopolitical dominance of source-language speakers over borrowing-language speakers or intimate contact in mixed households and/or other social settings" (Thomason – Kaufman 1988: 72; my numbering). The following is an abbreviated form of their typology:

(1) Casual contact: lexical borrowing only
Lexicon: Content words. For cultural and functional (rather than typological) reasons, non-basic vocabulary will be borrowed before basic vocabulary.

(2) Slightly more intense contact: slight structural borrowing
Lexicon: Function words: conjunctions and various adverbial particles.
Structure: [...]

(3) More intense contact: slightly more structural borrowing
Lexicon: Function words: adpositions (prepositions and postpositions). At this stage derivational affixes may be abstracted from borrowed words and added to

---

[16] The book is based on a study of very many contact situations throughout the world, and it is thus typological in the true sense of the word. Moreover, Thomason – Kaufman do not restrict themselves to lexical borrowings.

Lexical borrowing and the history of English

native vocabulary; inflectional affixes may enter the borrowing language attached to, and will remain confined to, borrowed vocabulary items. Personal and demonstrative pronouns and low numerals, which belong to the basic vocabulary, are more likely to be borrowed at this stage than in more casual contact situations.
Structure: [...]

(4) Strong cultural pressure: moderate structural borrowing
Structure: [...]

(5) Very strong cultural pressure: heavy structural borrowing
Structure: [...]   (Thomason — Kaufman 1988: 74-76)

Convincing as this typology is in the framework of Thomason — Kaufman's book, it is of little concrete use for the analysis of lexical borrowing, and this for two reasons. First, Thomason — Kaufman's emphasis is clearly on structural borrowing. Borrowings of content words are restricted to the category (1), "casual contact", with function words playing a certain role in categories (2) and (3). Since by general agreement the only function word and slight structural borrowings into English are from Scandinavian/ Old Norse (with some claims made for French, see Thomason — Kaufman 1988: 123-126), all other contact situations involving English as a borrowing language must be classed as "casual". Second, it needs to be emphasised that intensity of contact, Thomason — Kaufman's main criterion, is extremely hard to document independently for contact situations in the past. To take Scandinavian as an example again, we have no precise data for either the duration of contact[17] nor for the number of speakers involved, and we can only guess, for example, that Scandinavian-English language contact happened in either "mixed households and/or other social settings." The situation with regard to contact between speakers of French and English after the Norman Conquest is equally difficult to assess, although it is better documented. According to Berndt's (1965 [1969]) view, which is accepted by most linguists nowadays, the native speakers of French never amounted to more than ten per cent of the population and bilingualism was not widespread, especially after 1204. This version of events, in turn, fails to explain the massive influx of French loan-words between 1250 and 1400 (Baugh — Cable 1993: 173-174), for which other explanations have to be sought (language shift of native speakers of French, borrowing from written sources).

---

[17] See the interesting, but inconclusive studies by Ekwall (1930), Page (1971) and Samuels (1985).

4.3. Thus, sociolinguistic typologies are available, but their application to questions of lexical borrowing in the history of English is problematic. First, lexical borrowing, especially of content words, does not seem to be very indicative of the intensity of contact. Second, since there is so little information about the precise circumstances of contact situations in the past, circular arguments seem unavoidable: from the available linguistic evidence we reconstruct a picture of Scandinavian-English language contact in the Danelaw, and we then use this reconstruction to "explain" the linguistic evidence. The conclusions we arrive at in this way may not necessarily be wrong, of course, but one should always bear in mind that the socio-historical and the linguistic evidence in these cases cannot clearly be separated.

5. Conclusion

No radically new insights can be expected from an overview like the one just given, but some conclusions suggest themselves nevertheless. Of the three types of typologies discussed here, the third, although the most attractive because of its sociolinguistic orientation, is possibly the least useful for a study of lexical borrowing: lexis is an unreliable indicator of the precise nature of a contact situation, and the socio-historical evidence necessary to reconstruct the latter is often not available. Morphological and semantic typologies on the other hand, though seemingly more traditional, can yield a great deal of new information if an attempt is made to study a whole semantic domain and if such studies are truly comparative, comparing different text types, different source languages or different periods.

References

Algeo, John
    1989   "British-American lexical differences: a typology of interdialectal variation", in: Ofelia Garçía — Ricardo Otheguy (eds.), *English across cultures, cultures across English: a reader in cross-cultural communication*. (Contributions to the Sociology of Language 53.) Berlin — New York: Mouton de Gruyter, 219-241.

Baugh, Albert Croll — Thomas Cable
1993   *A history of the English language.* (4th edition.) Englewood Cliffs, N. J.: Prentice-Hall.
Benson, Morton — Evelyn Benson — Robert F. Ilson
1986   *Lexicographic description of English.* (Studies in Language Companion Series 14.) Amsterdam — Philadelphia: Benjamins.
Berndt, Rolf
1965   "The linguistic situation in England from the Norman conquest to the loss of Normandy (1066-1204)", *Philologica Pragensia* 8: 145-163.
[1969]   [Reprinted in: Roger Lass (ed.), *Approaches to English historical linguistics: an anthology.* New York: Holt, Rinehart and Winston, 369-391.]
1984   *A history of the English language.* (2nd edition.) Leipzig: Verlag Enzyklopädie.
Betz, Werner
1936   *Der Einfluss des Lateinischen auf den althochdeutschen Sprachschatz. 1. Der Abrogans.* (Germanische Bibliothek 2, Untersuchungen und Texte 40.) Heidelberg: Winter.
1949   *Deutsch und Lateinisch: die Lehnbildungen der althochdeutschen Benediktinerregel.* Bonn: Bouvier.
1951   "Lateinisch und Deutsch", *Der Deutschunterricht* 3: 21-36.
1974   "Lehnwörter und Lehnprägungen im Vor- und Frühdeutschen", in: Friedrich Maurer — Heinz Rupp (eds.). *Deutsche Wortgeschichte,* vol. 1. (3rd edition.) (Grundriß der germanischen Philologie 17.) Berlin — New York: de Gruyter, 135-163.
Bloomfield, Leonard
1914   *An introduction to the study of language.* New York: Holt.
[1983]   [New edition with an introduction by Joseph F. Kess. (Amsterdam Studies in the Theory and History of Linguistic Science 11, 3.) Amsterdam — Philadelphia: Benjamins.]
1935   *Language.* (Rev. [British] edition.) London: Allen and Unwin.
Bolton, Whitney French
1982   *A living language: the history and structure of English.* New York: Random House.

Bourcier, Georges
 1981 *An introduction to the history of the English language.* [*Histoire de la langue anglaise du Moyen Age à nos jours.*] English adaptation by Cecily Clark. Cheltenham: Thornes.

Carstensen, Broder
 1968 "Zur Systematik und Terminologie deutsch-englischer Lehnbeziehungen", in: Herbert Ernst Brekle — Leonhard Lipka (eds.), *Wortbildung, Syntax und Morphologie: Festschrift zum 60. Geburtstag von Hans Marchand am 1. Oktober 1967.* (Janua Linguarum, Series Maior 36.) The Hague — Paris: Mouton, 32-45.

Carstensen, Broder — Ulrich Busse
 1993-1996 *Anglizismen-Wörterbuch: der Einfluß des Englischen auf den deutschen Wortschatz nach 1945.* 3 vols. Berlin — New York: de Gruyter.

Ekwall, Eilert
 1930 "How long did the Scandinavian language survive in England?", in: Niels Bøgholm (ed.), *A grammatical miscellany offered to Otto Jespersen on his 70th birthday.* Copenhagen: Levin & Munksgaard, 17-30.

Fischer, Andreas
 1986 *Engagement, wedding, and marriage in Old English.* (Anglistische Forschungen 176.) Heidelberg: Winter.
 1989 "Lexical change in Late Old English: from *æ* to *lagu*", in: Andreas Fischer (ed.), *The history and the dialects of English: festschrift for Eduard Kolb.* (Anglistische Forschungen 203.) Heidelberg: Winter, 103-114.
 1992 "Laughing and smiling in the history of English", in: Wilhelm G. Busse (ed.), *Anglistentag 1991 Düsseldorf: Proceedings.* Tübingen: Niemeyer, 51-62.
 1994 "'Sumer is icumen in': the seasons of the year in Middle English and Early Modern English", in: Dieter Kastovsky (ed.), *Studies in Early Modern English.* (Topics in English Linguistics 13.) Berlin: Mouton de Gruyter, 79-95.

1997 " 'With this ring I thee wed': the verbs *to wed* and *to marry* in the history of English", in: Raymond Hickey — Stanisław Puppel (eds.), *Language history and linguistic modelling: a festschrift for Jacek Fisiak on his 60th birthday*, vol. 1. (Trends in Linguistics, Studies and Monographs 101.) Berlin — New York: Mouton de Gruyter, 467-481.

Gneuss, Helmut
1955 *Lehnbildungen und Lehnbedeutungen im Altenglischen.* Berlin: Schmidt.
1992 "*Anglicae linguae interpretatio*: language contact, lexical borrowing and glossing in Anglo-Saxon England", *Sir Israel Gollancz Memorial Lecture: proceedings of the British Academy* 82: 107-148.

Görlach, Manfred
1974 *Einführung in die englische Sprachgeschichte.* (UTB 383.) Heidelberg: Quelle und Meyer.
1978 *Einführung ins Frühneuenglische.* (UTB 820.) Heidelberg: Quelle und Meyer.
1990 "Heteronymy in international English", *English World-Wide* 11: 239-274.

Haugen, Einar
1950 "The analysis of linguistic borrowing", *Language* 26: 210-231.
1953 *The Norwegian language in America: a study in bilingual behavior.* 2 vols. Philadelphia: University of Pennsylvania Press.

Jespersen, Otto
1905 *Growth and structure of the English language.* Stuttgart: Teubner.
[1982] [9th edition. Oxford: Blackwell.]

Jorgensen, Peter A. — Barbara M. Ferré
1986 "Kategoriale Überlappungen im sprachlichen Entlehnungsprozeß", *Zeitschrift für Dialektologie und Linguistik* 53: 6-18.

Käsmann, Hans
1961 *Studien zum kirchlichen Wortschatz des Mittelenglischen 1100-1350: ein Beitrag zum Problem der Sprachmischung.* (Buchreihe der Anglia 9.) Tübingen: Niemeyer.

Kay, Christian J.
1994 "*Historical thesaurus of English*: progress and plans", in: Merja Kytö — Matti Rissanen — Susan Wright (eds.), 111-120.

Koziol, Herbert
1984 *Grundzüge der Geschichte der englischen Sprache*. (3rd edition.) (Grundzüge 9.) Darmstadt: Wissenschaftliche Buchgesellschaft.

Kytö, Merja— Matti Rissanen — Susan Wright (eds.)
1994 *Corpora across the centuries: proceedings of the first international colloquium on English diachronic corpora. St. Catherine's College Cambridge, 25-27 March 1993.* (Language and Computers: Studies in Practical Linguistics 11.) Amsterdam — Atlanta, Ga.: Rodopi.

Lass, Roger
1987 *The shape of English: structure and history*. London: Dent.

Leisi, Ernst
1985 *Das heutige Englisch: Wesenszüge und Probleme.* (7th edition.) (Sprachwissenschaftliche Studienbücher.) Heidelberg: Winter.

McArthur, Tom (ed.)
1992 *The Oxford companion to the English language*. Oxford — New York: Oxford University Press.

Page, R. I.
1971 "How long did the Scandinavian language survive in England? The epigraphical evidence", in: Peter Clemoes — Kathleen Hughes (eds.), *England before the Conquest: studies in primary sources presented to Dorothy Whitelock.* Cambridge: Cambridge University Press, 165-181.

Pfeffer, J. Alan
1987 *Deutsches Sprachgut im Wortschatz der Amerikaner und Engländer: vergleichendes Lexikon mit analytischer Einführung und historischem Überblick.* Tübingen: Niemeyer.

Roberts, Jane
1994 "*Old English thesaurus*: progress and plans", in: Merja Kytö — Matti Rissanen — Susan Wright (eds.), 155-161.

Roberts, Jane — Christian J. Kay — Lynne Grundy
1995 *A thesaurus of Old English.* 2 vols. (King's College London Medieval Studies 11.) London: King's College London, Centre for Late Antique and Medieval Studies.

Rynell, Alarik
1948 *The rivalry of Scandinavian and native synonyms in Middle English especially "taken" and "nimen".* (Lund Studies in English 13.) Lund: Gleerup.

Samuels, Michael Louis
1985 "The great Scandinavian belt", in: Roger Eaton — Olga Fischer — Willem Koopman — Frederike van der Leek (eds.), *Papers from the fourth international conference on English historical linguistics, Amsterdam, 10-13 April 1985.* (Amsterdam Studies in the Theory and History of Linguistic Science 4, 41.) Amsterdam — Philadelphia: Benjamins, 269-281.

Scheler, Manfred
1977 *Der englische Wortschatz.* (Grundlagen der Anglistik und Amerikanistik 9.) Berlin: Schmidt.

Serjeantson, Mary Sidney
1935 *A history of foreign words in English.* London: Kegan Paul, Trench, Trubner.

Thomason, Sarah Grey — Terrence Kaufman
1988 *Language contact, creolization and genetic linguistics.* Berkeley: University of California Press.

Ullmann, Stephen
1962 *Semantics: an introduction to the science of meaning.* Oxford: Blackwell.

Udo Fries (Zürich)
Foreign place names in the ZEN-Corpus

1. Introduction

ZEN, the *Zurich English Newspaper Corpus*, is a machine-readable corpus of London newspapers, covering the period from 1671 to 1791. It is not yet complete.

Only very few newspapers appeared before 1671. The first of these was *The London Gazette*, established from the very beginning as an official vehicle for news, and first published in 1665; it appeared twice weekly. *The Daily Courant*, the first daily newspaper, began to appear in 1702. During Queen Anne's reign (1702-1714), about four morning papers as well as the *London Gazette* were published on Tuesdays, Thursdays, and Saturdays, when "the post left the capital for all parts of the British Isles" (Harris 1978: 87). By around 1710, "twenty papers of two or four pages were appearing weekly, twice- and thrice-weekly ... in London" (Harris 1978: 83f.).

The ZEN-Corpus includes texts from *The Post Man, The Post Boy, The London Post, The English Post, The Daily Post, The Flying Post; Read's Weekly Journal, The Country Journal*, and *The Daily Journal*; furthermore, there are papers whose major concern was the publication of advertisements but which also carried foreign news, like *The London Morning Advertiser*, and *The General Advertiser*. Finally, sections from *The Morning Chronicle, The London Chronicle, Lloyd's Evening Post, The General Evening Post*, and *The Gazetteer* are also included.

Many of these papers were read in coffee-houses. In 1739, 559 coffee-houses were in business in London, "and most found it necessary to offer a selection of newspapers for their customers" (Harris 1978: 86).

For the ZEN-Corpus, texts from the beginning of 13 decades were collected: from 1671, 1681, 1691, 1701, and so on until 1791 in order to cover the intervening 120 years as fully as possible. We stopped in 1791, nine years before the end of the century, and six years after 1785, in which year *The Universal Daily Chronicle*, which changed its name into *The Times* a few years later, started publication. We have – so far – not included *The Times*, because it is easily available around the world on microfilm.

Udo Fries

2. Foreign place-names

A typical foreign news item looks as follows:

> Milan December 26. *The Count* Ercole Viscomti, *Envoye from His Catholick Majesty to the Great Duke of* Toscany, *is returned thence extreamely satisfied with his entertainment in that Court.*

This shows that foreign place-names may occur in two positions in the papers: first, as a sort of headline, an introductory line to the following text, usually printed in italics; and secondly, foreign names appear within the text itself, in italics or in normal print. For a foreign name to appear in an English text, there must have been some kind of language contact. The text, normally, arrived in London either in French or in English. In the former case, somebody at the London office of the newspaper worked as a translator, in the latter case, someone writing in English had to decide what to do with a foreign place-name.

No doubt, through these early newspapers, a larger public than ever before became acquainted in this way with places and place-names they had never heard before. On the other hand, many foreign place-names had been known to English-speaking people for many centuries.

In both instances, the old and the more recent acquisitions, at some point there must have been some contact with a foreign language, though not necessarily with the language in which the place-name in question was originally coined.

Today, most foreign place-names in English newspapers appear in just one form, there is hardly any variation in the spelling or in the form of foreign place-names, be it in the news media or in other printed texts. In this respect, late-seventeenth- and eighteenth-century newspapers clearly differ from present-day texts. Usage was not as fixed as it is today. Yet even today, political developments lead to new names for old places, and this can be seen in newspapers.

3. Stability of place-names

There is a group of place-names for which there seems to be no variation in the ZEN-Corpus – and which are spelled in the same way today; there has "always" been a separate English form:

Foreign place names in the ZEN-corpus

(1)   ZEN   Modern English   Italian
       Rome   *Rome*   *Roma*
       Florence   *Florence*   *Firenze*
       Milan   *Milan*   *Milano*
       Naples   *Naples*   *Napoli*
       Venice   *Venice*   *Venezia*

(2)   ZEN   Modern English   German
       Brunswick   *Brunswick*   *Braunschweig*
       Cologne   *Cologne*   *Köln*
       Hanover   *Hanover*   *Hannover*
       Vienna   *Vienna*   *Wien*

In all these instances, English does not use the standard Italian form of today. French can often be clearly seen as the mediator between Italian and English:

(3)   French   Modern English
       Florence   *Florence*
       Novarre   *Novarre*
       Venice   *Venice*

The *OED* contains two nineteenth-century quotations of the Italian form *Novara*; *Collins English Dictionary* (1991) gives *Novara* only.

French was also the mediator between Russian and English in

(4)   French   English
       Muscovie   *Muscovy*
       Flandres   *Flanders*

Muscovy refers to the principality of *Moscow*.

However, a lack of variation in seventeenth- and eighteenth-century newspapers does not mean that there was no variation before that period. Especially in the fifteenth and sixteenth centuries, there was much variation, some instances of which are given in (5):

(5) Older forms ZEN
Rom, Roome, Room Rome
Melayn(e), Mylleyn, Mil(l)ayn(e), Myllain, Milan
Millain(e), Millan(e), Melane
Venysse, Venise, Ven(i)ys, Vennys, Venice
Venes, Vennis, Venis

Earlier spellings are taken from recordings found in the *OED2* on CD-ROM.

Rome is used in Chaucer's famous verse where it rhymes with *to me*, a line that has been used as evidence of Chaucer's pronunciation of final *-e* in rhyme.

(6) *With hym ther rood a gentil Pardoner*
of Rouncivale, his freend and his compeer,
That sreight was come fro the court of Rome.
Ful loude he soong "Com hider, love, to me".

Chaucer also uses *Venyse, Melan,* and *Milayn.*

In Shakespeare's *Tempest*, we find both *Millaine* and *Milan*; a text quoted in the *OED* for 1623 gives *Millane*, referring to "those rediculous Italian Gazetts, that come from Rome, Millane, and Antwerp", but Massinger's *Duke of Milan* of the same year has only the modern form.

The Italian form *Milano*, which occasionally occurs in modern texts, does not refer to the city, but to *panettone* and to soccer. From the *OED*:

(7) 1938 Joyce *Let.* 20 April (1966) iii 420
I spent all Easter day with Lucia, ... eating panettoni di Milano and fooling generally

1978 S. Brill *Teamsters* ix. 327
Milano is most probably getting similar retainers from several Teamster locals.

The modern Italian *Venezia* is also occasionally recorded in the *OED*.

Only larger places which have acquired some additional meaning beside their reference to a town or country are, however, included in the *OED* with a headword of their own. Other place-names can only be found with a command to search the entire *OED*, but many place-names do not appear in the *OED* at all.

Foreign place names in the ZEN-corpus

Thus, the *OED* includes:
- *Florence*, because the word also refers to a gold florin (from 1400), to certain woven fabrics (1483), and a type of wine (from 1707).
- *Milan*, used attributively for textile fabrics and steel work (as in Milan bonnet, fustian, gloves, lace, sleeves, and Milan needles, points, and steel).
- *Leghorn*, as the name of straw plaiting for hats and bonnets and, from the nineteenth century onwards, as the name of a breed of domestic fowl.
- *Lisbon*, because of a white wine, soft sugar, and a kind of lemon using this name.
- *Naples*, for various things in some way connected or associated with Naples. Similarly, with *Venice*.

Basically, the first group of place-names, including *Rome*, *Venice* and *Florence*, have not changed their spellings for many centuries; their pronunciation follows the normal lines of phonological development in English.

We also come across early English forms that have survived until today and do not seem to follow the Italian pattern or pronunciation. Besides the standard language, one would, however, have to investigate the form of these words in local Italian dialects.

(8) English   Italian
     Rome   *Roma*
     Turin   *Torino*

Then, there are shortened forms in English: Italian towns ending in *-ano, -ino* appear as *-an, -in* in English.

(9) English   Italian
     Milan   *Milano*
     Urbin   *Urbino*
     Turin   *Torino*

The Italian *Toscana* is Modern English *Tuscany*. A form with <o> in the stem is used in the corpus, and recorded between 1617 to 1666 in the OED. This was clearly a marked form; the great majority of quotations in the *OED*, also beginning in the early seventeenth century, have the modern form *Tuscany*.

121

(10)   ZEN   Modern English   Italian
       Toscany ~ Tuscany   Tuscany   Toscana

To interpret changes correctly, we must take account of developments in the countries/languages of origin and in the form of the name at the time of borrowing. A well-known instance is the Modern Italian *Livorno*, which goes back to *Legorno*, a form used in the sixteenth and seventeenth centuries. In Modern English, the two forms *Leghorn* (mainly used by the older generation), and *Livorno* are used side-by-side.

(11)   ZEN   Modern English   Italian
       Legorn, Leghorn   Leghorn, Livorno   Livorno

The earliest quotation in the *OED* is from no. 16 of the *London Gazette*, in its first year of publication, 1665:

(12)   *A Legorn ship ... bound to Tunis with moneys to lade Corn.*

The modern form, *Livorno*, is recorded just once, for 1967, in the *OED*, with reference to a Socialist schism at *Livorno*. There is, of course, an emotional side to changing place-names: in a letter to the Editor of an English newspaper a few years ago, someone was bitterly complaining about this new name, *Livorno*, and pleading for the old version of *Leghorn*.

This is a common feature that can be found in other languages as well. Some twenty years ago *Die Presse* in Vienna had an article about *Kromeritz*, which they thought could be used as the German for the Czech *Kromeritze*, only to be told in a letter that there was already a German form, *Kremsier*, used until 1918 by the German-speaking population.

In spite of all this, there is a large number of place-names that have remained stable over the centuries, place-names that lack synchronic variety of forms in the ZEN-Corpus, and which have remained stable in the subsequent development, i.e. from the eighteenth to the twentieth century.

3. Synchronic variety in place-names

Place-names in the ZEN-Corpus may appear in different forms, even in the same year, and even in the same edition of a newspaper:

(13) ZEN
  Lisbonne   ~   Lisbon
  Wolfenbutel   ~   Wolfenbuttel   ~   Wolffenbuttel
  Muscovy   ~   Moscow   ~   Mosko

The form *-bonne* in *Lisbonne* is clearly a variant of *-bon*. The two forms are also found in German names:

(14) ZEN
  Bon   ~   Bonn
  Ratisbonne   ~   Ratisbone   ~   Ratisbon

The Portuguese form, *Lisboa*, does not occur, but the *OED* has more than a hundred instances of *Lisbon*, many of which, however, refer to the *Lisbon (Dakota) Star*, from 1883 and later.

The *OED* also records the German form *Wolfenbüttel* with reference to music and the famous library – with the umlaut – between 1611 and 1805.

Moscow is a special case. The modern form, *Moscow*, appears in the *OED* as the "name of the capital of the U.S.S.R." *Muscovy*, on the other hand, refers to the principality of Moscow, applied by extension to Russia generally. *Muscovy* is the normal form in the seventeenth century, but there were more variants than the ZEN-Corpus shows: (15) shows varieties in the seventeenth century:

(15)   *Muscovie*
     Muskevia
     Muscovia

Earlier, in the sixteenth century, forms include *Muskovie* and *Muskovia*, both spelled with a <k>, as in (16), from the *OED*:

(16)   c 1573
     c 1591
     The Companie of Merchauntes tradinge Muskovia havinge bene preiudiced by the errours.

Also, the variant *Mosko* is not a single instance, but has its ancestors in the sixteenth century:

123

Udo Fries

(17)   1591 G. Fletcher *Russe Commw.*
         One hundred rubbles of going money of Mosko.
         The Metropolite of Mosko.

The earliest instance of the spelling *Bonn* in the *OED* goes back to 1752, in a reference to the "courts of Mannheim and Bonn".
Ratisbonne takes up the Latin form of *Regensburg*, lat. *Ratisbona*.
The broad influence of French on foreign place-names was pointed out above. But the original forms are also used in the ZEN-Corpus:

(18)   ZEN   Italian   Modern English
       Mantoua   *Mantova*   *Mantua*
       Padoua   *Padova*   *Padua*
       Genoua/Genova   *Genova*   *Genoa*

The modern forms used in English are: *Mantua*, *Padua* and *Genoa*.
The modern inhabitants are either the *Genoese* or the *Genovese*.

5. Anglicisation

Just as in *Edinburgh*, German names ending in *-berg* or *-burg* take a final <h>:

(19)   ZEN   German
       Lunenburgh   - *Lüneburg*
       Heydelbergh   - *Heidelberg*
       Oldenburgh   - *Oldenburg*
       Brandenburgh ~ Brandenburg - *Brandenburg*

But note, (perhaps) on the analogy of *Luxembourg* or *Strasbourg* via French:

(20)   ZEN   German
       Hambourg   - *Hamburg*
       Presbourg   - *Pressburg*
       Mersebourg   - *Merseburg*
       Neubourg   - *Neuburg*

Foreign place names in the ZEN-corpus

Apart from the German form used in Austria, *Pressburg* is today *Bratislava*. In Austria, however, old names of the former Austro-Hungarian empire have survived two world wars, e.g., *Pressburg* for *Bratislava*, *Agram* for *Zagreb*, and *Laibach* for *Ljubljana*, which is difficult to pronounce for speakers of German.

Strasbourg occurs in two variants in the newspaper corpus, one French and one German:

(21)   ZEN   Modern English   French   German
      Strasbourg ~   *Strasbourg*   *Strasbourg*   *Strassburg*
      Strasburg

French influence can also be observed in Dutch names (which did not survive):

(22)   ZEN   Dutch
      Groningue   *Groningen*

There was also a French influence on Swiss names:

(23)   17th/18th century English   French   German
      Berne   *Berne*   *Bern*
      Basle   *Bâle*   *Basel*

The German umlaut is not adopted:

(24)   ZEN   German
      Aichstadt   – *Eichstädt*
      Dusseldorp   – *Düsseldorf*
      Lunenburgh   – *Lüneburg*
      Munster   – *Münster*
      Wolfenbutel   – *Wolfenbüttel*
      Zurick   – *Zürich*

It would, of course, be wrong to compare the English forms of eighteenth-century place-names with their present-day foreign forms, as there have been changes in other languages as well.

Innsbruck, to take an Austrian city, had spellings with one or two <n>, its bridge being spelled with either <b> or <p>. *Linz* was spelt without or alternatively with a <t>: *Lintz*.

Udo Fries

A few earlier forms are listed in (25).

(25) ZEN   Earlier foreign
    Aichstadt   *Aichstet*   1488
    Dusseldorp   *Dusseldorp*   1159
    Ratisbon   *Ratisbona*   704
    Wolfenbutel   *Wolffenbuttel*   1267
    Erfort   *Erfort*   1350
    Heydelbergh   *Heydelberga*   1184
    Dantzick   *Dantzkh*   1454

6. Changes after the eighteenth century

Frequently, there has been a change towards adaptation to the foreign form, for example by a change in the spelling of the stem vowel:

(26) ZEN   Modern English   Foreign
    Burdeaux   *Bordeaux*   *Bordeaux*

Chaucer has *Burdeux*.
Variation of stem and medial <-e->:

(27) ZEN   Modern English   Foreign
    Fountainbleau   *Fontainebleau*   *Fontainebleau*

Older <o> to <ou>:

(28) ZEN   Modern English   Foreign
    Lovain   *Louvain*   *Louvain/Leuven*

Double <ll>:

(29) ZEN   Modern English   Foreign
    Barcellona   *Barcelona*   *Barcelona*

Addition of an <h>:

(30)	ZEN	Modern English	Foreign
Tholouse	*Toulouse*	*Toulouse*
Thoulon ~ Toulon	*Toulon*	*Toulon*
Maltha	*Malta*	*Malta*

Diphthongal spelling <ey> instead of <ei>:

(31)	ZEN	Modern English	Foreign
Heydelbergh	*Heidelberg*	*Heidelberg*

Endings in <-er> instead of <-ar>:

(32)	ZEN	Modern English	Foreign
Gibralter	*Gibraltar*	*Gibraltar*

Endings in <-or> instead of <-ur>:

(33)	ZEN	Modern English	Foreign
Erfort	*Erfurt*	*Erfurt*

Ending <-y> instead of <-ey>:

(34)	ZEN	Modern English	Foreign
Turky	*Turkey*	*Turkey*

Sancto to *Santo*:

(35)	ZEN	Modern English	Foreign
Sancto Domingo	*Santo Domingo*	*Santo Domingo*

Seventeenth/Eighteenth-century form in <-ick>

(36)	ZEN	Modern English	Foreign
Zurick	Zurich	Zürich
Dantzick	Danzig	Danzig/Gdańsk

No change towards the foreign word took place in the following instances:

Denmark: occurs in at least three varieties in the seventeenth/eighteenth century:

(37) ZEN   Modern English   Foreign
      Danemark(e) ~ Denmark   Denmark   Danmark

Variation of final <-e>:

(38) ZEN   Modern English   Foreign
      Spaine ~ Spain   Spain   España
      Stockholme ~ Stockhome   Stockholm   Stockholm
      Ukrain   the Ukraine

English adds an <-s>:

(39) ZEN   Modern English   Foreign
      Lyons   Lyons   Lyon

The ancient *Lugdunum* is *Lyon* in Modern French, but *Lyons* [laiənz] in English.

Finally, there may have been a complete change of name:

Seventeenth- and eighteenth-century newspapers lead us to some places which have since changed their name:

(40) ZEN   Modern English   Foreign
      Adrianople   Edirne   Edirne
      Constantinople   Istanbul   Istanbul
      Presbourg   Bratislava/Pressburg Bratislava

7. Conclusion

In England, as elsewhere, changes in foreign names may be a very emotional matter, at least for some people. "To hell with Beijing," exclaimed Herbert Sandford, of the College of St. Mark and St. Peter, Plymouth, speaking for the geography section of the British Association. He complained, about two years ago, about the faceless people at the BBC and a group of experts at the United Nations (BBC-speak and UN-write), about "a veritable babel of meaningless foreign unpronounceables and unspellables." His rage was directed against the forms *Beijing* and *Guangzhou*, instead of *Peking* and *Canton*, as well as at the rapid changes from *Cambodia* to *Kampuchea* and back to *Cambodia*. These unnecessary changes, he argued, meant "that children were being increasingly burdened by unpronouncable and unspellable names". "Gothenburg, the Swedish city-port," he continued, "is so spelt and pronounced in English. But the UN is spelling it Goteborg, people are calling it Goatborg, and those hearing that are writing Goatborg. And now that the BBC is pronouncing it 'Yer-ter-borg' people think its spelt Yerterborg. Beginning with a Y, of course, they can't look it up in an atlas."

There are clearly two groups of foreign place-names: the familiar ones, which resist change, and the less familiar ones, which one can change around more easily. Part of the problem is that the position of the borderline between these two groups is different for different people. It is *Mailand, Venedig, Paris* und *London* for speakers of German, and *Peking, Shanghai, Burma*, perhaps even *Ceylon* for speakers of English. Spelling may indicate differences in pronunciation, but not necessarily so. Since Kenya (pronounced [kiːnjə]) became independent, it has been pronounced [kenjə].

Early English newspapers are an important source for the adaptation and the spelling of foreign place-names. For a more comprehensive study, they should be supplemented by sixteenth- and seventeenth-century travel reports and geographical/political treatises.

References

Collins
  1991   Collins English dictionary. (3rd edition.) Glasgow: Harper Collins.
Harris, M. – Alan Lee
  1986   The press in English society from the seventeenth to the nineteenth centuries. Rutherford: Fairleigh Dickinson.
OED
  1992   The Oxford English dictionary 2 on compact disk. Oxford: Oxford University Press and AND Software.

Raymond Hickey (Essen)
Language contact and typological difference: transfer between Irish and Irish English

1. Introduction

The purpose of the present paper is to examine the question of contact between two typologically quite different languages, Irish and English, and the role of this in the development of a specifically Irish variety of English since the beginning of the early modern period in the seventeenth century. Care should be taken to note that the concern in the present paper is with the origin of features of Irish English and not with their further development in latter-day non-contact situations.[1] There will be no definitive statement made as to whether contact is responsible for the many idiosyncratic characteristics of Irish English particularly in its syntax. Ultimately proof of contact as a source of non-standard features in a variety is never forthcoming no matter how convincing a case can be made for contact origins. In the final analysis it is a matter of belief whether one accepts a contact hypothesis or not. However, one can render contact more probable by presenting the case in a linguistically convincing manner.[2] One of the major difficulties is that the contact case for Irish English has been exaggerated by Celtic scholars, perhaps because of a false sense of loyalty to Irish. Indeed many linguistically sound analyses of Irish English have come from authors who do not favour contact as the default source for non-standard syntactic structures.[3] The time would appear to have come to redress the balance somewhat not simply by restating uncritically the stance of older authors but by offering a new analysis, above all by considering in more detail the structure of Irish and by taking into account all necessary factors, such as syntactic principles, morphological type, intonational patterns etc., and by assessing the systemic status of elements suspected of having their origin in contact.

---

[1] This has been done elsewhere, for instance by Kallen (1989, 1990) in his treatment of aspectual types in Dublin English.
[2] As has been done in more recent contributions such as Filppula and Ó Baoill, see other references to Filppula (1990) and Ó Baoill (1990).
[3] See the various publications by John Harris such as Harris (1983, 1984, 1986, 1991).

2. Preliminaries on contact

2.1. Contact-based explanations are as old as historical linguistics itself. Broadening out for a moment from Irish to the group of languages to which it belongs one sees that the possible influence of Celtic on neighbouring languages has had more than its fair share of attention from linguists. The reason for this lies not so much in an undue affinity with the Celtic world as in the doors it opens for contact theories seeing as how the Celts from their first appearance in history have been in contact with many peoples whose languages developed unexpected features. A comprehensive assessment of these is not the aim of the present study but it would be perhaps appropriate to mention the main instances with a view to arriving at a clearer picture of types of contact.

2.2. Contact with Germanic

One of the defining features of the Germanic branch of Indo-European is initial stress accent which separates it from other more conservative sub-groups of the family such as Slavic or Baltic. The fixing of stress can be postulated to have occurred by about around 500 B.C., for both Celtic and Germanic (Salmons 1984: 269ff., 1992: 87ff.). Salmons notes that accent shift, particularly a fixed stress accent, is a common feature in language contact situations and postulates that in Germanic it could have stemmed from Celtic going on the assumption (Salmons 1984: 274) that the Celtic group was dominant over the Germanic one.[4]

2.3. Contact with Western Romance

2.3.1. The development of /y/

The notion that the high front rounded vowel of French derived from Celtic contact was apparently an idea first proposed by G.-I. Ascoli. A good earlier examination of the arguments for and against this are to be found in Wartburg (1950) where he has a chapter "Die Auswirkung des Gallischen" ('The effect of Gaulish') in which he enumerates the changes which he sees as being due to Celtic influence such as the development of

---

[4] Salmons (1984: 118) is inconclusive on the direction of influence (Celtic to Germanic or vice versa) and just points to Celtic domination.

ct → xt → it (due supposedly to frequent palatalisation in Celtic languages). For u → ü in French he gives a very detailed treatment (Wartburg 1950: 36-51). Wartburg is a good example for an author who is sceptical about presupposing that contact is the source of a hitherto unexplained feature. Basically he says there are three objections to Celtic influence here: (1) sometimes /u/ causes velarisation and not fronting, (2) the shift to /y/ is not complete in French, and (3) its distribution is not coterminous with that of the Celtic area of Gaul.

Modern argumentation is based on the alternation of /i/ and /u/ in Welsh, also found in Celtic words such as *BRUS, BRIS* which give Old French *bruisier* (ModFr. *briser*) and English *bruise*. Note that /y/ is not found in Celtic anymore. The Welsh alternation is between /i/ and a centralised /ɨ/. In Scottish Gaelic and Ulster Irish there is a fronted realisation of /u/ [ʉ], but this is unconditional, i.e. not due to anything like umlaut.

2.3.2. Lenition in Western Romance

The particular case of /y/ in French is a weak case for contact origin.[5] Paradoxically, the more general assumption that the weakening of consonants in Western Romance could be due to substrate influence from Celtic speakers is on a much sounder footing. Lenition is a general feature of French and Iberian Romance and of course of all the Celtic languages. The arguments for Celtic origin have been put forward on a number of occasions, the best presentation being that of Martinet (1952) where he claims that both the historical loss of intervocalic and final consonants in French and the present-day alternations in Spanish are of the same kind as the lenition which has been morphologised in all the Celtic languages.

This line of argument has been taken up by many authors since, such as Delattre, who maintains (1964: 322ff.; 1969/1970) that the lax pronunciation of the Celts in Gaul had a significant effect on the pronunciation of Vulgar Latin in this region and compares Old French with modern English in the strong stress on a syllable in a word with

---

[5] This applies to similar cases such as tracing Spanish /h/ from Latin *F* back to a Basque substratum in Ibero-Romance. It is true that Basque originally did not have /f/ (only latterly in loan-words) but the /h/ from *F* might just as well have been an internal development in Spanish. Of course the articulatory motivation for this may well stem ultimately from lenition as a make-up feature of Spanish which in its turn may be a contact feature. This puts the contact source at one remove but in fact increases its plausibility.

accompanying weak stress on other syllables (the syllable-timing of Modern French is a relatively recent development). Delattre is sometimes vague in his statements and not all his arguments are equally convincing, see the opposing views of Fowkes (1966) who examines modern Welsh to partly dismantle some of Delattre's more extreme claims about features of Celtic and Old French.

The curious fact of this putative influence is that as a general directive in Celtic, consonant lenition may well have migrated to the Romance dialects with which it was in contact. This is quite clearly the case in Ireland where the same lenition as a general make-up feature of Irish has spread into English. The case for contact, however, breaks down if one tries to maintain a one-to-one correspondence of lenited elements on a segmental level.

2.3.3. Head marking

Nichols in her examination of typological features within a very large cross-section of languages suggests (1992: 273f.) that the head-marking (rather than dependent-marking) tendencies of colloquial French could be a response to contact in general, specifically the route of contact with Celtic or "the result of pure peripherality of location".

2.4. Types of contact

Any discussion of language contact and ensuing transfer must take the various types of contact[6] and the results for the languages involved in this contact into account. For the present discussion one must distinguish two basic types. The first is "direct transfer", where the effect is immediate, frequently with alteration in the structure of the recipient language. Immediate influence on closed classes of a language (morphology and syntax) presumes intensity of contact and a lack of external constraints such as a notion of standard, perpetuated by general education and a literate public.

The second main type can be termed "delayed effect" contact. The effect is not immediate. There is no structural upheaval in the recipient language but a gradual penetration due to prolonged exposure to another language by large bilingual sections of a community. Characteristic for such a

---

[6] See Appel — Muysken (1986) for a taxonomy of contact.

scenario is low-level influence in a general sense: "speech habits" migrate from one language to another. These may lead later to structural if not indeed typological change. The development of Gaulish French [y], if it has its origins in contact with Celtic (see above), must have arisen in this manner. This view of gradual change is of course more Neogrammarian than one which presupposes the sudden appearance of a contact phenomenon in a recipient language. If Celtic had /y/ at the time of the initial development of Latin to French in Gaul (which is by no means certain), then an abrupt appearance could only have occurred in a scenario which assumes lexical diffusion: the Romance speakers started borrowing words from the Celts and among these words would have been some with /y/ and this pronunciation would have then spread to encompass native sections of their vocabulary causing a shift of *U* to /y/.

2.5. Sources for features

When viewing the characteristic features of Irish English one can recognise three possible situations. It is by no means possible to assign features unambiguously to these positions as the evidence is undecisive.

2.5.1. Features which only have an Irish source

The view is commonly held that virtually the only Irish English structure which can definitely be attributed to direct Irish influence (Harris 1984: 319f.) is the use of *after* and present participle to express an immediate perfective aspect (see section 6. below). This is certainly the easiest case as there is no model in other varieties or archaic forms of English which could have been an input to Irish English in this case. In addition it represents a quite natural metaphorical extension of a locative statement to the temporal sphere (Hickey 2000) and hence could travel well across a language barrier. However, there are a number of other phenomena which can be traced back to Irish if they are analysed properly, see section 3.4. on convergence.

2.5.2. Features from Irish or dialect/archaic forms of English

This group is that which has provoked the contact sceptics to turn away from Irish as a source. The following are two clear examples. (1) The distinction between second person singular personal pronouns in Irish English (*you* for the singular and various forms for the plural: *ye* [ji], *youse* [juz], *yez* [jiz], the differential use of the latter is sociolinguistically significant). (2) The use of the word-order OV (NON-FINITE) to indicate the resultative perfective aspect: *I've the book read* 'I am finished reading the book' which contrasts with *I've read the book* 'I read it once'. The object-verb word-order has of course precedents in the history of English and corresponds to the original Germanic sentence brace which is still to be seen in German (*Ich habe das Buch gelesen.*). But equally it has an equivalent in Irish in which the past participle always follows the object: *Tá an leabhar léite agam*, literally 'is the book read at-me'.

2.5.3. Features which have no Irish model

Care should be taken here: some features which exist represent transfer on a general level and might not be recognised as such although they have a substratal source. A case in point is lenition in Irish English (see section 3.4.4.). Apart from such deceptive cases there are phenomena which do not have any parallel in Irish. Certain aspects of Dublin English can be treated under this heading, such as the breaking of long vowels or more recent changes such as the Dublin vowel shift (retraction of /aː, ai/ and raising of /ɒ, ɔː, ɔi/ (Hickey 1998, 1999)) which represent developments independent of Irish and not immediately traceable to the varieties of English imported into Ireland.

3. The speech community in early modern Ireland

3.1.1. Periodisation

The presence of English in Ireland can be divided into two main periods. The first is that from the initial invasion of Anglo-Norman overlords and the English in their retinue (1169 onwards) up until the renewed campaign of Anglification by the Tudors which culminated in the defeat of the Irish at the Battle of Kinsale in 1601 and the subsequent suppression of specifically Irish forms of social organisation. The second period begins after this and continues up to the present day. The external circumstances were provided by the plantations of large parts of the south of Ireland by English who were assigned land, largely as a reimbursement for services as mercenaries in the years of military strife during and after the Civil War (1642-1649).[7]

At the beginning of the seventeenth century, English was spoken in the towns of the east coast from Dublin in the centre down to Waterford in the south east and in Galway and Limerick in the west where English survived from the first period. There were also a few isolated rural pockets of English speakers. But in all these locations Irish was the vibrant language of the majority, in the countryside this was totally so, as the former rival here, Anglo-Norman, had long since been abandoned when the Norman settlers of the twelfth and thirteenth century assimilated to the numerically superior Irish who surrounded them.

3.1.2. Status of the early modern period

The critical period for the contact situation which obtained in Ireland is from the renewed settlement of Ireland in the seventeenth century to the wholesale decline of Irish as a first language after the Great Famine and the subsequent mass emigration in the nineteenth century (de Fréine 1966). For this time span one must bear other salient aspects in mind. The first is that there was no organised education in English for the Catholics, i.e. for the vast majority of the population, this having been prohibited by the so-called Penal Laws which started in 1695. Any formal instruction which the native Irish may have had was gained from so-called hedge-schools[8], an uncoordinated system of self-taught wandering teachers who eked out a living teaching the local population the rudiments of written English and a small selection of other subjects (Dowling 1936 [1968], 1971). This situation is interesting in itself. Bliss (1977) maintained that the hedge-school system led on the one hand to different stress patterns in Irish English with respect to standard English: either (1) through a lack of knowledge due to the sole experience of English in written form or (2) because of old stress patterns preserved as archaic features in Irish English and on the other hand to an abundance of malapropisms: confusion among phonetically similar words or wrong semantic application of a word. Indeed it is no coincidence that the term "malapropism" comes

---

[7] There were a number of attempted large-scale settlements in the late sixteenth century undertaken by the Tudors in the centre of the country and in the central-south in Munster (in 1585, Hayes-McCoy 1967: 183) but their demographic effect was marginal.
[8] This is a reference to the practice of instruction out of doors to facilitate quick dispersal if confronted suddenly with the authorities.

from a figure (Mrs. Malaprop) in the Irish play *The Rivals* (1775) by Richard Brinsley Sheridan. For the development of a unique feature of Irish English what is important to note is that no normative forces were operative in the period of greatest transition from Irish to English. The Irish were motivated to learn English as it led to social opportunity, or at least was the precondition for this arising. But the process of transition was a long one, which meant that there was a protracted period of better or worse bilingualism which obviously facilitated transfer from Irish to English.

3.1.3. Planters and their English

One can accept that at the beginning of the modern period, Irish was by far and away the dominant language in Ireland. This situation changed, initially on a legal level, when English was reinforced as the official language. By the middle of the seventeenth century the practical effects of English influence were beginning to be felt as large numbers of non-aristocratic English moved to Ireland to settle there permanently. These immigrants form the layer in Irish society known as planters. They would seem to have been socially identifiable as a group and in the early eighteenth century no less an author than Swift was to ridicule the English they spoke which had come under the strong influence of Irish. This he did in his *Irish eloquence* and *A dialogue in the Hybernian stile*, which exhibit lexical items as characteristics of planter English. Now note that these pieces were written just two generations after the Cromwellian plantations. One would indeed expect that lexical items would be the first forms of interference to appear in the English of these settlers.

The planters were dialect speakers mainly from the west and northwest of England. When in Ireland their contact with speakers of southeastern mainland English was negligible. One indirect piece of evidence of this is the fact that no bundles of dialect features corresponding to established features of mainland English dialects are to be found in Ireland. This would seem to point to an adoption of Irish speech habits and a levelling of dialect differences after settlement in Ireland.

Of greater significance is the considerable contact of the planters with their Irish servants and tenants, something which would suggest an influence of the speech of the latter on the former. This opinion is shared by most scholars on the subject. Bliss (1976: 557), for instance, supports the view that the planters were cared for by Irish nurses and had contact with

Language contact and typological difference

the children of the native Irish something along the lines of the model for the development of the English of whites in the south of the United States with respect to the African American population.

For early modern Irish English, it is not what the planters spoke that is of relevance but what type of English the Irish learned. True, the planters will certainly have passed on their variety of English to their offspring, but the sheer numbers of the local Irish would make their language the more likely of the two to have been formative in the genesis of a specifically Irish form of English.[9] One should add furthermore that the planters ceased to be an identifiable social group in the course of the eighteenth century. On a social level they doubtlessly merged into the urban Irish who were in somewhat better positions than their rural counterparts. On the other hand many planters who were affluent became part of that (Protestant) rural gentry which is known in Ireland as the ascendancy class, part of the lower aristocracy and better-off farmers of the eighteenth and nineteenth centuries, much admired by Yeats and much derided by later more realistic writers such as Louis MacNeice (Foster 1988: 167ff.).[10]

The main consideration remains the manner in which the overwhelmingly large Irish-speaking section of the population acquired English. Taking both extremes of the second period one can recognise that Ireland moved from an almost entirely Irish-speaking community to a largely English-speaking country in which there are some very small pockets of Irish speakers along the western seaboard and in which there is a general consciousness of the Irish language for vaguely nationalistic but non-linguistic reasons.

When examining the manner in which the Irish acquired English from the mid-seventeenth to late nineteenth centuries, it has been noted above that there was no means of formal instruction. The upshot of the lack of education and the obvious fact that the planters did not teach the Irish English in any systematic way is that the Irish acquired English of their own accord, to a limited extent through informal tuition but more often than not by picking up as much English as they could from those

---

[9] Thomason – Kaufman (1988: 43) assume that there was no emulation of English features of Irish speakers by the descendants of settlers but that given that the first group was very much more numerous their "speech habits prevailed anyway".

[10] There is arguably an accent of Irish English which is fairly close to standard forms of southern British English and which is known as the "ascendancy accent". However, this has never been linguistically defined and the features it exhibits are in the opinion of the present author more likely if anything to be discourse features and in general non-linguistic.

members of their community who could already speak English. This held for many urban dwellers and for those who had contacts with the English. The details of the process are not known as there are no historical documents on the subject. But it is known that the Irish soon realised that the acquisition of English was to their social advantage. Prominent political figures like Daniel O'Connell (early nineteenth century) urged the native Irish to acquire English as best they could in order to increase their chances of social advancement. MacNamara (1973: 36) says that the Irish have indulged in instrumental bilingualism[11] as there were distinct advantages to be accrued from a knowledge of English but were not motivated by the desire to integrate themselves into and identify with the community of English speakers. Furthermore it is remarkable that the numerous Irish emigrants to the United States in the nineteenth century abandoned their native language virtually instantaneously. There is no diaspora form of Irish,[12] contrast this with many other European ethnic groups which showed and to some extent still show a high degree of language retention.

The fact that the Irish learned English from each other has several consequences. One is that the switch-over to English was a very slow process. It really only got under way in the course of the nineteenth century when the native population was decimated by the Great Famine of the late 1840's and subsequent massive emigration set in, chiefly to America, often via England, this fact feeding the Irish section of the population in Merseyside, Liverpool being the natural port of arrival for the Irish on their crossing to Britain.

Before 1850 there are no reliable census data on the number of Irish speakers. The first census was in 1851 (Adams 1974, 1979), but this shows that for large sections of rural Ireland well over half the population[13]

---

[11] The distinction between "integrative" and "instrumental" bilingualism derives from Lambert, see Gardner — Lambert (1972). In essence the difference is that integrative bilingualism is practised by the speakers of a community who wish to partake in the social life of the second language's community. Instrumental bilingualism characterises those speakers who learn a second language for utilitarian reasons without any subjective commitment to it.

[12] Or practically none, see Kallen (1993).

[13] Estimates for the beginning of the nineteenth century (Hindley 1990: 9), i.e. in the pre-Famine period (Freeman 1957), suggest that over two thirds of the south was still overwhelmingly Irish-speaking, that is, the linguistic Anglicisation of the country did not tip the balance in favour of English until after the Great Famine in the latter half of the nineteenth century. Note that not all authors agree on these estimates; Ó Cuív (1986:

spoke Irish and this two centuries after the energetic re-Anglicisation of Ireland had begun. It is safe to assume that those Irish, above all in rural areas, who knew English did not have anything like native knowledge. Rather they would have typically used it in certain well-defined situations where they had dealings with English speakers: in contact with urban dwellers such as merchants, with native English-speakers such as bailiffs and tax collectors, or at least with those whose English was better than average and who would have, for reasons of social aspiration, chosen to neglect what Irish they had.

3.2. Possible creolisation

3.2.1. Preconditions

The scenario depicted above is one of imperfect bilingualism in which Irish would have been the vernacular in the sense of the most natural and vital of the two languages for the speakers in question. This type of situation is one in which there could well have been incipient creolisation of English as some authors have suggested going on linguistic evidence (Corrigan 1993). Before attempting a conclusion about the possible creole nature of early Irish English it would be instructive to consider just what features would be required to legitimately classify it as a creolised variety of English. For a much fuller discussion of the issues touched on here, see Hickey (1997a).

3.2.2. Phonology

Creoles have restricted phonological inventories and phonotactics. English-based creoles typically shift dental fricatives to stops and simplify clusters. Irish English has, it is true, plosive equivalents to English dental fricatives, i.e. [t̪, d̪] for /θ, ð/, but there is no marked simplification of clusters, for instance sequences of fricative and plosive are legal, e.g. *desk* is [dɛsk]. Clusters may be broken up by vowel epenthesis (see section 3.4.2.) but this is clearly a transfer phenomenon from Irish. Plosives for dental fricatives can also be interpreted in this light.

384ff.) is more conservative in his assumptions concerning the numbers of native speakers in the pre-1851 period.

3.2.3. Grammar

Restructuring of, and not just transfer from, the lexifier language is characteristic of true creoles, indeed this is their defining internal feature, analytic plurals being a clear example where, say, a deictic element in pre-head position is used as exponent of the grammatical category in question. *Them* has been a likely candidate as plural marker in the formation of English-based creoles. There is nothing like this shift to an analytic type in Irish English.

The verbal area in creoles is that which offers most similarity with Irish English. The first salient characteristic of the verbs in creoles, and one which is not found in Irish English, is serialisation where two or more verbs are concatenated in a sentence like *He went look buy sugar*. The second feature is the use of pre-verbal particles for modification purposes, typically to express tense, modality or aspect (Schneider 1990: 89ff.), e.g. *go* for the future, *done* for past state (as in *He done marry his sister* (Feagin 1991)).

Irish English offers two instances of this type of modification which are probably from Irish.

(1) *Do + be* to express habitual aspect as in *He does be in his office every morning.*
(2) *After + present participle* to express immediate perfective aspect as in *He is after drinking the beer.*

In this connection one should mention front-focussing structures like *it*-clefting which are characteristic of Caribbean creoles and Irish English.

(1) a. *It's to Dublin he's gone today.*
    b. *It's her brother who rang up this morning.*

Fronting is not solely a feature of creoles; furthermore Irish influence on the development of Caribbean forms of English cannot be excluded (Bailey 1982; Roy 1986; Williams 1986, 1988).

3.2.4. Lexicon

Given the initial impoverished lexicon of creoles, they tend to use polysemy, semantic extension or periphrasis to cover the lexical range required of them certainly in later creolisation, constructions like Tok Pisin *gras bilong fes* 'beard', *gras bilong hed* 'hair' being standard examples of this phenomenon. Again there is no evidence of this in Irish English.

Reduplication is found in creoles but is absent from Irish English with the exception of the phrase *at all at all* 'not at all' where it would seem to have derived from the rhythmic similarity with the semantically equivalent Irish phrase *ar chor ar bith* (Hickey 1990).

3.2.5. External circumstances

A number of facts would seem to point away from a creole interpretation of Irish English. Firstly, the Irish were too close to England and too often confronted with more or less standard forms of the language by speakers either originating from England or, in the case of urbanites from the east, indeed native speakers of the language. Secondly, the Irish shared a common native language. There was nothing like the mixture of (West African) backgrounds which characterised the forced immigrants in the Caribbean area in the early phase of the slave trade. Anything like abrupt creolisation is ruled out for Irish English as there was no shift in population.[14]

Imperfect acquisition of English would mean for a speaker of Irish that the latter would have had a continuing influence on the former. Indeed if this were not the case, if the Irish learned English and changed over to it fully within their lifetime, the country would have been English-speaking by the eighteenth century, which is quite obviously not the case.

In a situation of imperfect bilingualism structural transfer is at a premium (Haugen 1969; Weinreich 1953). Lexical transfer does not presuppose such a scenario. Just consider the many instances of cultural borrowing where the two languages in question are not in speaker contact (Modern German and English for instance).

Now looking at the deviant features of Irish English today, one finds a fair share of them in syntax, an area not affected by indirect language contact.

3.3. Code-switching

In recent years much attention has been paid to code-switching, for instance in the speech of Latin American (mostly Mexican) immigrants in

---

[14] Given the clear defining features of creoles the designation is treated as binary. Terms such as semi-creole or creoloid are not regarded as appropriate in the present context.

the south-west of the United States, and it has been examined as a possible locus of language change.[15]

The simplest kind of code-switching involves lexical items from one language strewn into sentences in another language. Certainly this is a phenomenon which is observable in present-day Irish vis-à-vis English (Stenson 1990, 1991). The motivation for this is usually extra-linguistic. A speaker of Irish referring to a phenomenon which he/she associates with the English-speaking areas of the country will probably use English keywords, for example *cancer, operation, therapy* when talking of disease, as this is something which would be associated with hospitals which are not located in the Irish-speaking areas. During the development of Irish English the situation would have been the reverse: speakers could be expected to have used code-switching for terms which they associated with their native Irish domestic background. Such code-switching may have been a temporary phenomenon during the period of transition from Irish to English, a pre-stage to lexical borrowing before the elements used in the second language are more or less permanently established. However there are no references to this code-switching and it is not parodied by those authors who chose to caricature Irish English at the beginning of the modern period (Bliss 1979).

Syntactic code-switching is a different matter. Basically there are three types: tag-switching, inter-sentential (Pfaff 1979), intra-sentential switching (Poplack 1980). The second is the most structurally dependent (Berk-Seligson 1986; Clyne 1987). It requires certain joints or crossover points at which one can dock onto the second language. Such points are typically clause breaks or major constituent boundaries.[16] The structure which follows must be in keeping with the syntactic expectations of the beginning of the sentence and be syntactically interpretable to other members of the bilingual community. Prosody plays an important role here as well: a large degree of agreement in the number of stressed (and

---

[15] See Hoffmann (1992: 109ff.) and Romaine (1989: 110ff.) for general discussions of the phenomenon. Further pertinent treatments of the subject are Eastman (1992); Kachru (1982); Mühlhäusler (1985); Myers-Scotton (1993) and Woolford (1983).

[16] Sankoff – Poplack (1981) visualise two constraints: (1) the free morpheme constraint (no switching at a point of bound morphemes) and (2) the equivalence constraint: switches tend to occur at points where the juxtaposition of elements from both languages does not conflict with a syntactic rule from either language. In their government and binding analysis DiSciullo – Muysken – Singh (1986) and Singh (1985) maintain that switching is only possible between elements which are not related by government, i.e. where the last element before a switch may be dominated by a previous category but does not itself dominate a further category.

perhaps) unstressed syllables favours syntactic code-switching. The syntactic contact phenomena of Irish English may have had their origin in this type of switching seeing as how the deviant structures often match the prosodic structure of the Irish original. The absence of such a match could be used to explain why certain central features of Irish were not transferred into English, such as the system of prepositional pronouns. Here the elements which correspond functionally to subjects are placed in clause-final position (see examples (16a-e)). There is no way, however, in which English could have a stressed element expressing subjecthood at the end of a clause or sentence so that transfer is excluded. Indeed none of the attestations of Irish English, even those showing heavily influenced syntax, involve translations of prepositional pronouns to indicate the subject of a sentence in English.

3.4. Convergence

3.4.1. General background

Among the possible scenarios for language contact there is that of language convergence where two or more languages come to resemble each other in the course of time. There are well-known cases such as that of the Indo-Aryan and Dravidian contact area (with retroflexion of consonants as a shared feature, for instance, cf. Emeneau 1956; Gumperz 1971) or that of the Balkans (for a comprehensive discussion of features, see Schaller 1975). These models are not, however, applicable to Ireland. The main reason is that English (in Ireland) has not been influenced in its morphology, if one neglects the retention of a distinction between singular and plural with second person pronouns which has had support due to the similar distinction in Irish. Furthermore, although Irish latterly has been influenced by English (Ó Cuív 1951: 54), there is a chronological staggering involved here: English came under the influence of Irish from the beginning of the early modern period onwards and Irish in the present century shows the effects of the decline in monoglot speakers in the transfer phenomena which appear. But this is not a convergence of linguistic systems on a more or less equal footing, and only this would lead to features arising comparable to, say, the post-positive article or the loss of the infinitive or the pleonastic use of personal pronouns found in most of the Balkan languages.

145

Raymond Hickey

If one moves from a consideration of structural elements to that of low-level phonetic phenomena then one finds that there are areal features in Irish and English in Ireland. Two examples of this are particularly salient. The first is vowel epenthesis and the second consonant lenition.

3.4.2. Epenthesis

This is determined in both Irish and Irish English by phonotactics. In Irish there are severe restrictions on the types of clusters which can occur, thus a sequence of two sonorants is impermissible, particularly one where the first is a liquid.

(2) a. *feilm*   [fʲɛlʲɪmʲ]   'farm'
    b. *arm*   [aɹəm]   'arm'

The distribution of epenthesis varies within the different dialects of Irish. Its scope is greatest in Munster Irish (in the south, cf. Breatnach 1947: 16f.), where it occurs, as elsewhere, in clusters of liquid and fricative. Here, however, the usual condition that the cluster affected by epenthesis be within a single syllable does not apply (example (3b)).

(3) a. *garbh*   [gaɹəv]   'rough'
    b. *dorcha* [dʌɹəxə]   'dark'

Now this epenthesis has spread to all varieties of (southern) Irish English, the latter adopting the phonotactic rules of Irish in this respect. The scope is not quite as great as in Irish, for instance clusters of liquid and fricative are not affected, but it applies universally to sequences of *l* and a following sonorant.

(4) a. *swerve* [swɛɹv]
    b. *film*   [fɪləm]
    c. *helm*   [hɛləm]

3.4.3. Metathesis

Connected with epenthesis is the metathesis of consonants which is widely attested in Irish (de Bhaldraithe 1945: 115f.; de Búrca 1958: 136f.; Ó Cuív 1944: 127f.) and also found to a limited extent in English in Ireland.

The motivation for the most common cases would appear to be the resolution of phonotactically unacceptable clusters. In Irish this is historically well documented. For instance in the Anglo-Norman period (late twelfth to fourteenth centuries) many loans from French entered Irish with affricates. Where these were in initial position they were simplified, in internal or final position the elements of the affricates were metathesised practically without exception.

(5) a. *chaumbre* → *seomra* /sjoːmrə/ 'room'
    b. *page* → *páiste* /paːsjtjə/ 'child'
    c. *college* → *coláiste* /kəlaːsjtjə/ 'college'

In Irish English metathesis has the frequent function of resolving heavy syllable codas consisting for instance of /r/ and a sonorant, a purpose which is fulfilled by epenthesis in the cases of /l/ and a sonorant, see above.

(6) a. *pattern* [pætrən]
    b. *modern* [mɒdrən]

3.4.4. Lenition

The relationship between Irish and Irish English with regard to lenition may at first appear tenuous as there is a lack of phonetic correspondence in its manifestation. In Irish lenition is a morphological phenomenon which has its origins in phonetic weakening of obstruents in intervocalic (sandhi) environments in pre-Old Irish. Its main realisation is (and has been for the entire period of English in Ireland) the fricativisation of stops and the glottalisation or loss of fricatives. This occurs under certain grammatical conditions.

(7) a. *ciall* [kiəɫ] 'sanity'
    b. *a chiall* [ə xiəɫ] 'his sanity'
    c. *saol* [siːɫ] 'life'
    d. *do shaol* [də hiːɫ] 'your life'

The existence of morphological lenition and the similar process of nasalisation in Irish represents one of the major typological differences in the grammars of Irish and English. There is obviously nothing similar to it in the morphology of Irish English. And yet if one considers lenition in a more general light as a phonological directive to weaken segments in inherently weak environments[17], then there is a good deal of common ground between both languages. In Irish English alveolars in intervocalic positions and before a pause are weakened to corresponding fricatives ([t̞] symbolises an apico-alveolar fricative (Hickey 1984)).

(8) a. *Italy* ['ɪt̞əli]
    b. *tight* [tait̞]

It would seem that the principle of lenition as a weakening process has been transferred to English in Ireland, but not of course its manifestation in the donor language. Here one can see that lenition as a directive has been able to straddle the typological divide between both languages.

The above considerations show that while one would look in vain for structural similarities between Irish and English, there are shared peripheral phenomena which are usually associated with the textbook linguistic convergence areas such as the Baltic with regard to tone[18] or the Balkans as mentioned above. If one also takes northern Ireland and Scotland into account, then more such features can be distinguished. For instance the high mid realisation of /uː/, i.e. [ʉ], is an areal feature which holds for most of Scotland, large parts of Ulster and notably for Ulster (Donegal) Irish. Another feature is the lack of phonemic vowel length. For many varieties vowel length is predictable on the grounds of the phonetic environment as specified in the Scottish vowel length rule (Aitken 1981). This also applies to Ulster Irish English showing a continuum from Scotland down the south-west into northern Ireland.

4. Characterising Irish syntax

4.1. Before examining structures in Irish English, a brief characterisation of Irish should be offered to set the framework for later discussion. Probably

---

[17] This would certainly apply to alveolars in English which are often tapped (in forms of American English) or reduced to a glottal stop (forms of urban British English) as in *writer* [raɪɾɚ] and [raɪʔə] respectively.
[18] Jakobson (1931 [1962]) mentions polytony as a typical phenomenon which occurs in language areas: the Baltic area with most Scandinavian languages, North German (dialectally), north Kassubian, Lithuanian, Latvian, Livonian and Estonian. In all these languages (except Lithuanian and Latvian) this is an innovation (Jakobson 1931 [1962]: 137f.). See Lehiste (1988: 65ff.) for a critical reappraisal of Jakobson's original ideas.

the most well known fact about Irish is that, like other Celtic languages, it has a verb-subject-object word-order as seen clearly in the next example.

(9)   *Chuaigh sé thar sáile bliain ó shin.*
      went he over sea year from this
      'He went abroad a year ago.'

This word-order is absolute in Irish. If one wishes to topicalise some element in a sentence by fronting (the normal means in Irish), then this is done by clefting with subordination of the main clause.

(10) *Is thar sáile a chuaigh sé bliain ó shin.*
      it-is over sea that went he year from this
      'He went abroad a year ago.'

4.2. Irish has roughly the same division of word classes as does English. However, the status of these is quite different. Irish is what one can call a noun-centred language. By this is meant that apart from denoting objects, entities, ideas etc., Irish uses nouns with great frequency when referring to actions as well. Here a support verb is found with a noun carrying the main semantic load.

(11) *Rinne sé dearmad faoi.*
      did he forget-NOUN under-it
      'He forgot it.'

In such structures what in English would be a direct object appears in the genitive in Irish.

(12) *Lig siad fios bhur rúin linn.*
      released they knowledge your secret-GENITIVE with-us
      'They revealed your secret to us.'

A further function of the genitive which testifies to its established position in Irish is its use after nouns where in English one would have an adjective.

(13) *Mo theanga dhúchais.*
      my tongue inheritance-GENITIVE
      'My native tongue.'

4.3. As might be expected Irish makes use of the various avenues opened up by this nominalisation. Thus with the word for 'length', *fad*, it applies in a number of situations in which English would have an adjective or adverb.

(14) a. *Tá a thrí fad ann.*
 is its three length in-it
 'It's three times as long.'
 b. *Chuir sé a fhad eile leis.*
 put he its length other-ADJ with-it
 'He made it as long again.'
 c. *Tá fad na teanga air.*
 is length the-GENITIVE tongue-GENITIVE on-him
 'He is very talkative.'
 d. *Fad is a bhí mé ann.*
 length and COMP was-REL I in-it
 'As long as I was there.'

4.4. It will have been noted that in the above sample sentences there occur forms which unite prepositions and pronouns in a single word. These are the prepositional pronouns of Irish which play a central role in the syntax of the language. There are sixteen of them in all and they can combine with any one of the seven personal pronouns; in (15) a selection is offered to give an impression of what these synthetic forms look like.

(15) Base form	1 singular/ plural	2 singular/ plural	3 singular m/ plural	3 singular f.
ag 'at'	*agam*	*agat*	*aige*	*aici*
	againn	*agaibh*	*acu*	
ar 'on'	*orm*	*ort*	*air*	*uirthi*
	orainn	*oraibh*	*orthu*	
faoi	*fúm*	*fút*	*faoi*	*fúithi*
'under'	*fúinn*	*fúibh*	*fúthu*	
le 'with'	*liom*	*leat*	*leis*	*léi*
	linn	*libh*	*leo*	

A central part of the semantics of Irish is realised by extended use of locative and deictic prepositions in combination with personal pronouns.

To convey some of the flavour of such structures consider the following sentences.

(16) a. *Tá dhá orlach agam air.*
 is two inch at-me on-him
 'I am two inches taller than him.'
 b. *Níl dul uaidh agat.*
 is-not going from-it at-you
 'You cannot escape it.'
 c. *Níl teacht agam air.*
 is-not coming at-me on-it
 'I cannot reach it.'
 d. *Níl an teanga ó dhúchas aige.*
 is-not the language from inheritance at-him
 'He is not a native speaker.'
 e. *Is fada liom uaim í.*
 is far with-me from-me she
 'I miss her a lot.'

Indeed the prepositional pronouns can express a subject relation with a verb (example (17a)) or subject relevance with a type of passive (example (17b)) when used with the common impersonal form of a verb.

(17) a. *Rugadh mac di.*
 born-IMPERSONAL son to-her
 'She gave birth to a son.'
 b. *Múchadh an tine orm.*
 extinguish-IMPERSONAL the fire on-me
 'The fire was put out on me.' i.e. 'Someone put out the fire which I had started, was attending, etc.'

Lastly one should mention that contrast is often realised by a different choice of prepositional pronoun as can be seen in examples (18a) and (18b) and that a verb is frequently missing with such constructions, the prepositional forms being sufficient (examples (18c) and (18d)).

(18) a. *Rinne sí gáire liom.*
did she laughter with-me
'She smiled at me.'
b. *Rinne sí gáire fúm.*
did she laughter under-me
'She laughed at me.'
c. *Seo chugainn í.*
here towards-us she
'Here she is coming towards us.'
d. *Fúithi féin atá sé.*
under-her herself that-is it
'It's up to herself.'

5. What gets transferred?

5.1. General aspects

The above very brief sketch of Irish is hopefully sufficient to convey an impression of just how typologically different Irish is from English without even considering the morphology, let alone the phonology of the language. The central question is now: which if any of the above structural types get transferred into English?

Nominalised syntax (Henry 1960) in English is typical for contact speakers in the western seaboard today and is regarded by the English monoglots of the rest of the country as rather antiquated nowadays, somewhat reminiscent of stage Irish and typical of writers such as Synge, who attempted to capture something of the flavour of Irish syntax in the English they employed, albeit with considerable poetic license.

Now consider in general what items of a language get transferred most easily to another. There is little doubt that free-standing discourse elements pass from one language to another quite easily, cf. German with *sure, sorry, ok, bye-bye* from English. In Irish one has *well* from English and in Irish English one had formerly the filler *arrah* from Irish with more or less the meaning of 'no matter, whatever'.

Looking at other elements in a language one can ascertain that idioms and individual lexical items travel well, as they too are stand-alone elements in a language and can be prised out of the donor language without any structural consequences. Contrariwise, elements which have their origin in semantically transparent regular paradigms do not move easily assuming a strong typological difference between donor and recipient language. Consider this latter fact in connection with Irish. The prepositional pronouns sketched above have not been transferred into English as they are part of the morphological and semantic lattice of Irish and obviously this cannot have been transferred to English. However, if a prepositional pronoun were separate from the others already within Irish, it could be regarded as susceptible to transfer as examples (19a, b) would appear to show.

(19) a. *'Twas a bad day that was in it.*
       B'é droch lá a bhí ann.
       was-it bad day COMP was in-it
   b. *That's all that's in it.*
       Sin an méad atá ann.
       that the amount COMP-is in-it

In Irish the third personal pronoun with the preposition *in* 'in', i.e. *ann* 'in-it', was hived off from the remaining elements of this paradigm and developed an existential sense from the originally locative one, much as German did with the verbal phrase *da sein* (literally: 'be there', i.e. 'exist'). After its detachment from the paradigm the existential indicator *ann* was free-floating and a prime candidate for transfer as can been seen from the above instances.

One should perhaps not be too dogmatic about transfer of a productive paradigm from donor to recipient language. There is one situation in which this is nonetheless attested in Irish English. What it involves is re-interpretation of structures which already exist in the target language. The case in point is the use of the preposition *on* with a personal pronoun to express the relevance of an action to an individual (much as the dative in German). This re-interpretation can be seen in example (20a), where the literal meaning is replaced in Irish English by an interpretation in which the relevance of the action to the subject expressed by the pronoun is indicated. In example (20b) this second interpretation is the only possible one.

(20) a. *Thit an dréimire orm.*
       fell the ladder on-me
       (cf. German *Mir ist die Leiter heruntergefallen.*)
           on-me is the ladder fallen-down
       'The ladder fell down on me.'

b.   *They crashed the car on me.*
     'They crashed my car, the car I was looking after, etc.'

The transfer of this kind of grammatical category is facilitated where certain conditions are met. Among these are the following: there should be an available morpheme in the recipient language which does not show too much allomorphy and should not be homophonous with markers of other categories, and the morpheme used should be immediately translatable, i.e. others in the contact situation should be able to identify and isolate it.[19]

5.2. The question of acceptance

A consideration which is important when examining what gets transferred in a contact situation is just how this transfer arises. I think it is fair to assume that transfer begins on an individual level. Somebody starts by shifting a structure from his/her first language to the second language. This single incidence of transfer must, if it is to establish itself in the contact variety of the second language, be interpretable for other contact speakers. This imposes a restriction on restructuring the second language. The higher the degree of transparency with an interference form the greater the chance of its being accepted or spontaneously arising on several fronts so to speak, i.e. of it occurring with large numbers of speakers.

This view of how transfer arises holds for a particular type of contact situation where there is no prestige group within the community of contact speakers. If this were the case, however, then an item of transfer could diffuse through the community from its origin within a prestige group, in an extreme case from a single individual of high status in this group. The latter kind of scenario applies to German today vis-à-vis English (where there is no speaker contact involved). Certain key sectors of society such as the visual media, fashion, technology, industry and science are usually the source of a loan which is often accepted generally due to the high status of such sectors of the community.

---

[19] See Holm (1988: 67) for a discussion of similar conditions for substratal influence in the genesis of pidgins and creoles.

Language contact and typological difference

5.3. Pressure for transfer

An additional question must be addressed within the sphere of transfer. This concerns the motivation for transfer. To answer this one must make a two-fold distinction among types of transfer. The first is what one can term low-level transfer which shows slight consciousness among speakers and which derives ultimately from speech habits, mainly phonetic, which are so ingrained as to be impossible to abandon when speaking a new language. The second type is systematic transfer the origin of which can be seen in the search of speakers, conscious or unconscious, for equivalence of categories between the outset and the target languages. This situation can lead to remoulding in the recipient language when categories from the donor language are found missing. The phenomenon can be clearly recognised with a switch-over from one language to another[20] by an entire community over a relatively long period of bilingualism as is attested for the history of English in Ireland from the seventeenth to the nineteenth centuries.[21]

If one bears in mind that pressure to transfer results from the desire to express a category not available in the target language, then one can explain why the canonical VSO word-order of Irish has not been carried over into Irish English.[22] This structure is of no informational value and hence does not represent a category for which contact speakers would have felt a need in English. VSO word-order is an aspect of Irish syntax but it does not represent a category for which there is no ready equivalent in English. However, there are several such categories a couple of which will be looked at briefly now.

---

[20] This type of language shift is dealt with by Edwards (1992), Romaine (1989: 38ff.) and by Hoffmann (1992: 186ff.). Hoffmann (1992: 190f.) sees the prestige of the second language as largely responsible for the shift and quotes the decline of the Celtic languages, as an instantiation of this.

[21] There is evidence that even in Dublin Irish was used side by side with English in a literary capacity as late as the eighteenth century. A group of poets, the most noted of whom was one Seán Ó Neachtain, had set up base in Dublin and were active writers and contemporaries of Swift (Ó Cuív 1986: 393f.). Ó Neachtain is the author of a burlesque *Stair Éamuinn Uí Chléire*, which is of linguistic interest because in it he ridicules the efforts of the Irish to speak English and gives examples which show a heavy influence of Irish syntax (Ó Cuív 1986: 393f.).

[22] The complex of syntactic influence of Irish on English has been treated in particular detail by Filppula, see Filppula (1982, 1986, 1990, 1991) as representative publications.

6. Perfectives in Irish and Irish English

There are two major aspectual types in Irish, the habitual and the perfective. Both occur frequently and show no restriction to any subset of verbs. The first is the "resultative perfective". This emphasises that the state which is envisaged as the end point of an action has indeed been reached. It is a telic construction (Dahl 1985) much like the telic *Aktionsarten*[23] of German as with verbs like *aufessen* 'eat up', *aufbrauchen* 'use up' which imply arriving at a pre-defined endpoint. The following are examples of the resultative perfective in Irish English (Hickey 2000).

(21) a. *He has the job done.*
       Tá an obair déanta aige.
       is the work done at-him
    b. *They have the house built.*
       Tá an teach tógáilte acu.
       is the house built at-them

Given the telic nature of the resultative perfective construction, it is not applicable to sentences with stative verbs (Harris 1984: 312), so that an example like the following is impermissible:

(22)   **I have something about Russian grammar known.*

The second type of aspect common in Irish is the "immediate perfective".[24] Here the stress is placed on the fact that the action denoted by the verb has just taken place.

---

[23] For a discussion of the different definitions of the terms aspect and *Aktionsart* in the linguistic literature, see Brinton (1988: 3ff.).

[24] The term "hot news" is frequently used by authors dealing with aspect in Irish English, see Kallen (1989: 11) and Harris (1991: 201) for recent examples. This goes back to McCawley (1971) in his classification of aspect types. It is the equivalent of the term "immediate perfective" which I use here. The designation "resultative perfective" is standard usage, again see Harris (1991: 202). With reference to the treatment of perfective constructions offered in Greene (1979) some authors speak of PII (= resultative perfective) and PI (= immediate perfective). As these cryptic labels are difficult to remember correctly they will not be used here. The resultative perfective with *after* is recorded for the first time quite early on (1690). Other typically Irish syntactic structures surface much later, for instance the habitual present with *do + be* appears in print only at the beginning of the nineteenth century (Bliss 1976: 558).

(23) a. *He's after breaking the window.*
　　　 Tá sé tar éis an fhuinneog a bhriseadh.
　　　 is he after (2 words) the window COMP break-NON-FINITE
　　b. *She's after eating her dinner.*
　　　 Tá sí tar éis a dinnéar a ithe.
　　　 is she after (2 words) her dinner COMP eat-NON-FINITE

The above sentences carry the implication that the action is very recent and would typically be used in a narrative situation. It is not the attainment of a goal which is stressed, but the fact that the action has just been completed.

Now consider the sentences in (21) for a moment. The Irish sentence in each case has the usual initial verb, the subject NP (functionally an object) followed by the non-finite verb form and closing with a prepositional pronoun. In the Irish English construction the only restructuring which has taken place is that the order of non-finite verb form and object is reversed. The prepositional pronoun is ignored. This is in keeping with the view that in the process of transfer an equivalence between major categories is enough, here verb and object. Indeed this stands to reason if one considers that transfer in a contact situation first arises on an individual level. Were too much restructuring undertaken then the likelihood of acceptance by others in the contact community would diminish. Furthermore in sentences of the kind in (21), any transfer of the prepositional pronouns 'at-him' or 'at-them' would not be of any informational value as the subject reference is conveyed in English quite simply by the finite verb form.

A further point should be made in connection with sentences of the kind in (21). This is that the intonational pattern of Irish is carried over into Irish English. In the Irish original there is strong stress on both the object and the non-finite verb form, i.e. *Tá an 'obair 'déanta aige.* In Irish English both these elements also carry strong stress so that (21a) is pronounced *He has the 'work 'done* and not **He has the 'work ˌdone.*

Indeed the role of intonation in the identification of possible transfer structures should not be underestimated. Consider the clause coordination below which is regarded by some authors, but by no means by all, as resulting from Irish.

(24)　　*Chuaigh mé ag siúl 'agus 'é ag cur 'báistí.*
　　　　 went I at walking and it at putting rain
　　　　 'I went for a walk 'and 'it 'raining.'
　　　　 * 'I went for a walk ˌand ˌit 'raining.'

What is important here is that the coordinating conjunction, the dummy subject and the non-finite verb form all carry primary stress in both Irish and Irish English, an indication that the construction in the latter is the result of transfer from Irish (Ó Siadhail 1984). Note that the unstressed elements *ag cur* (literally 'putting') are of no relevance here as it is the number of stressed syllables which must match, the length of a foot is immaterial. In addition, of course the meaning of the sentence in both languages is 'I went for a walk in spite of the fact that it was raining'.

7. Supportive transfer

With many structures in Irish English, syntactic or morphological, Irish may have played a supportive role but it is by no means the only source which needs to be considered. This may of course have been the case with the word-order object + past participle discussed above. There are many other features of Irish English which could be retentions from earlier forms of English, particularly if one bears in mind that the speakers of the varieties brought to Ireland in the early modern period were in the main from the west or north west of England (the south-west played a greater role in the settlement of Ireland at the end of the medieval period, as of the late twelfth century (Hickey 1997b)). Arguments for two sample features of Irish English are offered here to show what the correspondence between Irish and English would look like.

The first concerns the occurrence of *for* with an infinitive. The use of *for* as a general preposition of purpose in English as in *He asked for help* is quite well established. The use with the infinitive is of course non-standard but very much present in localised forms of Irish English. The argument for supportive transfer from Irish derives from the fact that here the preposition *chun*, which has the general meaning 'for', is found with infinitives.

(25)   He went to Cork for to buy a car.
       Chuaigh sé go dtí Corcaigh chun gluaisteán a cheannaigh.
       went he to Cork for car to buy

An instance from the area of morphology is the retention of a separate form for the second person plural personal and possessive pronouns. The actual manifestations of these vary greatly in Irish English and some of the differences are attributable to the linguistic distinction between the north and south of the country, the form *yez* being more typical of the north, the form *ye* more general in the south, while *youse* is stigmatised (in the south) as popular Dublin English. Whatever the form found, the distinction between singular and plural is one which is present and active in Irish as can be seen in the glosses for examples (26a) and (26b).

(26) a. *Have ye* [ji] *got yeer* [jir] *homework done yet?*
    An bhfuil bhur cleachtannaí déanta agaibh fós?
    INTERROG is your-PL exercises done at-you-PL already
  b. *What are the two of ye* [ji] *up to?*
    Cad tá an beirt agaibh ag déanamh?
    what is the two at-you COMP doing

8. What about the lexicon?

A brief glance at Irish English is enough to recognise that of all linguistic levels the lexicon is that which shows least transfer as an open class which enjoys high consciousness among speakers.[25] If large amounts of lexical borrowing is an indication of superstrate status (recall the examples of Modern German and English or Middle English and Central French), then the lack of this would point to the low status of the potential donor language. This would appear to be indeed the case. It is obvious that cultural borrowings from Irish into English are not to be expected given the discrepancy in status between the two languages which increased steadily in the second period of English in Ireland.

But there is a further twist to this. The lexicon of Irish is quite different from that of English. Irish has few borrowings from Latin compared to English and neo-classical neologisms exist in theory[26], but are scarcely used by native speakers. Furthermore the Irish lexicon, with its heavy reliance on nouns where English would use adjectives, often conveys to speakers of other languages a sense of pathos and pseudo-poeticness which is not felt by speakers of Irish. Take the sentence *Tá fad na teanga*

---

[25] Thomason – Kaufman (1988: 129) note the large amount of phonological and morpho-syntactic interference from Irish into Irish English and the comparative lack of lexical transfer and indeed postulate that the few items there are may well have been introduced by English speakers confronted with Irish rather than by speakers of Irish English themselves.
[26] These are recorded in the main Irish-English dictionary, see Ó Dónaill (1977).

air. (is length the-GENITIVE tongue-GENITIVE on-him) 'He is very talkative' mentioned above. If translated literally ('The length of the tongue is on him') it evokes a sensation of floweriness which results from the metaphoric use of 'length' and 'tongue'. Added to this is the fact that for many English-speaking Irish this heavily nominalised syntax is regarded as antiquated usage, smacking of stage Irish. With attitudes like this it is not surprising that those Irish striving to acquire English would not have wished to transport such lexical elements or structures into the new language.

9. Conclusion

In the present paper the main typological differences between Irish and English have been enumerated and discussed. It has become apparent that only some of these differences are of relevance to a consideration of transfer between the two languages. One can maintain that when a category is lacking in the target language there will be pressure to transfer from the source language and hence the possibility of interference is greatest. This perhaps why on the one hand there has been no transfer of VSO word order from Irish but there have been attempts to represent the aspectual distinctions of Irish not found in English. This is all the more understandable if one bears in mind that most of the Irish learned English in a situation of uncontrolled adult second language acquisition where categorial equivalence to structures found in the first language (Irish) would have been expected in English and constructed through transfer, if not already present, as with the different distinctions of verbal aspect. Furthermore, one must distinguish carefully between the transfer of a distinction in principle and the exponence used to realise this distinction in the target language. This perspective can help to explain how lenition in Irish English is a phenomenon linked to Irish although the phonetic manifestation varies greatly across the two languages.

References

Adams, George Brendan
    1974   "The 1851 language census in the north of Ireland", *Ulster Folklife* 20: 65-70.
    1979   "The validity of language census figures in Ulster, 1851-1911", *Ulster Folklife* 25: 113-122.

Aitken, Adam Jack
1981 "The Scottish vowel-length rule", in: Michael Benskin − Michael Louis Samuels (eds.), *So meny people longages and tonges: philological essays in Scots and mediæval English presented to Angus McIntosh.* Edinburgh: The Editors, 131-157.
Appel, René − Pieter Muysken
1986 *Language contact and bilingualism.* London: Edward Arnold.
Bailey, Charles-James N.
1982 "Irish English and Caribbean Black English: another joinder", *American Speech* 57: 237-239.
Berk-Seligson, Susan
1986 "Linguistic constraints on intersentential code-switching: A study of Spanish/Hebrew bilingualism", *Language in Society* 15: 313-348.
Bhaldraithe, Tomás de
1945 *The Irish of Cois Fhairrge, Co. Galway.* Dublin: Institute for Advanced Studies.
Bliss, Alan J.
1972 "Languages in contact: some problems of Hiberno-English", *Proceeding of the Royal Irish Academy, section C* 72: 63-82.
1976 "The English language in early modern Ireland", in: Theodore William Moody − F. X. Martin − F. J. Byrne (eds.), *A new history of Ireland*, vol. 3: *Early modern Ireland (1534-1691).* Oxford: Clarendon, 546-560.
1977 "The emergence of modern English dialects in Ireland", in: Diarmuid O'Muirithe (ed.), *The English language in Ireland.* Cork: Mercier, 7-19.
1979 *Spoken English in Ireland 1600-1740: twenty-seven representative texts assembled and analysed.* Dublin: Cadenus.
Breatnach, Risteard B.
1947 *The Irish of Ring, Co. Waterford.* Dublin: Institute for Advanced Studies.

Brinton, Laurel J.
1988 *The development of English aspectual systems: aspectualizers and post-verbal particles.* (Cambridge Studies in Linguistics 49.) Cambridge: Cambridge University Press.

Búrca, Seán de
1958 *The Irish of Tourmakeady, Co. Mayo.* Dublin: Institute for Advanced Studies.

Clyne, Michael
1987 "Constraints on code-switching: how universal are they?", *Linguistics* 25: 739-764.

Corrigan, Karen P.
1993 "Hiberno-English syntax: nature versus nurture in a creole context", *Newcastle and Durham Working Papers in Linguistics* 1: 95-131.

Crozier, Alan
1984 "The Scotch-Irish influence on American English", *American Speech* 60: 310-331.

Dahl, Östen
1985 *Tense and aspect systems.* Oxford: Blackwell.

Delattre, Pierre
1964 "German phonetics between English and French", *Linguistics* 8: 43-55.
1969/70 "La théorie celtique et les substrats", *Romance Philology* 23: 480-491.

Dietrich, Julia C.
1981 "The Gaelic roots of a-prefixing in Appalachian English", *American Speech* 56: 314.

DiSciullo, Anna-Maria — Pieter Muysken — Rajendrah Singh
1986 "Government and code-mixing", *Journal of Linguistics* 22: 1-24.

Dolan, Terence P. (ed.)
1990 *The English of the Irish. Commemorative issue of "The Irish University Review" for Alan J. Bliss.* Dublin [No indication of publisher.]

Dowling, Patrick J.
1935 [1968] *The hedge schools of Ireland.* London: Longman.
1971 *A history of Irish education.* Cork: Mercier.

Eastman, Carol M. (ed.)
1992 *Codeswitching.* Clevedon: Multilingual Matters.

Edwards, John
1992 "Sociopolitical aspects of language maintenance and loss", in: Willem Fase — Koen Jaspaert — Sjaak Kroon (eds.), *Maintenance and loss of minority languages.* (Studies in Bilingualism.) Amsterdam: Benjamins, 37-54.
Emeneau, Murray Barnson
1956 "India as a linguistic area", *Linguistics* 32: 3-16.
Feagin, Crawford
1991 "Preverbal *done* in Southern States English", in: Peter Trudgill — Jack K. Chambers (eds.), 161-190.
Filppula, Markku
1982 "VSO and SVO languages in contact: sentence-thematic peculiarities of Hiberno-English", in: Thorstein Fretheim — Lars Hellan (eds.), *Papers from the sixth Scandinavian conference of linguistics, Røro, June 1981.* Dragvoll [Trondheim]: Tapir, 50-59.
1986 *Some aspects of Hiberno-English in a functional sentence perspective.* Joensuu: University of Joensuu.
1990 "Substratum, superstratum and universals in the genesis of Hiberno-English", in: Terence P. Dolan (ed.), 41-54.
1991 "New models and typologies of language contact", in: Muusa Ojanen — Marjatta Palander (eds.), *Language contacts east and west.* (Studies in languages 22.) Joensuu: University Press, 5-31.
Foster, Robert F.
1988 *Modern Ireland, 1600-1972.* Harmondsworth: Penguin.
Fowkes, Robert A.
1966 "English, French, and German phonetics and the substratum theory", *Linguistics* 21: 45-53.
Freeman, Thomas Walter
1957 *Pre-famine Ireland: a study in historical geography.* Manchester: University Press.
Fréine, Séamus de
1966 *The great silence.* Cork: Mercier.
Gardner, Robert C. — Wallace E. Lambert
1972 *Attitudes and motivation in second-language learning.* Rowley, Mass.: Newbury House.
Greene, David
1979 "Perfects and perfectives in modern Irish", *Ériu* 30: 122-141.

Gumperz, John J.
1971 "Convergence and creolization: a case from the Indo-Aryan/Dravidian border", in: Dell Hymes (ed.), *Pidginization and creolization of languages: proceedings of a conference held at the University of the West Indies, Mona, Jamaica, April 1968*. Cambridge: Cambridge University Press, 151-168.

Harris, John
1983 "The Hiberno-English "I've it eaten" construction: what is it and where does it come from?", *Teanga* 3: 30-43.
1984 "Syntactic variation and dialect divergence", *Journal of Linguistics* 20: 303-327.
1986 "Expanding the superstrate: habitual aspect markers in Atlantic Englishes", *English World Wide* 7: 171-199.
1991 "Conservatism versus substratal transfer in Irish English", in: Peter Trudgill — Jack K. Chambers (eds.), 191-212.

Haugen, Einar
1969 *The Norwegian language in America: a study in bilingual behavior*. Bloomington, Ind.: University Press.

Hayes-McCoy, G. A.
1967 "The Tudor conquest", in: Theodore William Moody — F. X. Martin (eds.), *The course of Irish history*. Cork: Mercier, 174-188.

Henry, Patrick Leo
1960 "The Irish substantival system and its reflexes in Anglo-Irish", *Zeitschrift für celtische Philologie* 28: 19-50.

Hickey, Raymond
1984 "Coronal segments in Irish English", *Journal of Linguistics* 20: 233-251.
1985 "Salient features of Irish syntax", *Lingua Posnaniensia* 28: 15-25.
1986 "Possible phonological parallels between Irish and Irish English", *English World Wide* 7: 1-21.
1990 "Suprasegmental transfer: on prosodic traces of Irish in Irish English", in: Jacek Fisiak (ed.), *Further insights into contrastive analysis*. (Linguistic and Literary Studies in Eastern Europe 30.) Amsterdam: Benjamins, 219-229.

1997a	"Arguments for creolisation in Irish English", in: Raymond Hickey – Stanisław Puppel (eds.), *Language history and linguistic modelling. A festschrift for Jacek Fisiak on his 60th birthday.* Berlin: Mouton de Gruyter, 969-1038.
1997b	"Assessing the relative status of languages in medieval Ireland", in: Jacek Fisiak (ed.), *Studies in Middle English linguistics.* Berlin: Mouton de Gruyter, 181-205.
1998	"The Dublin vowel shift and the progress of sound change", in: Jacek Fisiak – Marcin Krygier (eds.), *English Historical Linguistics 1996.* Berlin: Mouton de Gruyter.
1999	"Dublin English: current changes and their motivation", in: Paul Foulkes – Gerry Docherty (eds.), *Urban voices.* London: Edward Arnold, 265-281.

Hindley, Reg
1990	*The death of the Irish language: a qualified obituary.* London: Routledge.

Hoffmann, Charlotte
1991	*An introduction to bilingualism.* London: Longman.

Holm, John
1988	*Pidgins and creoles,* vol. 1: *Theory and structure.* Cambridge: Cambridge University Press.

Jakobson, Roman
1931	"Über die phonologischen Sprachbünde."
[1962]	[Reprinted in: Roman Jakobson, *Selected writings,* vol. 1: *Phonological studies.* The Hague: Mouton, 137-143.]

Kachru, Braj B.
1982	"Toward structuring code-mixing: an Indian perspective", *International Journal of the Sociology of Language* 16: 27-47.

Kallen, Jeffrey
1989	"Tense and aspect categories in Irish English", *English World Wide* 10: 1-39.
1990	"The Hiberno-English perfect: grammaticalisation revisited", in: Terence P. Dolan (ed.), 120-136.
1993	"Language maintenance, loss, and ethnicity in the United States: perspectives on Irish", *Teanga* 13: 100-114.

Lehiste, Ilse
1988	*Lectures on language contact.* Cambridge, Mass.: MIT Press.

MacNamara, John
1973 "Attitudes and learning a second language", in: Roger W. Shuy — Ralph W. Fasold (eds.), *Language attitudes: current trends and prospects*. Washington: Georgetown University Press, 36-40.

Martinet, André
1952 "Celtic lenition and Western Romance consonants", *Language* 28: 192-217.

McCawley, John D.
1971 "Tense and time reference in English", in: Charles Fillmore — D. Terence Langendoen (eds.), *Studies in linguistic semantics*. New York: Holt, Rinehart and Winston, 96-113.

Montgomery, Michael
1989 "Exploring the roots of Appalachian English", *English World Wide* 10: 227-278.

Mühlhäusler, Peter
1985 "Patterns of contact, mixture, creation and nativization: their contribution to a general theory of language", in: Charles-James N. Bailey — Roy Harris (eds.), *Developmental mechanisms of language*. Oxford: Pergamon, 51-88.

Myers-Scotton, Carol M.
1993 *Duelling languages: grammatical structure in codeswitching*. Oxford: Clarendon.

Nichols, Johanna
1992 *Language diversity through space and time*. Chicago: University Press.

Ó Baoill, Dónall P.
1990 "Language contact in Ireland: the Irish phonological substratum in Irish English", in: Jerold A. Edmondson — Crawford Feagin — Peter Mühlhäusler (eds.), *Development and diversity: language variation across time and space: a Festschrift for Charles-James N. Bailey*. Arlington: University of Texas Press, 147-172.

Ó Cuív, Brian
1944 *The Irish of West Muskerry, Co. Cork*. Dublin: Institute for Advanced Studies.
1951 *Irish dialects and Irish-speaking districts*. Dublin: Institute for Advanced Studies.

1986	"Irish language and literature 1691-1845", in: Theodore William Moody — William E. Vaughan (eds.), *A new history of Ireland*, vol. 4: *Eighteenth-century Ireland: 1691-1800*. Oxford: Clarendon, 374-419.

Ó Dónaill, Niall
1977	*Foclóir Gaeilge-Béarla*. Dublin: Oifig an tSoláthair [Stationary Office])

Ó Siadhail, Micheál
1984	"*Agus (Is)/And*: a shared linguistic feature", *Celtica* 16: 125-137.

Pfaff, Carol W.
1979	"Constraints on language mixing: intrasentential code-switching and borrowing in Spanish/English", *Language* 55: 291-318.

Poplack, Shana
1980	"Sometimes I'll start a sentence in Spanish y termino en español: toward a typology of code-switching", *Linguistics* 18: 581-618.

Romaine, Suzanne
1989	*Bilingualism*. (Language in Society 13.) Oxford: Blackwell.

Roy, John
1986	"Tense and aspect in Barbadian English creole", in: Manfred Görlach — John Holm (eds.), *Focus on the Caribbean*. (Varieties of English around the World: General Series 8.) Amsterdam: Benjamins, 141-156.

Salmons, Joseph C.
1984	*The extent of language contact change: Germanic and Celtic*. [Unpublished Ph.D. thesis. Austin: University of Texas.]
1992	*Accentual change and language contact: comparative survey and a case study of early Northern Europe*. Stanford: Stanford University Press.

Sankoff, David — Shana Poplack
1981	"A formal grammar for code-switching", *Papers in Linguistics: International Journal of Human Communication* 14: 3-45.

Schaller, Helmut Wilhelm
1975	*Die Balkansprachen: eine Einführung in die Balkanphilologie*. Heidelberg: Winter.

Schneider, Edgar
1990 "The cline of creoleness in English-oriented creoles and semi-creoles of the Caribbean", *English World Wide* 11: 79-113.

Singh, Rajendrah
1985 "Grammatical constraints on code-switching – evidence from Hindi-English", *Canadian Journal of Linguistics* 30: 33-45.

Stenson, Nancy
1990 "Phrase structure congruence, government, and Irish-English code-switching", in: Randall Hendrick (ed.), *The syntax of the modern Celtic languages*. (Syntax and Semantics 23.) San Diego: Academic Press, 167-197.
1991 "Code-switching vs. borrowing in Modern Irish", in: Per Sture Ureland – George Broderick (ed.), *Language contact in the British Isles: proceedings of the eighth international symposium on language contact in Europe, Douglas, Isle of Man, 1988*. Tübingen: Niemeyer, 559-579.

Thomason, Sarah Grey – Thomas Kaufman
1988 *Language contact, creolization and genetic linguistics*. Berkeley, Cal.: University of California Press.

Tristram, Hildegard (ed.)
2000 *Celtic Englishes II*. Heidelberg: Carl Winter Verlag.

Trudgill, Peter – Jack K. Chambers (eds.)
1991 *Dialects of English: studies in grammatical variation*. London: Longman.

Wartburg, Walther von
1950 *Die Ausgliederung der romanischen Sprachräume*. (Bibliotheca romanica 1: Manualia et commentationes 8.) Bern: Francke.

Weinreich, Uriel
1953 *Languages in contact: findings and problems*. New York: Humanities Press.

Williams, Jeffrey P.
1986 "Hiberno-English and white West Indian English – the historical link", in: John Harris – David Little – David Singleton (eds.), *Perspectives on the English language in Ireland: proceedings of the first symposium on Hiberno-English, Dublin 1985*. Dublin: Centre for Language and Communication Studies, Trinity College, 83-94.

1988	"The development of aspectual markers in Anglo-Caribbean English", *Journal of Pidgin and Creole Languages* 3: 245-263.

Wolfram, Walt
1988	"Reconsidering the semantics of *a*-prefixing", *American Speech* 63: 247-253.
1991	"Toward a description of *a*-prefixing in Appalachian English", in: Peter Trudgill — Jack K. Chambers (eds.), 229-240.

Woolford, Ellen
1983	"Bilingual code-switching and syntactic theory", *Linguistic Inquiry* 14: 520-536.

Thomas Kohnen (Duisburg)
The influence of "Latinate" constructions in Early Modern English: orality and literacy as complementary forces

1. Introduction

One of the more conspicuous characteristics of English texts from about 1450 to 1650 is the increasing number of syntactic constructions which strike us as typically "literate". Readers need to cope with lots of participle, gerund, relative and adverbial clauses or with combinations and clusters of any of these, often "spiced" with so-called absolute constructions or relative connectives. The emergence and rise of these syntactic patterns is often attributed to Latin influence and consequently these constructions are viewed as "Latinate" constructions.

An interpretation along these lines seems plausible since Latin is generally seen as one of the prominent cultural forces during the period under consideration. Latin influence on English ranks among the standard topics in textbooks on Early Modern English.[1] Still, it often seems quite difficult to ascertain where exactly and to what extent syntactic constructions can be traced back to their Latin roots. In many cases intermediate or competitive influence of other foreign languages (above all French) cannot be ruled out.[2] Furthermore, certain "Latinate" constructions often turn out to be merely a special trait of one author or of some fashionable style, which was soon abandoned.[3] But the greatest difficulty one is faced with is to decide first of all whether a given syntactic structure is to be considered a Latin borrowing or a native development. Sørensen (1957) provides a clear description of the general situation:

> [T]o practically all supposed Latinisms in English (with the possible exception of the Old English absolute dative) there corresponded a set of similar native constructions, which were in existence – if only in embryo – before the influence

---

[1] See, for example, Barber (1997: 43-48, 56-62, 70-75) and Görlach (1991: 36-40, 100-101, 124-130); on Latin influence on English syntax see Behrens (1937); Blatt (1957) and Sørensen (1957).
[2] See, for example, with regard to fifteenth-century prose translations, Workman (1940: 67).
[3] A list of fashionable styles, most of them to a greater or lesser extent influenced by Latin, is given by Görlach (1985: 36-38).

from Latin set in, and which might have developed in the way they did without any foreign influence. Most English Latinisms, then, are Latinisms only in the sense that Latin has contributed – slightly or materially – to increasing their frequency or to extending their syntactical domains.[4] (Sørensen 1957: 133)

If this description is valid the main point about so-called "Latinate" constructions is not proving whether they in fact are "Latinate" or not, but ascertaining where and how the increase in their frequency and the extension of their syntactic domains happened and what factors contributed to this development.

In this paper I would like to show that the influence of so-called "Latinate" constructions in Early Modern English has to be understood within the context of a general tendency towards literacy and that the spread and syntactic extension of these constructions is mediated through text types and regulated by patterns of orality and literacy acting as complementary forces. In this investigation I will concentrate on constructions involving the present or first participle (which will also be called "P1"). Participle clauses are certainly among the major syntactic patterns whose spread may be due to Latin influence.[5] The development of P1-constructions will be investigated in three text types (statutes, private letters and narrative prose) during the period from about 1450 to 1700.

2. Data and classification of participle constructions

The data I use are mainly taken from the *Helsinki Corpus*.[6] However, in certain places I made some major alterations, excluding some texts and adding new files.[7]

---

[4] A similar view is expressed by Blatt (1957: 35) with regard to Latin influence on "European" syntax: "extended use of the participle", for instance, "may be only partly due to classical influence" since the construction is "inspired by Latin ... usage without being quite contrary to the linguistic tradition of the European language."
[5] For a detailed account of the development of participle constructions in the history of English as presented in handbooks and research articles see Kohnen (1996) and Kohnen (in press).
[6] See Kytö (1991) and Rissanen – Kytö – Palander-Collin (1993).
[7] The following changes were made: among the narrative prose texts Samuel Pepys's *Penny Merriments* were excluded because samples 1-3 do not contain any narrative prose sections. Extracts from Congreve's *Incognita*, on the other hand, were added to ensure that the data from Behn's *Oroonoko* do not just reflect her personal style. – Among the private letters a letter by John Shillingford (1447) was omitted since it is only a cross

The influence of "Latinate" constructions in Early Modern English

Among the participles found in the data only those were considered which clearly show the status of clause equivalents. I excluded all P1s in premodifying position (*the singing girl*) and the so-called continuous form (*The girl is singing*). Furthermore, all P1s that have changed word class were left out: P1s which are used as adjectives (*interesting*), prepositions (*touching*), adverbs (*notwithstanding*) or nouns (*the living and the dead*). Finally I excluded all *ing*-constructions that cover noun position in the superordinate clause, that is the constructions traditionally called gerund constructions.[8]

P1-constructions have been classified according to the position of P1 relative to the superordinate clause. If we take "subject-verb-(object)-(complement)-(adverbial)" (Quirk *et al.* 1985: 49) (with object, complement and adverbial as optional elements) as a basic clause pattern, then a P1 can be found either in front of the clause or after the clause or as a postmodification following any of its nominal elements. Thus we find three basic construction types:

When P1 is in front of the clause it is called a front construction:[9]

(1)   ... *bending* [P1] *his browes like the angry god of warr, he* [S] *went* [V] *his waies* ...   (Deloney, 794)

A P1 following the clause is termed a back construction:

(2)   *And sire, I* [S] *was* [V] *with my lady of Southfolke at this day hopyng* [P1] *that I myght have hade hyre at sume leysyre* ... (CMPRIV, E. Stonor, 111)

When P1 follows a nominal element of the clause in postmodifying function, I call it a postmodifying construction:

---

between private and official writing, with the official elements predominating. Since the file of personal letters called <CEPRIV2>, which – in the Helsinki Corpus – is supposed to represent the period from 1570 to 1640, in fact only contains letters from 1621-1632 (with the exception of one piece from 1587), I set up an additional file <CEPRIV2A> covering the years 1583-1606. Bibliographical details are given in the reference section.
[8] In Quirk *et al.* (1985: 1292) these constructions are subsumed under the heading "participle clauses".
[9] The following abbreviations are used in the examples: [S] – subject of the superordinate clause; [V] – finite verb of the superordinate clause; [P1] – *ing*-participle; [S1] – subject of P1 (if expressed separately). Abbreviations and contractions in the text have been expanded. References contain text file and line number.

173

(3)   Be it therefore enacted by the Authority aforesaid That every Distiller [S] having [P1] and keeping [P1] any Private Pipe or Stopcock ... shall [V] ... take up or demolish every such Pipe ...
(CELAW3, Statutes 1695-1699, 935)

Furthermore, those constructions were considered which are traditionally called "absolute constructions", that is where the subject of P1 is expressed separately. These constructions usually appear as front (4) or as back (5) constructions:[10]

(4)   I did thinke never to imploy him, but he [S1] being [P1] all your tailors I [S] have altered [V] my resolution;
(CEPRIV3, E. Oxinden, 26)

(5)   And yf it [S] happe [V] eny suche ownour ... to dye duryng the seid viage ... his heire [S1] being [P1] within age, That all suche feoffes or other suche persons ... shall have the same honours ...
(CMLAW, Statutes 1488-1491, 747)

3. P1-constructions in statutes

The so-called "Statutes of the Realm" might well be presented as a good candidate for tracing back "Latinate" constructions during the Late Middle English and Early Modern English periods. Statutes can be seen as typical instances or at least as immediate successors to what has been called "Chancery English" or termed "curial prose".[11] This type of language, going back to Anglo-French traditions, was moulded by multilingual clerks who used Latin and French in the same routine legal contexts of their profession. As a specialised language variety of the law and the royal administration it has been shown to attract formal, literate, especially

---

[10] In all, the following construction types are distinguished: front construction [P1f], absolute front construction [P1Sf], back construction [P1b], absolute back construction [P1Sb] and postmodifying construction [P1po]. Unless indicated otherwise, the terms "front construction" and "back construction" cover all instances of the respective patterns, including (front and back) absolute constructions.

[11] See Fisher (1977) for Chancery English and Burnley (1986) for curial prose. Both terms, of course, refer to different aspects of the statutes: whereas the former stands for several rather loosely delimited types of official writing, one of them being the statutes, the latter denotes a special prose style, which is – at least to a certain extent – characteristic of the type of language used in the statutes.

"Latinate" constructions,[12] but also to have been most influential in the formation of a modern written English standard, at least during the fifteenth century.

The data given in table 1[13] show that P1-constructions start off on a fairly high level with a frequency of 0.75 per 100 words; they go up only slightly (to 0.97) and then, by the end of the seventeenth century, drop to 0.51.

Table 1. P1-constructions in statutes

	date 1488-1491 11240 words CMLAW	date 1509-1543 11790 words CELAW1	date 1588-1604 11780 words CELAW2	date 1695-1699 13180 words CELAW3
P1	0.75 [84]	0.97 [114]	0.96 [113]	0.51 [67]

Table 2. Front constructions [P1f], back constructions [P1b] and postmodifying constructions [P1po] in statutes

	date 1488-1491 11240 words CMLAW	date 1509-1543 11790 words CELAW1	date 1588-1604 11780 words CELAW2	date 1695-1699 13180 words CELAW3
P1f	- - [-]	0.01 1% [1]	0.01 1% [1]	- - [-]
P1b	0.13 18% [15]	0.09 10% [11]	0.18 18.5% [21]	0.08 16% [11]
P1po	0.61 82% [69]	0.87 89% [102]	0.77 80.5% [91]	0.42 84% [56]

The distribution of the various P1-constructions and their respective proportions (as shown in table 2) presents a fairly static picture: a striking predominance of postmodifying constructions, which cover between 80.5% and 89%; a minor share of back constructions, which range between 10% and 18.5%, and almost no front constructions, with a proportion of 1% in two out of four text samples. This situation seems to prevail without any significant indication of change even when the total number of P1-constructions drops significantly.

---

[12] Burnley (1986: 593, 596-598), Fisher – Richardson – Fisher (1984: 4-6, 42)
[13] The numbers shown in the tables give the frequencies of P1-constructions per 100 words. Numbers in square brackets indicate their incidence. Figures (both frequency and percentage) are rounded off. The abbreviations referring to the text files are explained in the reference section. Note that the figures given in tables 1, 2, 4 and 5 contain all the P1-constructions found in the respective data; absolute constructions (see table 3) as well as so-called "transitional constructions" (see table 6) are included there.

Absolute constructions (table 3) are present right from the beginning, although the proportion is low and does not seem to rise; instead, the number of absolute constructions drops significantly in the last two text samples.

Table 3. Absolute front constructions [P1Sf] and absolute back constructions [P1Sb] in statutes

	date 1488-1491 11240 words CMLAW	date 1509-1543 11790 words CELAW1	date 1588-1604 11780 words CELAW2	date 1695-1699 13180 words CELAW3
P1Sf	- - [-]	0.01 1% [1]	- - [-]	- - [-]
P1Sb	0.04 5% [4]	0.03 3% [4]	0.02 2% [2]	0.01 1% [1]

In all, therefore, the statutes provide quite a static and inflexible scene in Early Modern English – after the end of the fifteenth century there does not seem to have been any significant spread of P1-constructions within this text type. Rather, the same status quo, a relatively high incidence of P1-constructions, a marked predominance of postmodifying constructions, a certain amount of back constructions and a more or less dwindling share of absolute constructions seem to have prevailed until the seventeenth century.

These results may be explained in terms of the specialised style which seems to have developed within the text type "statutes". Statutes, like most administrative and formal genres, prefer elaborate and compact means of expression. Here participle constructions are useful as means of syntactic compression. Furthermore, the predominance of postmodifications (which are often linked together in an additive and repetitive manner, without ever leaving the framework of the main clause) reflects a striving after precision in reference, which is a typical characteristic of curial prose (Burnley 1986: 596). The persons and situations to whom the laws apply need to be qualified unambiguously.

So, stylistic factors, the formal nature of the text type and, in addition, its Anglo-French or even Latin roots (Burnley 1986: 596, 599) may account for the fairly early and relatively high incidence of P1-constructions found here. However, seen from the perspective of contemporary Standard English, the statutes do not seem to contribute much to the dissemination and evolution of the syntactic possibilities of participle constructions. Rather, their development is static and backward, in quantitative and qualitative respects,[14] especially so when compared with the two informal text types, narrative prose and private letters.

4. P1-constructions in narrative prose and private letters

Private letters and narrative prose[15] may be called informal text types: private letters deal with personal affairs among relatives or intimate friends; the narrative prose texts tend to make use of short, fairly episodic "stories" and are situated within the tradition of popular literature, although there is a certain amount of internal variation.[16]

Here the data suggest a situation which is quite different (table 4). Starting from fairly low frequencies (below 0.2) the numbers rise dramatically in both text types – up to 1.49 in the narrative prose texts and up to 0.54 in the private letters; both decline by the second half of the seventeenth century.[17]

Table 4. P1-constructions in narrative prose and in private letters
Narrative prose:

	date 1481 words 8760 Caxton	date 1526 words 6430 Talys	date 1567 words 5120 Harman	date 1608 words 5170 Armin	date 1619 words 7320 Deloney	date 1688 words 5480 Behn	date 1692 words 7146 Congreve
P1	0.16 [14]	0.96 [62]	1.27 [65]	1.49 [77]	1.26 [92]	0.58 [32]	0.76 [54]

---

[14] This finding is in line with Nevalainen – Raumolin-Brunberg (1993: 68), who state that legal language "quite soon lost its norm status and became a special language".

[15] The term "fiction", instead of "narrative prose", is used throughout the Helsinki Corpus. See Kytö (1991).

[16] The "coherence" or "homogeneity" of text types is, of course, one of the vexing problems of corpus linguistics. For a treatment of this question with regard to the Helsinki Corpus see, for instance, Nevalainen – Raumolin-Brunberg (1989) or Rissanen – Kytö – Palander-Collin (1993). The variation which makes itself felt within the narrative prose texts (jest books as opposed to Behn or Congreve) seems to reflect a genuine evolution of the genre in the seventeenth century, with authors exploring larger and more elaborate narrative forms.

[17] Even if the sample of Congreve's Incognita shows more P1-constructions than Behn, it still bears out the decreasing tendency by the end of the seventeenth century.

Private letters:

	date 1448-1482 words 15950 CMPRIV	date 1502-1547 words 10640 CEPRIV1	date 1583-1606 words 10953 CEPRIV2A	date 1621-1632 words 10570 CEPRIV2	date 1662-1706 words 13129 CEPRIV3
P1	0.18 [28]	0.4 [43]	0.42 [46]	0.54 [57]	0.29 [38]

In both text-types, then, P1-constructions seem to spread later, showing a parallel and continuous development which differs only in the extent of the increase. These findings, together with the results from the statutes, suggest two things:

First, P1-constructions seem to develop differently within different text types, that is, their occurrence and frequency differs according to text types. Second, P1-constructions seem to develop across text types, that is they reach individual text types in a certain order.

Does this regular and parallel development in both informal text types also apply to the distribution of the individual constructions and their respective proportions? Table 5, in fact, presents a rather complicated picture. However, here again some fairly noticeable tendencies can be distinguished in both text types:

First: at the beginning we find only postmodifying and back constructions. Front constructions only appear in the first half and at the end of the sixteenth century respectively.[18]

Second: apart from the jest books (<Talys> and <Armin>, where postmodifications predominate), back constructions are the most frequent constructions found in the data. However, in the course of time their powerful position seems to be challenged, at least in the narrative prose texts. (See, for instance, the proportions in <Behn> and <Congreve>.)

Third: with the rise of front constructions postmodifications lose importance. Within the two jest books (the only samples where postmodifications predominate) their proportion goes down from 65% to 39%.

---

[18] There is one single instance of a front construction in a letter by Margaret Paston from 1448 (in <CMPRIV>), which, however, can be left out of consideration. It is the only example in almost 16,000 words, the next one turning up more than 140 years later. Moreover, since the letter is not an autograph and therefore "cannot be trusted for the details of form" (Davis 1954: 121), this construction can probably be regarded as an untypical formal pattern inserted by a scribe to whom this letter was dictated. It goes without saying that front constructions can be found earlier in other, probably more formal text types. However, they seem to be quite rare in the fourteenth and fifteenth centuries (see Behrens 1937: 40).

Table 5. Distribution of front constructions [P1f], back constructions [P1b] and postmodifying constructions [P1po] in narrative prose and private letters

Narrative prose:

	date 1481 8760 words Caxton	date 1526 6430 words Talys	date 1567 5120 words Harman	date 1608 5170 words Armin
P1f	- - [-]	0.08 8% [5]	0.39 31% [20]	0.41 27% [21]
P1b	0.1 64% [9]	0.26 27% [17]	0.61 48% [31]	0.5 34% [26]
P1po	0.06 36% [5]	0.62 65% [40]	0.27 21% [14]	0.58 39% [30]

	date 1619 7320 words Deloney	date 1688 5480 words Behn	date 1692 7146 words Congreve
P1f	0.29 23% [21]	0.26 44% [14]	0.25 33% [18]
P1b	0.61 49% [45]	0.26 44% [14]	0.29 39% [21]
P1po	0.36 28% [26]	0.07 12% [4]	0.21 28% [15]

Private letters:

	date 1448-1482 15950 words CMPRIV	date 1502-1547 10640 words CEPRIV1	date 1583-1606 10953 words CEPRIV2A	date 1621-1632 10570 words CEPRIV2	date 1662-1706 13129 words CEPRIV3
P1f	0.01 3% [1]	- - [-]	0.08 20% [9]	0.21 39% [22]	0.08 26% [10]
P1b	0.12 68% [19]	0.33 81% [35]	0.22 52% [24]	0.3 56% [32]	0.18 61% [23]
P1po	0.05 29% [8]	0.08 19% [8]	0.12 28% [13]	0.03 5% [3]	0.04 13% [5]

In both letters and narrative prose the frequency of postmodifications is lower towards the end of the seventeenth century than in the sixteenth century.[19]

So by the end of the seventeenth century a shift of relative dominance of P1-constructions makes itself felt: a slightly less important back con-

---

[19] Compare, for instance, 0.12 in <CEPRIV2A> with 0.04 in <CEPRIV3> or 0.27 in <Harman> with 0.07 in <Behn>. The frequency in <Congreve> is rather high but clearly less than in <Talys>, <Harman>, <Armin> and <Deloney>.

struction, a stagnating postmodification and a rising front construction, which is clearly in second place.

5. Transitional constructions: evidence for linguistic evolution

Is this fairly regular pattern which evolves in both informal text types simply due to an accidental constellation of certain statistical facts? Or is it an evolutionary pattern governing the spread and the extension of the syntactic domains of P1-constructions?

The key to this problem should be found in the emergence of front constructions. Front constructions do not simply turn up "out of the blue" but appear to be somehow connected with absolute constructions, which in turn may be associated with certain postmodifications. In the data absolute constructions (both front and back type) are hardly ever found earlier than (non-absolute) front constructions, that is earlier than <Talys>[20] and <CEPRIV2A>. This suggests, at least for the present data, that both absolute and front constructions seem to develop rather late and basically at the same time within a given text type. Closer analysis reveals a developmental pattern: front constructions may have developed from certain postmodifications via absolute constructions.

This hypothesis is based on the fact that front constructions, when they first occur, appear to a large extent as what I would like to call "transitional" constructions. They are transitional because on the one hand they can be traced back to postmodifications, on the other hand they can be identified as front constructions, most of them as absolute front constructions.

The data contain three major types of transitional constructions: in the first transitional construction ([P1fT1], examples (6) and (7)) the subject [S1], which is postmodified by a P1, is cut off from its respective finite predicate by a coordinator (*and*); the resulting construction is something like a mixture of a postmodifying and an absolute front construction: on the one hand P1 could be called a postmodification of a subject whose predicate was lost (but might be easily retrieved); on the other hand the

---

[20] The further discussion will show that all front constructions in <Talys> can be analysed as "transitional constructions" and therefore as absolute constructions. See table 6.

division marked by the coordinator *and* makes the unit "subject plus participle" to some extent independent or absolute.[21]

(6)   *This simple man* [S1] *beholding* [P1] *him wel, and* [∅ (S = *this simple man*)] *sawe he was of taule personage with a good quarter staffe in his hand* (Harman, 392).

(7)   *my wife ... hath provided for her with Mrs. Eppes, who* [S1] *haveing* [P1] *now but one mayede and servant, and* [∅ (S = *who*)] *desireth* [V] *to have her as soone as possible may bee* (CEPRIV2, V. Pettit, 113).

In the second transitional construction ([P1fT2], examples (8) and (9)) the subject postmodified by P1 is repeated or pronominalised before the finite verb of the main sentence, producing a construction with a double subject: a postmodifying construction is turned into an absolute front construction.[22]

(8)   *William* [S1], *seeing* [P1] *his head soft, What soft post is this? quoth* [V] *he* [S] (Armin, 366).

(9)   ... *my cousin and my selfe* [S1] *going* [P1] *to the steward of Lambeth, we* [S] *found* [V] *that by Judds and my vncles vnderhand dealing, the Jury haue given in ther verdict and found my vncle heier to the moitye* ... (CEPRIV2, Th. Knyvett, 162)

The third transitional construction ([P1fT3], examples (10)-(12)) originates in those environments where the subject of P1 and the finite verb is also the subject in the preceding sentence or even in the preceding paragraph and can therefore be deleted both in front of participle and verb. Apparently, such a construction is a front construction since (the verb of) the superordinate clause follows P1, but it could still be called implicitly

---

[21] There are two instances where the coordinator *and* is missing, which, however, were counted as [P1fT1] because of the large distance between [S1][P1] and [V].
[22] One might argue, however, that this is not an absolute construction because the subjects are identical. (For a discussion see Mitchell (1985: § 3804).) Whereas such a view might only be based on semantic grounds (referential identity) I would rather define "absolute construction" in a formal way as a P1-construction with [S1] expressed separately.

postmodifying: P1 is connected with a subject that can only be deleted because it is implicitly given and therefore this (implicitly given) subject can be said to be postmodified by P1.

(10) sone after commeth into the barne one of the good mans boyes [S], to fet some haye for his horse. And [0 (S)] fyndinge [P1] his maister lyinge faste bounde and greuouslye beaten with rodes, [0 (S)] was [V] sodenly abashed and woulde haue runne out agayne to haue called for helpe (Harman, 774).

(11) Doctor Junius [S] is a man of a marvellous high stomach, and [0 (S)] caring [P1] the profession of the better law, [0 (S)] will [V] not yield unto the lower (CEPRIV2A, H. Wotton, 271).

(12) in a low and undistinguishable Voice, he [S] let her know how much he [S] held himself obliged to her, and [0 (S)] avoiding [P1] as many words as handsomely he [S] could, at the same time, [0 (S)] entreated her to give him her Advice, toward the management of himself in this Affair (Congreve, 476).

The percentage of transitional constructions in the narrative prose texts (table 6) reveals an interesting development. At the beginning all front constructions are transitional constructions; their number subsequently drops. Above all the incidence of the first and second transitional constructions (those with coordination and double subject) decreases significantly.

Table 6. "Transitional" constructions [P1fT] in narrative prose

	date 1481 8760 words Caxton	date 1526 6430 words Talys	date 1567 5120 words Harman	date 1608 5170 words Armin
P1f	[-]	[5]	[20]	[21]
P1fT	[-] -	[5] 100%	[11] 55%	[10] 48%
P1fT1	[-] -	[2] 40%	[2] 10%	[-] -
P1fT2	[-] -	[3] 60%	[-] -	[2] 10%
P1fT3	[-] -	[-] -	[9] 45%	[8] 38%

	date 1619 7320 words Deloney	date 1688 5480 words Behn	date 1692 7146 words Congreve
P1f	[21]	[14]	[18]
P1fT	[5] 24%	[5] 36%	[5] 28%
P1fT1	[-] -	[-] -	[-] -
P1fT2	[2] 10%	[1] 7%	[-] -
P1fT3	[3] 14%	[4] 29%	[5] 28%

When assessing the percentage of the third transitional construction one needs to remember that in any portion of narrative prose there may be strings of sentences with an identical subject, which, of course, may be deleted. P1-constructions that are found here must be subsumed under the third class of transitional constructions, regardless of whether front constructions are already established or not. The significant statement supported by the data is that the proportion of transitional constructions in 1526 is 100% and in 1608 less than 50%.[23]

It can therefore be inferred from the data that transitional constructions, or rather postmodifying constructions, contributed significantly to the development of front constructions. Since front constructions develop in most cases first as absolute front constructions, this also explains why absolute and front constructions appear in quite close proximity.[24] Finally the evidence of the transitional constructions provides safe grounds for the assumption that the spread of P1-constructions is not due to an accidental combination of statistical facts, but to a structured pattern of evolution.[25]

---

[23] In fact, the third category of transitional constructions does not seem to be a transitional construction in the strict sense of the term (although it can still be analysed as "implicit postmodification"). Rather, it is its temporarily high frequency which seems to be a striking feature of a transitional development.

[24] It is worth mentioning that the only two instances of front constructions found in the statutes (see table 2) are a transitional construction [P1fT3] and an absolute construction [P1Sf]. The only front construction found in <CMPRIV> - see note 18 - is also an absolute construction.

[25] In the private letters the pattern of the three transitional constructions does not work out as well as in the narrative texts. Here an additional factor has to be taken into account. In private letters the emergence of front constructions seems to be closely linked to a specific structural position in the texts, the ending formula. In <CEPRIV2>, for instance, 55% of all front constructions appear in this section (apart from 18% "ordinary" transitional constructions in the rest of the texts.). So it is above all the temporarily high incidence of front constructions in a specific section peculiar to a certain text type which reflects the evolutionary pattern here.

6. The pattern of evolution: orality and literacy as complementary forces

The development of P1-constructions in the two informal text types outlined so far may be summed up as a hierarchy:

$$P1b > P1po > P1S > P1f$$

Read in descending order from the left to the right, this hierarchy entails the following hypotheses:
1. Occurrence: if a certain construction type occurs in a text sample, all the construction types higher in the hierarchy will occur as well.
2. Frequency of occurrence: constructions higher in the hierarchy show a greater frequency of occurrence and a larger proportion than constructions lower in the hierarchy.
3. Direction of development: the development of P1-constructions proceeds from left to right. The exact sequence of the postmodifying and back construction types cannot be determined exactly – in many texts both constructions occur from the start and in the two jest books there are more postmodifying constructions than back constructions.[26]

Although the development sketched out in this hierarchy seems quite plausible, with the data and arguments given above, one could still argue that it lacks precise statistical evidence since in the data presented so far absolute constructions were included in the respective numbers of back and front constructions. In order to fully vindicate the claim entailed in the hierarchy, the presentation of data should enumerate separately (non-absolute) back constructions, postmodifications, absolute constructions (with all the transitional constructions) and (non-absolute) front constructions, with their respective frequencies and proportions (table 7).

---

[26] A separate hierarchy for the jest books would be: P1po > P1b > P1S > P1f.

Table 7. Distribution of (non-absolute) back constructions [P1b*], postmodifying constructions [P1po], absolute constructions (with transitional constructions) [P1S*] and (non-absolute) front constructions [P1f*] in narrative prose texts

	date 1481 8760 words Caxton	date 1526 6430 words Talys	date 1567 5120 words Harman	date 1608 5170 words Armin
P1b*	0.1 64% [9]	0.23 24% [15]	0.59 46% [30]	0.5 34% [26]
P1po	0.06 36% [5]	0.62 65% [40]	0.27 22% [14]	0.58 39% [30]
P1S*	- - [-]	0.11 11% [7]	0.33 26% [17]	0.33 22% [17]
P1f*	- - [-]	- - [-]	0.08 6% [4]	0.08 5% [4]

	date 1619 7320 words Deloney	date 1688 5480 words Behn	date 1692 7146 words Congreve
P1b*	0.56 45% [41]	0.18 31% [10]	0.22 30% [16]
P1po	0.36 28% [26]	0.07 13% [4]	0.21 28% [15]
P1S*	0.12 10% [9]	0.18 31% [10]	0.15 20% [11]
P1f*	0.22 17% [16]	0.15 25% [8]	0.17 22% [12]

The modified array of data given in table 7 bears out all the hypotheses entailed in the hierarchy – with the qualifications made before.[27] It also shows that the hierarchy applies only to the spread of P1-constructions, that is until about the first half of the seventeenth century. From then on an increasing share of [P1f*] (see <Deloney> and <Congreve>) and [P1S*] (see <Behn>) disturbs the fairly regular pattern.

How can this hierarchy and the development it entails be accounted for? I suggest that it may be seen as a continuum of increasing literacy and decreasing orality. There are three main arguments which may be given in support of this interpretation:[28]

---

[27] The only exception is Harman, where [P1S*] surpasses [P1po] by about 4%. The jest books, of course, follow a different hierarchy, as pointed out in note 26.
[28] See also Kohnen (1996).

First: participle constructions are commonly held to be typically literate constructions.[29] They are literate because they make a text more elaborate and more compact: the addressee has to supply not only the elements implicit in the non-finite form (for example, subject, tense, semantic relation etc.), they also have to face the more basic problem of coping with two predications within the framework of one sentence.

Therefore it seems reasonable to assume that the spread and evolution of P1-constructions as testified by the data can be taken as a development towards literacy. Since near the lower end of the continuum there is a stereotyped formal pattern (the absolute construction), it seems reasonable to assume as well that the continuum proceeds from a lower degree of literacy towards a higher degree of literacy.

Second: in the history of English participle constructions make their first appearance almost exclusively in formal or literate genres. They are characterised by scholars as patterns typical of written language or as learned expressions.[30] When such forms spread to informal or more popular genres they can be assumed to settle down first in those contexts which correspond to the informal quality of these texts and only later on to develop as fully-fledged literal constructions. I suggest that this increase in literacy typical of informal text types is reflected in the present continuum by the transition from back and postmodifying constructions to absolute and front constructions.

The third argument proceeds from psycholinguistic considerations of language processing. It presumes that different ways of arranging the order of information units presuppose different settings of communication and that different serial arrangements may be more or less adequate to different communication settings. Given the typical setting of oral communication,[31] above all the high pressure on the participants to produce or decode messages on the spot and their total dependence on the limited scope of short-term memory, it seems likely that postmodifying and back constructions are those participle constructions which can be handled best in oral situations. P1 is simply attached to the preceding noun or clause. The necessary link can be easily established, since the relevant reference is close by; missing elements may be provided without difficulty since the words of the addressor are still ringing in one's ears. Apart from that, the combination "noun + P1" or "clause + P1" is a con-

---

[29] For an overview see Biber (1988: 233).
[30] See, for instance, Behrens (1937: 14) and Koziol (1938: 103).
[31] For a survey see Biber (1988: 28-46).

struction closely resembling the typical oral communication pattern "topic + comment".

Front constructions, on the other hand, seem to be a real hindrance in typical oral situations. The hearer has to face a predication without reference, a fairly indeterminate piece of information, which needs to be stored away until an appropriate candidate for the subject position is found. All this certainly requires greater effort and greater skill in language processing, or it may presume an addressee with a well-developed reading competence.

The additional effort required can be reduced if in front constructions the subject of P1 is made explicit and if it is coreferential with the subject of the main sentence, or if front constructions are only used in those contexts where the subject can be easily supplied since everybody knows who is being spoken about. And this fits the description of two of the transitional constructions ([P1fT2], [P1fT3]), which in this connection might also be called front constructions adapted to orality.

Seen from this perspective, back constructions, postmodifying constructions, absolute constructions (together with transitional constructions) and front constructions form a continuum of syntactic classes which are increasingly compatible with typical written settings and increasingly incompatible with typical oral settings.

To sum up the three arguments, the pattern of evolution shown in the hierarchy of P1-constructions may be explained as a development towards literacy, as a cline moving from contexts of "greater orality" towards "greater literacy". The remarkable thing about this pattern is that the construction which is expanding needs to take the route of "decreasing orality", that is the predetermined sequence proceeding from orality to literacy; it does not skip a construction to start with a thoroughly literate pattern. Thus the underlying process cannot simply be described as an increase in literacy or as a replacement of oral constructions by literate ones; instead, literacy and orality seem to act as complementary forces: the former provides the aim, the latter determines the sequence.

7. Conclusions

The outline presented so far, of course, cannot give a fully satisfactory account of the general influence of "Latinate" constructions, their rise in frequency and the extension of their syntactic domains. Still, with its restrictions to participle constructions and three text types, this investigation yields some tenable hypotheses which not only complement the history of participle constructions in English but may also cast fresh light on the way "Latinate" constructions spread and developed in the course of Late Middle English and Early Modern English:

First: the spread of P1-constructions (and perhaps also of other "Latinate" constructions) seems to be mediated through text types. P1-constructions have been shown to spread to a different extent in different text types and to affect different text types in a particular order (statutes vs. narrative prose and private letters). Apparently, more formal prestigious text types are affected first and to a larger extent.[32]

Second: text types may be static or dynamic with regard to the development of P1-constructions (and perhaps other "Latinate" constructions). The formal text type investigated does not seem to move far beyond the initial stage investigated: it does not show any significant quantitative or qualitative development (apart from the decreasing numbers at the end of the seventeenth century). The informal text types, on the other hand, change dramatically, both in quantitative and qualitative terms. They seem to be moving towards the full range of participle constructions found in Standard English, whereas the static statutes can hardly be said to act as trend-setters for later qualitative developments towards Standard English.

Third: apparently, such dynamic developments unfold in highly structured patterns of evolution and are accompanied by a marked increase in variability.[33]

One important developmental pattern can be described as a continuum regulated by the principles of orality and literacy, which act as complementary forces. Although there may be other patterns this one seems to be highly plausible. It accounts for the general tendency towards literacy, which stands out as a characteristic trait of the period under investigation. Furthermore, it reflects the special nature of the influence exerted by Latin. Latin was not spoken or acquired as a native language by native speakers. Rather, it was pronounced a model language by the prestigious cultural and educational elite and firmly entrenched in the formal teaching of grammar schools. So Latin and "Latinate" constructions were not

---

[32] However, such early developments do not necessarily guarantee further evolution later on, as has been shown in the case of the statutes.

[33] This increase in variability shows up in the transitional constructions, but also in a general trend towards "creative freedom" during this period (Partridge 1969: 81); Partridge's remark that "Tudor writers made freer use of participial constructions than is possible now" (1969: 82) is very much to the point.

internalised with everyday language, but rather along with stylistic norms of formal, prestigious texts, which later spread across the various functional domains of English, including informal colloquial text types. When these norms reached informal texts people seeking to fulfil the literate norm inserted literate constructions first in those contexts available to and compatible with their oral standards. This could account for the fact that the spread of P1-constructions (and perhaps also of other "Latinate" constructions) is mediated through text-types and seems to be a gradual evolution regulated both by orality and literacy.

Fourth: so far I have not yet commented on the significant drop in the frequencies of P1-constructions by the end of the seventeenth century, which is noticeable in all the text types investigated. I suggest that this decrease is due to processes of standardisation and codification.[34] If this is true, the development of P1-constructions outlined so far would correspond to what Einar Haugen has called the elaboration and codification stage of standardisation (1972: 107-110). A literate construction spreads to all functional domains of English and is established as an important part of the standard variety. The interesting thing to note here is that in the text types investigated the full range of participle constructions was only reached by a significant increase in frequency and variability. Where this dynamic pattern is not found we end up with a restricted inventory of constructions. The conclusion to be drawn is that standardisation, at least in its initial elaboration stage, entails increase in variation rather than reduction of variation. Apparently the complete range of forms and constructions to be codified later is only achieved by an initial pattern of increased variability.

So "Latinate" constructions, apart from their immediate import as an instance of language contact in the history of English, open up perspectives of more general linguistic interest as well. It is not so much the precise "genealogy" of these constructions as the way they spread and developed which point to recent issues in historical and socio-historical

---

[34] The drop in P1-constructions by the end of the seventeenth century does not contradict the explanation in terms of increasing literacy. A general increase in literacy does not necessarily imply an unlimited quantitative proliferation of a particular literate construction, nor does it exclude a restricted quantitative reduction of this. The main reason for attributing the present development to increasing literacy is the qualitative evolution of P1-constructions establishing their full constructional range as found in Standard English. The resultant inventory of participle constructions is maintained at the end of the seventeenth century, despite their decreasing frequency. The drop in frequency, on the other hand, may be due to the receding influence of Latin and the "fashionable styles" associated with it.

developed which point to recent issues in historical and socio-historical linguistics. Here at least factors like "orality" and "literacy", "standardisation", "structured variability" and "text type" would find a framework which might enable us to integrate the developments sketched out here into a more comprehensive account of language change.

Primary Sources[35]

1. Statutes

<CMLAW>	The Statutes of the Realm (III)
	Helsinki Corpus
<CELAW1>	The Statutes of the Realm (III; Henry VIII)
	Helsinki Corpus
<CELAW2>	The Statutes of the Realm (IV; Elizabeth I, James I)
	Helsinki Corpus
<CELAW3>	The Statutes of the Realm (VII; William III)
	Helsinki Corpus

2. Narrative Prose

<Caxton>	William Caxton, *The History of Reynard the Fox*
	Helsinki Corpus
<Talys>	*A Hundred Mery Talys*
	Helsinki Corpus
<Harman>	Thomas Harman, *A Caveat or Warening for Commen Cursetors*
	Helsinki Corpus
<Armin>	Robert Armin, *A Nest of Ninnies*
	Helsinki Corpus
<Deloney>	Thomas Deloney, *The Pleasaunt History of ... Iack of Newberie*
	Helsinki Corpus

---

[35] These references specify full details only for those excerpts which are not part of the Helsinki Corpus (for the texts of the *Helsinki Corpus* see Kytö (1991)). They indicate page and line numbers of the excerpts (6.1.-14.19 = excerpt from page 6, line 1 to page 14, line 19.).

<Behn>　　　　Aphra Behn, *Oroonoko*
　　　　　　　　Helsinki Corpus
<Congreve>　　(This text sample is not part of the *Helsinki Corpus*.)
Brett-Smith, H. F. B. (ed.)
　　1922　　　　*Incognita or Love and Duty Reconcil'd by William Congreve*. Oxford: Blackwell, 9.1-28.13.

3. Private Letters

<CMPRIV>　　*Paston Letters, The Stonor Letters, The Cely Letters*
　　　　　　　　Helsinki Corpus
<CEPRIV1>　　*Beaumont papers, Clifford letters, The correspondence of Sir Thomas More, Original letters (illustrative of English history), Plumpton correspondence*
　　　　　　　　Helsinki Corpus
<CEPRIV2A>　(This text sample is not part of the *Helsinki Corpus*, except for one letter from Jeayes 1906)
　　　　　　　　Correspondence of Robert Dudley, The letters and epigrams of Sir John Harington, The Chamberlain letters, The works of Sir Walter Ralegh, The life and letters of Sir Henry Wotton, Letters of Philip Gawdy
Bruce, John (ed.)
　　1844　　　　*Correspondence of Robert Dudley, Earl of Leycester, during his government of the Low Countries, in the years 1585 and 1586.* (Camden Society 27.) London, 150.1-151.34.
McClure, Norman Egbert (ed.)
　　1977　　　　*The letters and epigrams of Sir John Harington together with "The Prayse of Private Life"*. New York, 96.11-98.29; 109.23-111.35; 121.25-126.12.
McClure Thomson, Elisabeth (ed.)
　　1966　　　　*The Chamberlain letters: a selection of the letters of John Chamberlain concerning life in England from 1597 to 1626.* London, 45.18-46.5; 47.6-48.19; 50.16-53.7.
Ralegh, Walter
　　1829　　　　*The works of Sir Walter Ralegh, Kt., now first collected: to which are prefixed the lives of the author,* by Oldys and Birch, vol. 8: *Miscellaneous works*. Oxford. [Reprinted New York.] 654.1-654.16.

Smith, Logan Pearsall (ed.)
 1966 *The life and letters of Sir Henry Wotton*, vol. 1. Oxford, 227.1-236.12; 239.24-241.21.
Jeayes, Isaac Herbert (ed.)
 1906 *Letters of Philip Gawdy of West Harling, Norfolk, and of London to various members of his family, 1579-1616.* London, 24.18-27.5.
<CEPRIV2> *Barrington family letters, The correspondence of Lady Katherine Paston, The Ferrar papers, The Knyvett letters, Letters of the Lady Brilliana Harley; The Oxinden letters*
 Helsinki Corpus (one letter from Jeayes 1906 was moved to the additional text sample <CEPRIV2A>)
<CEPRIV3> *Correspondence of the Family of Haddock, Correspondence of the Family of Hatton, Diaries and letters of Philip Henry, Letters of John Pinney, Original letters of eminent literary men, The Oxinden and Peyton letters*
 Helsinki Corpus

References

Barber, Charles
 1997 *Early Modern English.* Edinburgh: Edinburgh University Press.
Behrens, Werner
 1937 *Lateinische Satzformen im Englischen: Latinismen in der Syntax der englischen Übersetzungen des Humanismus.* Münster: Buschmann.
Biber, Douglas
 1988 *Variation across speech and writing.* Cambridge: Cambridge University Press.
Blatt, Franz
 1957 "Latin influence on European syntax", *Travaux du cercle linguistique de Copenhague* 5: 33-69.
Burnley, John David
 1986 "Curial prose in England", *Speculum* 61: 593-614.

Davis, Norman
   1954   *The language of the Pastons.* (Sir Israel Gollancz Memorial Lecture: Proceedings of the British Academy XL.) London: Cumberlege.

Fisher, John H.
   1977   "Chancery and the emergence of standard written English in the fifteenth century", *Speculum* 52: 870-899.

Fisher, John H. — Malcolm Richardson — Jane L. Fisher
   1984   *An anthology of Chancery English.* Knoxville: The University of Tennessee Press.

Görlach, Manfred
   1985   "Renaissance English (1525-1640)", in: Sidney Greenbaum (ed.), 30-40.
   1991   *Introduction to Early Modern English.* Cambridge: Cambridge University Press.

Greenbaum, Sidney (ed.)
   1985   *The English language today.* (English in the International Context.) Oxford: Pergamon.

Haugen, Einar
   1972   "Dialect, language, nation", in: J. B. Pride — Janet Holmes (eds.), *Sociolinguistics: selected readings.* Harmondsworth: Penguin, 97-111.

Kohnen, Thomas
   1996   "Ausbreitungsmuster syntaktischer Standardisierung bei der Entwicklung englischer Partizipialkonstruktionen (Partizip Präsens) 1450-1700", *Anglia* 114: 154-201.
   (in press)   *Text, Textsorte, Sprachgeschichte: englische Partizipial- und Gerundialkonstruktionen 1100–1700.*

Koziol, Herbert
   1938   "Zur Syntax der englischen Urkundensprache des 14. und 15. Jahrhunderts", *Anglia* 62: 100-115.

Kytö, Merja
   1991   *Manual to the diachronic part of the Helsinki Corpus of English texts.* Helsinki: Department of English, University of Helsinki.

Mitchell, Bruce
   1985   *Old English syntax.* 2 vols. Oxford: Clarendon.

Nevalainen, Terttu — Helena Raumolin-Brunberg
   1989   "A corpus of Early Modern English Standard English in a socio-historical perspective", *Neuphilologische Mitteilungen* 90: 67-110.
   1993   "Early Modern British English", in: Matti Rissanen — Merja Kytö — Minna Palander-Collin (eds.), 53-73.

Partridge, Astley Cooper
   1969   *Tudor to Augustan English: a study in syntax and style from Caxton to Johnson*. London: Deutsch.

Quirk, Randolph — Sidney Greenbaum — Geoffrey Leech — Jan Svartvik
   1985   *A comprehensive grammar of the English language*. London: Longman.

Rissanen, Matti — Merja Kytö — Minna Palander-Collin (eds.)
   1993   *Early English in the computer age: explorations through the Helsinki Corpus*. (Topics in English Linguistics 11.) Berlin — New York: Mouton de Gruyter.

Sørensen, Knud
   1957   "Latin influence on English syntax", *Travaux du cercle linguistique de Copenhague* 5: 131-155.

Workman, Samuel K.
   1940   *Fifteenth century translations as an influence on English prose*. (Princeton Studies in English 18.) Princeton: Princeton University Press.

Lucia Kornexl (Greifswald/Rostock)
"Unnatural words"?
Loan-formations in Old English glosses

1. The problem

This paper aims to provide a critical examination of Old English glosses as a "languages in contact"-phenomenon. My object of study are loan-formations in continuous interlinear versions, especially those which are analysable in structure, but not equally so in meaning. Such material takes up considerable space in our Old English dictionaries, and it is from the preface of a late-nineteenth-century work of this type that the title phrase of this paper has been borrowed: in his *Student's dictionary of Anglo-Saxon*, Henry Sweet (1896: viii) decided to employ an exclamation mark in order to warn the reader against "unnatural words ... which are contrary to the genius of the language, some of them being positive monstrosities, the result of over-literal rendering of Latin words".

Sweet is not the only one who has commented on the alleged artificiality of quite a number of Old English gloss-formations. Nevertheless, comparatively little attention has been devoted to the question what exactly distinguishes this type of linguistic evidence from that contained in prose translations and in native texts independent of a foreign source.[1] Increased research efforts are needed in order to establish a sound basis that will enable us to properly evaluate the status and role of gloss materials within the corpus of Old English loans and in the broader context of Old English usage, linguistic development and language change.

---

[1] To some extent at least, this seems to be due to the fact that such "over-literal renderings" have frequently been put down to mechanical translation and a poor command of Latin, and thus did not qualify as material for serious study.

2. The peculiarities of glossing as a borrowing process

2.1. The contact situation

Vernacular glosses that were coined in order to gloss Latin lemmata no doubt form a distinctive category in the study of language contact. First, we have an exceptional contact situation, the donor being a "dead", though highly prestigious language, universally used in the church and functioning as the main idiom of learning in Anglo-Saxon England. The hierarchical relationship between the two languages involved is reflected in the layout of interlinear versions, where the Old English gloss is normally written in a much smaller script.[2] It seems quite clear that the decision to gloss a text instead of turning it into an Old English prose version was a deliberate and well-considered one, this method allowing for the preservation of the authoritative Latin original whilst procuring some help for those not or not yet sufficiently familiar with that language.[3]

The fact that for the student of a gloss, Latin functioned as the target language, puts the notion of "borrowing" into a different light. A genuine gloss-formation may at times have made its way into the common stock of the Old English language; what is, however, usually regarded as an essential motive for borrowing processes, namely the acquisition of new means of expression for active use in the vernacular, appears to have been subordinate if not irrelevant to the glossator as well as to the users of a gloss. It is hardly surprising that a large number of pertinent loans formed on a native basis belong in the category of low-frequency words and show a rather poor survival rate. Generally speaking, the Normans cannot be blamed for the loss and replacement of what seems to have been a vocabulary with very specific functions and a restricted range of application mostly outside the fields of common speech and ordinary prose writing.

---

[2] There are, however, a few notable exceptions, such as the continuous Old English glosses to the *Expositio Hymnorum* in MS B.L. Cotton Vespasian D. xii and to the psalter in MS Cambridge, University Library Ff. 1.23, which are both approximately equal in size to the underlying Latin texts.
[3] For the – mostly religious and liturgical – texts that received extensive glossing in Anglo-Saxon times, see Wieland (1984: 91, note 1) and Gneuss (1989: col. 1513).

2.2. The purpose and function of gloss-formations: problems of evaluation and typological deficiencies

We have no insight into the actual process of glossing and only vague ideas about when, how, and by whom such tools were used. The widespread notion, however, that glosses are nothing but an unconnected series of mechanical renderings by not always very competent translators has been disproved in a number of recent studies.[4] In various ways, interlinear versions shed light on "utramque linguam, uidelicet latinam et anglicam" – a methodological and didactic principle formulated by Ælfric in the Latin preface to his *Grammar* (Zupitza (ed.) 1880 [1966]: 1, 6-7) and reflected in several areas of Anglo-Saxon linguistic scholarship.[5] Continuous Old English glosses may not only to some extent have served as a substitute for bilingual dictionaries, then not yet in existence; depending on the translational technique adopted, they could equally offer comparative information on structural patterns and even a sort of practical grammar in contrastive form.

In Sweet's eyes, however, interlinear versions constituted the main breeding ground for what he regarded as "word monsters". Indeed such glosses often exhibit a marked tendency towards formal reproduction or at least close structural imitation that results in a great number of loan-formations showing full or at least partial morphosemantic correspondence to their models. Formally being compounds or derivatives, these loan-translations or loan-renditions belong to the most frequent word-formational patterns in Old English. Why are they nevertheless considered "unnatural"?

Here we have to take a critical look at the criteria applied to them. When Sweet, in a rather romantic vein, talked about composite terms "which are contrary to the genius of the language" (Sweet 1896: viii), he of course meant the English language, probably as found in well-written prose, whereas such gloss-formations were in fact mainly destined to bring out "the genius" of the Latin language. It is certainly a modern fallacy to measure a glossator's achievement by what Heinrich Götz in his study of Old High German glosses (Götz 1977: 111) calls "*morphologische Lateinferne*" 'morphological distance from Latin'.

---

[4] Cf. the publications listed by Gneuss (1993: 145, note 148). For a re-evaluation of Old High German interlinear versions cf. the seminal article by Henkel (1996).
[5] See the pertinent study by Gneuss (1990). On Ælfric's Latin-Old English *Grammar*, which was obviously a classic in Anglo-Saxon language teaching, cf. Gneuss (1990: 13-17).

What seems equally problematic for gloss material is the assumption that formal correspondence necessarily implies equivalence on the semantic side. Our natural preoccupation with meaning makes us suspicious of formations that render non-transparent, i.e. lexicalised Latin terms. Having to assess a loan-translation like *betwuxsendan* for Lat. *intermittere* 'to leave (sth.) undone, neglect, omit (sth.)', to which I shall return later, we usually see two basic options: either to save its linguistic credibility by postulating a case of semantic borrowing, or to classify it as a clumsy mistranslation, as Sweet did by supplying an equally simplistic literal definition, 'send between',[6] followed by his stigmatising exclamation mark. As I shall try to show, such imitative gloss-formations could, however, in various ways make sense to the contemporary user trained to interpret them.

In this connection, a word must be said about the typology of loans. The terms employed here are the English equivalents of the categories established by Werner Betz in his work on the Old High German version of the Benedictine Rule published in 1949,[7] and first applied to Old English gloss material by Helmut Gneuss in his investigation of the Old English psalter glosses (1955). This is not the place to deal with Betz's scheme in detail. I only mention two closely related points that are immediately relevant to our topic:

Betz's taxonomy of loans has been criticised for its methodological inconsistency, the divisions being based on either formal or semantic criteria.[8] In my opinion, the fundamental difficulty is not the diversity of criteria, which only reflects the complex nature of the material. Betz might rather be accused of having failed to establish them with sufficient clarity and to apply them consistently enough. What seems particularly problematic is his concept of "*Lehnbedeutung*" 'semantic loan', a term that stands for a specific category within his scheme (i.e. native words employed with a foreign meaning), and at the same time denotes a process

---

[6] The same rather unhelpful definition appears in Clark Hall – Meritt (1960) s.v. *betwuxsendan*, and Bosworth – Toller (1892-1898) s.v. *betwyx-sendan*. In such complex cases, defining strategies are crucial for the dictionary reader's adequate understanding of a term, both with regard to its meaning and its status within the Old English lexicon(cf. section 5.2.).

[7] For an outline and discussion of this scheme in the light of subsequent research into borrowing by Betz himself and other scholars, see Betz (1974).

[8] Cf. for example Öhmann (1951: 61) and Coleman (1965: 73). For a critical discussion of Betz's basic categories in the light of somewhat more recent research, see also Tesch (1978: 111-128). Scholarly doubts about the validity of Betz's system have, however, strongly been rejected by Schottmann (1977).

potentially involved in every loan-formation. The pertinent distinction between a 'semantic loan' and 'semantic borrowing' – *Lehnbedeutung* versus *Bedeutungsentlehnung* – was never fully worked out by Betz. He also failed to establish an analytical distinction that seems essential for the exploration of language contact via glossing: that between semantic and morphological "borrowing" in the special and restricted sense indicated above.

3. The basis of this study: the interlinear version to the *Regularis Concordia*

Old English gloss material has never been studied in a comprehensive way. With glosses having been provided for a variety of texts at different times and for diverse purposes, it seems best to examine each of them individually against its own specific background, bearing in mind that such an approach can produce results of limited applicability only.

The present investigation is mainly based on the gloss to the *Regularis Concordia*, the major document of the Benedictine Reform in England compiled by Bishop Æthelwold of Winchester in the early 970s. The Old English interlinear version preserved in the mid-eleventh-century manuscript British Library, Cotton Tiberius A. iii from Christ Church, Canterbury, consists of more than 10,000 lexical items. The underlying Latin text, though partly composed in the so-called "hermeneutic style"[9] and not always easy to construe, for the most part offers facts and gives instructions; there is no room for sophisticated interpretation as for example provided by the exegetical glosses in various psalter versions.[10] Certainly in the heyday of the Reform, and probably for some time after, the *Regularis Concordia* must have been part of the monastic curriculum. I very much doubt though that the Old English interlinear version as we have it was used in the classroom;[11] but, as I shall try to show, it reflects the different methods of analysing and interpreting Latin probably taught and practised there and presupposes a familiarity with such techniques on the part of the user.

---

[9] Cf. Lapidge (1975: 67 [1993: 105]): "By 'hermeneutic' I understand a style whose most striking feature is the ostentatious parade of unusual, often very arcane and apparently learned vocabulary."
[10] See Gneuss (1993: 146 and note 150).
[11] Cf. Kornexl ((ed.) 1993: clxxxvii-clxxxviii); for the controversial concept of the "classbook", see Lapidge (1982), Page (1982), Wieland (1985).

4. Translation technique and the handling of structural differences: the rendering of composite terms

To exemplify the problems of classification posed by formations induced by Latin models, and to demonstrate the explanatory potential of Sweet's "over-literal renderings", I will first deal with two variant patterns that have no fixed status in the typology of loans. Their Latin bases are not, as usual, complex lexemes, but syntactic groups composed of two elements and often institutionalised as technical terms. Such formations are of special interest because they attest to marked structural differences between Latin and Old English and enable us to study how the glossator handled such differences within the given limits.

4.1. "Natural" structural equivalents:
Old English compounds for Latin syntactic groups

In contrast to Old English, compounding is extremely rare in Latin.[12] Instead of compounds, Latin favours syntactic groups of adjective + noun or genitival noun + noun. Unique translation equivalents such as OE *æfenlof* for Lat. *uespertinalis laus* (line 1462) and OE *pancdæd* for Lat. *gratiarum actio* (line 544)[13] suggest that consciously, or guided by his *Sprachgefühl*, the *Regularis Concordia* glossator sometimes opted for what may be called the "natural" solution from an Old English point of view. A number of loan-formations also attested outside the *Regularis Concordia* gloss – most of them technical terms – demonstrate that we are dealing with a divergent pattern of a systematic nature.[14] A typology of loans should probably allow for such regular structural differences between model and replica language.[15] Old English compounds rendering

---

[12] See Hofmann – Szantyr (1965: 68*, § 58).
[13] These and all subsequent line numbers of the *Regularis Concordia* and its Old English gloss refer to my edition (Kornexl (ed.) 1993).
[14] Cf. for example OE *wucpen*, rendering *ebdomadarius minister* 'weekly servant', i.e. the monk or priest appointed for a week's duty, and OE *cildamæssedæg*, corresponding to Lat. *innocentium festiuitas* 'Childermas', i.e. the feast of the Holy Innocents. For a glossary of the monastic and ecclesiastical terms contained in the *Regularis Concordia* and their Old English equivalents in the interlinear version, see Kornexl ((ed.) 1993: ccxl-cclvi).
[15] An exactly parallel morphosyntactic relationship exists between Old High German and Latin; cf. Toth (1980: 15-16).

4.2. Practical constraints on idiomaticity: word-for-word renderings

In his choice of words, the glossator of the *Regularis Concordia* evidently felt quite often bound by the variable order of the elements in Latin. Thus, for instance, the seven occurrences of *matutinalis missa* 'morrow mass' (i.e. the first mass of the day as part of the monastic office) are all rendered idiomatically by OE *capitelmæsse* (lines 385, 387, 447, 841, 1327, 1567, 1604); by contrast, *missa matutinalis* is once glossed by *mæsse capitel* (line 733), with the elements of the compound appearing in reverse order.[16] Of particular interest for our purpose are two further unique *Regularis Concordia* renderings of the foreign term: *mæsse ærnemergenlicu* (line 382) and *mæsse dægrædlicu* (line 696). Both are cases of faithful imitation, but – as becomes immediately apparent when comparing the quotations below – in the second instance the Latin word-order simply left the glossator no choice:[17]

(1)   *þaræfter   mæssan   ærnemergenlice   wyrþian*
      382 *[I]nde   missam   matutinalem   celebrent*

(2)   *syþþan*   mæssan   *hi singan*   dægrædlice*
      696 *post   missam   celebrentur   matutinalem*

The same practical constraint on idiomaticity was at work in the following instance concerning the "principal mass" (i.e. the second mass of the day):

(3)   *mæsse   si sungen   seo ealdorlice*
      481 *missa   celebretur   principalis*

However, by inserting the definite article *seo* before *ealdorlice*, the glossator established an explicit syntactic and semantic tie between noun

---

[16] This peculiar glossing technique has also been observed by Ross (1934: 116-117) in the *Lindisfarne Gospels*.
[17] The square brackets in example (1) mark the scribal omission of the initial in Lat. *Inde*; the asterisks in example (2) are to indicate the divergence of the gloss from the syntax of the Latin text (cf. Kornexl (ed.) 1993: 261-262, note on line 696).

and adjective, thereby leaving no doubt that he had recognised and understood the syntactic group separated in the Latin text.

Ærnemergenlic, a hapax legomenon, as well as *dægredlic*, and probably also *ealdorlic*, are loan-translations; yet the status of the above combinations with *mæsse* is not altogether clear. There is no such category of composite terms in Werner Betz's scheme. Neither can we regard them as belonging to a special type of *Lehnformeln* or *Lehnfügungen* established by Eberhard Siebert (1971: §§ 18, 35) to denote vernacular equivalents of Latin nominal phrases such as ModHG *Heiliger Geist* for Lat. *Spiritus Sanctus*, because according to Siebert's definition, such "loan formulas" are accompanied by semantic borrowing. We rightly hesitate to make such a claim for our Old English word-for-word renderings, thereby attributing to them the status of idioms. Still, in the process of comparative analysis they deserve special consideration.

5. Possible modes of interpretation: the explanatory potential of "literal" renderings

5.1. Quasi-etymological translations

Compared to what we perceive to be a natural Old English prose style, "literal" renderings closely depending on the underlying Latin forms may seem awkward and unnatural, but seen from a different perspective, the immediate presence of the Latin model endowed them with an explanatory capacity not possible otherwise. This not only pertains to information on Latin word-formational patterns by way of reproducing the formal make-up of the item in question. Replications unidiomatic in their disregard of contextual meaning and proper syntagmatic arrangement from an Old English point of view could also be used to reveal the original idea behind a composite Latin term, if that was obscured by lexicalisation, or they could serve to display the underlying concept in the foreign language, if this did not, or not clearly enough, come out in what may be regarded as the usual and idiomatic Old English equivalent. After all, to take a rather straightforward example, Lat. *missa matutinalis* does mean the mass celebrated in the early morning, i.e. OE *ærnemergenlic* or *dægredlic*, a notion not contained in the established vernacular designation *capitelmæsse* 'mass immediately preceding the Chapter Office'.

Such a conscious endeavour to bring out the original semantics of foreign terms can, for instance, be observed in the *Regularis Concordia* glossator's rendering of various idiomatised combinations with Lat. *canonicus*:

(4) a. *tida domlice* – *horas canonicas* 'the canonical hours' (line 371)
 b. *tide rihtgesetre* – *hora canonica* (line 567)
 c. *gesettendlices rynes* – *canonici cursus* '(of) the Office of Canons' (line 500)
 d. *rihtlices rynes* – *canonici cursus* (line 907)

It seems remarkable that he chose no less than four different adjectives – *domlic, rihtgeset, gesettendlic* (a hapax legomenon) and *rihtlic* – to express the notion of 'rightly set, regular, according to the rule', on which Greek κανών and Lat. *canon* 'rule' are based.

As we know, Anglo-Saxon scholars showed a great concern for etymologies in the wake of the classical and the early medieval tradition. Thus it appears most likely that etymological techniques also had an impact on glossing.[18] The well-known method of motivating a term by associating it with real or – by modern scholarly standards – quite often alleged cognates[19] does of course not normally work across language boundaries. Essentially, however, there seems to have been the same explanatory principle behind the "literal" renderings cited above.

In this context, it seems useful to recall the fact that as an object of study and learning, the Latin text clearly took precedence over the accompanying interlinear version, and that the latter defies categorisation as a translation not only because of its lack of syntactic coherence. The usual concept of "translation" appears equally amiss for certain individual components of such a gloss.

This has important consequences for our evaluation: loan-formations that were intended to serve an independent function as vernacular equivalents to Latin technical terms must by definition have been accompanied by semantic borrowing. However, there obviously existed a second category of terms for which the relationship between model and replica was not one of semantic correspondence but rather a defining one. What we have in cases such as *tida domlice* for *horae canonicae* is, strictly

---

[18] See Gneuss (1990: 22-25, 1993: 147).
[19] Cf. for example Isidore of Seville's " 'reges' a [regendo et] recte agendo" and " 'homo', quia sit ex humo" (*Etymologiae*, Lindsay (ed.) 1911: 1.xxix), which found their way into Ælfric's *Grammar* (Zupitza (ed.) 1880 [1966]: 293.7-11).

speaking, not a translation in its own right but rather an explanation that can be properly understood only against the background of the underlying Latin. One would scarcely wish to assign this unique combination, which bears all the characteristics of a gloss, an independent status within the Old English lexicon,[20] especially as there is a sufficiently attested vernacular word for 'canonical hour', namely OE *tidsang*.[21] In a similar manner, we find Lat. *psalmi graduum* 'gradual psalms' in the *Regularis Concordia* both rendered by *sealmas stæpa* (line 304) – a "literal" translation that brings out the idea of spiritual progress on which the Latin designation is based –, and by the presumptively idiomatic OE *fiftyne sealmas* (line 833).[22] Such lexical variants, each with its own distinctive function, are of course not regularly represented in the Old English corpus, but we may quite confidently assume that Anglo-Saxon students were familiar with these different ways of "translating" a Latin term into the vernacular.[23]

As a consequence, the question of linguistic status needs to be applied systematically to the whole religious vocabulary in Anglo-Saxon sources, and it seems also worth asking with regard to Latin-induced technical terms in other fields. Ælfric's famous grammatical terminology, for

---

[20] In this case, the policy of the *Dictionary of Old English*, which records the occurrence in a special subentry s.v. *domlic* 3. (with the extract cited unfortunately distorting the Latin syntax), seems somewhat inconsistent. The extended definition: "canonical; *domlice tida* 'the canonical hours', rendering *horas canonicas* (referring to the divine office)" indicates that this is a loan-translation, but by rearranging the elements to make them conform to the Old English word-order, and by assigning to this group the meaning that would better go with the Latin term, probably too much "technical weight" is given to this imitative formation.
[21] Cf. the exhaustive list of occurrences in the *Microfiche concordance to Old English* (1980). The *Regularis Concordia* glossator was of course familiar with the term *tidsang* and uses it once to render Lat. *laus* 'canonical hour' (line 662).
[22] See Kornexl ((ed.) 1993: 205, note on line 304).
[23] The *Regularis Concordia* text contains a notable and rather curious example for the conscious application of the etymological method to a Latin technical term in anticipation of practical problems that might have resulted from taking the Old English equivalent too literally: readers are reminded that the *auditorium* is chiefly called by this name because in this room the monks are to listen to their master's commands ("qui et ab hoc maxime eo censetur nomine, quod ibi audiendum sit, quid a preceptore iubeatur", lines 1351-1353). Æthelwold's great concern with the true meaning of the word in its original (i.e. Latin) form becomes clear in the light of the corresponding native designation, later supplied by the *Regularis Concordia* glossator: the name *spæchus*, literally 'speechhouse', was apt to evoke associations that were definitely at variance with the ardent reformer's intention of reinforcing monastic silence (cf. Kornexl (ed.) 1993: 340-341, note on line 1351).

example, which largely consists of loan-formations of Latin terms that are usually placed side by side with their foreign models, or are clearly embedded in an explanatory context that refers to these models, has been credited with a degree of lexical and semantic autonomy by historical linguists and lexicographers that may neither do justice to the author's intentions nor to actual usage.[24]

5.2. "What every syllable signifies":
etymological explication by morphological imitation

In a different way, the concept of *translatio* intended as *interpretatio* seems to manifest itself in loan-formations on lexicalised Latin words of a non-technical sort. A pertinent example is the *Regularis Concordia* glossator's rendering of Lat. *intermittere* in the sense 'to leave (sth.) undone, neglect, omit (sth.)' by the loan-translation *betwuxsendan* (lines 148, 362, 440, 1371) and the unique loan-rendition *betwuxasendan* (line 253), with -a- perhaps having been inserted here for merely euphonic reasons.[25] Such items show that transparency of meaning was not always the prime concern of a glossator when opting for a loan-formation. It may be mentioned in this connection that in the Tiberius manuscript of the *Regularis Concordia* as well as in other glossed manuscripts Old English lexemes are frequently divided into their component parts by means of spacing, thus testifying to a remarkable sense of morphological structure among Anglo-Saxon glossators and scribes.[26]

---

[24] See for example the passages from Ælfric's *Grammar* cited in the *Dictionary of Old English* s.v. *dælnimend* 2. and the definition that accompanies them, suggesting a fully accepted technical term: "grammatical: participle". For the presumed currency of the Ælfrician linguistic terminology in Anglo-Saxon scholarly circles cf. Gneuss (1990: 15-16).
[25] The semantic diffuseness of the Old English prefix *a-* is usually ascribed to its different origins; cf. Lutz (1997: 276-277) and the bibliographical references given there.
[26] This splitting up of complex lexical items into morphological units can also be observed in the Latin section of the glossed *Regularis Concordia*, though in a far less systematic way. A comprehensive study of scribal word division in Old English manuscripts within the line is still a desideratum (cf. Gneuss 1973: 18, note 17; [1994: 19, note 19]), especially in the light of recent research into the principles of division at the end of lines, where normally syllabic rules obtained; cf. the pertinent investigation by Wetzel (1981) and the critical re-evaluation of Wetzel's material by Lutz (1986). The case for a syllable-based word division at the end of lines common to all Old Germanic languages has been strengthened by some recent studies; cf. Fix — Birkmann (1998) and the bibliographical references (Fix — Birkmann 1998: 24).

The motivating force behind what must chiefly be regarded as instances of "morphological borrowing" in glosses may be similar to the one that produced straightforward literal renderings of the type described in the preceding section. After all, in the Anglo-Saxons' understanding of etymology, the analysis of word-formation was an acknowledged means of revealing the true sense of a word.[27] This method of structural interpretation is probably alluded to at the beginning of an anonymous Anglo-Latin poem of the mid-eleventh century, in which a student is ordered by his master to analyse in depth a series of Greek medical terms, thereby explaining each word syllable by syllable:[28]

> Dic duo que faciunt pronomina nomina cunctis;
> omnia dic que sunt uerbi, que silliba signet.
> 'Tell me two nouns which provide synonyms for all (the following terms);
> tell all things about each word, what every syllable signifies.'

Though the poet talks about "syllables", the meaningful units to be explicated are of course the morphemes into which the words concerned can be divided. Michael Lapidge (1975: 84, note 1 [1993: 122, note 1]) has identified these lines as a verbatim quotation from an *enigma* of Alcuin; this suggests an established didactic technique practised on both sides of the Channel.

Structural replicas like *betwuxsendan*, which offer no immediate help on the semantic side, are extremely difficult to evaluate, as we have no precise idea about the "pragmatic dimension" of such formations, i.e. the exact way they were taught and used. It seems, however, a reasonable assumption that proper language learning and thorough text study did not stop at the analysis of morphological constituents but also involved elucidation of lexical meaning. The place where the techniques of analytical "translation" reflected in glosses were acquired and where the missing semantic links were provided, was no doubt the classroom, in the case of the interlinear version added to the *Regularis Concordia* pre-

---

[27] Cf. note 18.
[28] For this "glossarial poem of medical terminology from Canterbury" (MS CUL Gg.5.35, fol. 422ᵛ), see Lapidge (1975: 84 [1993: 122]); the whole poem is printed in Lapidge 1975 [1993], Appendix I (a).

sumably that of Christ Church, Canterbury.[29] That the glossator understood what he was translating can be shown by the fact that we find a further attestation of *intermittere* in the sense of 'to discontinue' being rendered by what might be called a "normal" lexical equivalent, OE *lætan* (line 1574), while the corresponding noun *intermissio* is glossed by the transparent *ad hoc*-nominalisations *geswicincg* (line 146) and *forlætincg* (lines 1484, 1594).[30]

There is even evidence that *betwuxsendan* had a semantic value attached to it. How else could we account for the fact that in the *Durham Hymnal* (MS Durham, Cathedral Library B. iii. 32, fols. 1-43), most probably written in the Christ Church scriptorium in the first half of the eleventh century, Lat. *interpolare*, meaning 'to interrupt', is glossed by OE *betwuxsendan* (Milfull (ed.) 1996: 201, hymn 39.27)? As no immediate morphosemantic connection can be established between **polare* and *sendan*, the sense 'to interrupt' shared by Lat. *intermittere* and *interpolare* must have functioned as a mediator. We may tentatively conclude then that in learned contexts, *betwuxsendan* had some currency in eleventh-century Christ Church, Canterbury. One would, however, hesitate to postulate unreservedly a case of semantic borrowing for this rare formation. Here, the policy adopted by the *Dictionary of Old English*, which in such cases usually attaches its definition to the lemma instead of the gloss, marks real progress in lexicographical method.

The concern for morphological correspondence evidenced in Old English interlinear versions led to perplexing results in yet another way – perplexing at least from a modern point of view. Let us compare the following selection of lexical equations produced by the glossator of the *Regularis Concordia*:[31]

(5) a.   *on-þeowian* – *in-seruire* 'busy o.s. with (sth.)' (line 187, 650)
    vs.   *þeowian* – *in-seruire* (line 167)
    b.   *under-ycan* – *sub-iungere* 'add' (line 402)
    vs.   *to-geycan* – *sub-iungere* (line 399)

---

[29] The linguistic and other arguments that strongly suggest a Christ Church origin for the *Regularis Concordia* gloss are laid out in the introduction to my edition (Kornexl (ed.) 1993).
[30] *Geswicincg* is a hapax legomenon, while *forlætincg* is restricted to three interlinear versions which are all connected with Christ Church, Canterbury (cf. Kornexl (ed.) 1993: 178, 366, notes on lines 146 and 1484).
[31] Similar attestations in other sources are treated by Gneuss (1955: 36 and *passim*), and Kastovsky (1992: 313-314).

c. *efen-þwære* – *con-cors* 'unanimous' (line 45)
d. *forþ-tihtincg* – *ex-hortatio* 'exhortation' (line 1660)

Complex formations such as *onþeowian, underycan, efenþwære*, and *forþtihtincg* constitute examples of what one might call "morphological overload": as far as we can judge, in all these cases the base alone would normally have sufficed to render the Latin lemma which, though morphologically complex, is no longer clearly analysable in terms of meaning. What makes these glosses fall under the category of loan-formations at all is the translation of the prefix. Though they may thus come unnaturally close to their Latin models, it seems scarcely convincing to class such formations as attestations of error or even ignorance, as has formerly been suggested,[32] because he who rendered *inseruire* by *onþeowian* could not possibly have done so without knowing the sense of the Latin word. We may rather see them as further manifestations of the putative classroom practice to tell "what every syllable signifies".

The morphological and, above all, the semantic logic behind the various kinds of real or pseudo-derivatives need to be examined in detail. In our random sample (5a-d), we find as translation variants both prefixed and unprefixed forms (*on-þeowian* vs. *þeowian*), and instances of apparently meaningless prefixation (*under-ycan*) side by side with replicas that appear to make sense (*to-geycan*, with intensifying *to-*). Correspondences can be more or less standard, such as OE *efen-* for Lat. *con-* in *efen-þwære* – *con-cors*, but are not always so easily predictable (cf. *forþ-tihtincg* – *ex-hortatio*).

6. Loan-formations and linguistic change

The morphological and semantic value of Old English prefixes proper as well as of adverbial elements that could serve as such and the conditions under which these entered into new combinations still await thorough investigation.[33] As Dieter Kastovsky has pointed out (1992: 313-314), such study must also include possible impacts of linguistic change in this field

---

[32] Cf. for example the pertinent comments on formations of this kind in the *Salisbury Psalter* by Sisam — Sisam ((eds.) 1959: 19) and in the *Stowe Psalter* by Kimmens ((ed.) 1979: 20).
[33] The complexity of the matter, which also comprises a highly complicated syntactic dimension, is reflected in the title of Bruce Mitchell's paper (1978): "Prepositions, adverbs, prepositional adverbs, postpositions, separable prefixes, or inseparable prefixes, in Old English?"; for further bibliographical references see Mitchell — Kingsmill (1980).

on glossing practice and vice versa. In this paper, I can touch on these aspects only very briefly. It seems doubtful to me if a significant correlation can be established between the progressive loss of meaning which is said to have affected many Old English prefixes and the productivity of supposedly "empty" prefix formations in glosses.[34] Examining for instance the Old High German interlinear version of the Benedictine Rule, which probably originated in the last decade of the eighth century (cf. Masser (ed.) 1997: 50) and has come down to us in an early-ninth-century copy (MS St. Gall 916), we find numerous formations of exactly the same make-up; we could, however, not reasonably assume a weakening process of similar extent to have been at work in that branch of the Germanic languages. By way of example, I have selected those Old High German items in the *Regula Benedicti* that parallel the Old English ones from the *Regularis Concordia* listed above (5a-c):[35]

(6) a. ana-deonon   – in-seruire
    b. untar-mahhon   – sub-iungere
    c. eban-keherzida – con-cors

In contrast to the situation in English, a considerable proportion of Old High German imitative gloss-formations have apparently made their way into the common language and survived into Modern German, thus completely losing any "artificial flavour" that may have accompanied them in their early days.[36] If we are to trust the textual evidence that has come down to us from Anglo-Saxon times, the effect of this glossing

---

[34] This possibility is suggested by Kastovsky (1992: 313-314).
[35] Masser ((ed.) 1997: 163 [52.10], 165 [53.5], 341 [141.5]); see also Betz (1949: 139 [A.VI.5], 170 [B.8.15], 120 [A.IV.10]).
[36] In his dissertation on *The systematic exploitation of the verbal calque in German*, Carpenter (1973) records quite a number of Modern German high-frequency verbs that have purportedly been modelled on Latin derivatives, such as *anfangen* < *incipere*, *unterbrechen* < *interrumpere*. Unfortunately, this work exhibits some serious methodological shortcomings. While clearly being aware that he is "treating a situation that is obviously the product of a lengthy diachronic development", Carpenter nevertheless aims at "a primarily synchronic study" (1973: 11). As for the most part he confines himself to merely listing Latin-Modern German correspondences without specifying how he arrived at his conclusions, his results can scarcely be regarded as reliable. Foreign influences on the Old High German lexicon are labelled systematically in Köbler's *Wörterbuch des althochdeutschen Sprachschatzes* (1993), apparently according to the categories established by Betz (cf. section 2.2.). For the careful user these labels can provide valuable information, but their uncommented application must be viewed with due caution.

technique on the decline of prefixations cannot in any way have been a fundamental one. Anglo-Saxon glossators no doubt made what seems to us a remarkably liberal use of derivational elements. The individual history of many of their prefixal derivatives indicates, however, that the language of the study and the living Old English language were to a great extent kept separate.[37] Because of the strong restrictions put upon the glossator by the nature and the rules of the genre, a structural change like the one from prefix verbs to phrasal verbs, which becomes apparent in Middle English, would in any case scarcely have had a chance to find adequate expression in a gloss.[38]

7. Conclusion

I hope to have shown that in warning his readers against what he regarded as "unnatural words and word-meanings" (1896: viii), Henry Sweet more than a century ago touched upon a whole range of problems that merit systematic study. In sum, we might say that "a healthy suspicion of glosses as indicators of normal usage", which Sir Christopher Ball expressed at a preparatory meeting for the *Dictionary of Old English*,[39] surely seems justified. To consider glossing as a method in its own right and in its own way, however, also requires us not to take too

---

[37] The nature and extent of translation-induced influences on the productive processes in Old English word-formation is still insufficiently explored (cf. Kastovsky 1992: 317). It will, for example, be important to ascertain whether a morphological pattern recurrent in loan-formations is matched by a substantial number of native innovations of the same type, or if a native pattern owes its productivity mainly to foreign instigation, as seems for instance to have been the case with derivatives prefixed by the adverbial particle *eft-* 'again', corresponding to Lat. *re-*. The editors of the *Dictionary of Old English* have recently devised "an improved system for treating verbs the component elements of which are calqued on Latin" (fascicle *E*, "Preface to readers", 1996: 1), so that such gloss formations are now given the status of "quasi-compounds" and appear as sub-entries under the initial adverbial element (cf. e.g. *efen~*, *efne~*, *eft~*). For the problems resulting from the *Dictionary*'s original resistance to recognise the existence of such verbal compounds cf. Kornexl (1994: 446-450).

[38] Traditional assumptions about the factors governing this process are conveniently summarised by Lutz (1997), who shows the destructive effects of phonotactically determined destabilisation on a considerable number of the Old English prefixes. Lutz also demonstrates that – contrary to established views – the Norman Conquest in fact exerted a stabilising influence on verbal prefixation.

[39] See Session III: "The form of the Dictionary of Old English" (Frank — Cameron (eds.) 1973: 6).

simplistic a view of so-called "over-literal renderings". Literalness need not necessarily be a sign of simple-mindedness, as we are wont to think. Due to a special training that to a considerable extent must have been based on oral instruction and thus escaped written notation, Anglo-Saxon students and glossators held the interpretative key, or rather the keys, to various types of vernacular glosses of different structure and function, but with the common purpose of elucidating the Latin text. It is our own preoccupation with lexical semantics that denies Old English glossformations closely modelled on Latin lemmata the full range of their explanatory potential; their "meaning" may in fact equally lie on the formal, i.e. structural and grammatical side. After all (to conclude with a somewhat anachronistic remark), the *Oxford English Dictionary* informs us that the term *artificial*, defined as 'opposed to *natural*', once carried a second sense: 'displaying special art or skill'.

References

Betz, Werner
- 1949   *Deutsch und Lateinisch: die Lehnbildungen der althochdeutschen Benediktinerregel.* Bonn: Bouvier.
- 1974   "Lehnwörter und Lehnprägungen im Vor- und Frühdeutschen", in: Friedrich Maurer — Heinz Rupp (eds.), *Deutsche Wortgeschichte*, vol. 1. (3rd edition.) Berlin: de Gruyter, 135-163.

Bosworth, Joseph — T. Northcote Toller
- 1892-1898   *An Anglo-Saxon dictionary based on the manuscript collections of the late Joseph Bosworth.* Edited and enlarged by T. Northcote Toller. London: Oxford University Press.

Brooks, Nicholas (ed.)
- 1982   *Latin and the vernacular languages in early medieval Britain.* Leicester: Leicester University Press.

Carpenter, Charles Whitney II
- 1973   *The systematic exploitation of the verbal calque in German.* (European University Papers, Series I: German language and literature 65.) Frankfurt/Main: Lang.

Clark Hall, John R. — Herbert D. Meritt
   1960   *A concise Anglo-Saxon dictionary.* (4th edition.) With a supplement by Herbert D. Meritt. Cambridge: Cambridge University Press.
Coleman, Evelyn Scherabon
   1965   "Zur Bestimmung und Klassifikation der Wortentlehnungen im Althochdeutschen", *Zeitschrift für deutsche Sprache* 21: 69-83.
Dictionary of Old English
   1986ff.   Angus Cameron — Ashley Crandell Amos — Antonette diPaolo Healey *et al.* (eds.), Toronto: Pontifical Institute of Mediaeval Studies.
Fix, Hans — Thomas Birkmann
   1998   "Die Worttrennung am Zeilenende in Handschriften der *Snorra Edda*", in: Hans Fix (ed.), *Snorri Sturluson: Beiträge zu Werk und Rezeption.* Berlin: de Gruyter, 23-33.
Frank, Roberta — Angus Cameron (eds.)
   1973   *A plan for the Dictionary of Old English.* Toronto: University of Toronto Press.
Gneuss, Helmut
   1955   *Lehnbildungen und Lehnbedeutungen im Altenglischen.* Berlin: Schmidt.
   1973   "Guide to the editing and preparation of texts for the Dictionary of Old English", in: Roberta Frank — Angus Cameron (eds.), 11-24.
   [1994]   [Reprinted with supplementary material in: Donald G. Scragg — Paul E. Szarmach (eds.), *The editing of Old English: papers from the 1990 Manchester conference.* Cambridge: Brewer, 7-26.]
   1989   "Glossen, Glossare: IV. Englische Literatur", in: *Lexikon des Mittelalters*, vol. 4. Munich: Artemis, cols. 1513-1514.
   1990   "The study of language in Anglo-Saxon England", *Bulletin of the John Rylands University Library of Manchester* 72: 3-32.
   1993   "*Anglicae linguae interpretatio*: language contact, lexical borrowing and glossing in Anglo-Saxon England", *Proceedings of the British Academy* 82: 107-148.

Götz, Heinrich
1977 "Zur Bedeutungsanalyse und Darstellung althochdeutscher Glossen", in: Rudolf Grosse — Siegfried Blum — Heinrich Götz (eds.), *Beiträge zur Bedeutungserschliessung im althochdeutschen Wortschatz.* (Sitzungsberichte der Sächsischen Akademie der Wissenschaften zu Leipzig: Philosophisch-historische Klasse 118/1.) Berlin: Akademie-Verlag, 53-208.
Henkel, Nikolaus
1996 "Die althochdeutschen Interlinearversionen: zum sprach- und literarhistorischen Zeugniswert einer Quellengruppe", in: Joachim Heinzle — Leslie Peter Johnson — Gisela Vollmann-Profe (eds.), *Übersetzen im Mittelalter: Cambridger Kolloquium 1994.* (Wolfram-Studien 14.) Berlin: Schmidt, 46-72.
Hofmann, Johann Baptist — Anton Szantyr
1965 *Lateinische Syntax und Stilistik. Mit dem allgemeinen Teil der lateinischen Grammatik.* (Handbuch der Altertumswissenschaft, 2. Abteilung, 2. Teil, vol. 2.) Munich: Beck.
Kastovsky, Dieter
1992 "Semantics and vocabulary", in: Richard M. Hogg (ed.), *The Cambridge history of the English language,* vol. I: *The beginnings to 1066.* Cambridge: Cambridge University Press, 290-408.
Kimmens, Andrew C. (ed.)
1979 *The Stowe Psalter.* (Toronto Old English Series.) Toronto: University of Toronto Press.
Köbler, Gerhard
1993 *Wörterbuch des althochdeutschen Sprachschatzes.* Paderborn: Schöningh.
Kornexl, Lucia (ed.)
1993 *Die "Regularis Concordia" und ihre altenglische Interlinearversion: mit Einleitung und Kommentar.* (Münchener Universitäts-Schriften: Texte und Untersuchungen zur Englischen Philologie 17.) Munich: Fink.
1994 "Progress in historical lexicography: *The Dictionary of Old English*", *Anglia* 112: 421-453.
Lapidge, Michael
1975 "The hermeneutic style in tenth-century Anglo-Latin literature", *Anglo-Saxon England* 4: 67-111.

[1993] [Reprinted with corrections in: Michael Lapidge, *Anglo-Latin Literature, 900-1066*. London: Hambledon, 105-149 and 474-479.]

1982 "The study of Latin texts in Late Anglo-Saxon England [1]: the evidence of Latin glosses", in: Nicholas Brooks (ed.), 99-140.

Lindsay, Wallace Martin (ed.)

1911 *Isidori Hispalensis episcopi etymologiarum sive originum libri XX*. 2 vols. Oxford: Clarendon.

Lutz, Angelika

1986 "The syllabic basis of word division in Old English manuscripts", *English Studies* 67: 193-210.

1997 "Sound change, word formation and the lexicon: the history of the English prefix verbs", *English Studies* 78: 258-290.

Masser, Achim (ed.)

1997 *Die lateinisch-althochdeutsche Benediktinerregel: Stiftsbibliothek St. Gallen Cod. 916*. Göttingen: Vandenhoeck und Ruprecht.

A microfiche concordance to Old English

1980 Compiled by Antonette diPaolo Healey — Richard L. Venezky. (Publications of the Dictionary of Old English 1.) Toronto: Dictionary of Old English Project, Centre for Medieval Studies, University of Toronto.

Milfull, Inge B. (ed.)

1996 *The hymns of the Anglo-Saxon church: a study and edition of the "Durham Hymnal"*. (Cambridge Studies in Anglo-Saxon England 17.) Cambridge: Cambridge University Press.

Mitchell, Bruce

1978 "Prepositions, adverbs, prepositional adverbs, postpositions, separable prefixes, or inseparable prefixes, in Old English?", *Neuphilologische Mitteilungen* 79: 240-257.

Mitchell, Bruce — Allison Kingsmill

1980 "Prepositions, adverbs, prepositional adverbs, postpositions, separable prefixes, or inseparable prefixes, in Old English? A supplementary bibliography", *Neuphilologische Mitteilungen* 81: 313-317.

Öhmann, Suzanne
1951 *Wortinhalt und Weltbild: vergleichende und methodologische Studien zu Bedeutungslehre und Wortfeldtheorie.* Stockholm: Norstedt & Söner.

The Oxford English Dictionary
1989 James A. H. Murray — Henry Bradley — William Alexander Craigie — Charles T. Onions (eds.). (2nd edition, prepared by John A. Simpson — E. S. C. Weiner. Combined with *A supplement to the Oxford English Dictionary*, ed. R. W. Burchfield, and reset with corrections, revisions and additional vocabulary.) 20 vols. Oxford: Clarendon.

Page, R. I.
1982 "The study of Latin texts in Late Anglo-Saxon England [2]: the evidence of English glosses", in: Nicholas Brooks (ed.), 141-165.

Ross, Alan S. C.
1934 "Notes on the method of glossing employed in the Lindisfarne Gospels", *Transactions of the Philological Society* 1933: 108-119.

Schottmann, Hans
1977 "Die Beschreibung der Interferenz", in: Herbert Kolb — Hartmut Lauffer (eds.), *Sprachliche Interferenz: Festschrift für Werner Betz zum 65. Geburtstag.* Tübingen: Niemeyer, 13-35.

Siebert, Eberhard
1971 *Zum Verhältnis von Erbgut und Lehngut im Wortschatz Otfrieds von Weissenburg.* Munich: Fink.

Sisam, Celia — Kenneth Sisam (eds.)
1959 *The Salisbury Psalter, edited from Salisbury Cathedral MS 150.* (Early English Text Society, Original Series 242.) London: Oxford University Press.

Sweet, Henry
1896 *The student's dictionary of Anglo-Saxon.* Oxford: Clarendon.

Tesch, Gerd
1978 *Linguale Interferenz: theoretische, terminologische und methodische Grundfragen zu ihrer Erforschung.* (Tübinger Beiträge zur Linguistik 105.) Tübingen: Narr.

Toth, Karl
1980 *Der Lehnwortschatz der althochdeutschen Tatian-Übersetzung.* (Epistemata: Würzburger wissenschaftliche Schriften, Reihe Literaturwissenschaft 6.) Würzburg: Königshausen und Neumann.

Wetzel, Claus-Dieter
1981 *Die Worttrennung am Zeilenende in altenglischen Handschriften.* (Europäische Hochschulschriften, Reihe XIV: Angelsächsische Sprache und Literatur 96.) Frankfurt/Main: Lang.

Wieland, Gernot
1984 "Latin lemma – Latin gloss: the stepchild of glossologists", *Mittellateinisches Jahrbuch* 19: 91-99.
1985 "The glossed manuscript: classbook or library book?", *Anglo-Saxon England* 14: 153-173.

Zupitza, Julius (ed.)
1880 *Ælfrics Grammatik und Glossar: erste Abteilung: Text und Varianten.* Berlin: Weidmann.
[1966] [Zweite, unveränderte Auflage mit einem Vorwort von Helmut Gneuss. Berlin: Weidmann.]

Manfred Markus (Innsbruck)
Duplications of vowels in Middle English spelling

1. The problem

Double spellings, both of vowels and consonants, were used in Middle English manuscripts with an apparent and irritating irregularity. Accordingly, researchers have reacted in various, partly contradictory ways. On the one hand, they have widely abstained from any attempt to explain the duplications, seeing them as a mere matter of coincidence and describing them eclectically[1]; on the other hand, scholars have resorted to arguing in terms of eccentric authors (like Orm, with his doubled consonants[2]) or even, more recently, to individual scribes, like the so-called "Scribe B" of Trinity College Cambridge MS R. 3.2, who – according to Samuels (1983: 20) – made a habit of using <ee> for Gower's typical <ie> spelling in *Confessio Amantis*. As regards the habits of scribes, however, Kristensson (1981: 161) has shown that even an individual scribe could be inconsistent when copying a source text repeatedly[3]. And with regard to consonantal digraphs, Scragg (1974: 50) has warned us that "[the] convention of repeating a consonant to indicate a preceding short vowel, which is so widespread in current English, has a long and complicated history".

Scragg refers to Latin and French spelling habits as influencing factors without fully making clear the extent of their influence. Concerning the vowels, his picture is at best fragmentary.[4] Generally speaking, little has been added to Luick's statement (1921-1940 [1964]: § 59)[5]:

> Vereinzelt schon im Frühmittelenglischen, vielfach schon im vierzehnten und fünfzehnten Jahrhundert versuchte man Vokallänge wenigstens in geschlossener Silbe

---

[1] Cf. e.g. Wild (1915: 8-10).
[2] For a critical explanation of Orm's spelling habits cf. Markus (1989).
[3] This inconsistency of scribes, adding to the inconsistencies caused by dialect mixture, is particularly common in manuscripts of the fifteenth century (cf. McIntosh – Samuels – Benskin (eds.) 1986, 1: 13).
[4] Apart from the valid points about doubled <u> (Scragg 1974: 47), nothing much is said by Scragg about vowel duplication; cf. Scragg (1974: 12, 37).
[5] Cf. Berndt (1960: 14); Fries (1985: 6); Mossé (1973: § 14); Robinson ((ed.) 1974: XXX); Sandved (1985: 20).

durch Doppelschreibungen zum Ausdruck zu bringen, am häufigsten bei e und o, demnächst bei a, seltener bei i und kaum bei u.

But while the acumen and patience of earlier scholars who studied Middle English spelling are admirable, I am glad to say that machine-readable Middle English texts, as they are now available, e.g. within the Innsbruck ICAMET project[6], give us a much better chance to answer the old questions of Middle English spelling habits in a new, data-based, and above all explanatory way. The questions are these:
1. Are the digraphs in Chaucer's English used systematically or not?
2. How can the different frequencies of the various double spellings be explained?
3. Is the difference between <ee> and <e> in Chaucer more than a mere marking of the opposition of vocalic length and perhaps one of vocalic or syllabic quality?[7]

In the following it will be shown that, due to the contact of Middle English scribes with the spelling habits in other languages, above all French, their vocalic double spellings were conditioned by a whole cluster of factors and, naturally enough, at different times and places in varying ways and degrees.

2. A new look at the material

The following explanation is based on the belief that there is some "orderly variation in (Middle English) spelling". In this belief and with this phrase we follow Milroy (1992: 134), who refers to the loss of the velar fricative in Middle English (as in *niʒt*) and takes spelling inconsistencies as evidence of the imminent loss of this fricative. Likewise, we try to find the systematic character of apparent inconsistencies of double spellings, combining quantitative evidence with arguments of language contact and an immanent change of the phonemic system.[8]

---

[6] *Innsbruck Computer Archive of Machine-readable English Texts*. For a general description of the concept of this project see Markus (1994).
[7] Cf. Sandved's assumption (1985: 20) that Middle English spelling reveals a "certain tendency" towards a marked distinction of open and closed syllables.
[8] Jeremy Smith (1994: 160) rightly said that handling languages in contact "you're dealing constantly with interaction between user and system, system and user."

Duplications of vowels in Middle English spelling

Table 1. Middle English occurrence of double spellings of vowels in dialectal arrangement

dialect	date	text	<aa>	<ee>	<oo>	<uu>	<ii>
EML	1121-1160	peterbor	(x)	(x)	–	(x)	–
EML	1200	oehom	(3)	(7)	(1)	(x)	–
EML	1225a (c1200)	vicesme	–	(x)	–	(x)	–
EML	1450	julian	(6)	x	x	–	–
EML	1500a (?c 1450)	merlin1	(x)	x	x	–	–
EML (East A)	1400 {-1450]	mandevil	(2)	x	x	–	–
EML (Linc?)	1400	abbey	–	x	x	–	–
EML (Norf)	1417ff.	paston2	x	x	x	(x)	(x)
EML (Norf)	1450	caplives	(1)	x	x	(x)	(x)
EML (Oxf)	1460c	oseney	x	x	x	(x)	–
EML (Norf)	1464a	capgra	(x)	x	x	(x)	(x)
EML/N	1415a	lantlit	x	x	x	(x)	(x)
EML/N	1430c (a1410)	mirbles	x	x	x	–	–
EML/N	1434 ( )	misyn	(4)	x	x	(x)	(x)
London	1390 ( )	persbla	x	x	x	–	(x)
London	1390 ( )	astce	x	x	x	–	–
London	1390 ( )	boesce	(x)	x	x	–	–
London	1390 ( )	melbla	x	x	x	–	x
London	1390 ( )	melsce	x	x	x	–	(x)
London	1450 {-1500]	boke	–	(5)	–	–	–
London	1480	wyclif	x	x	x	–	–
London	1480c	caxtquat	x	x	x	–	–
London	1485	caxtpar	x	x	x	(x)	(x)
London	1489	caxtdoc	x	x	x	(x)	(x)
N	1300 {-1325]	rolhor	x	x	x	(x)	(x)
N	1300 {-1325]	rollebok	x	x	x	–	(x)
N	1300 {-1325]	rollpros	x	x	x	–	(x)
N	1440c (?c1350)	mirredm	(2)	x	x	–	–
N	1450 {-1500]	craftdye	–	(4)	–	–	–
N	1450 {-1500]	solomon	(1)	–	(x)	–	–
N	1450c	alpha1	–	x	x	–	–
N (Yorks)	1450	liber	x	x	(x)	–	–
SE	1275c	kentserm	(2)	5	9	(20)	–
SE	1340 ( )	ayenbit	(4)	4	2	–	(x)
SML	1375 {-1400]	kingscam	(x)	x	x	(2)	–
SML	1375 {-1400]	kingsroy	x	x	x	–	–
SW	1225c (?c1200)	halibod	(x)	(3)	–	–	(x)
SW	1225c (?c1200)	halitit	–	–	–	(x)	–
SW	1225c (?c1200)	juliaroy	–	–	–	(x)	–
SW (Shrops)	1225c (?a1200)	ancrene (CC)	x	x	x	(x)	–
SW	1450	trevdial	(3)	x	x	–	–
SW	1450	trevmeth	(2)	x	x	–	–
WML (Here/Glou)	1200	kathroy	x	–	–	(x)	–

219

WML	1200c	sawleswd	(x)	–	–	(x)	–
WML (Here)	1225c (?c1200)	juliabod	–	–	–	–	–
WML	1400	mirror	x	x	x	–	–

Selecting from the Innsbruck Corpus[9] 46 Middle English texts which represent relatively purely the main dialects of Middle English, we get a table of vocalic double spellings (table 1[10]), which allows the following conclusions:

a. While <ee> and <oo> were used in all dialect areas at some stage, <aa> was sometimes used, but in most cases extremely rarely, and both double <u>'s and double <i>'s, where they occurred, were used with a deviant function explained below. In table 1 I have marked such rare or functionally deviant use by parentheses.[11]

b. Chronologically arranged[12], only the texts of the twelfth and thirteenth centuries and three late texts from the North (1450-1500)[13] reveal a clear abstention from vocalic double spellings.

In view of these observations a number of questions can be raised:
1. Why were duplicated spellings used at all?
2. What was the function of <aa>, <ee> and <oo>?
3. Why were <aa>-spellings, compared with <ee> and <oo>, so restricted?
4. Why were the <ii>- and <uu>-spellings almost non-existent?

3. Why duplicated spellings for monophthongs?

The general Middle English bias for using duplicated spellings can to some extent, but not fully, be explained from Old English, where the length of a vowel, like that of a consonant, was phonemically distinctive (cf. Pilch 1970: 62), though not graphically marked (*mĕtan* 'to measure' vs. *mētan* 'to meet'). Vocalic digraphs did not have the function of length, but were genuine diphthongs. As is well-known, these were mono-

---

[9] The fact that all the texts are in prose is quite an advantage, since many a stem vowel in verse literature may well have been quantitatively modified for the sake of rhyme or metre; cf. Frieshammer (1910: XIV).
[10] "x" means frequent occurrence. The abbreviations in the column "text" are those of ICAMET; the complete list of the Innsbruck texts is available from my department.
[11] Where in doubt I have given the frequency numbers. Of course, their weight depends on the length of a text.
[12] For lack of space the diachronic table is not printed out here, but it can easily be constructed from table 1.
[13] *boke, craftdye* and *solomon*.

phthongised in or by the eleventh century (Luick 1921-1940 [1964]: §§ 191; 358, 1; 359, 2), but the spelling survived for some time. Twelfth-century readers and scribes of English had thus to learn that digraphs like <ea>, <eo> and <io> marked monophthongs. This was the precondition for occasional cases of inverse spelling as in Middle English *people* <eo> = /eː/.

But the Old English spellings are hardly sufficient to explain the vocalic double spellings of the fourteenth and fifteenth centuries, the less so since this explanation would make the time gap of two centuries less plausible. I take it that other causes must have contributed to the habit of using digraphs for monophthongs in Middle English, such as[14]:

List 1. Vocalic digraphs in Middle English	
(a) <ie> for /eː/:	the spelling comes from Central French, its smoothed monophthongal value from Anglo-Norman. The spelling was even used in non-French words like *lief* and *priest* (both /eː/).
(b) <ui>, or allographic <uy>, for /üː/:	in words like *fruit*, the spelling reveals the former Central French rising diphthong /üí/. The Anglo-Normans, for their part a Germanic tribe, had not been used to rising diphthongs and therefore, in analogy to <ié> above, smoothed /üí/ to /üː/, which is the very vowel that then crossed the Channel.
(c) <ou> (or <ow>/<ov>[15]) for /uː/:	this is a clear case of borrowing from French (where it has survived up to now, like in *court*). <u> could not be used for /uː/ in Middle English, since it had taken over from French the function of marking the rounded vowel /ü/ (like in *virtú*).

As to the consonants, there is again quite a list of Middle English digraphs which signify monophthongs. The spelling was often a mere relic of Old English geminates. But since consonantal length was no longer phonologically distinctive, digraphs in Middle English could be used to mark specific monophthongs:

List 2. Consonantal digraphs in Middle English
<cc> for /k/ (cf. Lat.): *accorde*
<gh>, with allographs <gʒ, ʒh> for /χ/: *toughte, nyght*
<gn> for /ñ/ < OFr.: *resigne*
<jh> for /j/ in some foreign proper names: *Jhesus* (Fries 1985: 23)

---

[14] On these and other examples of the general influence of French spelling, cf. Bøgholm (1939: 158-161).
[15] The use of <v> or <w> as the second element of the digraph at issue is allographic and was common practice in mediaeval Latin. In Middle English the use of these graphs is connected with the so-called "minim" problem (cf. section 6.).

\<ng\> for /ŋ/ < OFr. velar nasal: *song*
\<ph\> for /f/ < Greek via Latin and Old French: *philosophye*
\<sc\> for /s/ < OFr. affricate /ts/: *science*
\<th\> for voiced and voiceless th. The spelling is based on Old French/ Latin, where it was needed for Greek *theta*; it superseded thorn in the thirteenth century.
\<wh\> for voiceless /w/ in line with Anglo-Norman.

4. The function of \<ee\> and \<oo\>

It is clear from lists 1 and 2 that digraphs were partly traditional, but in the main widely used in Middle English either for foreign phonemes, or for well-known native phonemes where traditional spellings (such as thorn) were abandoned in favour of foreign spellings that had been popularised by loan-words. Digraphs, then, were a multi-purpose instrument. It is, therefore, unlikely that \<ee\> and \<oo\> had a well-defined phonological function, least of all as a marker of length, since [length] was on its way to losing its phonemic quality, with other features going hand in hand with it, such as word stress or the closed type of the syllable.

The non-phonological role of the feature [length] results from the fact that the systems of the short and of the long vowels had split by the fourteenth century (cf. Fisiak 1968 [1977]: 41f.). The traditionally short vowels only had one row of a medium height; the long vowels had two. An opposition of length therefore implied an opposition of height (or openness), and tongue-raising of /ɛ/ or /ɒ/ would have implied a development to /eː/ and /oː/ respectively.

Considering now the empirical data available for \<ee\> and, for the sake of brevity, focussing on *The Parson's Tale*, we see that \<ee\> occurs in the following cases:

List 3. \<ee\> in *The Parson's Tale* (survey)

a. word initial position:	In word initial position only 106 cases of *eek* 'also' and two cases of *eet* 'ate' are to be found. *Eek* is also spelt \<eke\> in three cases. Both *eek* and *eet* had /æː/ in Old English (Anglian). On this basis the double spelling could have marked either length or openness of the vowel. But the double spelling is basically restricted to one or two favourite words and therefore not very conclusive.
b. word final position:	\<ee\> marks length and/or word stress.
	< OFr. -é,-ée: 22 (e.g. *priuee*)
	< OFr. -et: c100 (e.g. *secree*)
	< OE -ē,-ēo: 45 (e.g. *thee, knee, free, tree*)
	< OE -ǣ: 2 *see*

Duplications of vowels in Middle English spelling

c. word
medial
occurrences:　　Word medial occurrences allow the following conclusions:
1. In all cases the vowels concerned are etymologically long or, like the French vowels, interpretable as long.
2. There is a subset of Anglo-Norman words like *seculeer, degrees, encreeseth, contrees, natureel*, etc. Here, as in the frequent cases of word-final <ee> in (b), the <ee> spelling obviously marks length combined with word stress.
3. For the rest, most of the words concerned are closed monosyllabics.
4. If the words are plurisyllabic this is often due to a weak second syllable with a schwa or a mute *e*, sometimes followed by the final *-n* of the infinitive; in these cases the weak final syllable could be dropped and the stem syllables can be considered as closed: *beet(e), deel(en), deer(e)*, etc.
5. Another subset of words is that of derivations or compounds based on words as mentioned in (2) and (3) above. Thus, *beestes* derives from *beest, cleerness* from *cleer, feelynge* from *feel(en)* etc. In many such cases, derivations (and to some extent compounds like *heerof*) follow the rule of analogy and have the stem vowel spelling of their base words.

In sum, <ee> spelling was used in initial and final position (a and b) for both closed and open /e/, which was usually, but not always, etymologically long. In medial position <ee> was likewise used for etymologically long /e/ (or foreign /e/ wrongly interpreted as long; cf. c1) in closed monosyllables (c3); plurisyllables can be seen as "exceptions", explicable on phonological (c4), morphological (c5) or etymological (c2) grounds.

These observations allow the reverse conclusion that, with the exception of loan-words, there was no marking by double-spelt vowels in open stem syllables of di- or plurisyllabic words. This conclusion refers us to MEOSL, the Middle English open-syllable-lengthening rule, which is known to have applied to /e/, /o/ and /a/. The connection of the double spellings with this rule is the following: since non-high stem vowels in open syllables of disyllabic words were systematically lengthened in the thirteenth century (*mĕ-te* > *mē-te*), the CVCV-spelling could be interpreted as evidence of words affected by this rule. No extra marking was needed. The words on our list were all and sundry unaffected by the Middle English open-syllable-lengthening rule; their stem vowels had length *before* the thirteenth century, whether they were rooted in Old English or Old French, and no matter whether they were based on OE *ēa* or *ēo* or *ǣ* or *ē* or *ĕ* lengthened before homorganic clusters in late Old English. Table 2 lists a few examples of these types:

223

Table 2. Examples of <ee> in *The Parson's Tale* for assumed /eː/

< OE ēa	seeth /æː/, sleen, breed, greet, deedly, deeth, teerys
< OE ēo	been /eː/, weep, deere, preest, freend, seen, bees
< OE ǣ	nathelees; -lees, deel, heele, heeste, reed
< OE ē	teeth, feeld, heer(e), seed, feeleth, feet, feed
< OFr. e/æ	cleerness, beestes, feeste, irreguleer, releese, seculeer, natureel

What scribes obviously did by these markings of the long stem vowels of different etymology was to respond to the general disorientation concerning the feature of vocalic length. French loan-words like *irreguleer* had caused people to associate [length] with word accent. In addition, the Middle English open-syllable-lengthening rule had evoked further criteria for [length], namely the number and structure of syllables and the height of the tongue. The double spellings reveal the attempts of scribes to mark foreign or old long vowels in view of the new, allophonically variable criteria of [length].

Thus, in the fourteenth and early fifteenth century, the sense of English speakers of what [length] meant must have been unstable. The two principles of [length] mentioned were both valid: on the one hand the etymological concept, where length was defined by the traditional "outfit" of a vowel; on the other hand, what we may call a "prosodic" concept: the length of a vowel depended on its changing role within the syllable, the word or morpheme, and the sentence[16]. Due to the validity of both principles by the time of *The Parson's Tale*, the double spellings as evidenced in our list were challenged by spelling modernists. Accordingly, we have a number of counter-examples to list 3:

Table 3. <ee> spelling and "exceptions" in *The Parson's Tale*[17]

etymological spelling	alternative spelling
been 9	ben 380
(nathe)lees 5	natheles 2
greet 4	grete 17
beest(es) 5	bestes 3
leest(e) 1	lest 1
deeth 39	deth 7
freend(es) 10	frendes / frendshipe 5
cleernesse 1	clere 2

---

[16] For the many quantitative changes of Middle English vowels, see Markus (1990: 58).
[17] The figures given are based on the first 20% of the text.

| heer(e) 'hic' 5 | here 3 |
| weep 1 | wepe 7 |

Of the spelling alternatives we have the following types:

<e> in monosyllabic or, at least, non-disyllabic words, like in *ben, natheless, lest* and *deth*: *-les* as a suffix is likely to have had a quantitatively reduced vowel by 1400, due to its weak position in the word. <ben> also gives evidence of the allophonic quantitative reduction, this time due to the unstressed position in the sentence; the same goes for *lest*, which, at second sight, turns out to be a conjunction, i.e. a function word, whereas *leest* with double <ee> on the left is an adjective. As to *here* (in the sense of the deictic adverb), it must often have been in an unstressed sentence position and was thus bound to lose its original length.

In the other words on the right of table 3, the immanent spelling principle is that of the Middle English open-syllable-lengthening rule, where the length of a stem vowel is marked by <VCe> spelling. This is a case of inverse spelling. But then the average fourteenth-century speaker could not know whether words like *bestes* or *clere* had been borrowed early enough from French to be affected and lengthened by the Middle English open-syllable-lengthening rule. It so happened that words with a "long" vowel of a foreign origin were "nostrified" (to use an Anglicised Austrian term).

Deth and *frendshipe*, not fulfilling the conditions of the Middle English open-syllable-lengthening rule, are deviant in another way. The vowel in *deth* was shortened later, according to Pinsker (1974: § 63, A.2b) in the sixteenth century. It may well be that the stem vowel of *deth* allophonically lost its length earlier than this, perhaps due to contextual adaptation (like in *deth-bed*, etc.; cf. Pinsker 1974: § 63, A.1). *Frend*, at least in derivations and compounds, also ended up short in Early Modern English, like in *friendship* and *friendly*[18]; perhaps the alternative spellings with a single <e> can be interpreted as early evidence of that shortening.

5. <aa>

If we take it for granted that the <aa> spelling was similarly motivated as <ee> and <oo>, the vowels in question in the following list, again based on *The Parson's Tale*, should be etymologically long and positioned both in

---

[18] Cf. Bøgholm (1939: 171).

closed and stressed syllables of monosyllabic words. Table 4 gives the list, with alternative spellings added:

Table 4. <aa> in Chaucer's *The Parson's Tale*

double spelling	alternative spelling
caas 1 (< OFr. *cas*)	*cas* 2
estaat 10 (< OFr. *estat*)	*estat* 6
delicaat 3 (< OFr. *delicat*)	*delicat* 1
Canaan (proper name!)	–
chaast 1 (< OFr. *chaste*)	*chaste* 1, *chastity* (many)
curaat 2 (MLat. *curatus*)	*curates* 1
baar 2 (OE *bær*, /æː/ from *bæron*)	–
taak 2 (ON *taka*)	*take* 30

It can easily be seen that the three criteria are not entirely fulfilled: the words all have closed stressed syllables[19], but the /a/ is not etymologically long in all cases and the words concerned are not monosyllables.

This, however, accords fully with what was said above about the allophonic role of [length]. In words like *caas*, *estaat*, *delicaat* etc., the /a/ in the original French form had a medium length and was obviously, like Old French vowels generally, classified in English as long or short, according to syllabic and prosodic criteria. The most striking of these was the stressed final syllable, which caused its vowel to be interpreted as long[20].

The main point concerning <aa> is, of course, this: by the fourteenth century, long /aː/ changed to open /ɔː/ south of the Humber, so that the <aa> spellings on our list are extremely limited in number in comparison with <ee> and <oo>; and they concern words with long allophonic /a/ of various foreign origins.

There were several reasons for foreign stem vowels to be identified or interpreted as long. In *baar*, vocalic length can only be derived from the plural form OE *bæron* with the help of the analogy principle; the singular was *bær*. *Chaast* probably had a long stem vowel in Old French already, due to the loss of /s/ (Pinsker 1974: § 45, 1d). As to *taak* (from ON *taka*), this can only have received its length of the stem vowel from the Middle English open-syllable-lengthening rule, which is why the "alternative" spelling with a single <a> and a final <e> is by far more typical and the

---

[19] Except for *Canaan*, which is, however, a proper name.
[20] For further details see Pinsker (1974: § 45, B).

<aa> spelling must be considered as an aberrant case of inverse spelling. At a time when the reasons for vocalic length had become so complex, a small number of such "errors" should be allowed for.

6. <uu> and <ii>

<u> and <i> were obviously unaffected by the Middle English double spelling habits. In table 1 only a few texts were listed where <uu> and <ii> do occur, but on closer investigation they turn out to be used for anything but length. As to <uu>, it has the following functions:

Table 5. The function of <uu> in selected texts of ICAMET

<uu> =	
<w>:	peterbor
<o> + <w>:	paston 2/5, kingscam, caxtdoc
<uv> / <vu>:	oehom, vicesme
second part of <qu> + <u>:	capgra
Latin hiatus diphthong[21]:	/u-u/, as in *tu-us*
English hiatus diphthong:	/u-u/, as in *affectu-usly*: mysin

What this comes to is that double <uu> spellings, if they occur at all in Middle English rather than Latin words, do not represent monophthongs, but are, in line with Latin habits, either a transsyllabic combination of <u>'s, or allographs, at least partly, of <v> or <w>[22]. Double <ii> either takes us to Latin/ancient words/names directly, as in *Assyriis* 'to the Assyrians', or to hiatic diphthongs where the syllable border is crossed; this occurs rarely enough with derivations of verbs ending in <-i>, as in *fri-inge* 'frying' in *ayenbit*.[23] The far more frequent spelling, particularly in the special case of *-ing*-suffixes, is <iy> or <yi>. This is no doubt due to the "minim problem" (Markus 1990: 35; Wright 1960: XVII): with <i> surrounded by letters with vertical strokes the Middle English scribes generally preferred its substitution by <y> or <j>. Accordingly, we have further spelling variants of <ii>, namely <ji> and <ij>. All the variants are fairly frequent in connection with the expanded-form suffix and generally

---

[21] = so-called "internal open juncture" (Crystal 1991: 188).
[22] <uu> for /w/ was reintroduced in England, where it had been used before, in early Old English by Norman scribes of the eleventh century; cf. Bøgholm (1939: 161).
[23] Other extremely rare examples in *vicesme, rollebok* and *halibod* (some 10 examples altogether).

transsyllabic, but in *rolhor* there are also a few occurrences (17 altogether) of the words *lijk* and *lijf*.

Why were <u> and <i> less affected by graphemic duplication than the vowels of a lower tongue position? Again there is a connection with the Middle English open-syllable-lengthening rule in the thirteenth century: this had not reached the two vowels[24]. Accordingly the Old English complementary conditions of lengthening and shortening (*child* vs. *children*) were still relevant; the rule was, in fact, revitalised in the fourteenth century, affecting new words after the French intermezzo phase. The various quantitative changes, as listed by Pinsker for the late Middle English period (Pinsker 1974: § 50), are really a revival of the late Old English principle which redefined the length of a stem vowel by allophonically modifying it according to its surroundings.

7. Conclusions

Summing up: etymologically long vowels, whether of Old English, Old French, Latin or Scandinavian origin, were generally marked by duplicated spelling from 1300 to 1500 (North: to 1450). These were the stem vowels where the Middle English open-syllable-lengthening rule did not apply. The marking of length by double vowels did not prevent [length] from becoming a frequent allophonic feature in Middle English, so that etymologically long vowels were often phonetically shortened and Scandinavian or French vowels of a short or medium quantity were pigeonholed as long, for instance under the influence of word stress. <a> was less affected by duplication, since after the change of /a:/ to open /o:/ south of the Humber not many /a:/'s were left. <i> and <u> were lowered in time to avoid being affected by the Middle English open-syllable-lengthening rule.

As to <ee> and <oo>, this paper has tried to demonstrate that the "inconsistent" spelling practice of Middle English scribes is not simply due to general carelessness, but rather to the feature [length] having lost its former phonemic identity. Not always knowing for sure whether a stem vowel was to be classified as long or short (due to the many factors that had come to influence length), scribes tended to mark length by double spelling and often failed to do so correctly and consistently.

In this confusing picture we naturally have to cope with a great many "exceptions"; they can only be explained with the help of different types of

---

[24] On the reason for this, cf. Markus (1988).

linguistic rules, from misinterpreted etymology to the role of word accent or word families. But one intraphonological conclusion seems pretty stringent: with <e> and <o> there is a complementary distribution between traditionally long vowels (including the half-understood tradition from abroad), and the vocalic neo-lengths of the Middle English open-syllable-lengthening rule. Or were the latter, the MEOSL vowels, "tense" vowels rather than "long" ones?[25] This, however, is another story.

References

Berndt, Rolf
    1960   *Einführung in das Studium des Mittelenglischen.* Halle: VEB Niemeyer.

Bøgholm, Niels
    1939   *English speech from an historical point of view.* London: Allen and Unwin.

Chomsky, Noam — Morris Halle
    1968   *The sound pattern of English.* New York: Harper and Row.

Crystal, David
    1991   *A dictionary of linguistics and phonetics.* (3rd edition.) London: Blackwell.

Fisiak, Jacek
    1968   *A short grammar of Middle English.* (4th edition.) Warszawa: Polish Scientific Publishers.
    [1977]   [Reprinted London: Oxford University Press.]

Fries, Udo
    1985   *Einführung in die Sprache Chaucers: Phonologie, Metrik und Morphologie.* Tübingen: Niemeyer.

Frieshammer, Johann
    1910   *Die sprachliche Form der Chaucerschen Prosa.* (Morsbachs Studien zur englischen Philologie 42.) Halle.

Kristensson, Gillis
    1981   "Another piece of evidence for the study of Middle English spelling", *Neuphilologische Mitteilungen* 82: 159-161.

---

[25] Cf. Chomsky — Halle (1968).

Luick, Karl
1921-1940   *Historische Grammatik der englischen Sprache.*
             Bearbeited von Friedrich Wild und Herbert Koziol. 2 vols.
             Stuttgart: Tauchnitz.
[1964]   [Reprinted Oxford: Blackwell.]

Markus, Manfred
1988   "Noch einmal: zur mittelenglischen Dehnung des 13.
             Jahrhunderts", *Klagenfurter Beiträge zur Linguistik* 14:
             384-410.
1989   "The spelling peculiarities in the Ormulum from an
             interdisciplinary point of view: a reappraisal", in: Uwe
             Böker — Manfred Markus — Rainer Schöwerling (eds.),
             *The living Middle Ages: studies in mediaeval English
             literature and its tradition: a festschrift for Karl Heinz
             Göller.* Stuttgart: Belser, 69-86.
1990   *Mittelenglisches Studienbuch.* Tübingen: Francke.
1994   "The concept of ICAMET (Innsbruck computer archive of
             Middle English texts)", in: Merja Kytö — Matti Rissanen
             — Susan Wright (eds.), *Corpora across the centuries:
             proceedings of the first international colloquium on
             English diachronic corpora, St. Catharine's College,
             Cambridge, 25-27 March 1993.* Amsterdam: Rodopi, 41-52.

McIntosh, Angus — Michael Louis Samuels — Michael Benskin (eds.)
1986   *Linguistic atlas of late mediaeval English.* 4 vols.
             Aberdeen: Aberdeen University Press.

Milroy, James
1992   *Linguistic variation and change: on the historical socio-
             linguistics of English.* Oxford: Blackwell.

Mossé, Fernand
1973   *Mittelenglische Kurzgrammatik: Lautlehre, Formenlehre,
             Syntax.* Translated by Herbert Pilch und Ursula Siewert.
             München: Hueber.

Pilch, Herbert
1970   *Altenglische Grammatik: Dialektologie, Phonologie,
             Morphologie, Syntax.* München: Hueber.

Pinsker, Hans Ernst
1974   *Historische englische Grammatik.* (4th edition.) München:
             Hueber.

Robinson, M. L. (ed.)
1974 *The works of Geoffrey Chaucer.* (2nd edition.) Oxford: Oxford University Press.

Samuels, Michael Louis
1983 "Chaucer's spelling", in: Douglas Gray — E. G. Stanley (eds.), *Middle English studies: a festschrift for Norman Davis in honour of his seventieth birthday.* Oxford: Clarendon, 17-37.

Sandved, Arthur O.
1985 *Introduction to Chaucerian English.* (Chaucer Studies 11.) Cambridge: Brewer.

Scragg, Donald G.
1974 *A history of English spelling.* New York: Barnes and Noble.

Smith, Jeremy J.
1994 "Panel III, Languages in contact: discussion", in: Margaret Laing — Keith Williamson (eds.), *Speaking in our tongues: proceedings of a colloquium on medieval dialectology and related disciplines.* Cambridge: Brewer, 149-168.

Wild, Friedrich
1915 *Die sprachlichen Eigentümlichkeiten der wichtigeren Chaucerhandschriften und die Sprache Chaucers.* Wien: Braumüller.

Wright, C. E.
1960 *English vernacular hands from the twelfth to the fifteenth centuries.* Oxford: Clarendon.

EAST BATON ROUGE PARISH LIBRARY
MAIN LIBRARY
Expiry date: 28 May 2025

Borrowed on 01/29/2024 15:16 Till

1) Interlibrary Loan Book.
   Due date: 03/10/2024
   No.: 31659005013251

Total on loan : 1

To renew items call 225-231-3744
or visit www.ebrpl.com.
01/29/2024 - 15:16

Gabriella Mazzon (Naples)
Language contact in the history of Englishes, or the genesis of extraterritorial varieties

1. The problem

An even cursory review of the literature available on extraterritorial Englishes (ETEs), which I had the occasion to survey while working on Maltese English (Mazzon 1992) and on varieties of English in general (Mazzon 1994), reveals that extraterritorial Englishes have several features in common, although each possesses its own individuality.

This intriguing situation is bound to enfold interesting insight for "genetic linguistics". Our starting point is thus what could be called the extraterritorial Englishes' paradox: so different yet so alike.

2. The antefact

Weinreich's basic work on languages in contact (Weinreich 1953) represented a convincing introduction to the pervading impact of interlingual influence. This was so well depicted as to lend authority to a few decades of contrastive studies and to even the most overtly prejudiced characterisations of localised varieties, making us certain that yes, there is an Indian/Italian/German/Scottish way of interpreting English: we may laugh about it or be nationalistically proud of it, but it is there, a testimony of the struggle between our linguistic roots, our "mother tongue", and our social mask, the "other tongue".

Not very much later, however, another current in linguistic theory which was to become very influential was playing down the importance of language contact by suggesting that psycholinguistic processes are universal; the development of linguistic competence, though subject to external and idiosyncratic influences in its most superficial aspects, is common to all human beings, and yields similar results, even developed in comparably ordered sequences. In flowed the evidence in favour of such hypotheses, as abundant and convincing (in some cases even more so) as that in favour of mother-tongue influence. Thus, the field of second language acquisition theory, the main arena for this kind of debate, came to be dominated by the controversy between universalists and relativists (as many other fields did); the contendents were (and are) on the one side the pidginisation/ creolisation hypothesis, an extreme form of belief in language contact influence, and on the other side the Universal Grammar hypothesis in its stronger and weaker versions; each side accusing the other of excessive abstractness and of neglecting the data, each trying to generalise its explanations. The battle has been transferred to other arenas, though marginally; among these, historical linguistics is perhaps the most promising. Very few attempts have been made so far at exploiting the ample variation in forms and uses that the present spread of English affords us in the search for confirmation or falsification of these theoretical positions: the present paper is a modest contribution to this aim.

3. The object of the study

3.1. A major distinction has to be drawn between primary extraterritorial Englishes, or extraterritorial Englishes as L1 (both the "Atlantic" and the "Southern hemisphere" types) and secondary extraterritorial Englishes, or extraterritorial Englishes as L2 (the so-called new Englishes or non-native Englishes). There are both historical and formal differences between these two groups of varieties: in the latter group, English is mainly a second language; it is subject to constraints in its functions and domains of use as well as in the number of users; it is transmitted from one generation to another mainly via formal instruction and literacy; and it is imposed "from above" by a colonial government, with certain political aims, and not originated from the communicative needs of everyday life. In primary extraterritorial English communities, by contrast, English was the mother tongue of the majority of settlers over a vast area (with the natives more or less relegated into the background) and it has been used for all communicative purposes, by tacit consent, even by the millions of later immigrants. This is one external or historical difference which has had a considerable impact on the formation of new varieties. As examples of formal differences derived from this we can mention, first of all, the fact that the functional and social constraints are responsible for the reduced stylistic range of most non-native Englishes; moreover, the two other features mentioned produce, in secondary extraterritorial Englishes, the well-known tendency to the bookish, learned word, and the shifts in register whereby expressions which are highly formal in British English can be used in normal conversation while slang, colloquial phrases are often heard in formal exchanges and are sometimes used in writing even at high levels of formality.

3.2. Another, much deeper consequence of the difference in the way English was introduced and is transmitted concerns more directly the problem of assessing the role of language contact. Local languages had a very marginal role in the formation of primary extraterritorial English varieties, apart from the adoption of a few loan-words. Of course, mother-tongue impact also depends on the degree of vitality of the local languages, on their sociolinguistic role and on the numerical and socio-economic "weight" of their speakers: at the extremes of this scale we find the relative marginality of the contribution from Celtic languages to varieties of English in the British Isles, or of that from the Amerindian languages in North America, as opposed to the relatively important role of Afrikaans in the formation of South African English. A certain amount of foreign influence can be detected, e.g. in American English, from the adstrate, i.e. the languages of later immigrants, such as German, Italian and Yiddish, and from the contiguous use of Spanish in some regions, but this is again mostly confined to a few loan-words and calques in the mainstream varieties (i.e. not considering the mixed varieties employed by some ethnic groups, or the rather different and still partly controversial individuality of the Black English Vernacular).

In non-native Englishes, on the other hand, contact with the mother tongues has been much more influential, though precisely to what extent remains to be determined, as we shall see later on. This, however, does not mean that language contact was uninfluential in the formation of primary extraterritorial Englishes, but it was a contact of a different sort, i.e. the contact between different British varieties brought over by large numbers of immigrants, whose product will be briefly explored in the next section.

4. Dialect levelling in primary extraterritorial Englishes

4.1. It is to be assumed that each individual member of each colony brought over with him a repertoire of dialects and accents, with a dominant form depending on his provenance, and that only some settlers possessed a knowledge of the standardised, educated variety. The close-knit network of interaction between these different groups brought about the levelling of the more extreme dialect forms, since each speaker in contact with others probably tended to select from his repertoire the forms which were more likely to be widely understood.

We must hypothesise that a great deal of accommodation and quick koinéisation was going on in those early days, and it is interesting to note that the koiné that emerged was always basically of the same type, showing a predominance of one specific dialect form; the Law of Swamping, formulated by Roger Lass (1990c) on the evidence that "there is no ETE that is not a dialect of Southern English", yielded consistent results: a mixed dialect input in a colonial settlement, no matter what the demographic patterns and the geographic distribution may be, will always produce a variety of the Southern, and particularly South-Eastern, type, while contributions from other dialects will be submerged ("swamped") or will at best surface very marginally as relics.

4.2. All counter-evidence offered against this, claiming that specific features of old British dialects can still be traced in separate areas of the Old Commonwealth, is based on linguistic "islands" where the input was not really mixed, i.e. where there was a strong predominance of settlers from one dialect area; this can be seen in Canadian Newfoundland as well as in New Zealand's South Island, where typical features actually seem to correspond exactly to some old dialect features that could have been preserved, especially those that are characteristic of Scottish English and of the dialects of the English South-West. Since the premise of Lass's Law of Swamping is that the input be mixed, however, these relic areas do not really constitute counter-examples, since their settlement history is different. The controversy over such phenomena has raged particularly when specific and widespread extraterritorial English structures were observed that seem to mirror exactly specific dialectal British structures typical of an area. This has led some linguists to overgeneralise the conclusions derived from such observations. Thus, for instance, much has been made of the alleged "conservativeness" of American English; recent studies, however, warn us against hasty deductions (see e.g. Görlach (1987) and Lass (1990a: 137), who maintains that it is altogether impossible to speak in a global way of conservativeness or innovativeness for extraterritorial Englishes as we do for other languages).

The arguments in favour of "conservativeness" generally employ examples such as the occurrence of double modals of the type *might could* in southern United States English, a phenomenon which is rather narrowly confined, on the British Isles, to some varieties of Scottish English. The use of these forms in American English, though, does not match exactly their use in Scotland, so we cannot simply speak of conservation of these "transported" forms but have to investigate their spread, their frequency and the constraints on their use. Double modals have certainly been brought to America by Scottish settlers, but this does not explain their present distribution; we have to take into account the subsequent, internal development of the extraterritorial Englishes, which can include changes in meaning and constraints on occurrence beside geographical spread (Bailey — Ross 1992). It could thus be hypothesised that these double modals are one of the relic phenomena which survived the initial swamping and were later recovered, and that they gained different functional load and currency, probably thanks to local innovations.

4.3. There are other factors playing against hasty identification of extraterritorial Englishes' features with original British dialectal features; one is that the accommodation processes which eventually lead to swamping (and which take place in all dialect contact situations) are subject to constraints, such as those related to the relative salience of adopted elements. Among such constraints there are, for example, those induced by the variability enhanced by phonotactic mechanisms, possibility of homonymic clashes and other forms of overlapping, strength of the stereotypes linked to specific dialect forms; to this we must add the fact that different speakers follow, as it were, different routes towards accommodation in their adoption of phenomena pertaining to other dialects, and that very strong salience can have the rebound effect of inhibiting accommodation (Trudgill 1986: 11-31, 125). All of this contributes to the creation of varieties whose sources are not immediately identifiable (Trudgill 1986: 57-58):

> the linguistic form which is, as it were, transmitted from the originating dialect is not necessarily identical to the form that is eventually acquired by the receiving dialect. There may well be, that is, a certain amount of disturbance during the transmission, presumably because of different types of "imperfection" in accommodation. Some of these "imperfections" may well simply consist of incompleteness, bearing in mind that speakers during accommodation reduce dissimilarities with other speakers, not imitate them slavishly.

This confirms the futility of trying to reason in terms of global innovation or conservation in these cases, and is also a warning against generalisations about dialect levelling, since it must not be forgotten that the output of language contact is never equal to the algebric sum, or even the product, of the various components. This seems to speak in favour of the possibility to look for universal tendencies in the formation of new varieties, or at least to exclude direct filiations. Algeo (1992) has emphasised that the attempt made by some scholars at drawing very precise "family trees" for extraterritorial Englishes are misled by prejudice; it cannot be claimed nor implied, as sometimes is the case, that British English is the "mother" of American English, for instance - it would be more correct to say, Algeo maintains, that they are "sisters", since filiation should not be measured in terms of an absolute time scale but in terms of dominance, degree of overlap and individuability of boundaries between varieties. British and American English are certainly divergent varieties, but the interplay between them has varied in intensity over the decades, and after World War I they have tended to converge again, not only as a result of the increasing inflow of "Americanisms" into Britain but also as a consequence of the less studied flow of "Briticisms" to America. This example shows how difficult it is to pinpoint the lineage of a variety, especially since the degree of interaction between varieties made possible by the modern media system influences the process of divergence-convergence between the varieties themselves and even the rate at which they diverge or converge (Lass 1987: 10-11).

5. Second language acquisition theory and non-native Englishes

5.1. Although several phenomena typical of emerging varieties cannot be put down to any one specific extralinguistic cause, a number of studies are confidently turning to second language acquisition theory to explore the possibility of explaining some such phenomena in general sociolinguistic or psycholinguistic terms and in their connection with the process of language acquisition/learning.

The most controversial point within this approach concerns the predominance of "external" vs. "internal" factors in language development in contact situations, i.e. whether extralinguistic factors are decisive or whether the impact of strictly linguistic factors is overriding, as is widely held (Danchev 1988: 38-39). For those who work within a sociolinguistic framework, it is clear that external factors should predominate: Milroy (1992) points out that we ought to discuss these phenomena in terms of speaker contact rather than of language contact, since what is decisive in such situations is the adoption of features by more and more numerous speakers, which ultimately depends on their language attitudes; these, in turn, are obviously connected to social, extralinguistic factors, which are thus responsible for the spread of change.

On the other hand, internal factors should not be overlooked, since there seem to be some universal constraints on change, such as those subsumed by Labov (1965 [1986]) under the claim that "in contact situations mergers expand at the expense of distinctions" (quoted in Trudgill 1986: 119); this phenomenon could be ascribed to simplification, which is held to

be responsible for many second language acquisition general tendencies as well as for the elements in common between the structures of several pidgin and creole varieties. It must not be forgotten, however, that simplification cannot explain everything; on the contrary, it is often the case that finer distinctions are introduced, although they are not the same distinctions employed by the mainstream variety. One well-known example is the existence, in many pidgin and creole continua, of aspect and tense markers which cut into linear temporal sequences in a different way from similar markers used both in the L1 and in English. It is also well known that non-native extraterritorial Englishes develop their own vocabulary; speakers introduce such changes in order to "accommodate" their lexical inventory to the new extralinguistic context. Apart from the adoption of loan-words, there are also innovations in terms of semantic shift, and these do not include only "mergers", i.e. meaning extensions or spread of hyperonyms, but also "distinctions", i.e. meaning specialisations.
5.2. Recourse to second language acquisition theory could however be very useful to investigate some phenomena pertaining to secondary extra-territorial Englishes which are still, for the vast majority of the speakers involved, second or additional languages. Very few studies have so far taken advantage of this possibility; Sridhar (1989: 52-58) remarks that second language acquisition theory and Non-Native English studies often seem to be "at cross-purposes", while it is apparent that they would have much to gain from closer association. For instance, among the assumptions of second language theory we often find the following:
1. the target of acquisition is native-like competence;
2. the input is nearly unlimited;
3. the relative sociolinguistic "weight" and communicative functions of the L1 and of English are not to be taken into account;
4. motivation for the acquisition of English is integrative, i.e. it springs from the speakers' desire to be identified as part of a community or group of native speakers;
5. research should be concentrated on phonology and syntax.
In the contexts in which the "New Englishes" have developed, on the other hand, it is usually the case that:
1. the target of acquisition is intelligibility for intranational communication purposes;
2. the input is restricted and often non-native;
3. the complementarity of roles of the languages is very important, since it conditions acquisition and uses;

4. motivation is mainly instrumental, i.e. there is a desire to acquire English in order to use it as an asset (e.g. a professional or educational qualification) or as a tool (e.g. for business contacts or for interethnic communication);
5. the main innovations take place in the lexis and in the pragmatics of the language.

The observation of all these differences has produced a reciprocal distrust in the two fields of research, to the detriment of possible advantages, especially as regards contributions from second language acquisition theory to the hypotheses about the reasons for the existence of so many features in common between most New Englishes.

5.3. The notions developed within second language acquisition theory that seem most promising for the analysis of the role of language contact in the formation of Non-Native Englishes are several. First, there is the notion of "transfer", to a certain extent derived from that of "interference"; nowadays there is a wider perspective on the phenomenon of interlingual influence, explored in several of its aspects e.g. by Odlin (1989), and the term "transfer", which appears more comprehensive and devoid of negative overtones, is preferred. Evidence in favour of transfer, which seems to be facilitated by salience and other factors pertaining to attitudes, is not lacking in non-native Englishes. It has been noted that some features, especially features of pronunciation, vary consistently with the varying of the speaker's L1: this happens in the English of Nigerian speakers from different ethnic groups, as well as in the English of Indians living in South Africa, which in many respects looks remarkably like Indian English and rather different from all other types of South African English (Mesthrie 1987, 1991). In this case, the influence of the contact with the L1 is apparently stronger than the influence of the context, i.e. of the norms spread in the community at large.

Even more useful than the notion of "transfer" (which shows its limits in not so clear-cut situations) is that of "interlanguage", first developed by Selinker (1972) and then elaborated by other second language acquisition theorists. The appeal of this notion resides mainly in the fact that the "intermediate" systems developed by the learner while acquiring an L2 are seen not as immature approximations but as real "language", as a developmental product which shows extensive regularity but also intrinsic variability, and whose study can throw light on the process of language acquisition. Part of the success of this notion is due to the fact that it appeals both to "universalists" and to "relativists", since the results pointing to similar routes in acquisition could be explained in several

ways, and in particular either of the following, broadly formulated explanations could be correct:
(1) language contact and transfer may not be very important, since there are "universals" of language acquisition which tend to prevail, or
(2) language contact is such a pervasive phenomenon that it tends to yield similar results in whatever situation.
5.4. What is really in question here is not the existence or emergence of universal tendencies, but the level at which they apply, and whether they are significantly influenced by context and extralinguistic factors or not. As mentioned, the contrast is between cognitive approaches (maintaining that these tendencies emerge naturally in the individual's mind when confronted with language) and creolist approaches (claiming that these tendencies are triggered by the encounter between languages, as in a chemical reaction). Both approaches have convincing evidence on their side. In creoles there are items whose "source" cannot be pinpointed, which suggests that contact varieties are not just simplified versions of a system, nor just the sum of two different systems: "[t]his difficulty of apportioning items to their appropriate systems is inevitable, since not only are these forms which span both the putative "systems", but also "intermediate" forms which cannot be properly assigned to either" (Bickerton 1975: 13). The only way of explaining this for Bickerton (1975: 180) is that "the natural condition of an individual human speaker is to have a competence that embraces several possibly similar but certainly distinct varieties of language, plus all the intersystems that relate them to his own". This modernised version of the notion of linguistic repertoire, whose existence is referred to by some linguists as a possible explanation of the dynamics of language change (Weinreich – Labov – Herzog 1968: 104; Milroy 1992), is called by Bickerton "polycompetence", a term we will come back to.

Universalist approaches also employ evidence in favour of interlanguage systematicity, under the claim that L2 acquisition shows consistent similarity with L1 acquisition processes that are universal, with very limited contribution from the L1 in contact situations; this is maintained e.g. by Zobl (1984: 83): "The L1 furnishes a very abstract auxiliary evaluation measure which, under certain well-defined conditions, is activated and supplants that of the acquisition device". Convincing endorsement of these positions also came from error analysis, which did away with the prejudice that learners' errors are an automatic consequence of the clash between the "rules" which form their competence in the L1 and those pertaining to the new system which is being acquired.

Recent evidence, however, is often not so clear-cut, and although advocates of a strong cognitivist hypothesis still abound, scholars like Mairs (1989) and Ard (1989) now emphasise the fact that features shown by the learner seem to be produced by the interplay of several different factors.

5.5. Another reason why the notion of interlanguage appears to be useful in studies on non-native Englishes is the fact that interlanguage is subject to fossilisation, i.e. it can be "blocked" on its way towards the target norm (as opposed to L1 competence, which develops continuously). According to Corder (1981: 74-75), this happens for extralinguistic reasons:

> the learner continues ... to upgrade, or elaborate, his understanding of the target language only so long as he has a motive for doing so. When his ... grammar reaches that state of elaboration which enables him to communicate adequately for his purposes with native speakers, his motive to improve his knowledge or elaborate his approximate system disappears.

When fossilisation occurs, those which could seem errors or applications of remedial acquisition strategies become stable features of a variety; Non-Native Englishes could represent extreme cases of interlanguage fossilisation, since after the independence of the former colonies there has been less and less pressure to conform to an external standard, and in some cases there has even been a reversal of attitudes, with the more localised forms gaining the status of national standards, as repositories of a new identity which rejects the colonial past. It is clear that language contact plays a role in this process (even if we agree that most interlingual structures emerge as the product of the action of universal tendencies), since the features which are more likely to gain acceptance and to acquire identity value are those which are more immediately recognisable as "local", i.e. most probably influenced by the L1. According to Williams (1987: 162-164), the peculiarity of Non-Native Englishes resides in the fact that their fossilisation does not really represent an intermediate stage, but the fossilised form becomes a target in itself: the New Englishes are "frozen interlanguages". Can we subscribe to this view?

6. A look at some data

6.1. In the analysis of features of Non-Native English, the first difficulty is to get a clear picture of the diffusion and distribution of the language clines present in a specific community. If language contact certainly plays a relevant role in the formation of such varieties and, as just mentioned, in the institutionalisation and "autonomisation" of a variety, its impact on the definition and delimitation of varieties is not to be overlooked.

Here we could come back to the notion of polycompetence introduced by Bickerton (1975), which represents an internalised repertoire and seems to be an enlarged version of what Weinreich (1954) described as "dyasystem", a complex system made up of two or more varieties, that the multilingual or multidialectal speaker can use freely for his linguistic options. There is a deep difference between these sets of options in the so-called monolingual and in the bilingual speaker, since in the former the set is made up of variants which basically belong to "the same language", each specifically placed within a dialect and/or register and style range. For the bilingual speaker, the options come from (at least) two different systems, plus some varieties that are intermediate between them. This has a number of consequences, among which the following:
(1) The stylistic option range within English is restricted, since a portion of the cline is realised through L1 forms. As a result, many of the stylistic subtleties that the speaker may be exposed to from native sources are lumped together or undergo shift in style ranking.
(2) There is constant reinforcement from the substratum, either as a result of contact (whence the considerable attention granted by some scholars to the degree of typological "similarity" vs. "distance" between L1 and L2, often a modernised version of old contrastive studies) or through fossilisation in interlanguage development.

The influence of these factors on actual language production consists mainly in the emergence, on a wide scale, of two processes:
(1) reduction of target language inventories (sometimes hastily described as just "simplification"; we have already claimed that this is not the case);
(2) extensive contextualisation, mostly visible at the lexical and idiomatic levels, but also present elsewhere.

A brief analysis of some examples of these two tendencies will show how difficult it is to attribute such phenomena to "universal" or "relative" reasons; the examples below have been picked among the best known and kept to a minimum for reasons of space, but their number could easily be enlarged.

6.2. Although most Non-Native Englishes are non-rhotic, they often lack the centralised vowels and diphthongs typical of Standard Southern English, which brings about a drastic reduction in the vowel inventories of these varieties. This phenomenon often appears to be induced by L1 influence, but it is also typical of some primary extraterritorial Englishes and of English interlanguage (where it could be connected to different perception and discretisation of glides, whose exiguous phonic substance could detract from their being perceived as salient).

In the consonant system, a near-universal is the non-occurrence of /θ, ð/, variously substituted by [t, d] or [s, z] or [f, v]. This is a phenomenon that could most productively be studied in depth, since it is spread from British dialects to Hawaiian, Cameroonian and Japanese English. Most people seem to be glad to "get rid" of these fricatives: why? Are they characterised by intrinsic "difficulty" of articulation or are they just replaced because they are unfamiliar to speakers of English as L2? They can be easily reproduced by anyone, besides being commonly employed in standard L1 accents (as well as, for that matter, by speakers of Castilian Spanish), yet they are statistically rare in the world's languages. Is this rarity responsible for the consistent expunction? Providing an answer to this question goes beyond the scope of this paper, and does not seem to be easy.

6.3. The most widespread morphological features that characterise Non-Native Englishes, even in formal varieties, concern the use of verb and noun endings, which are often deleted. This leads to an overuse of the base forms, a tendency which could perhaps be ascribed to fossilisation, since it is the result of paradigm levelling and consonant cluster simplification, two phenomena that are typical of "intermediate" systems (also in L1 acquisition) and of "mixed" varieties (where however they tend to be counterbalanced by the introduction of separate tense and number markers). Other common phenomena concern deviations in the use of articles and quantifiers, as well as of prepositions. In these cases, though, contact with the L1 seems more directly responsible for the deviations: while these areas seem to be problematic for all L2 learners, the outputs differ considerably.

Syntax also appears influenced, at least in the most "deviant" patterns, by the L1 or the substratum (compare the widely quoted construction *I'm after having breakfast* of Hiberno-English, which has been preserved as part of the variety although this is no longer an L2 for the community at large). Not surprisingly, the constructions where deviation recurs more often are either those which are highly marked in native English (e.g. the inversions in the interrogative forms or the subsystem of question tags, the latter normally simplified through generalisation of a single form of the type *isn't it?/no?)* or those which involve thematisation and the organisation and distribution of information, which of course vary across languages and cultures (the most striking examples concern topicalisation with fronting of the type *Your sister, I haven't seen her,* and pronoun copying as in *My father he's retired.* In all these cases, phenomena which appear at best in the most colloquial styles of native English become generalised features of a whole cline of Non-Native English and even of local standards.

6.4. This is the level of most evident "deviation" of Non-Native Englishes, both because of the need for contextualisation and as a result of the bias in favour of formal, bookish terms typical of varieties spread through formal education. A whole set of innovations concerns the exploitation of word-formation patterns (e.g. use of "different" suffixes, conversion, formation of new compounds and phrasal verbs; compare Indian English *unemployee, a matured woman, to off the light, a shoebite* 'blister', *to neck so. out* 'to kick so. out'), while the direct and indirect contribution of loan-words should not be forgotten either; yet the bulk of innovations in Non-Native Englishes entails (some type of) meaning shift.

6.5. All these processes are perfectly natural and typical both of language change (with or without contact) and of language acquisition; but here again, the product of these processes varies widely: relativism and contextualisation are present, at this level, at a very high degree, and even higher is the degree of importance of these two factors in the pragmatics of Non-Native English varieties, e.g. in the use of terms of address, greetings etc. (for all these aspects see Mazzon 1994: chapter 6, Nihalani — Tongue — Hosali 1979, Platt — Weber — Ho 1984).

7. Conclusions?

The few examples quoted are certainly insufficient to draw any conclusion and may seem biased against the immense variety of such a large number of "Englishes". What has been said so far allows us, though, a few general comments:

(1) There is much more in the investigation of both primary and secondary extraterritorial Englishes that could be useful to the understanding of language contact phenomena at large, than what has been taken into consideration so far (in my opinion, this comment is not so trivial as it may sound at first, since the present "state of the art" reflects the widespread distrust of variation that has affected a large part of modern linguistics).

(2) Primary extraterritorial Englishes show levelling and accommodation while secondary extraterritorial Englishes show the interplay between a

(mostly standard) native variety and local languages, so their study should be kept separate to an extent.
(3) An examination of some widespread features of the main non-native Englishes seems to show that unconscious "pressures from below" tend to emerge in phonology and morphology, while conscious "pressures from above" (e.g. contextualisation) mainly emerge in vocabulary and pragmatics, with syntax somewhere in between.
(4) The question whether contact-induced processes prevail over universal tendencies or vice versa is not easily to be decided, even in the case of rather specific phenomena. While the similarity of some outputs across geographically, linguistically and culturally distant contexts seems to speak in favour of cognitive "fixed routes", the importance of extra-linguistic, contextual factors should not be demoted; these are responsible for the formation of speakers' attitudes, which pose heavy constraints on language behaviour, whose infinite variety reflects the infinite variety of the human mind.

References

Adamson, Sylvia — Vivien Law — Nigel Vincent — Susan Wright (eds.)
    1990   *Papers from the fifth international conference on English historical linguistics, Cambridge, 6-9 April 1987.* (Amsterdam Studies in the Theory and History of Linguistic Science 4, 65.) Amsterdam: Benjamins.

Algeo, John
    1992   "The birth of a language", paper presented at ICEHL7, Valencia.

Ard, Josh
    1989   "A constructivist perspective on non-native phonology", in: Susan M. Gass — Jacqueline Schachter, 243-259.

Bailey, Guy — Garry Ross
    1992   "The evolution of a vernacular", in: Matti Rissanen *et al.* (eds.), 519-531.

Bickerton, Derek
    1975   *Dynamics of a creole system.* Cambridge: Cambridge University Press.

Corder, Stephen Pit
    1981   *Error analysis and interlanguage.* Oxford: Oxford University Press.

Danchev, Andrei
 1988 "Language contact and language change", *Folia Linguistica* 22: 37-53.
 1989 "On global patterns of interlingual influence", in: *Proceedings of the 2nd symposium on English and Greek, 28-30 March 1988, Aristotle University of Thessaloniki,* 33-46.
Gass, Susan M. — Jacqueline Schachter (eds.)
 1989 *Linguistic perspectives on second language acquisition.* Cambridge: Cambridge University Press.
Görlach, Manfred
 1987 "Colonial lag? The alleged conservative character of American English and other "colonial" varieties", *English World Wide* 8: 41-60.
Harris, John
 1984 "Syntactic variation and dialect divergence", *Journal of Linguistics* 20: 303-327.
Kachru, Braj B.
 1986 *The alchemy of English.* Oxford: Pergamon.
Labov, William
 1965 "On the mechanism of linguistic change", *Georgetown University Monograph Series on Languages and Linguistics* 18: 91-114, 131-132.
 [1986] [Reprinted in: John J. Gumperz — Dell Hymes (eds.), *Directions in sociolinguistics: the ethnography of communication.* (2nd edition.) Oxford: Blackwell, 512-538.]
 1992 "Evidence for regular sound change in English dialect geography", in: Matti Rissanen *et al.* (eds.), 42-71.
Lass, Roger
 1987 *The shape of English.* London: Dent.
 1990a "Early mainland residues in Southern Hiberno-English", *Irish University Review*: 137-148.
 1990b "How to do things with junk: exaptation in language evolution", *Journal of Linguistics* 26: 79-102.
 1990c "Where do Extraterritorial Englishes come from? Dialect input and recodification in transported Englishes", in: Sylvia Adamson *et al.* (eds.), 245-280.

Mairs, Jane Lowenstein
 1989 "Stress assignment in interlanguage phonology: an analysis of the stress system of Spanish speakers learning English", in: Susan M. Gass — Jacqueline Schachter, 260-283.

Mazzon, Gabriella
 1992 *L'inglese di Malta*. Napoli: Liguori.
 1994 *Le lingue inglesi*. Roma: Nuova Italia Scientifica.
 1996 "Conservazione ed innovazione nelle varietá extraterritoriali d'inglese", *Quaderni di Lingue e Letterature Straniere* 16: 227-251.

McArthur, Tom
 1998 *The English languages*. Cambridge: Cambridge University Press.

Mesthrie, Rajend
 1987 "From OV to VO in language shift: South African Indian English and its OV substrates", *English World Wide* 8: 263-276.
 1991 "Syntactic variation in South African Indian English: the relative clause", in: Jenny Cheshire (ed.), *English around the world: sociolinguistic perspectives*. Cambridge: Cambridge University Press, 462-473.

Milroy, James
 1992 "A social model for the interpretation of language change", in: Matti Rissanen *et al.* (eds.), 72-91.

Nihalani, Paroo — R. K. Tongue — Priya Hosali
 1979 *Indian and British English: a handbook of usage and pronunciation*. Delhi: Oxford University Press.

Odlin, Terence
 1989 *Language transfer*. Cambridge: Cambridge University Press.

Platt, John — Heidi Weber — Mian Lian Ho
 1984 *The new Englishes*. London: Routledge and Kegan Paul.

Poussa, Patricia
 1990 "A contact-universal origin for periphrastic *do*, with special consideration of OE—Celtic contact", in: Sylvia Adamson *et al.* (eds.), 407-434.

Rissanen, Matti — Ossi Ihalainen — Terttu Nevalainen — Irma Taavitsainen (eds.)
1992   *History of Englishes: new methods and interpretations in historical linguistics.* (Topics in English Linguistics 10.) Berlin — New York: Mouton de Gruyter.

Selinker, Larry
1972   "Interlanguage", *IRAL* 10: 209-231.
1992   *Rediscovering interlanguage.* London: Longman.

Sridhar, Kamal K.
1989   *English in Indian bilingualism.* Delhi: Manohar.

Todd, Loreto — Ian Hancock
1986   *International English usage.* London: Routledge.

Trudgill, Peter
1986   *Dialects in contact.* Oxford: Blackwell.

Varvaro, Alberto
1984   "Sociolinguistica e linguistica storica", in: Alberto Varvaro (ed.), *La parola nel tempo.* Bologna: Il Mulino, 105-116.

Weinreich, Uriel
1953   *Languages in contact.* New York: Academic Circle.
1954   "Is a structural dialectology possible?", *Word* 10: 388-400.

Weinreich, Uriel — William Labov — Marvin I. Herzog
1968   "Empirical foundations for a theory of language change" in: Winfred Philip Lehmann — Yakov Malkiel (eds.), *Directions for historical linguistics.* Austin: University of Texas Press, 95-195.

Williams, Jeffrey P.
1987   "Non-native varieties of English: a special case of language acquisition", *English World Wide* 8: 161-199.

Zobl, Helmut
1984   "Cross-language generalizations and the contrastive dimension of the interlanguage hypothesis", in: Alan Davies — Clive Criper — Anthony P. R. Howatt (eds.), *Interlanguage.* Edinburgh: University of Edinburgh Press, 79-97.

Ruta Nagucka (Cracow)
Latin prepositional phrases and their Old English equivalents

1. Introduction

Linguistic influences are much easier to claim and speak of than to ascertain and justify.[1] The fascination of lexicographers with the subject could have positive and interesting results, but I am sceptical about effecting changes from a foreign language in the field of syntax. It is common knowledge that during the Old English period Latin was used for centuries for the recording of various matters, chiefly ecclesiastical, because it was regarded as being more suitable for the system of the Catholic Church. On the one side was things-as-they-were: Latin writings which documented great scholarly achievements, and whose weakness was merely that they were gradually ceasing to be of much use, simply, because they were hardly understandable. From what we read in Alfred's and Ælfric's prefaces few of their contemporaries were well acquainted with Latin. In the other scale was a necessity for the spread of learning and religion, for restoring education to its former position, which process was attempted by translating Latin texts into vernacular. According to Hogg (1992: 12), however, it is doubtful that the general ignorance of Latin was the only reason for Alfred's translating; it was also probably "pride in the vernacular language", as he says. Whatever the reason, Latin texts were translated into Old English.

2. Translation in Old English

The art of translation depended on the subject matter of the text which was translated and on the purpose of translating such a text. In the case of Ælfric's *Colloquy* the language of the Old English translation was

---

[1] This article is a revised version of a paper which I presented at the conference on language contact in the history of English, Tulln, Austria, July 1994. My research for a possible influence upon Old English prepositional construction was undertaken when I was working on a semantic interpretation of Old English prepositional phrases. The results of scholarship on this subject appeared in my recent publication (Nagucka 1997b); no part of this article is a straightforward extract from the book, although there are some similarities between the two.

conditioned by the mechanical methods of the glossator because it was as important for pedagogical purposes "to render each Latin word by its English equivalent, as to render the meaning of the sentence as a whole" (Coll 16), while in Ælfric's *De Temporibus Anni* no single Latin text can be provided as a sole source. In *De Temporibus Anni* Ælfric made use of different authorities, chiefly Bede, but he did not translate them literally and only rarely did he translate a phrase, "a sentence or a short paragraph more or less closely" (TA liv). There are also other types of translation, something in between, i.e. not mechanical glosses, neither individual, loose renderings of the contents based on various sources. Such is Alfredian prose, the translation of Bede's *Ecclesiastical History of the English People* and the Old English *Orosius*. They show a relationship between the Old English and their Latin originals close enough "to allow comparison between source text and translation" (Kilpiö 1989: 2). Still in neither case do we have absolute equivalence. I have purposefully chosen these four texts which display various techniques of translation in order to find out how Latin prepositional phrases are rendered into Old English and whether any influence of the source language on the English equivalent can be noticed.

3. Prepositional phrases in Latin and Old English

3.1. Theoretically speaking, the following correspondences between Latin prepositional phrases and their Old English renderings are possible:

Latin PP – Old English PP: one-to-one correspondence, the two phrases are not only semantically equivalent but often also structurally congruent;

Latin PP – Old English non-PP: the equivalence, i.e. the semantic identity, or at least close affinity is expressed by other means than the PP.

Since all the texts under analysis have Latin works as sources or models one would believe that the influence of Latin must have been significant, particularly in prepositional phrases of the one-to-one pattern. But it need not be so, as is spelt out by Godden (1992: 514) when he says that in the "use of language there is evident in most writers a striking resistance to Latin prose as a model, and a preference for developing a learned vocabulary from native resources and building a system of syntax and sentence structure based on native idiom." Various hints in the literature on the subject as well as the views expressed by the translators warn a linguist to be cautious when interpreting even formally identical constructions as direct products of the language influence. Let us examine a few examples of prepositional phrases in the two relevant languages:

(1) a. Lat. *ad campum*
       OE *to felda* (Coll 24)
           to field
           'to the field'
    b. Lat. *In dormitorio cum fratribus*
       OE *On slæpern mid gebroþrum* (Coll 304)
           on dormitory with brothers
           'In the dormitory with the brothers (brethren)'
    c. Lat. *in hieme*
       OE *on wintra* (Coll 138)
           on winter
           'in the winter'
    d. Lat. *inter artes seculares*
       OE *betwux woruldcræftas* (Coll 217)
           between secular-crafts
           'among the secular occupations'

For obvious reasons the one-to-one correspondence is found as a rule in glosses; in Ælfric's *Colloquy* most prepositional phrases are glossed as English prepositional phrases. But they are also met in other texts, although their number is much smaller, probably because of greater independence of the translator when compared with a glossator. In the text of *De Temporibus Anni* we come across such phrases as:

(2) a. Lat. *a tenebris*
       OE *fram ðam þeostrum* (TA 2/3)
           from the darkness
           'from the darkness'
    b. Lat. *subter terras*
       OE *under ðysse eorðan* (TA 8/18)
           under this earth
           'under the earth'
    c. Lat. *in pennis eius*
       OE *on hire fiðerum* (TA 14/34)
           on her wings
           'on its wings'

Such equivalent and often congruent pairs of prepositional phrases occur in earlier translations as well. In *Orosius* there are

(3) a. Lat. *ab occasu*
    OE *on westhealfe* (Or 12/13)
    on western-side
    'on the western side'
 b. Lat. *apud Aegyptum*
    OE *mid Egyptum* (Or 32/25)
    with Egypt
    'among the Egyptians'
 c. Lat. *in Hispania*
    OE *on Ispanium* (Or 192/18)
    on Spain
    'in Spain'

and finally, in Bede's *Ecclesiastical History* we find:

(4) a. Lat. *in oceano* (Bae I 26/17)
    OE *on garsecge* (Bede 30/19)
    on ocean
    'in the ocean'
 b. Lat. *post Brettones et Pictos* (Bae I 18/18)
    OE *æfter Bryttum 7 Peohtum* (Bede 28/24)
    after Britons and Picts
    'after the Britons and the Picts'
 c. Lat. *inter septentrionem et occidentem* (Bae I 10/18)
    OE *betwyh norðdæle and westdæle* (Bede 24/30)
    between north and west
    'between the north and the west'

It is worth noting that in most examples the prepositions govern their nominal objects which take the same inflectional case in both languages, be it dative/instrumental in Old English and ablative in Latin, or accusative, e.g.

(5) a. Lat. *cum fratribus*   ablative
       OE *mid gebroþrum*   instrumental
    b. Lat. *inter artes seculares* accusative
       OE *betwux woruldcræftas*   accusative

3.2. Thematic relations of the Latin PPs and their Old English equivalents are the same but the affinity of these structures in the two languages under consideration seems to stem from other sources than the impact of one language upon another. The structural similarity or identity is conditioned here by the same typological characteristics. Latin and Old English are prepositional languages, both are highly inflected and both possess basically the same specifications, which according to Hawkins (1983: 331, 335) are:

    Latin (= oldest attested Italic language)
                                   (Indo-European: Italic) SOV; Pr; AN, GN/NG, NRel;
    Old English (Indo-European: Germanic: West)
                                     SOV/v-1, V-2; Pr; DN, PossN, AN, GN/ng, NRel.

Latin and Old English contain the same Indo-European genetic classification, the same basic word-order SOV, both are prepositional and both have the same word-order of modifier and head categories: AN, GN, NRel. The differences between these two languages refer to nonbasic word-orders which are postulated for Old English and not for Latin: Latin of that period had no native speakers, it was employed throughout Europe as a *Schriftsprache*, to use Hogg's term, and was not undergoing substantial modifications of a nonbasic type, which was the case in Old English. These characteristics allow us to assume that Latin and Old English, although of different Indo-European subgroups, had basically the same word-ordering, and it is therefore not very likely that Latin style influenced the native language of the English. Both used the same language means for prepositional relations and if anything, one could consider the possibility of mutual support rather than influence.

3.3. Additional evidence can be provided from Old English poetry which is of native origin and Germanic in character. Although the Old English poetic diction has been ascribed a "specialised grammar and syntax, extending all the way from matters of inflexion and the use of demonstratives to the structure of the sentence" (Godden 1992: 504), the prepositional phrase does not show striking features of positional distortion, except, perhaps, the fact that the preposition occurs more often postpositionally in poetry than in prose. Consider the following prepositional phrases taken from *Beowulf*:

(6) a. *to waroðe* (Beo 234)
 to sea
 'to the sea'
 b. *mid gesiðum* (Beo 1313)
 with companions
 'with his comrades'
 c. *on Heorote* (Beo 497)
 on Heorot
 'in Heorot'
 d. *fram mere* (Beo 855)
 from mere
 'from the sea'
 e. *under heofenum* (Beo 52)
 under heavens
 'under the sky', etc.

These and similar examples show that prepositional phrases were widely used both in prose and in poetry.
3.4. Finally, even if there were some directly supportive (not necessarily decisive) influence of Latin on the Old English prepositional phrase it would have been implied (even negatively) by Ælfric in his *Grammar*. My own suspicion is that, since Ælfric carefully chose relevant Old English examples to illustrate such aspects as Latin grammatical tense as shown by Traugott (1992: 180-181), or lexical compounds as discussed by Kastovsky (1992: 312), he must have done the same in the case of the preposition. Ælfric chose the most appropriate Old English prepositional constructions to illustrate their Latin equivalent counterparts and they are exactly the same structurally as has been so far presented. If this reasoning is correct, Old English prepositional phrases for Latin prepositional phrases seem to be optimal renderings.

4. The semantic interpretation of prepositional phrases

4.1. The semantic interpretation of Latin prepositional phrases and their Old English equivalents is very complex. Superficial observations corroborate a generally accepted view that the spatial meaning of prepositions is extended to temporal relations. Whether each of the two languages developed resourcefulness of this kind independently (with which I would intuitively agree), or whether it was due to the Latin influence upon Old English remains to be determined. The subject requires a thorough semantic examination of comparative character since the concepts underlying prepositional structures and thematic relations in general are subject to subtle constraints on the nature of human perception. An analysis of this kind will not be pursued in this paper.

4.2. In the preceding sections I discussed simple relationships between Latin and Old English prepositional phrases in which there was a one-to-one correspondence. Another group of correspondences with prepositional constructions comprises Latin prepositional phrases which in Old English translations appear as nonprepositional structures, e.g.

(7) a. Lat. *Narbonensis provincia habet ... a meridie mare Gallicum, habens in fronte qua Rhodanus fluvius in mare exit*
OE *Be suðan Narbonense is se Wendelsæ,*
by south Narbonensis is the Mediterranean-Sea
OE *þær þær Rodan seo ea utscyt* (Or 22/28)
there where Rhône the water flows-out
'In the South of Narbonensis there is the Mediterranean Sea into which the Rhône flows'

b. Lat. *In his finibus India est*
OE *þæt sint India gemæro* (Or 10/15)
that are India boundary
'These are the frontiers of India'

c. Lat. *Insula Creta ... habet in longo millia passuum centum septuaginta duo, in lato quinquaginta*
OE *Creto þæt igland ... hit is an hund mila long,*
Crete that island it is one hundred miles long
OE *7 hundsyfantig 7 fiftig mila brad* (Or 26/32)
and seventy and fifty miles broad
'Crete, that island, is a hundred and seventy miles long, and fifty miles broad'

This method of translating is quite frequent; the Latin text is loosely transformed into Old English *andgit of andgi[e]te* 'according to the sense' as Alfred says in his preface to Gregory's *Pastoral Care*. Native grammatical constructions are preferred here; the avoidance of compact, synthetic structures typical of Latin is noticeable in Old English versions not only in the case of prepositional phrases. Unsurprisingly, I have not come across examples of this type in Ælfric's *Colloquy*; nonprepositional

257

Ruta Nagucka

structures can hardly be used by the glossator for "glossarial restrictions" if we can use that term. To say that the following expressions

(8) a. X has ... miles in length
        Lat. *habet ... in longo*
and
    b. X is ... miles long
        OE *is ... mila long*

mean the same is correct from the semantic point of view; structurally, however, they are different, and that difference would be immediately displayed by the glossator. Literal translations, i.e. *word be worde* 'word by word' are possible as has been amply shown, but if there is a choice between a one-to-one translation and a more idiomatic one of native origin it is the latter that would be preferable: the same meaning would be expressed by different grammatical structures. There are other interesting examples of this way of translating, e.g.

(9) a. Lat. *patri in regnum successerit* (Bae I 46/20)
       OE *æfter his fæder to rice feng* (Bede 42/18)
           after his father to kingdom seized
           'succeeded his father in the kingdom (on the throne)'
    b. Lat. *Anno ab Urbe condita CCXLV*
       OE *Æfter þæm þe Romeburg getimbred wæs*
           after the that Rome built was
       OE *twa hunde wintra 7 iiiix* (Or 78/1)
           two hundred winters and iiiix
           'Two hundred and forty five years after Rome was built'
    c. Lat. *Ante annos Urbis conditae mille trecentos*
       OE *Ær ðæm ðe Romeburh getimbred wære*
           before the that Rome built were
       OE *þrim hund wintra 7 þusend wintra*
           three hundred winters and thousand winters (Or 28/25)
           'A thousand and three hundred years before Rome was built'

d. Lat. *Huic mortuo Semiramis uxor successit*
   OE *7 æfter his deaðe Sameramis his cwen*
       and after his death Semiramis his queen
   OE *fengc ægþer ge to þæm gewinne ge to þæm rice* (Or 30/14)
       seized both to the war and to the kingdom
       'and after his death Semiramis, his wife, took to the war and to the kingdom'

Typically Latin absolute constructions are substituted by native structures, sometimes they are idiomatic like OE *fon to rice* 'succeed to the kingdom', sometimes they are regular, stylistically unmarked constructions like OE *æfter þæm þe Romeburh getimbred wæs* 'after Rome was built'. Such language data seem to support a view, already mentioned elsewhere, that there was some reluctance on the part of the translator to mechanically copy Latin structures; he would rather avoid those that sounded alien to his ears, replacing them with a native idiom. This is the most natural reaction: why should a recipient language change any of its own constructions and adapt them to a foreign model if native resources are capable of expressing the same meaning? If there is any historical reality to these general reflections, theoretical speculations or interpretations (and I hope that there is), it is possible that they may translate into predictions about concrete language situations. For example, since Old English prepositional phrases are more commonly used than prepositionless oblique cases, it is natural to expect Old English prepositional phrases for Latin inflected forms of nouns without prepositions. What I mean is a correspondence between a Latin noun in the ablative, sometimes the dative, and its translational equivalence in Old English in the form of a prepositional phrase. Take the following as illustrative examples:

(10) a. Lat. *arte mea*
        OE *mid cræfte minon* (Coll 170)
            with skill my
            'with my skill'
    b. Lat. *accipitre*
        OE *mid hafoce* (Coll 126)
            with hawk
            'with a hawk'

c. Lat. *multis modis*
   OE *on feala wisan* (Coll 124)
   on many ways
   'in many ways'
d. Lat. *domi*
   OE *æt ham* (Coll 25)
   at home
   'at home'
e. Lat. *diluculo*
   OE *on dægræd* (Coll 23)
   on dawn
   'at dawn'
f. Lat. *nostra hyeme media*
   OE *on middan urum wintra* (TA 40/53)
   on middle our winter
   'in the middle of our winter'
g. Lat. *die septimo*
   OE *on ðam seofoðan dæge* (TA 8/16)
   on the seventh day
   'on the seventh day' etc.

These numerous examples are of particular interest and ask for some comments, especially those that appear in glosses where direct structural correspondence is more common. I am inclined to argue in this case that motivations for changing a Latin inflectional form into an Old English prepositional phrase could be considered in terms of analogical extension of one pattern over another, taken from among the variants permitted by the language as a whole. At that time of the history of English the inflectional system of the noun was ceasing to be the most important indication of conceptual relations such as temporality, locality, instrumentality, and the like. The inflectional forms which did exist did not even seem to compete with prepositional phrases; if they were used they had an archaic flavour as in the following line from *Beowulf*

(11)   *He þæm batwearde bunden golde swurd gesealde* (Beo 1900)[2]
        he the boatkeeper bound gold sword gave
        'He gave the boatkeeper a sword adorned with gold'

or are idiomatic in character as in

(12) a. *7 þy geare slogon East Engle Beornwulf* (SC 823 A)
         and the year killed East Angles Beornwulf
         'And in this year the East Angles killed Beornwulf'
     b. *þa sende he Æþlwulf his sunu ... to Cent*
         then sent he Æthelwulf his son to Kent
         micle werede (SC 823 A)
         great army
         'Then he sent Æthelwulf, his son, to Kent with a great army'

Since Latin oblique cases were often translated into Old English prepositional phrases (even in glosses where the author could have used a parallel inflectional form[3] but avoided it and used prepositions instead), the impact of Latin in this respect is rather insignificant. The native idiom was definitely preferred, which often resulted in the restructuring of the whole sentence, for example

(13)   Lat. *syluas fertur habere pomis onustas*
        OE *heora orcyrdas mid æpplum afyllede* (TA 40/53)
            their orchards with apples filled
            'their orchards were filled with apples'

Each sentence represents different grammatical structures showing their own characteristics; in spite of this formal difference the meaning of these two sentences is very close.

---

[2] The word *golde* is used in the function of the instrumental, cf. a prepositional phrase in
   Eall mid readum golde his cynestol
   all with red gold his throne
   geworhte (Bosworth)
   wrought
   'He wrought his throne all with red gold'.

[3] For Old English inflectional forms of nouns with or without prepositions used to express instrumental and sociative functions see Nagucka (1997a).

261

4.3. Although I have not come across a reverse structural correspondence where a Latin prepositional phrase would be translated as an inflected nominal without a preposition, i.e. something like

(14) Lat. *cum accipitre
 OE hafoce
 'with a hawk'

which situation is theoretically possible, there are occasional occurrences of inflected nominals in both languages. I see no reason in principle why such equivalences cannot be treated on the same grounds as the other types. However, in this case I feel that Latin is to a certain extent responsible for such a translational solution. Having admitted that there is a low frequency of inflected nouns without prepositions in Old English when the relations of time, space, instrument etc. are expressed, and given two languages in contact, Latin and Old English, some Latin infiltration into Old English is very likely. For example:

(15) a. Lat. *Quin et iisdem temporibus fames Constantinopolim invasit*
 (Bae I 64/12)
 OE *Swylce eac þissum tidum com mycel hungor*
 also (2 words) these times came great hunger
 OE *on Constantinopolim* (Bede 48/13)
 on Constantinople
 'Also, at this time, there arose a great famine at Constantinople'
 b. Lat. *His temporibus regno Nordanhymbrorum praefuit rex fortissimus* (Bae I 178/4)
 OE *Ðyssum tidum forewæs Norðanhymbra*
 these days was-before Northumbrians
 OE *rice se strongesta cyning* (Bede 92/3)
 kingdom the strongest king
 'At this time there ruled over the kingdom of Northumbria a very brave king'
 c. Lat. *dulcibus uerbis*
 OE *swæsum wordum* (Coll 256)
 suave words
 'with mild words'

The presentation of these data (examples in (15)) is some evidence that Latin might have had some influence on Old English translational "techniques" and that analogical extensions of certain structural types, however obsolete in Old English, might have been activated on a very limited scale.

But as we have already seen, translating Latin inflectional forms into their Old English equivalent prepositional phrases is much more common and more natural than translating Latin inflectional forms by Old English inflectional forms. This is probably the reason why the translator used Old English prepositional phrases for Latin settings which could have been rendered by Old English prepositionless expressions[4] as in

(16) a. Lat. *Misit idem beatus papa Gregorius eodem tempore etiam regi Aedilbercto epistolam* (Bae I 168/9)
  OE *Sende eac swilce se eadiga papa in þa*
  sent also (2 words) the noble pope in the
  OE *ilcan tid Gregorius Æþelberhte cyninge*
  same time Gregory Æthelberht king
  OE *ærendgewrit somed* (Bede 90/6)
  letter together
  'The blessed pope Gregory sent also at the same time a letter to king Æthelberht, along (with many worldly gifts)'
 b. Lat. *His etiam temporibus adeo jugis et gravis aestus incanduit*
  OE *On þære tide wæs sio ofermycelo hæto*
  on the time was the great heat
  OE *on ealre worulde* (Or 40/3)
  on all world
  'In those times there was great heat in the whole world'

5. Conclusion

I have examined the claim that Latin–Old English translational equivalences are to a considerable degree independently motivated, and argued that the view that Latin might have exercised structural influence

---

[4] What I want to emphasise is, that in spite of the fact that Old English prepositionless phrases were used in native literature without Latin sources and could have been used (and actually were used) in the Old English texts based on Latin models, this type of equivalence and congruence (i.e. Latin inflectional form – Old English inflectional form) is not a rule and seems to be restricted to some collocations only.

upon Old English within the prepositional construction does not seem to be supportable. Structural options for Old English native patterns are obvious in the texts under analysis; it is the Old English prepositional phrase which is preferred for the Latin inflectional form or other prepositionless structure. The fact that in both languages there are equivalent and congruent prepositional phrases is irrelevant in this respect; this fact is explained by genetic and typological characteristics. In effect, our observations eliminate Latin as an influential factor on the structure of the Old English prepositional phrase. Perhaps the prepositional phrase is a structural and conceptual area which does not tolerate influence or borrowing.

Abbreviations and sources of examples

Bae = *Bædae Opera Historica* with an English translation by J. E. King. 1954. *Ecclesiastical History of the English Nation* based on the version of Thomas Stapleton, 1565. 2 vols. London: Heinemann.

Bede = Thomas Miller (ed.), 1890/1891 [reprinted 1959]. *The Old English version of Bede's Ecclesiastical history of the English people*, part I, 1. and 2. (EETS OS 95 and 96.) London: Oxford University Press.

Beo = Fr. Klaeber (ed.), 1950. *Beowulf and The fight at Finnsburg*. (3rd edition.) Boston: Heath.

Bosworth = Joseph Bosworth, 1898 [reprinted 1954]. Edited and enlarged by T. Northcote Toller. *An Anglo-Saxon dictionary*. London: Oxford University Press.

Coll = G. N. Garmonsway (ed.), n.d. *Ælfric's Colloquy*. University of Exeter.

Or = Henry Sweet (ed.), 1883 [reprinted 1959]. *King Alfred's Orosius*, part I: *Old-English text and Latin original*. (EETS OS 79.) London: Oxford University Press.

TA = Heinrich Henel (ed.), 1942 [reprinted 1970]. *Aelfric's De Temporibus Anni*. (EETS OS 213.) London: Oxford University Press.

SC = *Two of the Saxon Chronicles*, 1899 [reprinted 1980]. Ed. by Charles Plummer on the basis of an edition by John Earle. 2 vols. Oxford: Clarendon.

References

Godden, Malcolm R.
1992 "Literary language", in: Richard M. Hogg (ed.), 490-535.
Hawkins, John A.
1983 Word order universals. New York: Academic Press.
Hogg, Richard M.
1992 "Introduction", in: Richard M. Hogg (ed.), 1-25.
Hogg, Richard M. (ed.)
1992 The Cambridge history of the English language, vol. 1: The beginnings to 1066. Cambridge: Cambridge University Press.
Kastovsky, Dieter
1992 "Semantics and vocabulary", in: Richard M. Hogg (ed.), 290-408.
Kilpiö, Matti
1989 Passive constructions in Old English translations from Latin: with special reference to the OE Bede and the pastoral care. (Mémoires de la Société Néophilologique de Helsinki 49.) Helsinki: Société Néophilologique.
Nagucka, Ruta
1997a "The instrumental in Old English", in: Raymond Hickey — Stanisław Puppel (eds.), Language history and linguistic modelling: a festschrift for Jacek Fisiak on his 60th birthday. Berlin: Mouton de Gruyter, 153-166.
1997b The language-to-cognition interface: the Old English prepositional phrase and the four-dimensional continuum. Kraków: Księgarnia Akademicka.
Traugott, Elizabeth Closs
1992 "Syntax", in: Richard M. Hogg (ed.), 168-289.
Zupitza, Julius (ed.)
1880 Ælfrics Grammatik und Glossar. Berlin: Weidmannsche Buchhandlung.

Gabriele Rinelli (Brindisi)
Scandinavian and native social terms in Middle English: the case of *cherl/carl*

1. The sociohistorical approach

This paper sets out to analyse the semantic evolution of OE *ceorl* > ME *cherl* and to compare it with the cognate ME *carl*, of Scandinavian origin (ON *karl*). Some terms indicating social status underwent a semantic change in Middle English. In the particular case of *cherl* two types of semantic change can be observed: a semantic pejoration and, at nearly the same time, a shift from a definite referential sphere to a connotative and less specific one.

First, it is useful to consider the evolution of the term as shown by the *Oxford English Dictionary*. In Old English *ceorl* meant 'a male human being, a man' (until the fourteenth century) and, in Old English laws, 'a member of the third or lowest rank of freemen' (only Old English). In Middle English, according to the *Oxford English Dictionary*, since 1225 it had taken on the meaning of 'a tenant in pure villeinage, a serf, bondeman', the position to which most of the Anglo-Saxon churls were reduced after the Norman Conquest. They in fact had lost the status of free men and were bound to certain services to a lord. The semantic pejoration is thus considered a consequence of the gradual depression of the socioeconomic status of churls. This interpretation is based on a sociohistorical approach, which considers some linguistic changes as signals of social changes. It is agreed, in particular, that "the lexical semantics of social terms is directly affected by social history" (Dahlgreen 1985: 111).

Another phenomenon in the semantic evolution of ME *cherl* is the shift from a mere referential sphere, indicating a specific class in society, to a level that could be defined evaluatively and connotatively – designating, in general terms, a man of rude manners.[1] As a matter of fact, from the thirteenth century onwards *cherl* is found with the meaning of 'countryman, rustic, boor' and, as a term of disparagement or overt contempt, with the meaning of 'base fellow, villain'. In Leith's words: "by about 1300,

---

[1] Leith (1986: 70-73) distinguishes various types of meanings and also different criteria for relating words to referents. In the case of ME *cherl* the criteria shifted from the functional to the evaluative, i.e. the meaning gradually shifted from a referential sphere to a connotative one.

churl had lost its technical sense as a term of rank, and indicated low breeding in general; from that point connotations of 'rudeness' gradually became criterial" (Leith 1986: 80).

It will be argued here that not only the semantic pejoration of the term but also its shift to the evaluative and connotative sphere can be tackled according to a sociohistorical approach.

To this purpose it may be useful to make two preliminary remarks. One concerns the difference in structure between English medieval society and the Anglo-Saxon one. The latter consisted broadly of three main strata – nobility (*eorls*), peasantry (*ceorls*) and slaves (*theows*) – which were well defined and easily distinguishable according to certain peculiarities (Runciman 1984: 5). Churls, for example, had a definite wergild,[2] possessed a certain amount of land, enjoyed the status of free men and paid food-rent to the king. Instead English society at the end of the Anglo-Saxon period and in the Middle Ages presented, especially in the lower strata, a far more subtly differentiated range of roles and positions. They often mixed up, coinciding in some aspects, and were not distinguishable according to any specific parameter. This situation obviously affected the social vocabulary which became variable.[3] Class boundaries were no more well-delimited as they were in the Anglo-Saxon period. Not only was a churl's status socially depressed, but it was also remarkably blurred, so it can be assumed that it did not any longer retain the peculiarities of an actual social class. Churls had in fact lost their juridical defining marks and only preserved such stereotypical features as rude manners.[4] For this reason, ME *cherl* must have gradually lost its technical denotation to acquire a general contemptuous connotation.

The second remark involves the changes occurring in a class which was below the churl's in the social scale. From the close of the twelfth century onwards the social condition of slaves improved slowly, though un-

---

[2] "In the eyes of the law, the chief mark that distinguished one class of society from another was the price that had to be paid in compensation for the slaying of one of its members, which was called the wergild" (Whitelock 1974: 83).

[3] "[T]he fluidity of the verbal distinctions drawn in the sources demonstrates the continued absence of any clear differentiation among the classes; and this fluidity not only implies ease of movement between one class and another, but also suggests that the verbal distinctions themselves may be changing in meaning between one generation and the next" (Runciman 1984: 5).

[4] In all social terms it is possible to make a distinction between constitutive information – i.e. objective parameters – and cultural stereotypical information (Dahlgreen 1985: 110-111).

steadily.[5] As the feudal system gradually declined the ties between the lord and the serfs became loosened and the latter less subject to servile work.[6] The descendants of the Domesday slaves probably crept up into the lower ranks of the rural population (Stenton 1965: 138). Therefore while churls were receding in the social scale, the lowest ranks of society were advancing. This turned out in a common movement of churls and slaves towards a shared servile condition, midway between freedom and slavery.[7]

Therefore it is possible that the semantic pejoration of ME *cherl* was not a mere and direct consequence of the social decline of churls, but that it was also due to the progressive advancement of slaves in the social scale. It can in fact be assumed that ME *cherl* came to be applied also to former slaves, placed now on the same level as churls. Progress on the social level, then, may have influenced the semantic pejoration of a word. Such a hypothesis is based on a sociohistorical consideration, in that it tries to explain a linguistic change as being affected by social change, but the process is thought to be different from the simple social pejoration → semantic pejoration evolutionary pattern. Social receding of churls and advancing of slaves may be considered as interacting in the semantic evolution of ME *cherl*, although the former process had a major and more direct influence.

This hypothesis is also useful for a further explanation of ME *cherl* shifting from the referential sphere to the evaluative one: former slaves advancing in the social scale no more belonged to the class of slaves but still retained rude manners.

2. The linguistic approach

The semantic change of social terms can also be tackled according to a linguistic approach. Words can be considered not as single units but as parts of a lexical network. Whenever a linguistic change takes place, the internal balance of the network is modified: this happens when loan-words are introduced in a language. In fact loan-words that don't bring with them new meanings create a superposition with native words and

---

[5] "[T]hroughout the period [early Middle Ages] there was a movement, slow perhaps, and to contemporaries not always perceptible, towards freedom" (Stenton 1965: 138).
[6] Marsh (1974: 96-103) shows interestingly how this process began at the close of the twelfth century and continued all over the Middle Ages.
[7] "[T]here was ... a progressive assimilation of former slaves and depressed *ceorls* into a more or less uniform serfdom" (Runciman 1984: 26).

this turns out in a situation that is linguistically uneconomic, causing instability in the network. As time passes a new balance is achieved through reduction of synonyms or semantic differentiation among them, with a consequent creation of a more close-meshed network (Gusmani 1987: 104-105). Therefore the semantic change of social terms is not exclusively a consequence of social changes.[8]

In the case of social terms in Middle English, as well as in other semantic fields, the internal balance of the lexical network was modified by various French and Scandinavian loan-words. Of course, it is difficult to find out all the reciprocal effects that the introduction of borrowings had on the network of social terms in Middle English. A hypothesis has been suggested by Leith about ME *cherl*, whose shift to the evaluative sphere is compared with that of *gentle*, and whose semantic pejoration is considered in relation to *villein* and *peasant*.[9]

Here only a comparison of the semantic evolution of ME *cherl* and the cognate *carl* will be made. Such a comparison is particularly interesting because both the linguistic and the sociohistorical approach prove to be useful.

3. *cherl* vs. *carl*

Scandinavian *carl* underwent a semantic change similar to that of the native term, so they could be considered, broadly speaking, as synonyms like many other Scandinavian and native doublets.[10] Nevertheless different usages of the two words can be noticed when both are found in an individual text or used by the same author.

In order to show how different but coexisting connotations can be inferred from an individual text, it may be useful to analyse both words as they occur in *Havelok* (Smithers (ed.) 1987) and in *Cursor Mundi* (Morris (ed.) 1893).

---

[8] "[W]e must beware of assuming that any change in meaning automatically accompanies, or signals, a change in social structure. The relationships between the words and the things they signify are more complex than that" (Leith 1986: 80).

[9] "If *gentle* denoted good breeding and gracious behaviour, then words like *churl* could be associated with coarseness. ... Interestingly enough, the French borrowing *villein* dropped even further. ... While both these words have become pejorised, a later borrowing, *peasant*, has retained its early meaning, 'one who works the land', as well as a later pejorative one" (Leith 1986: 80-81).

[10] The *Oxford English dictionary* and the *Middle English dictionary*, which have separate entries for *carl* and *cherl*, report for both of them similar meanings.

Cherl is found five times in *Havelok*, in the acception of 'serf, villain', twice with a strong contemptuous connotation (lines 685 and 2534). Interestingly it is used, in four cases, along with other social terms.

Here are such instances:

(1) a. *Louerd, we aren boþe þine –*
 þine cherles þine hine (lines 620-621)
 b. *And seyde "wiltu ben erl?*
 Go hom swiðe, fule drit-cherl!
 Go heþen and be eueremore
 þral and cherl als þou er wore" (lines 682-685)
 c. *Of Cornwayle þat was erl*
 (þat fule traytour, þat mixed cherl) (lines 2533-2534)

In the quotations above the meaning and connotations of the word can be inferred through a comparison with other terms of contempt, with which there is a relation of semantic association, and with some terms indicating higher social status, with which there is a relation of semantic opposition. There is a semantic association in the pairs *cherles – hine, þral – cherl* and *fule traytour – mixed cherl*. The *þral – cherl* cooccurrence is particularly interesting in that it shows the two terms in synonymic association while in Old English this same phrase was a typical example of semantic opposition. A contemptuous connotation is expressed in the compound *drit-cherl* 'scum' (ON *drit* 'excrement'), further strengthened by the adjective *fule* 'dirty'. The last two quotations show *cherl* in opposition to *erl*, collocated in the previous line. The rhyme strengthens the opposition, also evident from the context: a man that bears the title of earl unjustly is, in reality, nothing but a churl. In the following quotation, instead, there is a completely different situation: Havelok was thought to be a churl's son while he was the king's son:

(2) *For he wende þat Hauelok wore*
 Sum cherles sone and no more. (lines 1092-1093)

Also here *cherl* is to be considered a synonym of *thrall*, as few lines after we find:

(3) *He wende þat Hauelok were a þrall* (line 1100)

In the following quotation *cherl* is collocated within a rather conventional sequence of social terms:

(4)   *Of erles, baruns, lef and loth*
      Of knictes, cherles, fre and þewe. (lines 261-262)

Here *cherl*, as in the previous quotation, is used as a rank term, showing no actual shift to the evaluative sphere.

Now let us consider the Scandinavian form *carl*. It is found in the text in one occurrence, as a term of address, with the meaning of 'fellow'[11]:

(5)   *"We shole at þis dore gonge*
      Maugre þin, carl, or outh longe!" (lines 1789-1790)

This menacing sentence is uttered by a member of a gang and is addressed to a *greyue* – an administrative official of a town – whose social status was by no means a low one. The word comes from a despicable character and is addressed to a fully respectable person, i.e. Bernard the greyue, who had given hospitality to Havelok. A few lines later Havelok himself kills the assaulters.

Therefore *cherl* in *Havelok* is a near-synonym of *thrall*, used both as a description of a low social status and as a disparaging address-term, while *carl* is not descriptive of a social status but is simply an address-term whose connotation is less contemptuous.

Cherl and *carl* are also found in *Cursor Mundi* of which four manuscripts are available: Cotton, Fairfax, Göttingen and Trinity.[12] The first three manuscripts belong to a northern linguistic area and are dated back to the fourteenth century, while the Trinity manuscript is written in a hand of the first quarter of the fifteenth century in a south-western dialect. Therefore the Trinity manuscript has southern or native forms whereas the other manuscripts have northern or Scandinavian forms. For this reason in the Trinity manuscript we find *cherl* instead of *carl*.

In *Cursor Mundi* we find *carl* and *cherl* used nearly always with reference to biblical characters.

Carl occurs three times. In the following quotations it bears the meaning of 'man, fellow':

---

[11] Smithers ((ed.) 1987: 180), in the glossary to the text, translates it as 'fellow'.
[12] Cotton manuscript Vesp. A 3 in the library of the British Museum. Fairfax manuscript 14 in the Bodleian Library. Manuscript Theol. 107 in the Göttingen University Library. Manuscript R.3.8 in the library of Trinity College, Cambridge.

(6) a. With hething all þai him answard
And said, "qui es þis carl sua ferd?" (Cotton: lines 1735-1736)
b. "Þou carl, qui brekes þou vr lau,
Þi bedd noiþer aght þou ber ne drau" (Cotton: lines 13808-13809)

In the first quotation there is a context of derision: the word is in fact used in reference to Noah by people mocking him because he was building the Ark and warning them of the imminent Flood.

In the second quotation – only in Cotton and Göttingen – *carl* is used, as a term of address, in a context of accusation: it refers to a paralytic man who has just been miraculously healed by Jesus Christ and is accused by the Jews of violating their religious laws. A few lines later Christ himself is accused of the same sin: working on the Sabbath.

It is interesting to note that in these two instances *carl* is used with reference to persons who, even if they are considered either fools or sinners in the other characters' opinion, nevertheless are neither mad nor guilty, but actually righteous and, in the case of Noah, the instrument of divine plans. The real despicable character is not the man offended but the offender, and this situation is very similar to that noticed in *Havelok*.

In the following couplet *carl* bears the meaning of 'servant' without any particular connotation and is not used as a term of address:

(7) Þe toþer es woman, carl o feild,
And child þat es wit-in eild (Cotton: lines 29444-29445)

The case of *cherl* is quite different. It is found in the Fairfax manuscript:

(8) kniht and chorle and fre-man.
oute of þer þre breþer bigan.
Of sem freman. of Iaphet kniht
of cham chorle come ful riht (lines 2133-2136)

Here *cherl*, in the form *chorle*, bears the meaning of 'serf, slave' and is used as a rank term along with *kniht* and *freman*. In particular the passage is about the three different social ranks that derive from Noah's sons, Shem, Japheth and Ham – i.e. freemen, knights and churls respectively. Here the word is equivalent to *thrall* 'slave', as Ham is defined some lines before:

(9)   *cham he sayde wiþ-out doute*
sal be his brethers vnderloute
vnderlowte to ham and tralle (Fairfax: lines 2053-2055)

In Cotton, Göttingen and Trinity *thrall* is found instead of *cherl*, the two terms being actually synonyms.[13] It is interesting to note that, unlike *carl* in the occurrences quoted above, here *cherl* refers to an actually despicable character and sinner.

Different usages of the two terms can also be found in Chaucer. Interestingly in this case the northern form is the one that has a more remarkable contemptuous connotation. According to Burnley (1983: 150):

> To refer to another as a *cherle* may be simply socially-descriptive; nevertheless, it may also be an expression of abuse. The English-derived form of the word, *cherle*, was that used in London in the generation before Chaucer, but immigration had, by the later fourteenth century, introduced another form, *carl*, derived from Norse. Although in northern Middle English this was the normal referential form, in London and also perhaps more widely, the selection of this form seems to have carried an extra charge of contempt: it was perhaps a deliberately coarse form and, with one exception ... Chaucer restricts its use to the description of the coarsest of his characters and the most socially-depressed.

Chaucer's usage therefore appears to be rather different from that found in a northern text such as *Cursor Mundi*. It is interesting to note that in the London area *carl* was coarser than *cherl* while in northern Middle English the situation was exactly the opposite. As to the reason why, in London Middle English, certain Scandinavian words had social stigmas associated with them, Smith (1992: 64) asserts that this was not because of their etymology – in fact Scandinavian words had become fully naturalised – but because they originated from distant parts of the country that, coincidentally, had been exposed to a close social contact with Norse speakers.

Besides, as to the different usage of *cherl* and *carl* in the northern area a sociohistorical explanation is also possible. This involves a consideration of the social structure in the Danelaw and, subsequently, in the regions once inhabited by the Danes. In these regions the rural population enjoyed more freedom than elsewhere. Danish peasants, who had come to England as soldiers, took up again their work in the fields, retaining the status of free men. Stenton (1965: 138) describes peasants in the Danelaw as a

---

[13] The cooccurence *thrall* – *cherl* in line 685 in *Havelok* has already been noticed.

peasant "aristocracy" not found in other parts of England. Therefore while in other parts of England the native *cherl* underwent a semantic pejoration due to the worsening of the social conditions of peasants, in northern England the semantic path of the Norse *carl* seems bound to a different social evolution.

It is largely agreed among scholars that in the regions once occupied by the Danes a sort of hybrid language had risen from the close contact between Scandinavian and native speakers. There were small differences in vocabulary between Old English and Scandinavian dialects, and this fact rendered the amalgamation of the two languages easier. The English and the Northmen could very easily understand each other in their own languages, and the close connection between both nationalities must have caused numerous words to be introduced into one language from another one, even without either side noticing that the words adopted did not belong to their original vocabulary (Björkman 1969: 8). Words that were synonyms, or quasi-synonyms, and bore a strong phonological resemblance – due to a common etymology – were often used without distinction, and influenced each other on the phonological level as well as on the semantic one. Therefore a phonological and/or semantic hybridism often took place, owing to whether a word could have an English form but a Scandinavian meaning or vice versa. This is, just to quote a few examples, the case of words like ME *drem* 'dream' (< OE *dream*, but the meaning < ON *draumr*), ME *frend*, in the acception of 'relative' (< OE *freond*, but the meaning < ON *frændi*) or the ME verb *bitaken* 'hand over to' (< ME *bi-* + ON *taka*, but the meaning < OE *betæcan*). It is interesting to note that Middle English doublets like *chiste/kiste, chepe/coupe, chirche/kirke, bitechen/bitaken* have /tʃ/ and /k/ as allophones since no semantic difference is evident, while in the case of *cherl/carl* the opposition /tʃ/ vs. /k/ can be considered phonemically distinctive. In the latter case in fact a semantic difference is noticed, due to sociolinguistic factors, while in the doublets quoted above, in which no sociolinguistic factor is evident, a complete semantic blending takes place. In other words, the phonological resemblance and the social factors played opposite roles in the *cherl/carl* doublet. This fact must have prevented the semantic blending of the two forms, which therefore had different usages, at least until the fourteenth century. In this view the language contact and its dynamics are not considered in purely linguistic terms, but as processes also influenced by external factors, which in the case of social vocabulary appear to be relevant.

References

Björkman, Erik
    1969   *Scandinavian loan-words in Middle English.* New York: Haskell House.

Burnley, John David
    1983   *A guide to Chaucer's language.* London: Macmillan.

Dahlgreen, Kathleen
    1985   "Social terms and social reality", *Folia linguistica historica* 6: 107-125.

Gusmani, Roberto
    1987   "Interlinguistica", in: Romano Lazzeroni (ed.), *Linguistica storica.* Roma: La nuova Italia scientifica, 87-114.

Kurath, Hans — Sherman S. Kuhn (eds.)
    1956ff.   *Middle English dictionary.* Ann Arbor, Mich.: University of Michigan Press.

Leith, Dick
    1986   *A social history of English.* London: Routledge and Kegan Paul.

Marsh, Henry
    1974   "The end of serfdom", *History Today* 24: 96-103.

Morris, Richard (ed.)
    1893   *Cursor Mundi.* (Early English Text Society.) London: Kegan Paul, Trench Trübner and Co.

Murray, James A. H. — Henry Bradley — William Alexander Craigie (eds.)
    1989   *The Oxford English dictionary.* (2nd edition prepared by John A. Simpson — E. S. C. Weiner.) 20 vols. Oxford: Clarendon.

Runciman, Walter G.
    1984   "Accelerating social mobility: the case of Anglo-Saxon England", *Past and Present* 104: 3-30.

Smith, Jeremy J.
    1992   "The use of English: language contact, dialect variation, and written standardisation during the Middle English period" in: Tim William Machan — Charles T. Scott (eds.), *English in its social contexts.* New York — Oxford: Oxford University Press, 47-68.

Smithers, Geoffrey Victor (ed.)
    1987   *Havelok.* Oxford: Clarendon.

Stenton, Doris Mary
- 1965 *English society in the early Middle Ages.* London: Penguin.

Toller, T. Northcote (ed.)
- 1972 *An Anglo-Saxon dictionary.* Oxford: Oxford University Press.

Whitelock, Dorothy
- 1974 The beginning of English society. London: Penguin.

Nikolaus Ritt (Vienna)
The spread of Scandinavian third person plural pronouns in English: optimisation, adaptation and evolutionary stability

1. Prolegomena

This paper deals with the spread of Scandinavian third person plural pronouns in English from an unabashedly functional perspective. It views the evolutionary competition between Scandinavian and English pronoun forms against the background of the efficiency with which they fulfilled their respective communicative functions for the human organisms and societies in which they were embedded. In this sense, my paper proposes a "functional explanation" of a particular change. At the same time it deals with the theoretical status and explanatory value of such accounts and thus takes up a topic that has been a hotly debated subject for more than a decade at least.[1] The particular view on the problem presented here rests on concepts derived from evolutionary biology (particularly Dawkins 1982, 1989) and the study of so-called Complex Adaptive Systems (for instance Gell-Mann 1992, 1994). My paper is as much about those theoretical issues, therefore, as about the pronouns themselves.

2. The "facts"

As is well known, the Modern English system of third person personal pronouns is etymologically speaking a hybrid. It incorporates both Anglo-Saxon and Scandinavian elements, as well as one which is of Anglo-Saxon albeit not quite determinable origin. The picture is conveyed in table 1.

As already stated, my focus will be on the Scandinavian element in the English pronoun system[2]. It represents quite an extraordinary pheno-

---

[1] For a few contributions to this debate and for further rererences, see, for example, Campbell – Ringen (1981); Dressler (1981, 1985: 266ff.); Lass (1980, 1987a, 1987b) or Samuels (1987a, 1987b).

[2] The possible origin of *she* is not at issue here. It suffices to point to Mossé (1952), which contains a good summary of the problems related to it:

> The schō type which is that of the Northern dialect and of Scots. It is perhaps the form taken in Middle English by the Old English feminine demonstrative sēo, then sīo with a displacement of accent (s'io > si'o > sj'o) which would have facilitated the palatisation of s to sh [s to š]. But it must be remarked that the Old English

menon, because the borrowing of closed-class items is rare compared to the frequency with which open-class elements such as nouns or verbs are normally borrowed. Languages hardly ever seem to allow it. Being such a special phenomenon, the intrusion of Scandinavian pronouns into the English system has become a fixed point in accounts of the history of English, and there will be hardly any introductory course on the topic that does not deal with it.

(1)

	singular			plural
	m	f	n	
nominative	he	she	it	they
genitive/possessive	his	her	its	their
objective	him	her	it	them

Legend: xxx .... Scandinavian
xxx .... Anglo-Saxon
xxx .... disputed

While missing from no major handbook[3], however, the story how Scandinavian pronouns found their way into English has not been subject to as many detailed and systematic treatments as one would expect. As I would like to argue in the following, there is a reason for that which epitomises, to my mind, a typical and annoying problem historical linguists often have to encounter: phenomena that seem intuitively plausible and easy to account for turn out to be extremely complex and almost impossible to formalise within classical scientific explanation schemes. Therefore, most accounts of the problem have preferred to remain informal.

3. Two types of account

The story of Scandinavian personal pronouns and how they found their way into English is typically told from one of two perspectives. Either it

---

feminine pronoun hēo could very well, by an analogous process (h'io > hi'o > hj'o > hj'o), arrive at the same palatization. Beside the spellings shō, schō we have chō (Morte d'Arthur, 20/715, etc.) and ȝhō in the Ormulum (a spelling which notes perhaps more precisely the stage [hjō]. This type, native of the North, was extended, however, to the North West Midlands (Sir Gawain: *scho* 18/1259 but normally *ho*) as well as the North East Midlands (Ormulum, Havelok).
(Mossé 1952: 56)

[3] See, for instance, Baugh − Cable (1993: 100); Blake (1993: 120); Jespersen (1978: 66); Strang (1970: 266f.) or Sweet (1892: 100).

comes as a socio-geographical story about speech communities, or it comes as a story about rivalling linguistic systems.

3.1. A story of speech communities

In the first case, the focus is normally on the relationship between the Anglo-Saxons and the Danes, on the degree to which one or both speech communities were bilingual, and/or on the status one language had within the other community. Such accounts will normally contain maps illustrating the original concentration of the Scandinavian forms in an area largely isomorphic with the Danelaw (see (2)),

(2)

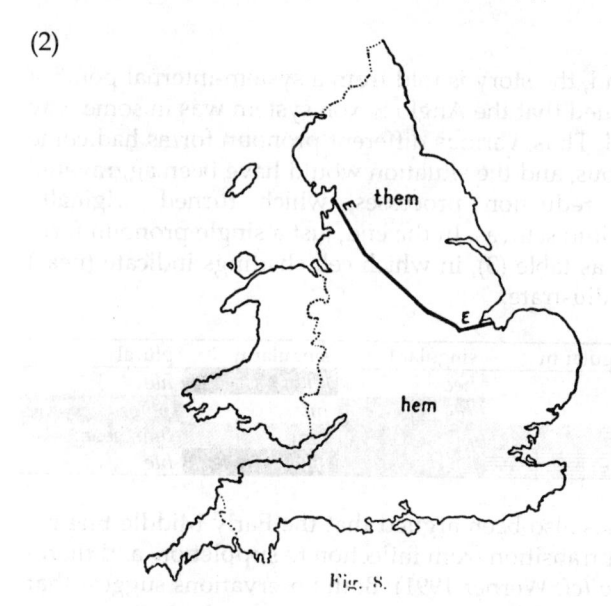

Fig. 8.

(Map taken from Mossé 1952: 57) and highlight that special extralinguistic conditions must have obtained for the borrowing of pronouns to have taken place. In *The Shape of English*, to give a typical example, Lass argues that

> there's no doubt that at least the later stages of the Danish presence ... were characterised by fairly extensive bilingualism [although ... u]nfortunately we have no direct evidence for what the contact situation was like. ... The intimacy of the

281

> contact between the two languages is ... suggested by the borrowing of rather unlikely items. In general, the most borrowable are members of the major "open" lexical classes, like nouns, verbs, adjectives; primarily structural forms, like pronouns and prepositions are normally resistant. Yet in English the Scandinavian pronouns *they, their, them* (ON *þeirr, þeirra, þeim*) have replaced OE *hie, hiera, hem*. (Lass 1987c: 53)

In such geo- and sociolinguistic accounts, system-internal aspects of the morphological borrowing are normally backgrounded – while it is of course not denied that there will have been some reasons for the pronouns to be borrowed since otherwise this wouldn't have happened.

3.2. A story of systems

When, on the other hand, the story is told from a system-internal point of view, it is normally argued that the Anglo-Saxon system was in some way deficient, or suboptimal. Thus, various different pronoun forms had come to be nearly homonymous, and the situation would have been aggravated through phonological reduction processes, which turned originally different vocalic nuclei into schwas. In the end, just a single pronoun form was left unambiguous, as table (3), in which cell shadings indicate (near) homonymies, serves to illustrate.

(3)

3rd	singular m	singular f	singular n	plural
nominative	he	heo	hit	hie
genitive	his	hiere	his	hiera
dative	him	hiere	him	him, heom
accusative	hine	hiere	hit	hie

Apart from that, it has also been argued that the Early Middle English system was in a state of transition from inflection to suppletion, and therefore inherently unstable (cf. Werner 1991). Both observations suggest that the import of the Scandinavian items served to create a "better" pronoun system.

The term "better" is crucial here, because it points to the inherent "incompleteness" of such purely formal accounts. After all, the "quality" of a system is normally, albeit often implicitly, measured in terms of benefits and costs to the speakers who employ it in actual communication. However, accounts that pretend to deal with linguistic (sub)systems often background the role of the speakers to such an extent that they do not even explicitly figure in them. At least on the surface, such stories want to be read as being about "systems as such" – continuing the traditional Saussurean approach to language: their message is that a better system ousted a less optimal one.

3.3. A combined story

3.3.1. Now, it seems obvious that the two stories, while making good sense for themselves and being intuitively appealing, somehow yearn to be put together. That this has not really been done in an explicit, confident manner may be not so surprising, if one takes into account that much of twentieth-century linguistic research has been based on the formalist view that language should indeed be studied by itself and particularly independently of its social context. As mentioned above, however, it is nevertheless interesting because, viewed from a naive, common-sense perspective, the situation seems to be pretty straightforward.

The inherited personal pronouns systems had become deficient and did not function well in communication anymore. Most speakers must have felt this somehow, and there will have evolved a general desire to do something about this. The chance offered itself when the contact between the Danish and the Anglo-Saxon speech communities had become so friendly and close that bilingualism became widespread and the respective other language ceased to be marked, socio-psychologically, as the language of the enemy. Playing around with bits and pieces of Danish, Anglo-Saxon speakers would have soon learned that the Scandinavian pronouns came in handy where the English ones were ambiguous, and would have begun to use them even in communication with English speakers, as long as they could be sure that the latter knew as much Danish as necessary to recognise the pronoun forms used. At first such uses will of course have been marked as highly exceptional, maybe as jocular or whatever, but sooner or later speakers must have got used to them to such a degree that Scandinavian forms lost their foreignness and might instead have been felt to be typical of a regional variety. Seeing how well their countrymen from the north-east fared with the new pronouns, then, the rest of the speech community soon followed suit and adopted the new streamlined pronoun system, until after some time only the most parochial speakers from the most backward communities stuck to their unwieldy heritage: the new Scandinavian pronouns had thus replaced the English ones.

While, informally speaking, this makes for a nice story, and sounds appealing, plausible and straightforward enough, it soon turns out to be highly problematic, if one attempts to translate it into a serious scientific hypothesis. Its mechanistic nature, which works fine in an informal metaphor, turns out to rely on untenable assumptions concerning the relation between linguistic systems and their speakers. For example, the informal account presented above tacitly assumes that speakers act by and large as rational agents on the market of linguistic systems, so to speak. When an opportunity to make a linguistic system more efficient offers itself, such rational speakers are expected to take it, so that speakers' behaviour is treated as if it were more or less predictable from the properties of the systems they employ. Conversely, it is also assumed that linguistic systems reflect the experiences speakers make when using them, and adapt themselves so that they best serve their speakers' purposes. The shape of linguistic systems is viewed to be straightforwardly predictable, on this assumption, from the nature and the purposes of their speakers.[4]

3.3.2. Although these assumptions are not often pronounced, they underlie many neo-functional approaches to linguistic change. Often they are implicit to theory specific terms such as "universal preferences" (cf. Vennemann 1988), "naturalness" (cf. Dressler 1985), or "optimality" (cf. Prince — Smolensky forthc.). These do bring the speaker in, but they do not always make this as explicit as it might be made. As is well established, however, the claim that speakers will prefer optimal to non-optimal systems, or that systems optimise themselves to suit their speakers does not hold much explanatory/predictive power. On the one hand, unless system-external criteria for measuring optimality are provided, such a statement may simply turn out to be circular and meaningless, reading 'whatever exists is optimal because if it were not, it wouldn't exist'. On the other hand, practically all criteria that have been proposed for measuring "optimality" have had to face counterevidence that would seem to falsify them at least by Popperian standards.

This problem is so well known that it might seem superfluous to present any examples. Let me nevertheless point out two, which suit the particular case studied here. If the success of Scandinavian pronouns is explained by a "necessity" to avoid ambiguity, how would one then account for the fact that ambiguity of the very same type was later reintroduced into the system, when the second person plural pronoun came to assume singular meaning, and *you* came to stand for nominative and

---

[4] The fact that these are not really understood very well yet is a different matter.

object case alike? Or how would one account for such pronoun systems as the Modern High German one, where *sie* stands for third person singular feminine, third person plural, as well as for the polite variant of the second person singular? Of course, explanations come to mind, but that doesn't alter the fact that the "avoid-ambiguity" rule cannot really be viewed as a general law of any predictive power, because hardly any violation of it could not be accounted for by "explanations" drawn from different levels in an *ad hoc* way.

3.3.3. The problem of putting systematic accounts and psycho-sociological ones together is related to the ontological status attributed to language, or rather to linguistic systems. On the one hand, they can be thought of as having psychological, or, more radically speaking, physiological, i.e. neuronal reality. Taking such a position implies the assumption of a material, though possibly very complex, relationship between internalised "systems" and their manifestations in actual use. It encourages the hope that this material relationship might be understood and modelled. It suggests, in other words, that although our observations and measurements are by far not exact or numerous enough, it should in principle be possible to show that some linguistic systems, i.e. cognitive schemata or neuronal configurations, are more easily accommodated in the central nervous systems of real speakers than others, so that it should eventually be possible to formulate sufficiently detailed hypotheses for putting them to experimental tests of some sort. Linguists taking such a view thus find themselves motivated to search for rules – even if only soft, probabilistic ones – relating formal properties of a system to the chance of their actual occurrence in human speech communities.

On the other hand, one may assume that linguistic systems are autonomous, albeit virtual objects, having ontological status in Popper's world three, for example.[5] Under this view, no material relationship between such virtual systems and their manifestations in speakers' brains or in actual utterances needs to be assumed. This view suggests that language systems can and ought to be analysed and dealt with independently from their neuronal, psychological or social bases. It is under this, essentially structuralist, approach, that language systems have the status of objects worth being studied in their own right. Such descriptions are externally unconstrained. Since criteria for evaluating rivalling accounts are not derived from language external "realities", they will often tend to be

---

[5] Roger Lass being the most prominent advocate of this view.

"aesthetic" and include internal adequacy, descriptive elegance or economy.

3.3.4. As I hope to have made clear already at the outset, this paper subscribes to the first of these two views. Admittedly, the reasons for this are partly personal. I simply find functional explanations extremely appealing, and they often seem to make so much sense that I simply don't want them to be useless. Personal preferences aside, however, it seems that recent developments in post-classical science, particularly in evolutionary biology or the study of Complex Adaptive Systems (cf. Gell-Mann 1992, 1994) might provide theoretical frameworks more fruitfully transferable to the study of human language than the scientific paradigm of classical physics, and – as far as I understand their theoretical implications – it seems that they are capable of clarifying at least some of the reasons why functional explanations of linguistic phenomena often appear so plausible when applied retrospectively, while not admitting of other than very probabilistic predictions at best.

In the next two sections of my paper, I would then like to do the following. First, I will present a functional, or if you will, "natural", explanation why the Scandinavian pronouns ousted the Old English ones – without problematising its predictive power. Having done so, I will embed the explanation within a generalised theory of Complex Adaptive Systems, and show that thus an understanding can be gained why the functional explanation proposed does make sense without necessarily having much of a predictive value in the classical sense.

4. A functional explanation of the success of the Scandinavian pronouns in Middle English

4.1. Outline

In the following I am going to compare the two rivalling third person plural subsystems of the personal pronouns within the framework of natural morphology as outlined in Dressler *et al.* ((eds.) 1987), or Dressler (1985). It is, in my eyes, the most refined and detailed framework for measuring the efficiency of morphological systems. In particular, I shall employ the parameters of iconicity, transparency, bi-uniqueness and figure-and-ground, or sign prominence.[6]

Diachronically, I will make a major distinction between the Old English period, with adjectival and pronominal inflection in general still relatively intact, and a Middle English stage, for which I assume adjectival and pronominal inflection to have mostly disappeared. As I will show, it is crucial to distinguish the two phases when dealing with the intrusion of Scandinavian forms into the English personal pronoun system, because the common view that Scandinavian pronouns were imported in order to reduce ambivalences in the pronoun system works only for the Middle English phase, and can therefore only explain the spread of the borrowed forms, not so much the borrowing itself. As for the latter, which must have occurred during the Old English period, I will show that phonological "advantages" of the Scandinavian forms were a much more likely motivation than other factors.

4.2. Stage I: Old English

4.2.1. Iconicity

It is normally argued that in the third person the original Anglo-Saxon pronouns are more iconic or speaking more exactly are more "diagrammatic" than the new Scandinavian ones.

The argument is as follows. As Lass (1994) and Werner (1991), for example, have mentioned correctly, the third person plural pronouns were analysable as "*h* + inflectional component". The inflectional components -*s*, -*m*, -*ne*, -*re*, -*ra* could also be found within the paradigms of the demonstrative pronouns and adjectives. Therefore, as table 4 shows, the morphotactic compositeness of the pronouns diagrammatically reflects their morphosemantic structure.

---

[6] Werner's paper (1991) on the topic works also within a functional framework, although he expresses doubts about the natural model as outlined by Mayerthaler (1981). His interpretation differs from this one mainly in that he attributes much more importance to suppletive tendencies. He argues that they were inherent to the Old English pronoun system already and that they were the main motivation for importing the Scandinavian items. As I will argue below, there are some arguments that speak against this view.

(4)   morphosemantic level: | {3rd person} | + | {case/number/gender} |

   morphotactic level: | /hV/ | + | /s/ /m/ /ne/ /re/ /ra/ |

Things are not quite as simple, however. In particular, two points have to be made: first, even in the Old English system, iconicity was not maximal, because the morphosemantic categories case and number are not expressed separately – as they are in agglutinating languages, where morphological diagrammaticity is greatest. Therefore, in perfect harmony with the fact that Old English inflectional morphology altogether was of the inflecting type, it seems that "iconicity" cannot have been a very important parameter for the Old English pronominal system either.

Second, and more importantly, the imported pronouns were in principle open to the same iconic analysis as the old ones. After all, their rhyme consonants integrate nicely into the Old English inflectional system of demonstratives and adjectives. The crucial difference would just be that the initial element was not *h*, but rather *þ*. Of course, the introduction of /ð/ as a new allomorph would increase ambivalence and would weaken, at the same time, the recognisability of the /h/ as a third person marker, but this would have affected iconicity only indirectly. What suffered primarily was bi-uniqueness or transparency. What all this amounts to, however, is that iconicity cannot have played any role worth mentioning in the competition between the Anglo-Saxon and the Scandinavianised pronoun systems.

4.2.2. Transparency

As just argued, the inherited Anglo-Saxon system third person pronouns fared better with regard to this parameter than the Scandinavianised one. It needs to be added, however, that on the whole, its morphotactic transparency was not very high in the first place. Particularly the vowels following the /h/ would have been difficult to assign to a specific morpheme.

Most plausibly, one could group them together with the /h/, which would result in a situation where a great number of different /hV/ sequences would stand for {third person}. This would offend the semiotic principle of maximising bi-uniqueness. The following table shows the results of an analysis along that line.

(5)
3rd	singular m	singular f	singular n	plural
nominative	hē + ∅	hēo + ∅	hi + t	hīe + ∅
genitive	hi + s	hie + re	hi + s	hie + ra
dative	hi + m	hie + re	hi + m	hi + m, heo + m
accusative	hi + ne	hie + re	hi + t	hīe + ∅

Under such an analysis, the allomorphs of the {third person} morpheme would include *hē, hi, heo, hie,* and *hīe*: clearly not an optimal situation. Still, if one assumes that the pronouns would have occurred often in unstressed positions, so that the contrasts would be diminished through phonological reduction processes anyway, this disadvantage would be neutralised to some degree.

Alternatively, the vowels could be grouped together with the {case/number/gender} morphs. The bi-uniqueness thus gained for the {third person} morpheme, which would always be represented by /h/, would have to be paid for, though, with an increased number of ambiguities among the {case/gender/number} morphemes present within the overall adjectival and pronominal inflectional system. Thus, if one assumes – as one must if one wants personal pronouns to have any transparency at all – that personal and demonstrative pronouns were analysed as containing the same inflectional morphemes, the following allomorphies would be created .

	singular m	singular f	singular n	plural
nominative	ē	ēo	it, æt	īe, ā
genitive	is, æs	iere, ǣre	is, æs	iera, ǣra
dative	im, ǣm	iere, ǣre	im, ǣm	im, ǣm
accusative	ine, one	iere, þā	it, æt	īe, ā
	PP* DP**	PP* DP**	PP* DP**	PP* DP**

* stem {h} ** stem {þ} or {s}

Holding the personal pronouns against strong adjectives would yield a similar, though less striking picture. However, the fact that the strong adjectival endings for the feminine genitive and dative singular as well as for the feminine genitive plural were rather unambiguously *-re*, rather than *-Vre*, seem to provide additional arguments against the plausibility of treating the vowels in the personal pronouns as part of the inflectional endings.

Finally, one could analyse the vowels as separate morphs altogether and argue that they stood for {gender}, for example, as in

(6)

	h	i	s
	third person	male	genitive

,

and interpret the final morphs to represent {case/number} only. Although such an analysis would be more diagrammatic than the other two, it is implausible and rather *ad hoc*, because /i/ was not a common marker for {masculinity} or {non-femininity} in Old English, and would most probably not have been recognised as one at all.

As far as my own intuition goes, the first analysis strikes me as the most plausible one of the three, but for the present argumentation it is in fact as pointless to make any decision as it may have been for the speakers of Old English themselves. What matters is that the morphotactic transparency of the Anglo-Saxon third person pronouns would not have been high under any analysis – which makes the choice between them difficult for speaker and linguist in very much the same way.

4.2.3. Bi-uniqueness

The commonplace view is that the Scandinavianised system fared much better with regard to bi-uniqueness (or minimal ambivalence) than the Old English one, because it contained fewer ambiguous forms.

However, things are again not quite what they seem. First of all, note that under the analysis proposed above for the Anglo-Saxon system, the Scandinavian imports did not decrease, but actually increase ambivalence. As argued above, the morphosemantic category {third person} could now be represented by two types of morphs, namely /hV/ and /ðV/, instead of just one, as originally. As far as the {case/number/gender} morphs were concerned, on the other hand, the new pronouns were very similar to the old ones, having an /r/ in the genitive and an /m/ in the dative/oblique.

It is true, of course, that the native Old English system was not very good with regard to biuniqueness. As already indicated, the morphosemantic category third person was represented by a variety of different /hV/ sequences, if the vowels were analysed as belonging to the first rather than to the second morph, and if they were not, increased ambivalence in the pronominal system as a whole was the consequence.

All that being as it may, however, the point remains that the Scandinavian pronouns were not better with regard to this, but rather worse, due to the additional ambivalence inherent to the /ð/-/h/ allomorphy. In Old English times, therefore, a preference for less ambivalent pronoun forms would have spoken against, rather than for the Scandinavian imports.

4.2.4. Sign prominence

As I will argue, this parameter is – in contrast to the common view – the only one where the Scandinavian imports clearly beat the Anglo-Saxon system. It is in fact more phonological than morphological and concerns, again, the difference between /ð/ and /h/. The former is definitely a stronger consonant, and more prominent than /h/, which lacks oral articulation and does not contrast well with the following vowels. In addition to that, /h/ was a typical victim of weakening processes and it is easily conceivable that in unstressed position it might have been reduced, which is tantamount to saying it was deleted because one cannot possibly reduce /h/ much further. Since in syllable onsets strong consonants are universally preferred to weak ones (cf. Vennemann 1988: 13ff.), the Scandinavian pronouns might have been preferred to the Anglo-Saxon ones on that account.

4.2.5. Summary

As has become evident, the only respect in which the Scandinavian pronouns were clearly superior to the Anglo-Saxon ones was the phonological structure of their onsets. Since this contradicts the standard view, it would be nice if there were some real textual evidence to support the claim that phonology might have been more important for the selection of /ð/ as against /h/ forms than normally assumed. Incidentally, such evidence can indeed be found, although the only piece that I have come across dates from a later period than the one under discussion here.

As is well known, Orm uses both *þeȝȝm* and *hem* forms to express the third person plural oblique (cf., already, Wright – Wright 1928). However, he does not vary them randomly but rather according to quite obvious phonological criteria. Table 7 shows the distribution of *hem* as against *þeȝȝm* forms in the second volume of Holt's (1878) edition of the Ormulum.

(7) a. *þeȝȝm*:
(21 tokens; numbers are line numbers in Holt's (1878) edition)

7 swa þeȝȝ didenn, 7 he þeȝȝm	10279
Þatt icc þeȝȝm muȝhe lesenn	10399
Þatt newe steorrne ȝaff he þeȝȝm,	11094
7 he þeȝȝm shollde frofrenn,	11243
Forr ȝiff þe Laferrd haffde þeȝȝm	12972
Þatt wære alls he þær haffde þeȝȝm	12974
Þatt he þeȝȝm þurrh hiss are	14463
Þatt he þeȝȝm ȝaff þurrh Haliȝ Gast	15344
Forr þu þeȝȝm birrlesst lifess drinnch	15406
Þatt he þeȝȝm hafeþþ filledd all	15444
Þatt he þeȝȝm hateþþ oppnedd.	15447
7 he þeȝȝm nemmneþþ swa þurrh hiss	15698
Forr þatt he þeȝȝm i Crisstenndom	15700
7 he þeȝȝm jaff annswere onnȝæn	16218
Settnessess bi þeȝȝm sellfenn.	16853
7 all swa summ he þeȝȝm bigrap	18150
7 all swa summ he kidde þeȝȝm	18152
He talde þeȝȝm, 7 kidde þeȝȝm	19405
He talde þeȝȝm, 7 kidde þeȝȝm	19405
He talde þeȝȝm hu mann maȝȝ Godd	19409
Ne flen fra þeȝȝm off tune,	19663

b. *hemm*:
(20 examples from a total of 305 tokens)

7 he till hemm; naȝȝ, namm I nohht	10285
7 he þa seȝȝde þuss till hemm;	10289
Hemm þuhhte þatt he mihhte ben	10299
Ja þurrh hemm baþe tacnedd;	10343
7 Sannt Johan hemm dide wel	10348
Þatt Sannt Johan hemm seȝȝde þuss,	10352
Forr min fulluhht ne maȝȝ hemm nohht	10358
Acc wel itt maȝȝ hemm brinngenn onn	10360
To ȝarrkenn hemm onnȝæness Crist	10362
To fullhtnenn hemm þurrh Haliȝ Gast	10364
To clennsenn hemm off sinness.	10365
Þatt tær wass hemm bitwenenn	10369

To clennsenn hemm *off sinness,*	10373
Þatt stod tær hemm *bitwenenn;*	10375
Off hellepine, 7 ȝifenn hemm	10400
To clennsenn hemm *off sinness,*	10447
Þatt clennseþþ hemm *all þwerrt út*	10460
Upp inntill heoffne, 7 ȝifenn hemm	10500
7 forþ wiþþ hemm *ec all þatt flocc,*	10566
Forr rihht iss þatt hemm *lasste wa*	10570

Clearly, while *hemm* occurs after consonants, practically all instances of *þeȝȝm* occur after vowels. It is in exactly such environments, however, that the relative consonantal strength of /ð/ as against /h/ will have mattered most. If the pronoun was preceded by a consonant, that consonant might have taken over the function of onset of the pronoun syllable. However, after vowels this possibility did not exist, so that the only way to have a good onset was to use the Scandinavian, rather than the Anglo-Saxon pronoun.[7]

It may well be, therefore, that the Scandinavian pronouns got their foothold in English not so much because they removed ambiguities, but simply because they were easier to perceive.

4.3. Stage II: Middle English

When adjectives ceased to inflect during the Middle English period, this had striking consequences for the status of the third person plural pronouns, because their rhymes were no longer recognisable as morphs well known from the adjectival paradigm. As I will show, it was this essentially independent development, which altered the odds in the competition between the Anglo-Saxon and the Scandinavianised pronoun systems in favour of the latter.

4.3.1. Iconicity and transparency

When their inflectional endings were no longer recognisable as such, the pronouns lost even the little morphotactic transparency they had had during the Old English period. Consequently, they were no longer diagrammatic either. This applies both to the Anglo-Saxon and to the

---

[7] This phenomenon is commonly known as "avoidance of hiatus", of course.

Nikolaus Ritt

Scandinavianised systems. Thus, without any dramatic changes in the pronoun system itself, it had become suppletive, so to speak. The pronouns had become monomorphemic, and all phonemes equally contributed to signalling a morphosemantic complex comprising the categories {person/number/case/gender}.

4.3.2. Bi-uniqueness

Once each of the different forms of the third person pronoun stood for its own {person/number/case/gender} morpheme, though, the homonymies, which were enhanced through the vowel reductions typical of minor stressed items, became a decisive factor in the competition between Scandinavian and Anglo-Saxon forms. Only now, when the initial consonant had ceased to represent exclusively {person}, the Scandinavianised pronoun system contained clearly fewer ambiguities than the one inherited from Anglo-Saxon. The following table makes this clear:

(8)

form: inherited system	function	form: Scandinavianised system
hN*	nominative singular m	hN*
	nominative plural	ðN
hNr	genitive singular f	hNr
	oblique singular f	
	genitive plural	ðNr
hNm	oblique singular m	hNm
	oblique plural	ðNm

*N stands for nucleus

4.3.3. Prominence

In very much the same way as in the Old English period, the Scandinavian pronouns were superior with regard to this parameter to the inherited ones.

4.4. Summary

Summing up, one can say that parameters of morphological naturalness provide a good framework for the explanation of the exceptional success of Scandinavian forms within the English system of third person pronouns. During the Old English period, the Scandinavian pronouns were superior to the Old English ones with regard to morphonological sign prominence, which seems to have allowed them to gain a foothold within the language. In a second stage, the parameter of bi-uniqueness, or ambivalence, became additionally relevant, due to the loss of (adjectival) inflections, which occurred at the beginning of the Middle English period and which was crucial – albeit in an indirect manner – for the ultimate success of the borrowed forms.

Both phonological prominence and the resolution of the particular ambiguities that characterised the Early Middle English third person pronouns represent advantages for sign processing by recipients rather than speakers. As far as the actual implementation of the change from the inherited to the Scandinavianised pronoun system is concerned, it would therefore seem plausible to assume that it occurred first in listener-centred variants of English, that is to say in careful speech and in formal registers. This would support Theo Vennemann's proposal (in his conference contribution) that the Scandinavian pronoun variants came to be interpreted as the phonologically foregrounded variants of the Anglo-Saxon ones, because, as is well known, careful speech and formal register favour phonological foregrounding.

5. The story of Scandinavian pronouns as a story of Complex Adaptive Systems

5.1. Basic assumptions: linguistic systems as "genotypes"

5.1.1. Having established the greater "naturalness", "optimality" or communicative efficiency of the Scandinavianised system, let me return to the question whether this is enough to actually explain why it ousted the inherited English one. Put differently, the problem is how to bridge the theoretical gap between saying that a potentially better system was in the air and explaining how it actually made its way into general usage, so that eventually it became the norm.

Before tackling that problem, let me outline the basic assumptions on which the present argumentation rests. First, linguistic systems are assumed to have psychological, or, if you will, neuronal reality. I do not consider them as mere abstractions derived from the observation of actual discourse, but rather as the neuro-psychological configurations that give rise to actual utterances. Languages are thus not virtual systems, but actual systems existing in the brains of real speaker-listeners – although they are difficult to observe.

5.1.2. Employing a biological metaphor, linguistic systems can thus be compared to genotypes. In the same way as the latter give rise to living organisms as their phenotypical expressions, so do linguistic systems give rise to actual texts, or utterances. In the same way as the qualities of individual organisms are co-determined by environmental factors, so are the qualities of actual utterances. Neither are exclusively determined by their underlying "genotypes".

Apart from having phenotypical expressions, linguistic systems share with genomes the quality that their survival on the stage of history crucially depends on their capacity of replicating. Genes do this through creating copies of themselves in sexual or a-sexual reproduction, the details of which need not concern us here. Constituents of linguistic systems, on the other hand, create copies of themselves through communication and language acquisition in the widest sense.

While genetic replication involves the straightforward copying of DNA strands, the replication of linguistic elements works through their phenotypic expressions. However, since the latter are strongly co-determined by environmental factors, including actual articulation and perception processes, as well as by contextual factors in a wider sense, it is obvious that the exactness with which the elements of linguistic systems are actually copied must be much smaller than the high copying fidelity characteristic of genetic replication. The consequence of this will be a high degree of variation within the speech community, or, if you will, the pool of linguistic elements. This is in perfect keeping with everyday experience, of course.

Crucially, however, not all variations of a linguistic element present in a speech community will be equally successful in it. Some will disintegrate before being able to produce copies of themselves that can be regarded as sufficiently exact to count as such. Linguistic systems are thus subject to a process of automatic selection that is analogous, in principle, to that which works on the level of genetic variation. Since the storage capacity of human brains must be regarded as equally limited as material resources in nature, it follows that in the long run variants of linguistic system constituents that are more efficient in replicating themselves will oust those which are less so. Furthermore, and in the same way in which a gene whose carrier dies before being able to reproduce does not get a foothold in the gene-pool of a species and can be considered as irrelevant from the evolutionary biological point of view, so any constituent of a linguistic system whose expression in an utterance does not create copies in the brains of other speakers before the brain in which it is itself encoded decays, cannot be regarded as an actual part of the language shared by a speech community and must be considered as uninteresting for linguists dealing with that language.

5.1.3. When dealing with the properties of linguistic systems at any particular historical point, one is therefore dealing with systems of such constituents as have proved themselves stable in time through successful replication in the speech community. When the biologist Richard Dawkins dealt with the questions why there are human beings and why they are the way they are, the perspective just outlined permitted him to come up with the ingenious answer that we exist because the genes that happen to produce human beings have been able to replicate before decaying (cf. Dawkins 1989: 1-12). In a similar manner, one might therefore argue that languages are the way they are because the neuro-psychological configurations underlying them have been able to replicate before disintegrating.

Since the constituents of linguistic systems replicate through communication and language acquisition, their successful reproduction, and thus their stability, depends on the costs and benefits their phenotypical expressions have for the organisms, or speaker-listeners, through which they come to be uttered and/or perceived. Elements whose phenotypical expressions are either too costly, or earn their producers no advantages, will stand a small chance of being uttered often enough to get acquired by other speakers. A stable linguistic system will thus be one creating utterances that cost human organisms sufficiently little – in terms of energy expended for articulation and sign-processing, for example – while carrying sufficient benefits in terms of enhancing speaker-listeners' well-being in the widest sense. It will be adapted, in other words, to the way in which its host works and to the functions it is required to fulfil.

5.2. Application

5.2.1. A language can therefore be regarded as a complex system capable of adapting to its environment. As scientists of various fields have recently become increasingly aware of, such Complex Adaptive Systems seem to exist in many different phenomenological domains ranging from evolutionary biology over immune systems to ecosystems, economies, cognition, human societies and cultures or artificial life. The increased interest in the properties of such systems has even lead to a research institution dedicated to their study, namely the Sta. Fe Institute in New Mexico.

One of the founding members of that institute, the particle physicist Murray Gell-Mann, has suggested the following rough schema to represent Complex Adaptive Systems (CAS) (cf. Gell-Mann 1992: 11):

(9)

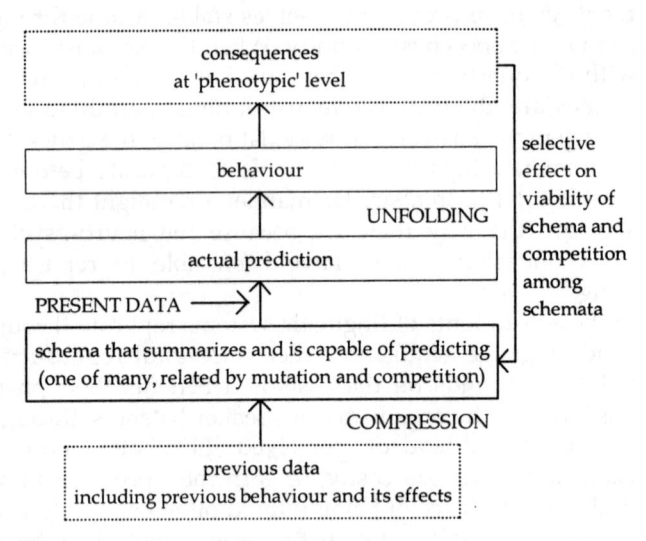

```
                    consequences
                  at 'phenotypic' level
                          ↑
                                                    selective
                      behaviour                     effect on
                          ↑                         viability of
                                    UNFOLDING       schema and
                   actual prediction                competition
                          ↑                         among
    PRESENT DATA ──→                                schemata
         schema that summarizes and is capable of predicting
         (one of many, related by mutation and competition)
                          ↑
                                    COMPRESSION
                    previous data
         including previous behaviour and its effects
```

Applied to the subsystem of English pronouns, in this particular case to the pronoun {ðem}, the schema would look as in (10).

The system proper, i.e. the scheme in the big box, has two aspects. Materially, it may consist of elements of all sorts. It may, for example, be the system of nuclear acids making up the genome of an organism, or it may, in our case, be neuronal configurations in a human central nervous system. Whatever its material makeup is, though, the system can also be viewed as information, because it predicts, or models, certain aspects of its environment relevant to its survival. An animal's genome can be read as a model of the environment in which it survives, containing information about factors ranging from climate, temperature or colour to the possibly threatening presence of predators. The neuronal configurations in the human brain that are relevant for language would similarly contain information about the makeup of human articulatory and auditory hardware, aspects of the world that are relevant for human survival to be communicated, as well as aspects about the way in which humans interact socially.

(10)

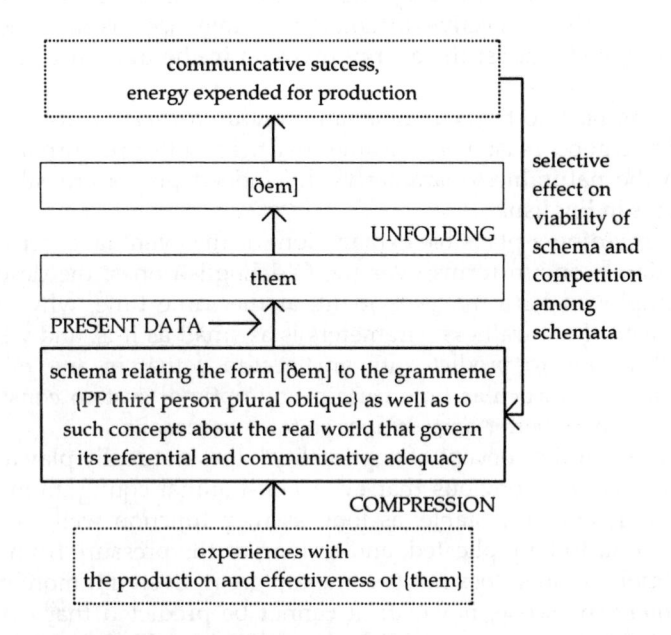

```
┌─────────────────────────────────────────┐
│     communicative success,              │
│     energy expended for production      │
└─────────────────────────────────────────┘
                    ↑
┌─────────────────────────────────────┐   selective
│              [ðem]                  │   effect on
└─────────────────────────────────────┘   viability of
                    ↑    UNFOLDING       schema and
┌─────────────────────────────────────┐   competition
│              them                   │   among
└─────────────────────────────────────┘   schemata
PRESENT DATA ──→    ↑
┌─────────────────────────────────────┐
│ schema relating the form [ðem] to the grammeme │
│ {PP third person plural oblique} as well as to │
│ such concepts about the real world that govern │
│ its referential and communicative adequacy     │
└─────────────────────────────────────┘
                    ↑    COMPRESSION
┌─────────────────────────────────────┐
│        experiences with             │
│ the production and effectiveness ot {them} │
└─────────────────────────────────────┘
```

Complex Adaptive Systems interact with their environments. Under relevant environmental conditions some of their elements will unfold and lead to phenotypic effects. A genome, for example, unfolds through developing an organism under favourable circumstances and through making the organism prone to behave in certain ways. The neuronal system responsible for language will unfold essentially through utterances. The phenotypic effects feed back on the system through selection among competing structures. A neuronal configuration unfolding as an incomprehensible utterance will earn no benefits from its environment while generating at least the cost of wasted articulation energy, so that its chances of remaining stable are relatively small. On the other hand, an utterance which successfully solves problems for speakers will be repeated, so that there is a good chance that copies of its underlying systematic configuration will arise in the minds of other speakers.

It is likely that the chances of an element within a linguistic system to survive will often be related to the frequency with which it unfolds itself and leads to an utterance. At the same time, the chance of such an element to unfold will be influenced by the number of times in which its unfoldings – i.e. the respective utterances – have been communicatively effective weighed against the energy invested in the utterance acts themselves.

5.2.2. Coming back to the Scandinavian pronoun forms, the above implies that we have good reasons to assume that their better performance with regard to the naturalness parameters did indeed play a crucial role for their success in English.

While providing a plausible explanation for the eventual success of the Scandinavian pronoun forms over the Old English ones, the descriptive schema employed here makes evident, at the same time, why the predictive power of naturalness parameters is as small as it is, and why they do not allow one to predict with any accuracy that in any particular language at any particular time certain non-optimal system constituents will be replaced by better ones. It does so for two reasons.

First, note that the concept of "optimality" does not really play a role in it. Rather, it is quite obvious that even non-optimal configurations may prove to be historically stable, as long as they function well enough in communication to be replicated, and as long as the pressure from clearly more optimal variants does not become too great. Even if a more optimal variant enters the scene, however, it cannot be predicted that it will inevitably be able to oust the established variant. And this brings me to the second, and central point.

It is obvious, of course, that the success of an utterance depends on the linguistic systems inherent to the speakers towards whom it is directed. Thus, an English utterance will have little success in an Italian-speaking community, just to give a very obvious example. Therefore, the evolutionary stability of any individual linguistic system will also depend on how well it predicts the structure of other systems in its environment, how well it adapts, in other words, to the systems already there.[8] On the one hand, this predicts that not every potentially "more optimal" variant will be able to get a foothold in a speech community; on the other hand, it makes plausible why it is practically unpredictable when it will. Since any system not only attempts to adapt to other systems, but represents at the

---

[8] A traditional way of referring to a system's "intuition" about the other systems in its environment would be by the term "norm". This term obscures, however, the elusive and essentially dynamic nature of the phenomenon.

same time part of the environment to which those systems themselves adapt, one must expect feedback and feedforward phenomena to occur. The behaviour of any system depends on what it expects the other systems to be like, and its own, tentative unfoldings in turn influence the expectations of other systems in that respect. Such a mechanism closely resembles the stock market, whose unpredictability is well known, even though such objectifiable parameters as the actual value of a company or the quality of its products do play their roles. In the same way, then, as the price of shares does not simply reflect a company's value at a given point of time, but to a great extent also investors' prediction of the behaviour of other investors, so does the status of any particular linguistic constituent within a speech community not reflect merely its communicative efficiency and effectiveness, but to a great extent also the status of that particular item within other systems.

What implications does all this have, then, for the way in which one ought to tell the story of the Scandinavian pronouns and their way into English?

5.2.3. Through contact with Scandinavian utterances, a configuration will have established itself within Anglo-Saxon brains that associated {third person plural oblique} with the articulatory gestures and acoustic impressions representable as /ðəm/. There, it would have competed with an alternative configuration in which the same concept bundle was associated with /hem/. Certainly, in this competition, all the factors adduced by naturalists, or more broadly speaking, functionalists, would have played their part: iconicity, prominence, transparency, bi-uniqueness, and what have you – in the same way as the quality of the products and the productive efficiency of a limited company will influence the price of its shares. What would have been equally important, however, was the number of copies of the competing configurations Anglo-Saxon systems predicted to be around in their relevant environments, and this estimate would have been constantly checked against actual utterance experiences. In order for the new imports to have been eventually successful, a critical number of systems must have been confident that others would share their trust in the innovations. Only if this condition was fulfilled, the semiotic superiority of the imports would have been able to unfold its impact and forced the main bulk of English linguistic systems to follow suit. In the same way as it is impossible to predict the stock market, however, it is impossible to predict when the critical point for such an event to take place would be reached.

Such account makes obvious why in the case of the *they/them* vs. *hee/hem*, as in probably most instances of linguistic innovation, parameters for the intersubjective determination of sign efficiency or any comparable parameters can never be sufficient to explain or to predict what actually takes place in the complex world of actual language systems. At the same time, this does not mean that their study is irrelevant. On the contrary, it is essential. Denying this would be tantamount to saying that an investor on the stock market should not bother at all with the fundamental economic condition of the companies s/he intends to invest in.

References

Baugh, Albert Croll — Thomas Cable
    1993   *A history of the English language.* (4th edition.) London: Routledge.

Blake, Norman
    1993   *The Cambridge history of the English language*, vol. 2. Cambridge: Cambridge University Press.

Campbell, Lyle — John Ringen
    1981   "Teleology and the explanation of sound change", *Phonologica* 1980: 57-68.

Dawkins, Richard
    1982   *The extended phenotype.* Oxford: Oxford University Press.
    1989   *The selfish gene.* (2nd edition.) Oxford: Oxford University Press.

Dressler, Wolfgang U.
    1981   "Outlines of a model of morphonology", *Phonologica* 1980: 113-122.
    1982   "A semiotic model of diachronic process phonology", in: Winfred Philip Lehmann — Yakov Malkiel (eds.), *Perspectives on historical linguistics.* (Amsterdam Studies in the Theory and History of Linguistic Science 4, 24.) Amsterdam: Benjamins, 93-132.
    1985   *Morphonology: the dynamics of derivation.* Ann Arbor, Mich.: Karoma.

Dressler, Wolfgang U. — Willi Mayerthaler — Oswald Panagl — Wolfgang U. Wurzel (eds.)
   1987   *Leitmotifs in natural morphology.* Amsterdam: Benjamins.

Eaton, Roger — Olga Fischer — Willem Koopman — Frederike van der Leek (eds.)
   1987   *Explanation and linguistic change.* Amsterdam: Benjamins.

Gell-Mann, Murray
   1992   "Complexity and complex adaptive systems", in: John A. Hawkins — Murray Gell-Mann (eds.), *The evolution of human languages.* Redwood, Cal.: Addison-Wesley, 3-18.
   1994   *The quark and the jaguar.* New York: Freeman.

Holt, Robert
   1878   *The Ormulum.* Oxford: Clarendon.

Jespersen, Otto
   1978   *Growth and structure of the English language.* (9th edition.) Oxford: Oxford University Press.

Lass, Roger
   1980   *On explaining language change.* Cambridge: Cambridge University Press.
   1987a   "Language, speakers, history and drift", in: Roger Eaton *et al.* (eds.), 151-176.
   1987b   "On sh*tting the door in Early Modern English: a reply to Professor Samuels", in: Roger Eaton *et al.* (eds.), 251-256.
   1987c   *The shape of English: structure and history.* London: Dent.
   1994   *Old English: a historical linguistic companion.* Cambridge: Cambridge University Press.

Mayerthaler, Willi
   1981   *Morphologische Natürlichkeit.* Wiesbaden: Athenaion.

Mossé, Fernand
   1952   *A handbook of Middle English.* Baltimore: Johns Hopkins.

Prince, Alan — Paul Smolensky
   1993   *Optimality theory: constraint interaction in generative grammar.* (RUCS Technical Report 2.) Rutgers Center for Cognitive Science, Rutgers University and Department for Computer Science, University of Colorado.
   [to appear]   (Linguistic Inquiry Monographs.) Cambridge, Mass.: MIT Press.]

Samuels, Michael Louis
    1987a   "The status of the functional approach", in: Roger Eaton *et al.* (eds.), 239-250.
    1987b   "A brief rejoinder to Professor Lass", in: Roger Eaton *et al.* (eds.), 257-259.
Strang, Barbara Mary Hope
    1970   *A history of English*. London: Methuen.
Sweet, Henry
    1892   *A short history of English grammar*. Oxford: Clarendon.
Vennemann, Theo
    1988   *Preference laws for syllable structure and the explanation of sound change*. Berlin — New York — Amsterdam: Mouton.
Werner, Otmar
    1991   "The incorporation of Old Norse pronouns into Middle English: suppletion by loan", in: Per Sture Ureland — George Broderick (eds.), *Language contact in the British Isles: proceedings of the eighth international symposium on language contact in Europe, Douglas, Isle of Man, 1988*. Tübingen: Niemeyer, 369-401.
Wright, Joseph — Elizabeth Mary Wright
    1928   *An elementary Middle English grammar*. London: Oxford University Press.

Herbert Schendl (Vienna)
Code-switching in medieval English poetry

1. Introduction

In many or even most countries of the world, at least part of the population is bilingual or multilingual, since the proportion of nation states to languages is about one to thirty (Romaine 1995: 8). A widespread linguistic phenomenon in such bi- and multilingual societies is code-switching, i.e. the change from one language to another within one act of communication.

The present paper is intended as a contribution to the study of code-switching in the history of English, a still largely neglected field of historical linguistics (for a brief discussion of research in this area cf. section 3.). The extralinguistic aspects of the various waves of immigration to Britain have been extensively studied, and so have the linguistic *results* of language contacts on the linguistic system of English (cf. the work on lexical borrowing from French, Latin and Scandinavian, or on the morphological and syntactic influence from these languages). But research into the dynamic *process* of earlier code-switching is still scarce, though during much of its history Britain has been a multilingual country with an ethnically and linguistically mixed population. This is particularly true of the Old and Middle English periods, with their mixing of different West and North Germanic speech communities (including the often neglected Low German immigrants in the Middle English period) as well as their contact with Celtic, French and Latin. In this complex linguistic situation, code-switching must have been an important discourse strategy in Britain, as in many other European nations.

2. Some theoretical concepts and definitions

2.1. We will follow Gumperz in defining code-switching "as the juxtaposition within the same speech exchange of passages of speech belonging to two different grammatical systems or subsystems" (Gumperz 1982: 59). This comprehensive definition covers both switching between different languages and switching between varieties of the same language, including styles and registers. However, the present paper will concentrate on switching between the three main languages of medieval England, namely English, Latin and French, and disregard the mixing of regional dialects in medieval manuscripts.[1] The term code-mixing, which has been used synonymously with as well as differently from the term code-switching, will not be used here. For a discussion of these terms see Myers-Scotton (1988: 158); Pfaff (1979: 295).

2.2. Various types of switching have been distinguished in the literature (cf. Romaine 1995: 122ff., 162f.). A widespread *functional* distinction is that between (i) situational and (ii) metaphorical (or conversational) switching (Gumperz 1982: 60-61). The first type, in which changes of code coincide with situational changes such as new participant, new topic, etc., is especially frequent in diglossic communities, where each language is typically related to specific functions. In metaphorical switching, on the other hand, switching fulfils a variety of discourse functions and stands in a complex relation with social variables (cf. Romaine 1995: 161-165, and section 5.3.); it occurs within the "same minimal speech act" and the syntactic, semantic and textlinguistic relations between the switched elements are "equivalent to those that join passages in a single language" (Gumperz 1982: 61).

Another typology of switching is based on syntactic criteria, with categories such as "tags" or "tag-like switches", "intrasentential", "extrasentential" and "intersentential" switching, though definitions of these terms vary.

Tags and tag-like switches such as interjections are rather freely moveable constituents and can thus be inserted at various places in an utterance (cf. Poplack 1980: 605).

Intersentential switches have been variously defined in the literature. While Romaine referring to Poplack (1980) interprets them as switches that involve "a switch at a clause [!] or sentence boundary" (Romaine 1995: 122), Poplack herself lists "independent clause" and "subordinate (and relative) clause" under "intrasentential categories" (Poplack 1980: 602, 605), while "freely distributable categories" such as "sentence, filler, interjection, idiomatic expression, quotation, tag" are subsumed under "extrasentential". For Myers-Scotton (1993: 3f.) "intersentential" only refers to switching between "a whole sentence (or more than one sentence)", while switches at clause boundaries and within a clause are classified as "intrasentential". We will basically follow Myers-Scotton and classify switches of both finite and non-finite dependent clauses as intrasentential; switches

---

[1] For the nature of such so-called *Mischsprachen* and *Pseudo-Mischsprachen* and their importance for the study of Middle English cf. Benskin – Laing (1981).

between two independent clauses, however, will be defined as intersentential. This seems justified for two reasons: firstly, the syntactic relation between two coordinated independent clauses is clearly less close than that of a dependent clause to its matrix clause; secondly, the distinction between two single sentences and two paratactically coordinated independent clauses forming a single sentence is often impossible in medieval English. In other words, intrasentential switches will be defined here as switches between or within constituents of a sentence, including dependent ("embedded") clauses. Such intrasentential switches evidently require a greater degree of bilingual competence, since the bilingual speaker has to match the syntactic and semantic rules of two or more languages involved in the speech act.

2.3. A much discussed problem in studies of code-switching is the distinction between switching and borrowing, particularly in the case of single lexical items (see e.g. Myers-Scotton 1992: 31-37, 1993: 20f.; Poplack 1980: 585; Romaine 1995: 142-144). Social factors and frequency of occurrence have increasingly been used for differentiating between switching and borrowing (cf. Myers-Scotton 1992: 36f.; Pfaff 1979: 297-298; Poplack 1988: 220), but these criteria are extremely difficult to apply for older data. In the present paper single items are only classified as switches if they are not morphologically integrated and occur in a text which has also larger switched constituents. However, since the linguistic integration of a lexical item is continuous rather than discrete, we often find borderline cases, especially with French language material, where a decision is to a certain extent arbitrary.

2.4. Frequently one of the languages involved in a bi- or multilingual discourse has been regarded as the base (or "matrix") language, while the language of the switch is the embedded language. The criteria used for distinguishing these two categories comprise morpheme frequency, language of the verb, underlying word-order, etc., though none of these seems to be fully convincing for a hierarchical ordering of the languages (cf. the critical overview in Romaine 1995: 145-149). In spite of the growing importance of this differentiation in recent research, especially in Myers-Scotton's Matrix Language Frame Model (cf. e.g., Jake 1994; Myers-Scotton 1992, 1993), this distinction will not be made in the present paper, i.e., the languages used in mixed texts will be treated as functionally and hierarchically equal.

3. Multilingualism, code-switching and historical linguistics

The interest of historical linguistics in bilingualism has so far been largely restricted to those aspects of language contact which are relevant for the explanation of linguistic change (cf. also Romaine 1995: 7). However, there is a small number of studies which have dealt with the "mixing" of languages in older texts, both for English and for other languages. The following survey will briefly present some of the most important ones, without aiming at anything like completeness.

An early study is Sullivan's detailed analysis of *The Latin insertions and the macaronic verse in Piers Plowman* (1932), which classified the Latin material on the basis of syntactic criteria. In view of the early date of this study, it is hardly surprising that the author regards the mixing of English and Latin in Middle English texts as a "literary peculiarity" and speaks of the need for further research to get more insight into "the medieval literary vogue of interpolating Latin quotations or macaronic verbiage in vernacular verse" (Sullivan 1932: x). More recently, Langland's use of Latin insertions has been studied by Nolan (1985) and again by Machan (1994).

A pioneering work integrating the early literature on code-switching and bilingualism is Stolt's (1964) study of the mixing of Latin and German in Martin Luther's table talks from the 1530's, which have been preserved as a result of note-taking by some participants (for a review of Stolt's study cf. Penzl 1966). Luther, like many of his educated contemporaries, was clearly proficient in Latin as the language of religion and scholarship and extensively switched from German to Latin in his (rather informal) conversations with scholars and students. The talks constitute a unique record of "the actual speech of educated men of the period" (Penzl 1966: 814) and their code-switching behaviour between the vernacular, the "low" variety, and Latin as the "high" variety in a basically diglossic situation. It becomes evident from Stolt's study that German-Latin code-switching "was not considered careless or sloppy" (Penzl 1966: 816) in educated circles in early-sixteenth-century Germany, and it certainly was not the result of insufficient education. Apart from Luther's talks, which form the main part of her study, Stolt also analyses Swedish-Latin records from 1593 and two texts with Old High German-Latin switching by Notker and Williram. She concludes from these that the basic tendencies observed in Luther also apply to the other texts. Stolt's study not only provides careful syntactic analyses of the data, but also looks for explanations for switching (Stolt 1964: 252f.).

The widespread use and function of code-switching in a very different genre, namely in early English business texts, has been impressively shown in a number of papers by Wright (1992, 1994). In view of Wright's contribution to the present volume, her work will not be discussed here any further.

Hunt's detailed investigation of vernacular and Latin glossing of medieval Latin school texts (Hunt 1991) should also be mentioned here even if his material does not contain code-switching in the strict sense of the word. Its relevance for the present topic lies in the regular use and mixing of all three main languages, Latin, French and English, in the glosses of these manuscripts, "often with a degree of interchangeability which betokens a not inconsiderable linguistic virtuosity" (Hunt 1991: 16; cf. also 434f.).

Wenzel's (1994) introduction to his edition of late medieval macaronic sermons provides detailed information on the syntactic and textual structures of these sermons. He closely investigates the different patterns of mixing Latin and English, using some insights from modern code-switching research. His analysis is, however, not linguistic in the strict sense of the word.

The considerable functional range of code-switching in medieval drama is the topic of a recent paper by Diller (1997/1998). Diller shows, among other things, that Latin tended to be used in drama as the "divine" language in the religious sphere, though exceptions do occur.

Though these studies provide valuable information on various aspects of language mixing, they make at the same time clear how little we still know about this field.[2] Furthermore, we still lack comparative analyses of a variety of historical text types and genres which might enable us to gain deeper insight into the code-switching behaviour in older periods of English and other European languages.

4. Code-switching in medieval England

4.1. The linguistic situation in medieval Britain was certainly a rather complex one: Latin as the "high" variety of religion, scholarship and partly of literature fulfilled many functions of a modern standard language,

---

[2] At the time of revising this paper, a workshop on code-switching in the history of English is being prepared by the present author together with Laura Wright for the ESSE 2000 conference in Helsinki. Furthermore, a number of papers are in preparation or in print.

followed by French in regard to its level of prestige, while English, though clearly spoken by the great majority of people, was the "low" variety well into the thirteenth century. (Subsequent changes in the relative status and functions particularly of English and French will not be discussed here, but cf., e.g., Berndt (1965); Kibbee (1991); Short (1980)). Though there is agreement today that French was never adopted by the vast majority of the population, especially the lower classes (cf. Berndt 1965; Short 1980), bi- or even trilingualism was by no means an unusual achievement with the literate part of society (cf. Short 1980: 474, 478f.). With much of the higher and the educated middle ranks of society a kind of diglossic situation must have existed, a situation which clearly favours code-switching. Richter gives a number of cases where bi- or multilingual speakers are reported to have used different languages according to situation and participants (Richter 1979: 55f., 61ff., 69, 73, 83, *et passim*). His detailed study of the languages used by more than 200 witnesses from different social classes in the canonisation procedure of Thomas Cantilupe, bishop of Hereford, in 1307 (Richter 1979: part 4, esp. section 3.) shows that a surprisingly large number of these witnesses were able to use a more prestigious language than their L1 on this occasion, a clear case of the situational code-switching typical of a diglossic situation.[3] A different kind of evidence can be seen in the fact that medieval manuscripts frequently contain texts in Latin, French and/or English in no apparent order, which clearly points to the multilingualism of the scribes and the users of these manuscripts (e.g., London, BM Harley 2253, [1]14th c.). These facts together with the material presented above prove that there were "continuity and complex interrelations of three languages" (Hunt 1991: 16) at least throughout the twelfth and thirteenth centuries, but most likely even later. While the role of French became increasingly restricted to a small number of functions such as law, Latin remained the high variety in most of its functions throughout the Middle English period.

4.2. In spite of the obvious existence of code-switching as a discourse strategy in medieval England, at least in certain social groups and specific communicative situations, information on the syntactic types and communicative functions of early switching is still fragmentary. This has two main reasons. Firstly, the history of English has too often been written as the history of the literary language as evidenced by monolingual sources – possibly due to the "monolinguistic and anglocentric per-

---

[3] According to Richter, the statements of the witnesses were in general made in the language representing the highest level of their linguistic competence (Richter 1979: 176).

ceptions of British culture" (Laing — Williamson (eds.) 1994: 10). Even large modern computer corpora of historical English texts tend to neglect the non-English language material from mixed texts and often only indicate its omission (cf. the *Helsinki Corpus* and the *Innsbruck Corpus*, which only provide information such as "6 lines Latin omitted"). Secondly, there are obviously no linguistic data of spontaneous speech from the medieval period (transcriptions of court proceedings and similar communicative events are almost non-existent before the Early Modern English period). However, a number of texts from a variety of literary and non-literary text types and genres show the mixing of two or more languages, and thus could supply us with essential information on the types, functions and strategies of code-switching in early written texts. In particular, switching is found in: (1) mixed or "macaronic" poetry, (2) drama, (3) business accounts, (4) sermons, (5) legal texts, (6) medical texts, (7) letters and diaries. All these – except diaries – are found both in the Middle and Early Modern English periods. If we follow Labov's well-known postulate to "use the present to explain the past", i.e., take modern insight into code-switching behaviour into account, and compare the strategies employed in various older texts and text types, we may even arrive at some guesses on code-switching in earlier spoken communication.

4.3. The present paper will concentrate on medieval poems using more than one language within its text, a poetic form also known as macaronic poetry. In its narrow sense, the term "macaronic poem" refers to the combination of a vernacular root with a Latin inflexion in an otherwise Latin text (cf. Morgan 1872: ix). This clearly artificial and often ridiculed poetic form seems to date from the late fifteenth century, and was not uncommon in Europe in the following centuries (cf. Genthe 1836). Since language switching between root and inflexion does not resemble naturally occurring code-switching, such poems will not be discussed here. In a wider sense, however, the term "macaronic" has been applied to any poem regularly using more than one language, a form which is found in England – as well as in Germany (cf. Genthe 1836: 13ff.) – throughout the Middle Ages well into the modern period.[4] The literary aspects and the

---

[4] The first English examples date from the Old English period, where we find two poetic texts with systematic switching between Old English and Latin. Apart from the last few lines of the *Phoenix*, it is above all the so-called *Macaronic Poem* (or *Call to Prayer*, Cambridge, Corpus Christi College 201) which deserves to be mentioned in this context. Its 31 lines show a regular change between the Old English on-verse and the Latin off-verse, with regular alliteration; cf. the first two lines below:

Herbert Schendl

history of "macaronic poems" will not be discussed here any further, and the term will be avoided in the following as a literary one, which even in its wider definition would not cover all the material of the present corpus. Though the present text corpus consists of literary texts, the focus of this study is entirely on the linguistic aspects of code-switching, in particular on the structural switching patterns used, though their textual function will also be briefly touched upon.

It is likely that code-switching in poetry follows at least partly specific poetic conventions and cannot be fully equated with code-switching in speech. In spoken discourse a variety of extralinguistic factors motivate or trigger the switches, while the number of such variables is certainly smaller in written texts, particularly in poetry. Poetic switching is most likely a deliberate and conscious choice of the author, while code-switching in spontaneous conversation is often unconsciously carried out by the speaker (Gumperz 1982: 69); while spontaneous switching seems to be more frequent in informal situations (Gumperz 1982: 68), medieval poetry can only rarely be classified as informal, though it may be less formal than prose writings. Finally, it is clearly not the result of insufficient competence in one of the languages involved, but often presupposes a high degree of competence in all languages. This last factor, however, also applies to much of spontaneous switching in speech, especially to intrasentential switching (cf. Poplack 1980: 588, 615). But even if code-switching strategies in poetic and other written texts differ from those in speech, it is worth looking at this kind of written data from a purely linguistic point of view. These mixed texts constitute an important, though neglected text type, and deserve to be studied in their own right. There seems to be more linguistically relevant information in these texts than one would expect at first sight. It is not unlikely that some of the authors of such poems would have used the same strategies in spoken discourse.[5]

þænne gemiltsað þe, N., <u>mundum qui regit</u>,
'Then (he) shows mercy to you, N., who reigns the world,'
ðeoda þrymcyningc <u>thronum sedentem</u>
'the king of people, sitting on the throne.'

[5] For the late fifteenth/ early sixteenth century, Skelton provides some proof for this claim. Skelton freely mixed Latin and English in his poetry, for example in *Colin Clout*. But he evidently appreciated Latin as a "living" language and objected to the study of Greek because it could not be used in conversation, cf. Lewis (1954: 133f.).

5. Empirical study

5.1. Corpus

The corpus of the present study consists of 80 poems dating from the thirteenth to the fifteenth centuries, though the majority are from the fifteenth century. Longer verse pieces, such as *Piers Plowman* or Chaucer's longer poetry, have not yet been analysed systematically, though they provide a lot of relevant material (for *Piers Plowman* cf. Sullivan 1932). The total number of switches amounts to about 1,800.[6]

The majority of the poems, namely 67, switch between English and Latin; in 55 of these, the English material clearly predominates over the Latin, while in 12 pieces English and Latin are distributed equally; 4 poems switch between three languages, namely English, French and Latin; 5 switch between English and French; finally, another 4 poems show Latin-French switches.

5.2. Structural patterns

The first part of the analysis will investigate possible switch boundaries and structural patterns of switching. In poetry, we have to differentiate between metrical factors and syntactic ones, though the two types are certainly interdependent and cannot be neatly separated. However, it seems justified to assume that metrical considerations would not regularly override possible syntactic constraints, even if individual cases of "poetic freedom" may occur.

5.2.1. Metrical patterns

The following selected examples should illustrate both the different languages used in macaronic poems and the correlation between metrical patterns and switching boundaries. Neither here nor in the following discussion will the material be classified according to date or languages. It is obvious that these parameters will have to be taken into account in further research.

---

[6] Since no distinction between matrix and embedded language has been made, every change from one language to another and back again has been counted as a separate switch.

Herbert Schendl

The early fourteenth-century *On the times* (*Proverbia trifaria*) switches in a very artistic way between the three languages Latin, French and English, see the extract under (1). In the first two lines, the first half line is in French, the second in Latin. In the following two lines, Latin comes first, followed by English.[7] Lines 5 and 6 change language from one half line to the next. All lines show both internal rhyme and end rhyme.

(1) *Quant houme deit parleir, videat que verba loquatur;*
 'When a man has to speak, let him take heed what words he utters;'
 Sen covent aver, ne stulcior inveniatur.
 'It is necessary to have understanding, lest one be considered more foolish.'
 Quando quis loquitur, bote resoun reste þerynne,
 'When anyone speaks, unless there be reason in it,'
 Derisum patitur, and lutel so shall he wynne.
 'He meets with scorn, and so he shall gain little.'
 En seynt' eglise sunt multi sepe priores;
 'In holy church many are often superior in position;'
 Summe beoþ wyse, multi sunt inferiores.
 'Some are wise, many are inferior.'
 (*On the times* (London, BM Royal 12 C xii ("R"), [1]14th c.); French-Latin-English)

A simpler pattern is provided by the regular switches from one line to the next, either between the three languages as in (2), or between two of them as in (3):

(2) *Soyez permenant et leal!*
 'Be constant and faithful!'
 Loue me so þat I it fele,
 'Love me so that I feel it,'
 Requiro.
 'I ask (you).'
 (*De amico ad amicam* (Cambridge, Univ. Gg. IV.27, [1]15th c.); French-English-Latin)

---

[7] For a discussion of the metrical pattern, including a different analysis of the metrical form cf. Aspin (1953: 160).

(3) Jhesus, almyghty Kyng of Blys,
Assumpsit carnem virginis.
'Took flesh by the Virgin.'
As Holy Kyrke makys mynd,
Intravit ventris thalamum,
'He entered the chamber of her womb,'
Fro heyuyn to erthe to saue monkynd,
Pater misit Filium.
'The Father sent his Son.'
(*Nativity* (Edinburgh, Advocates 19.3.1, ²15th c.); English-Latin)

Example (4), *On the King's breaking of the Magna Charta,* is a late thirteenth-century instance of French (Fr.)-English (E) switching. Here one language is maintained over two lines, the pattern being Fr.-Fr.//E-E//Fr.-Fr./E-E, etc.

(4) *Rome poet fere e defere,*
'Rome can do and undo,'
Si fet ele trop sovent;
'She acts thus full often;'
þat nis noþer wel ne veyre,
'That is neither good nor becoming,'
For þi is holy cherche ysend.
'For this holy church is put to shame.'
Merewele est de Deu vykere,
'Merewell is vicar of God,'
Ki a tel conseil consent.
'Who agrees to such counsel.'
þe man nis naʒt worþ þre eyre
'The man is not worth three eggs'
þat wel doþ and suþþe went.
'Who does right and afterwards changes.'
(*On the King's breaking of the Magna Charta* (Cambridge, St.John's College 112 ("J"), late 13th c.); English-French)

The next piece, the fourteenth-century poem *Against the king's taxes* (5), is an example of French-Latin switching. The first four lines of the five-line stanza form a quatrain, and switches occur after every half-line. The fifth line is in Latin. Without going into any phonological details, it is evident "that the accentual system of alternating stressed and unstressed syllables frequently does violence to the normal accentuation of French words in the whole poem" (Aspin 1953: 108).

(5) *Dieu, roy de magesté, ob personas trinas,*
'God, King of majesty, for the sake of the threefold persons,'
Nostre roy e sa meyné ne perire sinas.
'Let not our king and his household be undone.'
Grantz mals ly fist aver gravesque ruinas
'He brought great ills and grievous ruin on him'
Celi qe ly fist passer partes transmarinas.
'Whoever caused him to cross the seas.'
Rex ut salvetur falsis maledictio detur.
'May deceivers be accursed, so that the king may prosper.'
(*Against the king's taxes* (London, BM Harley 2253, [1]14th c.); French-Latin)

A rather frequent pattern is a Latin motto, burden and/or refrain, normally from a religious text (Bible, hymn, etc.) in an otherwise monolingual text, see example (6) with its refrain from the *Song of songs*, iv.8.

(6) *Ivy, chefe off treis it is,*
Veni, coronaberis.
'Come, you shall be crowned.'
The most worthye she is in towne –
He that seyth other do amysse –
And worthy to bere the crowne.
Veni, coronaberis.
(*In praise of ivy* (Oxford, Bodl. Lib. Eng. poet. e.1, [2]15th c.))

There are a number of subtypes of these main patterns, which will not be illustrated here. The essential point with the examples presented so far is that there is a clear correspondence between the placing of the switches and metrical considerations, which accounts for the regularity of the switching, e.g., in every second line or half-line, in the refrain, between title and main text, etc.

But there is also a number of poems where switching occurs irregularly – though not randomly – in an otherwise monolingual poem, see examples (30), (31). In many of these cases, a functional explanation can be found, see the discussion under 5.3.2.

5.2.2. Syntactic patterns

While the influence of metre on switching boundaries is obvious in the examples discussed above, syntactic factors have been claimed to play a major, possibly universal role in regard to possible switching patterns. The following discussion will look at the occurrence of the types introduced in 2.2., namely tag-like switches, inter- and intrasentential switches; in particular, however, the constituent structure of the intrasentential switches will be analysed in some detail.

5.2.2.1. Tag-like switches: there are neither tags nor tag-switches in the corpus. The nature of tags as rather freely moveable constituents makes it unlikely that there were any specific syntactic restrictions on tag-switching in Middle English or French. Their absence from the corpus may be pragmatically motivated, i.e., their rather informal nature may be less compatible with the poetic form or genre. However, there is a small number of interjections, all of them French phrases in an otherwise English utterance, which can be regarded as tag-like switches, cf. under (7):

(7) *Sire Emer de Valence, gentil knight and free,*
Habbeth y-suore oht that, par la grace Dée!
He wollith ous delyveren of that false contree, yef hii conne.
(*Execution of Sir Simon Fraser*, London, BM Harley 2253, ¹14th c.)

5.2.2.2. Intersentential switches (see section 2.2.) involve syntactically rather independent units and thus do not seem to present particular difficulties to the bilingual speaker.[8] Example (8) provides an instance of switching between two unconnected independent clauses, while in (9) the two are connected by a coordinating conjunction in the language of the second clause:

(8) *Vous estes ma morte et ma vye.*
'You are my death and my life.'
I praye you for youre curteisie

---

[8] This seems to contradict Romaine's statement that they require "greater fluency in both languages than tag switching since major portions of the utterance must conform to the rules of both languages" (1995: 123). However, Romaine also includes dependent clauses in the category of "intersentential" switches, for which this statement certainly applies; cf. the definitions discussed in section 2.2.

Amate!
'Love!'
(*Responsio,* Cambridge, Univ. Gg. IV.27, [1]15th c.)

(9) The shepeherdes ran to Bedleme than
Et invenerunt puerum.
'And found the boy.'
(James Ryman, *Now the Most High is born,* Cambridge, Univ. Lib. Ee. 1.12, end 15th c.)

Linguistically less interesting are the intersentential switches between the main body of a poem and its title, burden, motto or refrain, since here the syntactic and semantic relations (cohesion and coherence) between the utterances in the two languages are often weak, see the example under (10), which celebrates the English victory at Agincourt:

(10) *Deo gracias, Anglia,*
'Thanks to God, England,'
Redde pro victoria.
'Return for victory.'
Owre kynge went forth to Normandy
With grace and myght of chyualry.
Ther God for hym wrought mervelusly,
Wherfore Englonde may calle and cry.
'*Deo gracias*'.
(*A carol of Agincourt,* Oxford, Bodl. Lib. Arch. Selden B. 26, middle 15th c.)

The present corpus does not contain any examples of longer inserted passages, which are, however, frequent in Middle English verse pieces like *Piers Plowman* (for examples see Sullivan 1932: 36ff.).

5.2.2.3. Intrasentential switches have been defined here as switches between or within constituents of a sentence, including dependent ("embedded") clauses (cf. section 2.2.). The attempt to discover possible syntactic surface-structure constraints on this type of code-switching and on the syntactic constituents involved in switches was a central concern of much research in the 1970s and 1980s (cf. Timm 1975; Poplack 1980), and both language specific and more general constraints have been proposed (for a survey cf. Romaine 1995: 125-130). For most of the language-specific constraints counter-examples have been found and extensively discussed

Code-switching in medieval English poetry

in the literature; in spite of this, there seems to be a clear preference for certain patterns, while others are only rarely attested and often judged as unacceptable by informants.[9] In the case of corpus languages such syntactic constraints on switching are obviously impossible to discover, but the establishment of a frequency hierarchy of specific switches in the corpus should provide interesting results to be compared with statistics available for living languages.[10]

Intrasentential switches are widely used in the poems of the corpus and are more frequent than the other two syntactic types. They occur both between and within the major sentence constituents, such as NP and VP.

Noun phrase: there are no switches involving a pronominal subject or object in the corpus. The absence of this type of switch may reflect the strongest of the five syntactic constraints on switching established by Timm (1975: 477-480), though there are clear counterexamples even of this constraint (cf. Sankoff — Poplack 1981); pronominal switches of the type *íh consentiebam* have also been found in Old High German-Latin mixed texts, cf. Stolt (1964: 279).

Unambiguous single noun switches are less frequent in the present corpus than one would expect from modern research (cf. the results of Poplack's study, 1980: 603).[11] There are a few cases of single switched nouns in subject or object position, mainly personifications, cf. (11):

(11) a. <u>Tempus</u> ys come falshede to dystroy. (*Ballade set on the gates of Canterbury*, John Speed Davies MS, middle 15th c.)
 b. <u>Nego</u> is pouer clark in store. (*The song of Nego*, London, BM Harley 913, [1]14th c.)

Switches involving subject or object NPs are less rare when the NP is syntactically complex, cf. (12):

(12) a. *Jhesus, almyghty Kyng of Blys,*

---

[9] Constraints based on acceptability judgements by informants on made-up sentences (as, e.g., in Timm 1975) have, however, also been criticised on theoretical grounds. But even natural examples of switching are not always judged as grammatical by informants, as Myers-Scotton and Jake have shown in a paper delivered at NWAVE 28, Toronto 1999.
[10] For such a comparative statistical analysis of switches in a number of medieval and modern texts see Schendl (forthc.).
[11] For the difficult distinction between switching and borrowing see section 2.3. In Middle and Early Modern English business texts, on the other hand, the most common basic units of switching are words and morphemes, cf. the work by Wright referred to in section 3.

319

<u>Assumpsit carnem virginis</u>.
'Took flesh by the Virgin.'
(*Nativity*, Edinburgh, Advocates 19.3.1, [2]15th c.) (= example (3))

b. <u>Dieu, roy de magesté, ob personas trinas</u>,
'God, King of majesty, for the sake of the threefold persons,'
<u>Nostre roy e sa meyné ne perire sinas</u>.
'Our king and his household let not be undone.'
(*Against the King's taxes*, London, BM Harley 2253, [1]14th c.) (= example (5))

c. *Mayden moder milde*
<u>Oiez cel oreysoun</u>!
'Hear this prayer!'
(*Oreysoun*, London, BM Harley 2253, [1]14th c.)

Switches within the NP are attested a number of times between an English noun and a postmodifying Latin element; the postmodifier is a genitive (13), a non-finite clause (14) or a finite relative clause (15).

(13)   *Þuster nyth, and comth þe day*
'Dark night, and comes the day'
<u>Salutis;</u>
'Of salvation;'
Þe welle springeth ut of þe
<u>Uirtutis.</u>
'Of virtue.'
(*Hymn to Mary*, London, BM Egerton 613, middle 13th c.)

(14)   *A sterne forth ladde þeis kyngis all,*
<u>Inquirentes dominum.</u>
'Seeking for the Lord.'
(*Nativity*, Edinburgh, Advocates 19.3.1; [2]15th c.)

(15) a. *Drede ye nothing, grete joy I bringe,*
<u>Quod erit omni populo.</u>
'Which shall be for all men.'
(James Ryman, *Now the Most High is born*, Cambridge, Univ. Lib. Ee. 1.12, end 15th c.)

b. *he hus all to heuyn bryng,*

Code-switching in medieval English poetry

Qui mortem Cruce voluit.
'Who willed his death on the Cross.'
(*Nativity*, Edinburgh, Advocates 19.3.1; ²15th c.)

However, there are no unambiguous instances of switches between an attributive adjective and a noun. *Deifere* 'God-bearer', a designation for Mary, is listed in Latham – Howlett as a late Latin formation (1975: *s.v.*), but not recorded in the *MED*; nevertheless, it might also be a borrowing into Middle English religious language. The post-position of the modifying adjective is not infrequent in this particular poem.[12]

(16)   '*O! Deifere* delicate and doghter divine,
       Mother of mercy and meiden mellefluous,'
       (London, BM Addit. MS 20059, 1 c.)

The position of the object noun phrase can also be filled by a switched dependent object clause; in (17a) the English clause is dependent on a Latin finite verb, in (17b) on a French one:

(17) a.   *Ysayas cecinit*
          'Isaiah sang'
          Þat a chylde schalle be borne.
          (*A maid hath borne* ..., Oxford, Ashmole 189, 15th c.)
     b.   *Sachez bien, pleysant et beele*,
          'Know well, pleasant and fair,'
          That I am right in good heele.
          (*De amico ad amicam*, Cambridge, Univ. Gg. IV.27, ¹15th c.)

Verb phrase: there are a few cases where the switched VP is a single Latin finite verb, cf. the examples under (18); in (18a) both the preceding subject noun phrase and the following adverbials are in English, in (18b) the Latin verb is preceded by the French subject noun phrase and two English adverbial phrases:

(18) a.   Aungellys *EXALTANT*, bothe lowde and hih,

---

[12] In non-poetic language, the pre-position of the English adjective does not correspond to that in Latin and French, where the adjective is frequently placed after the noun; this would lead to the violation of the "equivalence constraint", which claims that code-switches "tend to occur at points in discourse where juxtaposition of L1 and L2 elements does not violate a syntactic rule of either language (Poplack 1980: 586; cf. also 581)."

(*Coronation of the Virgin II*, London, BM Harley 2255, 15th c.)

b. *Ma tresduce et tresamé*
    'My most sweet and most beloved'
    Night and day for loue of þee
    Suspiro!
    'I sigh!'
    (*De amico ad amicam*, Cambridge, Univ. Gg. IV.27, [1]15th c.)

Within the verb phrase, switches occur between the auxiliary and the infinite verb (cf. (19)), though this is only rarely attested (3 instances); this low occurrence again corresponds to constraints proposed for modern languages (cf. Sankoff — Poplack 1981; Timm 1975):

(19) *I wil in time whan I may*
      Venire.
      'Come.'
      (*Responsio*, Cambridge, Univ. Gg. IV.27, [1]15th c.)

Equally rare are switches between a finite verb and its infinite complement, see (20):

(20) *The Sinagoge hath hit in memorye,*
      'The Synagogue remembers it,'
      Yyt neuer he lynneth maliciusly
      'But, wickedly, never ceases'
      Esse ceca.
      'To be blind.'
      (Oxford, Bodl. Lib. Arch. Selden B. 26, middle 15th c.)

The greatest number of intrasententially switched constituents, however, are adverbials. In this function a variety of English and Latin constructions occur. Latin and English prepositional phrases are illustrated under (21):

(21) a. *He loste the blysse of paradys,*
         Pro sua superbia.
         'For his pride.'
         (London, BM Sloane 2593, [1]15th c.)

     b. *To London fro Kent*

<u>Sunt predia depopulantes</u>.
'Are the estates depopulated.'
(*Rebellion of Jack Straw*, 1381, Cambridge, Corpus Christi College 369, 14/15th c.)

c. *Of Mary mylde Cryste wolde be borne,*
<u>Sine virili semine</u>.
'Without a man's seed.'
(*Nativity*, Edinburgh, Advocates 19.3.1; ²15th c.)

In general, the whole prepositional phrase is in one language; switches within the prepositional phrase, i.e. after the preposition, only occur when the dependent constituent is a quotation, a title, etc., cf. (22). This would be in accordance with a claim made for modern languages that prepositions "tend strongly to remain in the language of the head element on which they depend" (Sankoff – Poplack 1981: 34; cf. also Timm 1975: 480; for a different view see Pfaff 1979).

(22) *Aske foryeuenes of thi trespas,*
With <u>Parce mihi, Domine</u>.
'With Spare me, o Lord.'
(Oxford, Bodl. Lib. Eng. poet. e.1, ²15th c.)

Besides prepositional phrases, Latin adverbial phrases in an oblique case occur, cf. (23):

(23) a. *Loue me well er I deye*
<u>Dolore</u>.
'For grief.'
(<u>De amico ad amicam</u>, Cambridge, Univ. Gg. IV.27, ¹15th c.)
b. *To saue monkynd, that was forlorne*
<u>Prime parentis crimine</u>.
'By the offence of our first parent.'
(*Nativity*, Edinburgh, Advocates 19.3.1, ²15th c.)

Equally well attested are both Latin and English finite adverbial clauses (24) and Latin non-finite adverbial clauses (25); the latter are functionally not always clearly distinguishable from postmodifiers:

(24) *When in suche wise founde him they had*

Ut dictum est per angelum,
'As was said by the angel,'
Ayene they came, being full glad.
(James Ryman, *Now the Most High is born*, Cambridge, Univ. Lib. Ee. 1.12, end 15th c.)

(25)   To seche that chylde thei toke tho wey,
Portantes sibi munera.
'Carrying presents for him.'
(*Nativity*, Edinburgh, Advocates 19.3.1, ²15th c.)

The isolated examples given above do not always adequately reflect the often very complex switching patterns, cf. the example under (26), where we find multiple switching in a single line:

(26)   *Regnum Anglorum regnum Dei est,*
'The kingdom of the angels/the English is the kingdom of God'
As the Aungelle to seynt Edward dede wyttenesse.
Now *regnum Sathane*, it semethe, *reputat best.*
'Now the kingdom of Satan, it seems, accounts best.'
For *filii scelerati* haue broughte it in dystresse.
'For the accursed sons have brought it in distress.'
(*Ballade set on the gates of Canterbury*, John Speed Davies MS, middle 15th c.)

The above discussion of syntactic switching patterns is in no way complete, but it should have illustrated the wide range of constructions involved in poetic switching. The investigation of both the relative frequencies of the various types and of the validity of more general constraints like the equivalence constraint seem rewarding topics for further research (cf. Schendl forthc.).

5.3. Functional aspects

5.3.1. General remarks

There is general agreement today that, even if there should be universal syntactic constraints on switching, functional factors are fundamental for an explanation of code-switching strategies. Much recent research has focused on the discourse functions and the pragmatics of switching (cf. Romaine 1995: 161ff.), aspects which take extralinguistic variables into account (Timm 1975: 474). Since these are often not applicable or recoverable in the case of early poetry, there are evidently limits to such a functional-pragmatic approach with older written data: firstly, poetic texts are in most cases monologues, not conversations between actual speakers; secondly, we frequently have no information on extralinguistic factors such as author, addressee, situation, event, etc., factors which clearly influence switching behaviour; thirdly, there are general functions of poetic switching which may override more specific functions. In spite of these limitations, certain textual functions of code-switching can be detected also in our material.

A basic function of switching in poetry is a poetic or artistic one. The conventionality of this form and the author's pleasure in playing with language are aspects which must have had considerable influence on switching patterns. This is particularly likely in the case of regular alternation between languages, as in the examples quoted under (1) to (5). Here traditional metrical patterns and poetic forms obviously influence the location of switches. There is also evidence that such poems were sometimes written by students and monks as a sort of poetic practice. Another general function of Latin-English switching in medieval poetry may have been to indicate membership of the educated social class; i.e., switching may have been used for reasons of prestige, much in the same way as conversational switching can function as "an overall discourse MODE" (Poplack 1980: 614). Poplack (1980: 614) has emphasised that "[t]he very fact that a speaker makes alternate use of both codes, itself has interactional motivations and implications, beyond any particular effects of specific switches ... It is then the choice (or not) of this mode which is of significance to participants rather than the choice of switch points." (Poplack 1980: 614).

But poets evidently have used switches also for more specific functions within a poem, such as to set a specific scene or to make the extralinguistic context more obvious. There are a number of examples of this use in the corpus. Some of the types discussed below have been established for code-switching in living languages, cf. Gumperz (1982: 75-84).

Herbert Schendl

5.3.2. Quotations

Switching language for quotations is a frequent code-switching strategy (cf. Gumperz 1982: 75f.), and most switches in the corpus are quotations, especially from the Bible or from religious texts. They sometimes set the motto for an otherwise monolingual poem, or provide the refrain, etc. and thus make the intended intertextual relations more explicit, cf. (27):

(27) *Make we mery in this fest,*
 For <u>verbum caro factum est</u>.
 'For the word has become flesh.'
 Godes sonne for the loue of mane,
 Flesshe and blode of Mary he nam,
 As in the gospell seyth Sent Johan,
 <u>Verbum caro factum est</u>.
 (Oxford, Bodl. Lib. Eng. poet. e.1, [2]15th c.)

However, biblical quotations in Middle English poetry are also frequently in English (cf. Smyth 1911), and there are even intrasentential switches within quotations, cf. (28):

(28) *"<u>Omne regnum in se divisum</u>", sayethe dyuyne Scrypture,*
 'Every kingdom divided in itself,'
 "Shall be desolate";
 (*Ballade set on the gates of Canterbury*, John Speed Davies MS, middle 15th c.)

Sometimes the source of the quotation is explicitly stated, as in the biblical quotations under (28) and (29); in (29) the insertion *scripture seith thus* has a function similar to the editing phenomena in "flagged codeswitching" (cf. Poplack 1988: 229f.):

(29) *For love of us (scripture seith thus)*
 <u>Nunc natus est Altissimus</u>.
 'Now the Most High is born.'
 (James Ryman, *Now the Most High is born*, Cambridge, Univ. Lib. Ee. 1.12, end 15th c.)

Quotations can, however, also be used to set a certain scene or background, and thus imply information not explicitly expressed. In the following poem on *The death of the duke of Suffolk*, seven of Suffolk's ecclesiastical supporters are introduced and characterised by the first words of various parts of the Office of the Dead (i.e. psalms, antiphons, and responses), which they have to perform on the execution of the duke. (No translation of the Latin phrases will be provided, since these would not make much sense without the larger context.)

(30) *Bisschopes & lordes, as grete reson is,*
Monkes, chanons, prestes, & other clergie
Pray for this dukes soule þat it might come to blis,
and let neuer suych another come after this!
...
"Placebo", begynneth the bisshop of Herford.
"Dilexi, for myn auauncement", saith þe bisshop of Chestre.
"Heu mei", saiþ Salisbury, "this goth to ferre for the".

"Ad Dominum cum tribularer", saiþ þe abbot of Gloucestre.
"Dominus custodit", saiþ the abbot of Rouchestre.
"Levaui oculos", saiþ frere Stanbury, "Volaui".
"Si iniquitates", saiþ þe bisshop of Worcetre,
"For Iac Nape soule, de profundis clamaui".
(*The death of the duke of Suffolk* (London, BM Cotton Vesp. B.xvi, middle 15th c.))

A subtype of "quotations" are those instances of direct speech which are used to characterise a speaker as a member of a certain linguistic or social group. In the *Song on the Flemish resurrection*, for example, some French knights begin with a French sentence or phrase, but then switch over to English, cf. (31). Though this is not done consistently throughout the poem, it clearly sets the French knights and court apart from the Flemish rebels. The inconsistency of the language used in quotations corresponds to observations in living languages that quoted messages are not always reported in the original language.

(31) *"Sire Rauf Devel", sayth the Eorl of Boloyne,*
"Nus ne lerrum en vie chanoun ne moyne,
'We will not let alive chaplain nor monk,'
Wende we forth anon ritht withoute eny assoygne."
(London, BM Harley 2253, [1]14th c.)

327

Herbert Schendl

5.3.3. Reiteration

Reiteration, i.e., the translation into or paraphrasing in another code, is an attested function of switching in speech, either for the sake of clarification, or to "amplify and emphasize a message" (Gumperz 1982: 78). A number of examples of this discourse strategy are also found in the corpus, cf. under (26) and (32).

(32)   *The sunne þat euer shyneþ bryȝt,*
         the sterre þat euer yeueth his lyȝt
         <u>Semper clara.</u>
         'Always bright.'
         (Oxford, Bodl. Lib. Arch. Selden B. 26, middle 15th c.)

A considerably longer instance of reiteration is John Lydgate's *Advice to the several estates* (33), where the whole English poem is a free paraphrase of the preceding Latin one.

(33)   <u>Rex sine sapiencia, Episcopus sine doctrina.</u>
         <u>Dominus sine consilio. Mulier sine castitate.</u>
         <u>Miles sine probitate. Iudex sine iusticia.</u>
         <u>Diues sine elemosina. Populus sine lege.</u>
         <u>Senex sine religione. Seruus sine timore.</u>
         <u>Pauper superbus. Adolescens sine obediencia.</u>

         Goo forth, kyng, reule the by sapyence;
         Bysshop, be able to mynystre doctryne;
         Lord, to treu counceyle yeue audyence;
         Womanhed, to chastyte euer enclyne;
         Knyght, lete thy dedes worshyp determyne;
         Be rightuous Iuge, in sauying thy name;
         Ryche, doo almes, lest thou lese blys with shame.

         People, obeye your kyng and the lawe;
         Age, be thou ruled by good religion;
         True seruaunt, be dredfull & kepe the vnder awe;
         And thou, poure, fye presumpcyon;
         (de Worde, *The temple of glas*)

5.3.4. Interjections

The syntactic status of switched interjections as tag-like elements has already been mentioned in 2.2. and 5.2.2.1., but interjections can also fulfil specific discourse functions (cf. Gumperz 1982: 77f.). The few interjections in the corpus are French ones in an otherwise English text. Similar to the quotation in (31), they help to characterise a speaker, such as the French knight talking to the French king in *The song of the Flemish resurrection* in (34):

(34)   Tho suor the Eorl of Seint Poul, *Par la goule Dé!*
       We shule facche the rybaus wher thi wille be,
       Ant drawen hem with wilde hors out of the countré.
       (London, BM Harley 2253, [1]14th c.)

5.3.5. Word-play

Switches may function as word-plays or provide a kind of folk etymology for a preceding word (cf. also McClure — McClure 1988: 41f.). In a Latin poem from the middle of the thirteenth century (London, BM Harley 978), the names of four greedy Anglo-Norman brothers, *Robert, Richard, Gilbert* and *Gefrei* are "explained" by the phonetically similar French words and phrases *robbur*, *riche hard*, *gilur* 'guiler', and *jo frai* 'I will do'.

5.3.6. Rhyming function

In many cases the switch provides a rhyme, though this is hardly independent from the functions mentioned so far — at least from the analyst's point of view. The corpus provides instances of poems which rhyme only within the same language, as in (1), as well as cases of rhyming across languages, as in (4). In the two Old English macaronic poems (see note 4), the Old English on-verse and the Latin off-verse are linked by alliteration.

6. Conclusion

One of the aims of the present study has been to show that even consciously structured written texts such as poems provide interesting material for the study of code-switching behaviour in medieval England. If we take recent approaches to code-switching into account and do not restrict our analyses to a single text type and style, we may be able to reconstruct at least some aspects of earlier code-switching behaviour. The data analysed for this paper have provided some indication that switching behaviour and patterns show a surprising correspondence to tendencies observed in modern studies of code-switching, both in regard to structural and to functional criteria. A detailed study of all the mixed language data available from earlier periods and a critical comparison of the results from various text types with those of modern studies of code-switching will greatly further our knowledge of bilingual speech behaviour in medieval and early modern England. Such an approach should add an important dimension to the study of the history of English – a history which is much less monolingual than older research has often made us believe.

References

Archibald, Elizabeth
    1992   "Tradition and innovation in the macaronic poetry of Dunbar and Skelton", *Modern Language Quarterly* 53: 126-149.

Aspin, Isabel S. T.
    1953   *Anglo-Norman political songs.* (Anglo-Norman texts XI.) Oxford: Blackwell.

Benskin, Michael — Margaret Laing
    1981   "Translations and *Mischsprachen* in Middle English manuscripts", in: Michael Benskin — Michael Louis Samuels (eds.), *So meny people longages and tonges: philological essays in Scots and medieval English presented to Angus McIntosh.* Edinburgh: Middle English Dialect Project, 55-106.

Berk-Seligson, Susan
    1986   "Linguistic constraints on intrasentential code-switching: a study of Spanish/Hebrew bilingualism", *Language in Society* 15: 313-348.

Berndt, Rolf
    1965   "The linguistic situation in England from the Norman Conquest to the loss of Normandy (1066-1204)", *Pragensia Philologica* 8: 145-163.

Clyne, Michael
1987 "Constraints on code-switching: how universal are they?", *Linguistics* 25: 739-764.
Delepierre, Octoave
1852 *Macaronéa ou mélange de littérature macaronique des différents peuples de l'Europe.* Paris: Gancia.
Diller, Hans-Jürgen
1997/1998 "Code-switching in medieval English drama", *Comparative Drama* 31: 506-537.
Genthe, F. W.
1836 *Geschichte der Macaronischen Poesie, und Sammlung ihrer vorzüglichsten Denkmale.* Leipzig: Meissner.
Gumperz, John J.
1982 *Discourse strategies.* Cambridge: Cambridge University Press.
Harvey, Carol J.
1989 "Intertextuality in the Anglo-Norman lyric", *Journal of the Rocky Mountain Medieval and Renaissance Association* 10: 17-28.
Heller, Monica (ed.)
1988 *Codeswitching: anthropological and sociolinguistic perspectives.* Berlin — New York: Mouton de Gruyter.
Hunt, Tony
1991 *Teaching and learning Latin in thirteenth-century England.* 3 vols. Cambridge: Brewer.
Jacobson, Rodolfo (ed.)
1990 *Codeswitching as a worldwide phenomenon.* (American University Studies 13, 11.) New York: Lang.
Jake, Janice L.
1994 "Intrasentential code switching and pronouns: on the categorical status of functional elements", *Linguistics* 32: 271-298.
Kibbee, Douglas A.
1991 *For to speke Frenche trewely. The French language in England, 1000-1600: its status, description and instruction.* (Amsterdam Studies in the Theory and History of Linguistic Science 3, 60.) Amsterdam: Benjamins.

Laing, Margaret — Keith Williamson (eds.)
  1994   *Speaking in our tongues: proceedings of a colloquium on medieval dialectology and related disciplines.* Cambridge: Brewer.
Latham, Ronald E. — David R. Howlett
  1975 -   *Dictionary of medieval Latin from British sources.* Oxford: Oxford University Press.
Lazzarini, L.
  1982   "Aux origines du macaronique", *Revue des Langues Romanes* 86: 11-33.
Lewis, Clive Staples
  1954   *English literature in the sixteenth century excluding drama.* (The Oxford History of English Literature 3.) Oxford: Clarendon.
Machan, Tim William
  1994   "Language contact in *Piers Plowman*", *Speculum* 69: 359-385.
Mahootian, Shahrzad
  1996   "Codeswitching and universal constraints: evidence from Farsi/English", *World Englishes* 15: 377-384.
McClure, Erica — Malcolm McClure
  1988   "Macro- and micro-sociolinguistic dimensions of code-switching in Vingard", in: Monica Heller (ed.), 25-51.
Morgan, James A.
  1872   *Macaronic poetry.* New York: Hurd and Houghton.
Myers-Scotton, Carol M.
  1988   "Codeswitching as indexical of social negotiation", in: Monica Heller (ed.), 151-186.
  1992   "Comparing code-switching and borrowing", in: Carol M. Eastman (ed.), *Codeswitching.* Clevedon: Multilingual Matters, 19-37.
  1993   *Duelling languages: grammatical structure in code-switching.* Oxford: Oxford University Press.
  1995   "A lexically based model of code-switching", in: Lesley Milroy — Pieter Muysken (eds.), *One speaker, two languages: cross-disciplinary perspectives on code-switching.* Cambridge: Cambridge University Press, 233-256.

Nolan, P. E.
1985 "Beyond macaronic: embedded Latin in Dante and Langland", in: Richard J. Schoeck (ed.), *Acta Conventus Neo-Latini Bononiensis.* (Medieval and Renaissance Texts and Studies 37.) Binghamton, N. Y.: Center for Medieval and Renaissance Studies, State University of New York, 539-548.

Penzl, Herbert
1966 Review of Stolt 1964. *Language* 42: 814-817.

Pfaff, Carol W.
1979 "Constraints on language mixing: intrasentential code-switching and borrowing in Spanish/English", *Language* 55: 291-318.

Poplack, Shana
1980 "Sometimes I'll start a sentence in English y termino en español: toward a typology of code-switching", *Linguistics* 18: 581-618.
1988 "Contrasting patterns of code-switching in two communities", in: Monica Heller (ed.), 215-244.

Richter, Michael
1979 *Sprache und Gesellschaft im Mittelalter: Untersuchungen zur mündlichen Kommunikation in England von der Mitte des elften bis zum Beginn des vierzehnten Jahrhunderts.* (Monographien zur Geschichte des Mittelalters 18.) Stuttgart: Hiersemann.

Romaine, Suzanne
1995 *Bilingualism.* (2nd edition.) (Language in Society 13.) Oxford: Blackwell.

Rothwell, William
1991 "The missing link in English etymology: Anglo-French", *Medium Aevum* 60: 173-196.

Sankoff, David — Shana Poplack
1981 "A formal grammar for code-switching", *Papers in Linguistics* 14: 3-46.

Schendl, Herbert
1996 "Text types and code-switching in medieval and Early Modern English", *Vienna English Working PaperS (VIEWS)* 5: 50-62.

1997	" "To London fro Kent / Sunt predia depopulantes": code-switching and medieval English macaronic poems", *Vienna English Working PaperS (VIEWS)* 6: 52-66.
forthcoming	"Syntactic constraints on code-switching in medieval texts", in: Matti Rissanen et al. (eds.), *Papers from the ICEHL, Helsinki 1996.* Berlin: Mouton de Gruyter.
2000	"Linguistic aspects of code-switching in medieval English texts", in: David A. Trotter (ed.), *Multilingualism in later medieval Britain.* Rochester: Boydell and Brewer, 77-92.

Short, Ian
1980	"On bilingualism in Anglo-Norman England", *Romance Philology* 33: 467-479.

Smyth, Mary W.
1911	*Biblical quotations in Middle English literature before 1350.* (Yale Studies in English 41.) New York: Holt.

Stolt, Birgit
1964	*Die Sprachmischung in Luthers Tischreden: Studien zum Problem der Zweisprachigkeit.* (Acta Universitatis Stockholmiensis, Stockholmer germanistische Forschungen 4.) Stockholm: Almqvist & Wiksell.

Sullivan, Sister Carmeline
1932	*The Latin insertions and the macaronic verse in Piers Plowman.* Washington, D. C.: Catholic University of America.

Timm, L. A.
1975	"Spanish-English code-switching: el porqué y how-not-to", *Romance Philology* 18: 473-82.

Wehrle, William O.
1933	*The macaronic hymn tradition in medieval English literature.* Washington, D. C.: The Catholic University of America.

Wenzel, Siegfried
1994	*Macaronic sermons: bilingualism and preaching in late-medieval England.* Ann Arbor, Mich.: University of Michigan Press.

Wright, Laura Charlotte
1992 "Macaronic writing in a London archive, 1380-1480", in: Matti Rissanen — Ossi Ihalainen — Terttu Nevalainen — Irma Taavitsainen (eds.), *History of Englishes: new methods and interpretations in historical linguistics.* (Topics in English Linguistics 10.) Berlin: Mouton de Gruyter, 762-70.
1994 "Early Modern London business English", in: Dieter Kastovsky (ed.), *Studies in Early Modern English.* (Topics in English Linguistics 13.) Berlin: Mouton de Gruyter, 449-65.

Zumthor, Paul
1960 "Un problème d'esthétique médiévale: l'utilisation poétique du bilinguisme", *Le Moyen Age* 66: 301-336, 561-594.

Robert P. Stockwell & Donka Minkova (Los Angeles)
The partial-contact origins of English pentameter verse: the Anglicization of an Italian model[1]

1. Introduction

By "partial contact origins" we do not mean quite the same as "partial learning" or "imperfect acquisition" as in almost all language learning at any significant remove from the cradle. By "partial contact origins" we intend to say that the donor system was understood pretty well, but constituents of the donor system, even though understood, failed to be embraced, consciously and by choice of the borrower. As far as we know, there is no established term in the literature for this sense that we intend our phrase to convey. Pieces of the system, we will claim, were borrowed and pieces were left out, but there is every reason to believe they were not left out through ignorance, which is what usually happens in borrowing. There were substantial differences between the prosodic systems of English and the Romance languages, differences of a type that made Chaucer's new focus on duple rhythm possible. Romance is overwhelmingly polysyllabic, while English, even in Chaucer's time, had a much higher proportion of actual or potential monosyllables. Iambic verse was in a sense "easier" in English because only the stressed syllables of polysyllables are metrically fixed: monosyllables go along in whichever way is required by convenient fit to the metrical template.

The issue of Chaucer's prosodic originality and the metrical components of his pentameter have received considerable attention lately; the

---

[1] We would like to thank our LACHE hosts, the editors of this volume, for organising the conference for which this paper was prepared. For useful questions and comments we are grateful to David Burnley, Hans-Jürgen Diller, Udo Fries, Manfred Markus. For especially careful and critical readings, we are much in debt to Steve Barney and Christopher McCully, though we absolve them of responsibility for the content, in particular the scansions, many of which they may disagree with. We gratefully acknowledge travel support from the UCLA Academic Senate Research Committee.

We won't repeat the complicated publication history of this volume; we want to point out, however, that the intervening five years of Chaucerian verse studies, ours and other people's, have forced substantial revisions of the original contribution, shifting the focus away from "partial-contact" and "origins" towards the linguistic components of English which enabled the transfer and Anglicization of the Continental model. The new focus accounts for our subtitle.

two studies that approach most closely our target are by Duffell (1996) and Youmans (1996). Duffell (1996: 218) pointed out that when Chaucer created iambic pentameter in English, he made one major innovative artistic cut against the grain of the contact system from which he derived it, namely to confine himself mostly to duple rhythm at the cost of triple rhythm. Dealing only with the mature and sophisticated pentameter lines of Chaucer's latest works, the *Troilus* (1382-1385), and the *Canterbury Tales* (1388-1400), we draw our historical background – the Romance progenitors – and therefore part of the "contact origins", from Duffell (1991, 1996, n.d.). For the English ingredients of the new entity, however, we differ both from Duffell and from Youmans. Youmans demonstrates convincingly that the principles of gradient metricality defined for Shakespeare's and Milton's verse are applicable to Chaucer's pentameter, and that in spite of the predominance of alternating stress patterns in Chaucer, a foot-based account of his meter covers more adequately the syntactic and stylistic variations of his pentameter line. Our focus is on elucidating the difference between metrical and rhythmic units in Chaucer and on the linguistic components of rhythmic variability. While Youmans's conclusions are fully compatible with our principles, his observations on scansion intersect ours only marginally. We differ from Duffell in that he does not discriminate between the notions of "triple rhythm" and "triplets" (= trisyllabic feet). He does not discuss the latter, only the former. We will argue that they are distinct notions, and we will claim that the Chaucerian metrists who allow occasional trisyllabic feet are mistaken – most of the famous metrists from Skeat (1894) and Ten Brink (1885 [1901]) to Schipper (1895), F. N. Robinson (1933 [1957]), Baum (1961), Gaylord (1976), and Barney (1993), but not including Smithers (1983), Bischoff (1897/1898), and Wild (1915), with whom we are in agreement.[2] The establishment position – but incorrect, in our opinion – is succinctly stated by F. N. Robinson (1933: xxxii) thus:

> Like most English poets, [Chaucer] not infrequently has an extra light syllable in a line (a trisyllabic foot in place of the regular iambus), though in such cases it is

---

[2] Youmans (1996: 186-187) dismisses the possibility of genuine anapestic foot substitutions within the iambic pentameter line because "it fails to account for the metrical constraints on extra syllables in these lines"; instead, he refers to the possibility of a "split position" analysis of sequences such as *Of a* in
   GP 364: Of a *solempne and a greet fraternytee*
He does not comment on either the frequency of such "split" positions, nor on constraints on them in Chaucer.

often impossible to determine whether to resort to apocopation. The extra syllable seems to have been most frequent in the caesural pause.

And Skeat (1894: lxxxviii):

[H]e freely accepted the principles of adding a syllable at the end of the line [on this there is no question, and everyone agrees – RS and DM] and at the end of the half-line. He also allowed himself to accept the principle of dropping the first syllable of the line.

Placement of an extra syllable just before the caesura is known as an "epic caesura" in the literature on the Romance progenitors of Chaucer. We find that in Chaucer's verse the examples are virtually all in the form of a final unstressed vowel followed by an initial unstressed vowel after the caesura, and that this fact strongly suggests that Chaucer actually avoided the epic caesura in favour of some sort of vowel merger to maintain the ten-syllable count. Even lines that seem to contain an epic caesura in some manuscripts, like this one from Hengwrt,

(a)   *We moste endure it; this is the short and playn.*
        (*KT* 1091, Hengwrt)

are commonly non-epic ordinary caesuras in other good manuscripts:

(b)   *We moste endure, this is the short and playn.*
        (*KT* 1091, Ellesmere)

In order to appreciate properly what Chaucer was doing in his borrowing and restructuring of the Romance decasyllabic model, it is absolutely crucial to discriminate between rhythmic triples and metrical triplets. Our reasons for viewing this distinction as being of central importance will gradually emerge: we emphasise the matter now in order to shape the forest which is easily obscured by underbrush.

We will discuss the pedigree of three separate components of Chaucer's iambic pentameter: (a) number of syllables, (b) caesuras and ictus, and (c) rhythmic grouping within the line – duple or triple or mixed. Terminologically, we shall distinguish between "meter" (by which we mean the abstract conceptual pattern of measured repetition of some phonological unit – number of syllables and relative prominence, or both) and "rhythm", by which we mean the actual periodicity and relative strength of beats that occur in a particular instantiation of a metrical pattern, i.e. a line of verse in a natural and unforced rendition. Meter is a design which exists independently of the properties of the linguistic material fitted into it, while rhythm is derived from the lexical and morphosyntactic properties of that linguistic material.

2. The basic template: syllabic count

Before Chaucer, there was no hendecasyllabic verse in English. There was proper syllabic verse, but it was predominantly octosyllabic, i.e. verse written in common time, everything divisible by two.[3] One point in which Chaucer is completely indebted to the Continental tradition is the syllabic count in his iambic pentameter: the syllable count must have been borrowed, because it did exist in France, Italy, Portugal, and Spain, and it did not exist in England prior to Chaucer. (1) represents the basic template of the French *décasyllabe*, the Portuguese *decasílabo*, the Italian *endecasillabo* and the Spanish *endecasílabo*, with the positions that count, counted:

(1) • • • 1 (•) -- • • • • • 1 (•) (•)
     1 2 3 4   5 6 7 8 9 10

Parenthesised syllables at the caesura and the line-end are extrametrical. French metrists indicate the presence or absence of the extrametrical (unstressed) syllables by distinguishing M(asculine) and F(eminine) verses. The template (1) is thus 4M/F + 6M/F. Since the M/F distinction is always extrametrical, the template may be called the 4+6 template of the decasyllabic line, or simply the X(I).[4] Except for caesural

---

[3] Even Orm's septenarius was common time, four feet + three feet + one foot, where the final foot was always a rest foot.
[4] This label, which is Duffell's, has certain advantages and may well become standard. It captures the central fact that the line must have exactly ten syllables, with a possible eleventh syllable, always weak, at the right edge; it ignores the position of the caesura, which became variable in Italian though not in French; and it ignores the extra syllable that can appear in Romance at the caesural juncture. The feminine caesura (the "epic caesura") does not appear in Chaucer, if – as we believe – Bischoff (1897/1898) was right; Smithers (1983) has correctly pointed out that Baum (1961: 21, footnote) erred in reading Bischoff as having asserted "that Caucer admitted no trisyllabic feet except at the so-called caesura"; scanned in accord with the X(I) template – and we agree – all trisyllabic feet disappear by virtue of the standard rules of elision, apocope, syncope, and synizesis (two vowels coalescing without forming a pre-existing diphthong). But triple rhythm

placement, the template is fundamentally the same, no matter the language.

3. Caesura and ictus in earlier verse: French vs. Italian

It has been persuasively argued in Duffell (n.d.) that the model that led to Chaucer's innovation – the pure five-beat iambic line – was Boccaccio's verse and not the French *vers de dix*. Duffell's argument is based primarily on freedom of caesural placement. He demonstrates that Chaucer "employs all Boccaccio's variants [of caesural placement], and his only debt specifically to France seems to be a few epic caesurae, borrowed from an earlier age. He also borrowed the Italians' favourite rhythm, because French lines have no rhythm within the line."

In the early French syllabic template, beginning with the *Vie de Saint Alexis* (shortly after 1020), only two syllables are ictic, bearing prosodic prominence by virtue of the verse template itself, the first one originally in position 4. The second ictic position was always in position 10. If medieval Romance verse is read with a modern French accent, these are the phrasal stress positions and the syntax must correlate accordingly. In hendecasyllabic verse all the other positions were, in principle, variably stressed, i.e. could be filled by weak or by strong syllables. Note well: it was not required, as part of the template, that these syllables alternate regularly between weak and strong. No one can be sure just when word stress was lost in French, but we are not satisfied that "word stress was much stronger in Old French than it is in the modern language" (Duffell 1991: 252). For this kind of verse to take root, lexical stress must have been ignored because it was very weak, or it was non-existent as in Modern French. In the early 4+6 template, among the non-ictic syllables after the caesura position 7 was commonly strong, i.e. coincided with a second phrasal stress. The most frequent original rhythm, therefore, was

(2)   de-de-de-DUM -- de-de-DUM de-de-DUM

---

does not disappear, because that depends on so-called "inversions", as we shall see in section 8.

Perhaps the most important detail to note about (2) is the presence of triple rhythmic groups after the caesura. This is extremely rare in Chaucer[5], though it occurs:

(3)   0   1- 0   1   - 0   0 1- 0   0 1
      And lif is lost -- but ye wol on me rewe   (TC 1.462)[6]

4. Notation of meter and rhythm

Our notation, in particular the bracketing that we have indicated by the single hyphens in the row above the words, is intended to represent the rhythm of the line. The meter, on the other hand, is always the same, namely

(0) 1 0 1 0 1 0 1 0 1 (0)

This is the metrical template. Since it is always the same we do not need to represent it overtly. We indicate the caesura with a double hyphen [--], writing it within the line of verse itself and often in the rhythmical notation, especially where it is crucial to the reading. The dash has a meaning in ordinary orthography which exactly fits the notion "caesura"

---

[5] Most potential examples, like this one, depend on the assumption that some weak monosyllables are weaker than others, discussed below under principle (e.iii.).
To avoid editorial interference we have taken all our *Canterbury Tales* examples from the Hengwrt manuscript (Ruggiers (ed.) 1979), except that we have replaced the letter thorn with <th>. (Of course, the Hengwrt scribe may have favored metrical regularities. If that be so, at least we did not select it for that reason; it is generally regarded as the best text. See Baker's defense of this view in Ruggiers ((ed.) 1979: xvii-xviii).) For *Troilus and Criseyde*, we have relied on the *Riverside* (Benson (ed.) 1983).

[6] This scansion is also Guthrie's (1988: 41), we infer (he does not actually scan it, but his discussion leaves no doubt). The only plausible alternative scansion strikes us as excessively forced:

0   1- 0   1   --   0   1 -0 1-   0 1
And lif is lost   but ye wol on me rewe   (TC 1.462)

By "forced" we mean: if we promote *ye* to ictus, then we also have to promote *on*. Thus we will have favored, for promotion, a pronoun and a preposition. To us it seems better to favor, in this string of three weakish syllables, the verb, hence our scansion. We discuss below this matter of what to do with three weak syllables in a row. A scansion with a single triple, contrastive emphasis on *ye* and *me*:

0   1- 0   1   --   0   1 -0   0-   1 1
And lif is lost   but ye wol on me rewe   (TC 1.462)

is marginally possible, though fourth-foot trochaic reversal is highly suspect, and we see no good reason to create stress clashes where not strongly required by the morphology and syntax.

and we therefore prefer it to arbitrary symbols like [|] or [/] which do not have a common language interpretation. We follow Duffell in using the notation [1] for [strong] and [0] for [weak]. Thus [1 0] notation is transparently [STRONG weak]. By other scholars it has been written [S w], [s w], [s x], [/ x], [/ ⌣], [x .], and no doubt still others. These can be used to notate feet: [1 0] = trochee, [0 1] = iamb, [0 0 1] = anapaest, [1 0 0] = dactyl, but whether [0 0 1] actually notates an anapaest (for example) in a particular instance depends on the metrical environment in which it is embedded. That is to say, there is a difference between an actual anapaestic foot, embedded in otherwise anapaestic lines, and the perceptual effect of a triplet, which would echo a single anapaestic foot in a non-anapaestic line. The effect may be achieved, for example, by trochaic reversal in the first foot: [01] + [01] may become [1 0] + [0 1], which may create an anapaestic rhythmic effect as though it were bracketed [1] + [0 0 1]. But an actual anapaestic foot, replacing an iamb in iambic pentameter, would necessarily move the final ictus to position 11, which does not happen in Chaucer's verse (we deal, below, with examples that have been claimed to have this property).

5. The insular components of the Continental model

We now turn to some specifically English elements in the tradition which originated with Chaucer. The home-grown elements of the template in (1) are, we claim, indispensable ingredients of the previously unattested iambic pentameter form. The ingredients are here expressed in terms of principles of line organisation and scansion; some of the principles listed below may be shared by other historical metrists, others not, hence the use of the first person pronoun.
(a) If there are 10 or 11 syllables, we take the tenth as always ictic, no exceptions.[7]

---

[7] The principle of an inviolable tenth syllable ictic realisation makes this position unique in the verse line in that it allows violations of the lexical stress of the word which appears in that position. The following lines from *The Wife of Bath's Prologue* illustrate the point:
  17: *Thou hast yhad fyve hóusbondes, – quod he,*
  18: *– And that ilke man that now hath thee*
  19: *Is noght thyn hóusbonde, – thus seyde he certeyn.*
  20: *What that he mente therby, I kan nat seyn;*
  29: *That gentil text kan I wel understonde.*
  30: *Eek wel I woot, he seyde myn housbónde*

(b) We do not allow more than 11 syllables and the eleventh must be weak. If there appear to be more than this number in the line, we omit from scansion excess weak syllables in accord with standard assumptions about elision, syncopation, apocopation of line-internal *-e*'s, and synizesis. Our practice is in substantial accord with that of Kökeritz in scanning Shakespeare, except that provable instances of trisyllabic feet are much more frequent in Shakespeare[8]: "Only when a line cannot be made to scan by the application of all known devices of word reduction are we entitled to consider the possibility of trisyllabic substitution" (Kökeritz 1961 [1969]: 224). We believe the meter respects most organic final *-e*'s, as nicely summarized, for example, in Barney (1993: 94-104), e.g. in infinitives, past participles, datives, monosyllabic weak adjectives; but occasional metrical deletion of organic *-e*'s is necessary.[9]

(c) If there appear to be only nine syllables, we try to find an appropriately realisable vowel or other syllabic unit to make a tenth. We of course allow headless lines, which are nine-syllable lines that conform to the decasyllabic template except that an initial weak syllable is missing. Headless lines must start with an unarguably strong syllable. So-called "Lydgatian lines" also have only nine syllables, but they start with an unarguably weak syllable, they have a stress clash usually across the caesura, and they are really bad lines. We would like to believe that Chaucer did not write them. This line from the *Troilus* (I.496) is scanned as indicated by Barney (1993: 109):

Such violations of the linguistic input in favor of the metrical template are not part of the borrowed model; they are a consequence of a well-understood principle of rhyming on stressed syllables, followed loyally by English versifiers from the time of the earliest specimens of rhymed verse.

[8] In the late plays especially, where there are lines like this from *Antony and Cleopatra*, cited by Kökeritz: "*Like to a Vagabond Flagge upon the Streame*" (1961: footnote 27). As Smithers (1983: 219) says, "[o]ne reason for doubting in practice that the author [Chaucer] used these two rhythmic patterns [either one, or two, successive weak syllables – RS and DM] is the fact that there are relatively so few secure examples of them."

[9] Elision of the infinitival final *-e* in hiatus is very common (e.g. *General Prologue* 34, 94, 95, 96, 106, 144 *et passim*), which suggests that the *-n*-less infinitive does not behave differently from other parts of speech. Barber – Barber (1991: 77) report that in the 382 infinitives they examined in non-elision environment, 40% lose their *-e*, while 60% keep it. By comparison, weak monosyllabic adjectives (445 examined) keep *-e* 84% of the time, 5% are ambiguous, and most of the 11% that lose the *-e* precede a noun that is not stressed on the initial syllable by virtue of its structure or position in the X(I), or both, e.g. *GP* 316 "*For his science and for his heigh renoun*".

(4) 0 1 - 0 1 -- 0 1 - 1 0 - 1 (0)
     Ne semed it -- that she of hym roughte

We would force the stress clash to straddle the caesura, but the resulting scansion is only slightly better:

(5) 0 1 - 0 1 -- 1 0 - 1 0 - 1 (0)

The fact is that Lydgatian lines are almost always emended (see Barney 1993: 111), a practice with which we are in full agreement. If not emended, they certainly must be discounted for the purpose of describing Chaucer's metrical system.[10]
(d) We let the strong syllables fall where they may, not demoting in violation of the regular stress-assignment rules of Middle English, except that more than a total of five strong syllables are never allowed (four indeed are possible, just as in the Shakespearean lines cited above). If appearances seem to suggest there might be more than five, something has to be demoted.
(e) We use the following three principles for the resolution of metrical ambiguity (i.e. instances where a scansion that violates the metrical template might be preferred):
(e.i) Lexical stress in polysyllables, especially disyllables, is inviolable. Thus words like *again, before, between, among, compel, rely* can only be scanned [0 1], and words like *battle, fancy, roses, angel* can only be scanned [1 0]. Where lexical stress is certain, it is never to be demoted in scansion and cannot be shifted for metrical convenience (but see (f) below for words that might have had variable stress, where therefore lexical stress is not certain).
(e.ii) Equally, a lexically weak stress cannot be promoted for metrical reasons except in rhyme position: this is the weak counterpart of (e.i). Together, principles (e.i.) and (e.ii.) are functions of the fact that stress in polysyllables is, in general, lexically fixed - almost totally, in native words, though somewhat more variable in recent French borrowings, in Chaucer's time.

---

[10] There is a long tradition of suspicion concerning the authenticity of Lydgatian lines, most colorfully expressed by Saintsbury (1923: 175, note): "I am perfectly certain that Chaucer's prosodic wit was never so thin, and his ear never so thick, as to write [My take is doon for my wit is thinne MT 438] ... [which is] easily mended with syllables that copyists were quite likely to slip, especially when Lydgate himself had misled them".

(e.iii) When no lexical stress is available to force the issue, there is always a basis for disagreement, and here English differs greatly from the Romance languages, where polysyllabicity is the norm, with regular and invariant stress assignment, and where ictus is usually determined by that fact (except, of course, in French after lexical stress was lost). Our view about scanning a language with many monosyllables - Chaucer's as well as Modern English - is that the best scansion is the one which assigns low prominence to the inherently weakest syllables, and promotes only those syllables to high prominence which are morphologically more salient. This is not easy to determine and may vary from case to case, but generally, we would rank function words lowest, with simple determiners (*a, the*) at the very bottom, monosyllabic prepositions and pronouns next, light auxiliaries (*be, wol*) next, monosyllabic conjunctions next (*and, or*), complementisers next (*for* in *for to*), and above that it doesn't seem to matter: heavy conjunctions, verbs, adjectives, nouns.

(f) Central to our practice in scanning polysyllabic words, including French loans, we assume that the stress accords with that of Modern English, wherever that reading is possible metrically.[11] There is a good argument to be made that by the second half of the fourteenth century disyllabic Romance nouns and most adjectives were stressed initially. This is especially true for nouns with a heavy first syllable: *auctor, fortune, garden, languour, mantel, martir(e), servise*. By that time the matching of syllable weight with stress in the native vocabulary was complete through vowel lengthening in open stressed syllables. Through early loss of final vowels in Anglo-Norman, as in *porpos(e)* 'purpose', the pattern of initial stress in disyllabic words was reinforced in the borrowed vocabulary. Some independent evidence is provided by sound change: in order to account for the semantic split later, we have to assume pronunciations of both *person* and *parson*, following the native pattern of *darling, harbour, harvest*. This change occurs only in stressed syllables where the *-r-* is tautosyllabic. So unless there is some overwhelming evidence to the contrary, or when there is indisputably trisyllabic realisation in rhyme, we count monomorphemic words of Romance origin in the same way we count native vocabulary. The following line from *The Miller's Tale* is a case in point:

(6)   1 - 0   0   1   -- 0   1 -   0   1 -   0   1 (0)
         Ialous he was, -- and heeld hire narwe in Cage (*MT* 3224)

---

[11] The arguments for this position are presented in Minkova (in press).

The partial-contact origins of English pentameter verse

One can insist on the (certainly possible) reading without the initial triple (i.e., without the initial trochaic inversion), but since this inversion, and this pronunciation, were both surely possible in Chaucer's time, why force this strict iambic reading:

(7)   0   1-   0   1   --   0   1-   0   1-   0   1
            Ialous he was, -- and heeld hire narwe in Cage (*MT* 3224)

Clearer still, if one is worried about the potential stress ambiguity of *Ialous*, are instances involving line-initial inversion of non-Romance single-root disyllables (the examples below are from *Troilus and Criseyde*):

(8)   *Liggyng abedde, and make hem for to grone;*   (*Tr.* I.915)

(9)   *Wolden as now do this honour to me,*   (*Tr.* II.1432)

(10)   *Stynteth right here, and softely yow pleye.*   (*Tr.* II.1729)

(11)   *After thiself next heried be she –*   (*Tr.* III.1256)

Some headless lines, which always start trochaic, remain trochaic throughout. Such lines are arguably best analysed as having a silent initial weak syllable.

(12)   (0)   1 - 0   1 -   0   1 - 0   1 - 0   1   (0)
           Heren noyse of reynes nor of thonder?   (*Tr.* III.662)

(g) We disallow, both metrically and rhythmically, sequences of three fully unstressed syllables anywhere in the line. We would have to, and do, scan lines such as:

(13)   *For al the tresour in the town of Troie* (*Tr.* III.874)
         I nolde setten at his sorwe a myte   (*Tr.* III. 900)
         But this thyng stant al in another kynde (*Tr.* III. 903)
         For I wol speke, or ellis go my wey.   (*CT* 3133)

347

as containing only duples.[12]

(h) To repeat a point which may already have been made sufficiently, we disallow all "trisyllabic" feet, even under the heading "slurring" (*Verschleifung*), whereby the final [1] would be forced into position 11. The term has been used ambiguously either (1) to mean "reduction to non-syllabic status" (Bischoff 1898: II.368, quoted in Smithers 1983: 230), which is fine by us: it means "elision" or "synizesis" under that reading; or (2) as Baum, Schipper and Skeat and most other metrists use it, to mean having full syllabic status, just, as it were, danced over lightly, a "compromise which hurries over one syllable" (Saintsbury 1923: 172).[13] This latter notion we view as psychologically unsound: if it is a syllable it has syllabic status. So while we do not object to the term "slur", a slur is always but one syllable, not two.

It should be clear that the bracketing we place above the line (indicated by hyphens) does not refer either to metrical template bracketing or to syntactic phrasing; it represents rhythmic grouping. Rhythmically, we believe, weak syllables are always proclitic to a strong syllable (barring a major caesura, which is syntactically determined). This is totally obvious in a phrase like, say, *bread on his plate*; we believe it is perceptually also true of a phrase like *working the crowd*: they are both [DUM – de de DUM]. The basis for this proclitic perception of the relation between duple meter and triple rhythm is probably the fact that the verse always ends [0 1], which forces the [1] in position 8 to seek, perceptually speaking, to the left for its partner(s); and so on back to the beginning of the line. Only the first [1], therefore, is stranded without a rhythmic partner in a trochaic reversal.

In their aggregate, the principles and constraints defined in (a)-(h) give a specific shape to the Continental decasyllabic line which is known as the "iambic pentameter". Some aspects of the borrowed model facilitated the introduction of these principles. The Italians (especially Boccaccio) had

---

[12] Our practice in scanning Chaucer differs, therefore, from some analyses of later verse, e.g. Tarlinskaja (1992: 173) considers the line
  To be the curtain of the inmost soul
from Frost, "The Fear of God" (13), as containing a mixture of disyllabic and trisyllabic feet. We would have to promote *of*, only a slight promotion if you will, but a reading without this promotion is in our opinion impossible, because it is neither duple nor triple rhythm within the phrase *curtain of the inmost*.
  0 1 0 1 0 0 0 1 0   1
  To be the curtain of the inmost soul

[13] Saintsbury writes that "with so excellent an English word as "slur", which exactly expresses the English practice, I can see no excuse for Synizesis (1923: 172, note)."

developed a unified line of ten syllables with a mobile caesura (Duffell 1991: 252). Their metrical template allows a much more flexible placement of the original two peaks, which puts in jeopardy the two-peak earlier template and paves the way for the iamb. In other words, if the first two peaks can come at positions 4, 5, 6, or 7, albeit in different lines, the model of a di- or tri-apical line (and salience of phrasal stress only) will easily be superseded by a model involving more peaks, matching not phrasal, but word stress. Gasparov (1980), quoted in Tarlinskaja (1987: 7-8), has shown that even early Italian verse has an iambic tendency, never less than 33%.[14]

6. On rhythmic duples and triples in the model and in English

The next question is, if the modified metrical template allows more than two ictus, and if the poet uses both word and phrasal stress to fill those ictic positions in a consistently decasyllabic verse, where do these stresses fall within the line? There were two extreme rhythmic possibilities: alternating stress as in (14)

(14)   0   1 - 0   1   --   0   1 - 0   1 - 0   1

which is duple time throughout; and

(15)   1 - 0   0   1   --   0   0   1 - 0   0   1

which is triple time throughout. A decasyllabic verse completely in triple time necessarily has only four stressed positions (since triples use up nine of the ten available positions). Early in the history of the Romance decasyllable it became possible to mix duple and triple time to produce what was called a *fragmento adónico*. The most common analogue of a rhythmically mixed line in Chaucer results from the familiar "trochaic reversal" in the first foot, which creates (rhythmically speaking) a triple. The traditional trochee + iamb analysis is given as (16), the rhythmic effect of the triple is given as (17):

(16)   1   0 - 0   1   --   0   1 - 0   1 - 0   1

(17)   1 - 0   0   1   --   0   1 - 0   1 - 0   1

---

[14] Pure syllabicity, on the other hand is only 4% - in 12 Italian poets from Dante to D'Annunzio (Saintsbury 1923: 172, note).

Duffell counts a string as duple time or triple time only if either rhythm recurs at least twice and preferably three times: thus [10101] and [1001001] are the minimal duple-defining and triple-defining sequences in his X(I) system. While we do not disagree with this as a basis for statistical generalisations, it is the matter of defining rhythmic variation and rhythmic redefinition across two languages, from Romance to English, that interests us here. The syllable-timed nature of the Romance languages will tend to blur and even conceal the effect of a single triple in the line. The reason why it is not necessary to have two consecutive occurrences of a duple or a triple to perceive and define rhythmic variation in a language like English is that English is roughly beat-isochronous: that is, strong stresses tend to be spaced equidistantly. The "speeding up" of syllable rates that goes with even a single triple among duples is rhythmically striking.[15] Since a triple rhythm is such a distinctive rhythmic variant from duple when it occurs even once, we count any trochaic substitution as a mixture of duple and triple, while Duffell would not. We therefore cannot compare his statistical generalisations with ours, though it appears that our definition roughly doubles the number of triples.

Elaborating further on point (b) in section 5.: our "triples" (and Duffell's) stay strictly within the X(I) syllabic count; they are to be distinguished from the "triplets" (tri-syllabic feet) commonly referred to in Chaucerian metrical studies, which editors allow – usually reluctantly and with reservations about textual corruption – as an additional weak syllable between ictus, creating a line with the fifth ictus on position 11. For example, within the context of a discussion of the syllabic count of the past participle of *come*, Barney (1993: 92-93) allows ... *come that* ... to be taken as an instantiation of a "triplet" in (18), arguing from the huge number of strong past participles that clearly have *-e(n)* compared with those that don't, as found in the manuscripts.

(18)   0   1   - 0   1 -   0   1 0 --   0   1 -   0   1 (0)
        For sith the day is come -- that I   shal dye   (KnT 1732)
        1   2   3   4   5 6 7   8   9   10 11 (12)

---

[15] Youmans (1996: 204-209) offers a convincing account of the syntactic and stylistic effects of metrically motivated inversions and concludes that though Chaucer was more sparing in their use, his verse obeys the same constraints on such inversion as can be found in Shakespeare and Milton.

The partial-contact origins of English pentameter verse

This produces a serious violation of the ten-syllable template. Indeed Barney asserts that the scansion above "may exemplify a recognised license in Chaucer's lines, an occasional "triplet" or extra light syllable between stresses" (Barney 1993: 93). His key argument depends on the undeniable fact that past participles frequently scan with the final *-e* pronounced. But sometimes they don't, as Barney recognises. For us *come* in the above line would have to be monosyllabic; the line cannot contain real triplets that increase the syllable count and force the final ictus to appear on position 11. That begins to happen much later in English verse (not really commonly until the nineteenth century); we believe it does not ever happen in Chaucer, properly scanned; and if it appears to be supported textually, then syncopation, elision, synizesis, or textual corruption is to be suspected.[16]

Below we illustrate these concepts with examples from Shakespeare and Milton, these examples having properties which differ superficially from Romance decasyllabic verse only in that Renaissance verse does not allow extrametricality within the verse.[17]

(19)   0   1- 0   1   --   1-   0   0   1- 0   1
        A wretched soul, -- bruised with ad ver si ty
                                                    (*Comedy of Errors*, II.1)

(20)   1- 0   0   1- 0   1   --   1- 0   0   1
        Be, as thy presence is, -- gracious and kind   (Sonnet 10)

(21)   1   0   0 1 0   --   1 0 1 0   1 0
            Holds in perfection -- but a little moment   (Sonnet 15)

---

[16] We are not convinced that Chaucer's metrical practice, abstracted away from errors in textual transmission, includes any lines that would be metrically parallel to modern examples with triplets (placing position X on the eleventh syllable) such as:
    0   1- 0 0   1- 0   0   1- 1 0   1
    Were really as dark as I hear sages tell ...
(Frost, "The Lesson of Today", quoted in Tarlinskaja 1992: 179). Such lines contain true anapestic feet replacing iambic feet. Indeed, this particular verse is so anapestic that probably the last three syllables should be scanned [0 0 1] to make a third anapest, with "sages" demoted, which also avoids an ugly stress clash.

[17] All the Italian examples are selected and scanned by Duffell. The English ones have been randomly selected by us from Shakespeare. There can certainly be some legitimate disagreements about particular scansions, but we feel confident that enough of them will withstand scrutiny to establish our main points.

(22) 1 - 0   0   1 - 0   --   0   1 - 0   0   1
      (OR [1 0   0   1 - 0   --   1   0   1   0   1])[18]
      Prouder than rustling -- in unpaid- for silk
      (Cymbeline 3.3)

(23) 1 -   0   0   1   - 0   1   --   1 - 0   0   1
      Brought on his way with joy; -- hee unobserv'd
      1 -   0   0   1 - 0   1   --   1 - 0   0 1
      Home to his Mothers house -- private return'd
      (Paradise Regained 638-639)

Below from Boccaccio's *Filostrato*, Book I, are lines 2, 4, and 6, chosen to illustrate triples:

(24)   0   0   0 1 0   0 1   0   0 1 (0)
         ne lor principii -- pietosi invocare

(25)   1   0   0 1 0   0   1 0   0 1 (0)
         io di Parnaso -- le Muse pregare

(26)   0 1 0 1   0   0   1 0   0 1 0
         novellamente -- m'ha fatto mutare

Italian verses show plenty of instances of duple alternating stress, of course. Boccaccio's verses, according to Duffell, are about 70% duple time. The proportions of duple to triple that Duffell cites for the Continental sources, and for Chaucer's contemporary Gower writing in French, are these (Duffell 1996: 217):

(27)   Boccaccio 70%   Thibaud 56%   Orleans 60%
       Petrarch 79%   Machaut 61%   Villon 48%
       Gower 78%   Froissart 60%   Ronsard 50%.

---

[18] The latter scansion is strict but somewhat forced; the scansion with three triples is the more natural reading.

7. From the Continent to Chaucer

Chaucer's first extended composition in decasyllabic verse, the *Parliament of Fowls*, written after his second trip to Italy in 1378, when he became familiar with Petrarch and Boccaccio, fully supports Duffell's claim that Chaucer invented[19] the iambic pentameter by the simple device of "excluding triple-time" rhythm in the close of the line. Iambic feet, which correspond to duple rhythm, are nothing new in the language at the time Chaucer started writing, of course: the Ormulum is monotonously and almost exceptionlessly duple. Chaucer's real originality lies in producing a new entity, the iambic pentameter, out of two familiar patterns: the iambic foot and the decasyllabic line.

Duffell's view is that Chaucer's verse is essentially constant, in the proportion of duples to triples, from the very beginning. Summing up all his sampling of Chaucer's decasyllabic verse, his figure is that Chaucer had only 3% triple rhythm. Even with a much less limiting definition of "triple", we have found only about 6%. Chaucer thus went far beyond his Italian models in his duplicity.

8. On duples and triples in Chaucer

What are the possibilities of "duple" vs. "triple" - or rather, where do the triple rhythmic groupings commonly show up? As has often been recognised, they are most likely to appear to the left of the caesura; most frequently the triple rhythmic effect is the result of trochaic reversal at the left edge of the line.

(28)   1 - 0   0   1   -- 0   1 - 0   1 -   0 1
        Whit was hir smok,   and broyden al   bifoore
                                            (*Miller's Tale* 3238)

(29)   1 - 0   0   1   -- 0   1 -   0   1 - 0   1
        Pipen he koude   and fisshe and nettes beete
                                            (*Reeve's Tale* 3927)

---

[19] According to Saintsbury (1923: 161-162) there might be some slender evidence that Chaucer did not invent the decasyllabic line in English. We do not think that the appearance of occasional earlier examples of decasyllables within predominantly octosyllabic verse damages in any way the claim that Chaucer was the originator of the iambic pentameter in English, and anyway, in Barney's nice phrase (personal communication), Chaucer "was certainly the founder of it in deed".

If the verse contains two triples, it necessarily has only four strong syllables, and the triple may cross the caesural boundary:[20]

(30) 0   1 -   0   0   1 - 0   0   1 -   0 1
      With hym ther was dwellynge -- a poure Scoler[21]
      (*Miller's Tale* 3190)

(31)   1 -   0   0   1   0   1 -   0   0   1
      (OR [1   0   0   1   0 --   1   0   1   0   1])[22]
      Which that he louede -- moore than his lyf
      (*Miller's Tale* 3222)

In the *General Prologue* (858 lines), there are 52 solid triples[23], most often at the left edge and lexically fixed, e.g.

(32) 1 - 0   0   1 - 0 -- 1 - 0   1 - 0   1
      Redy to weenden -- on my pilgrymage   (*Prologue* 21)

---

[20] The question of whether the existence of lines with two triples justifies the description of Chaucer's verse as a four-stress meter – the so-called "four-beat heresy" stated perhaps best by Lewis (1938) and defended by Southworth (1954) and Frye (1957) – is outside the scope of this paper, but we should note that the obvious origin of this view is indeed the existence of triple rhythm in duple-based verse, since two triples in a verse necessarily leaves room for only four beats.

[21] The rime is with *carpenter*. This fact forces the scansion as indicated. One more clarification related to our principle (d) should be added. Adjacent non-lexical items are assigned prominence on the basis of context and relative communicative weight. We promote *of* in:

Ne of *his wo ne dorste he nat byginne* (*Tr.* I.503)

but we count it as the first syllable of a triple in:

Or *feynede hire she nyste, oon* of *the tweye* (*Tr.* I.494)

It will be unstressed in:

And *lat me sterve, unknowe,* of *my destresse* (*Tr.* I.616)

[22] Since there is nothing metrically extraordinary about a headless line, or about a monosyllabic <moore>, we prefer the scansion with two triples because it preserves the prosodic prominence of *moore* over *than*.

[23] Lines 21, 46, 91, 93, 96, 99, 105, 106, 114, 125, 136, 145, 166, 195, 250, 275, 281, 307, 314, 324 (Ellesmere reading), 347, 352, 358, 393, 405, 421, 440, 447 (?), 458, 489, 491, 532, 541, 556, 579, 603, 621, 658, 680, 687, 745, 750, 755, 764 (?), 774, 788, 790, 798, 800, 833.

(33)   1 - 0   0   1   - 0   1 -   0   1 - 0 1
          Vnder   his   belt   he   bar   ful   thriftily   (*Prologue* 105)

(34)   1 - 0   0   1   -- 0   1 -   0 1 - 0   1
          Hardy   he   was   and   wys   to   vndertake   (*Prologue* 405)

While our conservative count shows triples appearing about twice as often as by Duffell's criteria, it is still the case that the overall figure of triples is not impressive. The same is true of our sampling of *The Knight's Tale*, *The Wife of Bath's Prologue*, and *The Miller's Tale* – the figures stay below 7%:

text	number of lines	lines containing triples	% of lines with triples
The General Prologue	858	50	5.8
The Knight's Tale	350	23	6.5
The Wife of Bath's Prologue	388	18	4.6
The Miller's Tale	161	10	6.2

The numbers are somewhat higher (up to around 7%) if we include in our counts not only the very clearly lexically and morphosyntactically forced triples, but also more ambiguous ones, for which the realisation of the rhythmical triple might well depend on the performer's or reader's own priorities. This means that for some lines such as these from the *General Prologue* (examples in (35))

(35)   But, for to tellen yow of his array   GP 73
        That of hir smylyng was ful symple and coy;   GP 119
        This is to seyn, a Monk out of his Cloystre   GP 181
        Noon of his bretheren cam ther in his haunt   GP 252[c]
        And rounded as a belle out of the presse.   GP 263,

one could imagine two acceptable scansions: one, placing morphosyntactic weight above the predominant pattern of duple alternation, resulting in triple rhythm for the embolded groups, and one, often ridiculed as "mechanical", which respects the duple alternation which is the metrical core of the poem in violation of the relative morphosyntactic prominence of the monosyllables, thus:

(36)   But, FOR to tellen yow of his array   GP 73

> That OF hir smylyng was ful symple and coy; GP 119
> This IS to seyn, a Monk out of his Cloystre GP 181
> Noon OF his bretheren cam ther in his haunt GP 252b
> And rounded AS a belle out OF the presse. GP 263

It would be impossible to incorporate such rhythmically, but not metrically, ambiguous lines into a strict template. The way to deal with them, we believe, is to assign the ambiguous sequences an intermediate position between the clearly alternating duple rhythm and the clear triples as in (32)-(34) above. In allowing for this "middle" rhythmical ground, we extend Youmans's idea that metricality is a gradient concept (Youmans 1996: 190) to the rhythm associated with a particular meter. Thus we would argue that Chaucer, in creating the iambic pentameter, allowed both clear triples and "fuzzy" triples; the latter did not have Continental predecessors.

9. Alternative analyses of triples

The occurrence of triples in Chaucer's best verse has led some metrists[24], especially Lewis (1938) and including Frye (1957), I. Robinson (1971), Southworth (1954), to claim a degree of similarity, even possible influence and "continuity", between earlier accentual verse and the mature Chaucerian line. We believe that this is truly a red herring and that there is no connection. Weismiller (1989), in a study where he brilliantly links the occurrence of triples in English Romantic and Victorian verse to what he calls the "freeing" of verse, though we agree with most of the rest of his argument, is in our opinion wrong to claim that "genuine triple rhythms in English originate in strong-stress meter" (Weismiller 1989: 282). Chaucer's pentameter derives from Italian models which had a characteristic mix of duples and triples. The earlier Middle English traditions, on the other hand, were either almost entirely duple (like Orm's 4 / 3+1) or predominately triple (like many of the popular ballads) – that is, there were true trisyllabic feet, which is not the case in Chaucer.

The impression that Chaucer wrote a five-foot four-stress hybrid is based on the relative frequency of triples (including trochaic inversions) in Chaucer's verse. As Weismiller (1989) argues, this lead to substantially greater freedom in the form of later pentameter. Unlike the metrists who

---

[24] The issue is insightfully discussed in Baum (1961), esp. appendix 1, though he fails to get to the heart of the matter, the occurrence of triple rhythm amongst underlying duples.

adhere to the Halle — Keyser tradition, Weismiller analyses initial inversions as triples in a basically duple rhythm (1989: 281)[25], citing examples like (32)-(34) above.

One might expect that the particular kinds of variation on strict iambicity that Chaucer confronts us with are largely in those instances where triples may be viewed as calques of the Romance triples. This claim has been made by Weismiller (1989) for post-Renaissance verse, as we noted above; but also specifically for Chaucer, by Guthrie (1988). Guthrie claims: "The key to Chaucer's variational system lies in its permission of weak-position ternary prominences (anapests), and the majority of these involve purposeful violations of the stress maximum principle" (1988: 31). But there are problems with Guthrie's analysis, much as we agree with the principle. Here are two examples illustrating his claim: note that the conflict appears in *his help in,* where there is a strong syllable assigned to a weak position even though it has weak syllables on both sides. Though we prefer this reading, it does not involve a lexically fixed stress. One could just as well read it [01 01 01 01 01]. The other reading is not forced (whereas a lexical stress forces a reading). In (38) the first triple is forced by the lexical stress in *preieth.* The second triple is reasonable but not forced.

(37)   0 1 - 0   1 -- 0   0 1 - 0   0   1   rhythm
        0 1 - 0   1 -- 0   1 - 0 1 - 0   1   meter
        So fer am I -- from his help in derknesse   (*TC* I.18)

(38)   0   1 - 0   0   1 --   0   1 - 0   0   1
        And preieth for hem -- that ben in the cas   (*TC* I.29)

We conclude from these examples that Guthrie's analysis is wrong: he finds numerous triples where they are not forced by lexical stress and where the scansion is therefore dubious. Guthrie senses part of the correct generalisation: "Chaucer's stress contour ... suggests the background presence of the Romance line, with its characteristic tendency toward stress at positions 4, 7, and 10" – de-DUM de-DUM, de-de-DUM de-de-DUM. While Guthrie claims that certain examples favour "a rising rhythm in the phrase as a whole", the claim does not stand up to close inspection

---

[25] Weismiller also lists several other metrists over the past 200 years (1989: 266) who have found it useful to characterise initial inversions and seventh position weak syllables in duple/ triple terms rather than as "mismatching" within a purely alternating stress template.

because it entails a predominance of triples after the caesura, whereas our scansions show triples much more frequently before the caesura. Even his best cases, like the ones below, do not in general "rise" in the manner of the Romance *fragmento adónico*:

(39)   0   1 - 0   0   1 -- 0   1 - 0   0   1
       And preieth for hem -- that ben in the cas   (*TC* I.29)

(40)   0   0   1   0   0   1 -- 0   1   0   1
       ffor whan men han wel cryde -- than wol they rowne
                                                                 (*TC* IV.587)

(41)   0   1   0   0   1 -- 0   0   1   0   1
       And that ye me wolde -- as youre brother trete
                                                                 (*TC* V.134)

The fact is that triples occur before the caesura overwhelmingly more frequently than elsewhere in Chaucer's verse, and Guthrie's "rising rhythm" does not exist in Chaucer – it is not part of his "Romance hangover". His "Romance hangover" is the possibility of allowing triples; but it is not a tendency for triples to occur after the caesura. These are relatively rare. In the *Prologue*, with 50 triples, only seven appear after the caesura:

(42)   46: *Trouthe and honour, FREdom and CURteisye*
       100: *CURteys he WAS, LOWely, and SERuysable*
       195: *And for to festne his hood VNder his CHYN*
       324: *That from the tyme of kyng WILliam weere FALle.*
       393: *Aboute his nekke, VNder his ARM adown.*
       680: *But hood, for Iolitee, WEred he NOON.*
       774: *To ride by the weye DOUMB as the STOON*[26]

It is important to note, however, that Guthrie is right in claiming that a strict "matching" analysis cannot cope with triples, because it makes them all appear to be grossly deviant where a rhythmical reading with triples in the right places gives variation without a sense of violation though certainly deviant from the norm. A closely similar argument is made by Weismiller (1989: 265-267) with respect to the opening lines of Donne's "Twicknam Garden":

---

[26] Cited from the Ellesmere manuscript. In Hengwrt, this line lacks the second *the* and should therefore preserve the *-e* on *weye*, producing a perfectly regular duple scansion across the line.

(43)   Blasted with sighs, and surrounded with teares,
         Hither I come to seeke the spring.

A simple "matching" analysis (following Halle — Keyser) also does not capture the correct generalisation.

10. Conclusions

10.1. Fourteenth-century English had a lexical and phrasal tendency toward iambic rhythm (*above, below, beyond, on top, annoy, the boy, my word, compel,* ...). This shows up in iambic verse long prior to Chaucer.
10.2. But iambic rhythm did not exist as a design feature in the Romance X(I).
10.3. Chaucer imposed, as a design feature, his native iambic rhythm onto his English version of the X(I).
10.4. But underlying rhythms of the Romance X(I) sometimes surfaced, with about 6% triples throughout his work.

References

Barber, Charles — Barber, Nicolas
    1991   "The versification of the *The Canterbury Tales*: a computer-based statistical study", *Leeds Studies in English* 22: 57-84.
Barney, Stephen A.
    1993   *Studies in* Troilus: *Chaucer's text, meter, and diction.* East Lansing: Colleagues Press.
Baum, Paul
    1961   *Chaucer's verse.* Durham, N. C.: Duke University Press.
Benson, Larry D. (ed.)
    1983   *The Riverside Chaucer.* (3rd edition.) Boston: Houghton Mifflin.
Bischoff, Otto
    1897-1898   "Über zweisilbige Senkung und epische Caesur bei Chaucer", I, *Englische Studien* 1897: 353-392, and II, *Englische Studien* 1898: 339-398.

Duffell, Martin J.
1991 The Romance hendecasyllable: an exercise in comparative metrics. [Unpublished Ph.D. dissertation, Queen Mary and Westfield College, University of London.]
1996 "Chaucer, Gower, and the history of the hendecasyllable", in: Christopher McCully — John M. Anderson (eds.), 210-219.
n.d. "The craft so long to lerne": Chaucer's invention of the iambic pentameter. [Unpublished manuscript.]

Frye, Northrop
1957 *Anatomy of criticism.* Princeton: Princeton University Press.

Gasparov, M. L.
1980 "Ital'janskij stix: sillabika ili sillabo-tonika? (Opyt ispol'zovanija verojatnosntyx modelej v stixovedenii)" [Italian verse: syllabic or syllabo-tonic? (An experiment in using probability models in metrics)], in: P. V. Grigor'ev (ed.), *Problemy strukturnoj lingvistiki 1978.* Moskva: Nauka, 199-218.

Gaylord, Alan
1976 "Scanning the prosodists: an essay in metacriticism", *The Chaucer Review* 11: 22-81.

Guthrie, Steven R.
1988 "Prosody and the study of Chaucer: a generative reply to Halle — Keyser", *The Chaucer Review* 23: 30-49.

Halle, Morris — Keyser, Samuel Jay
1971 *English stress: its form, its growth, and its role in verse.* New York: Harper and Row.

Kökeritz, Helge
1961 "Elizabethan prosody and historical phonology", *Annales academiae regiae scientiarum Upsaliensis* 5: 79-102.
[1969] [Reprinted in: Roger Lass (ed.), *Approaches to English historical linguistics: an anthology.* New York: Holt, Rinehart and Winston, 208-227.]

Lewis, Clive Staples
1938 "The fifteenth-century heroic line", *Essays and Studies* 24: 28-41.

McCully, Christopher — John M. Anderson (eds.)
1996 *English historical metrics.* Cambridge: Cambridge University Press.

Minkova, Donka
   1997   "Constraint ranking in Middle English stress-shifting", *English Language and Linguistics* 1: 135-175.
   in press   "Middle English prosodic innovations and their testability in verse", in: Irma Taavitsainen et al. (eds.), *Placing Middle English in context*. Berlin: Mouton de Gruyter.

Pearsall, Derek
   1992   *The life of Geoffrey Chaucer: a critical biography*. Oxford: Blackwell.

Robinson, Fred Norris
   1933[1], 1957[2] *The complete works of Geoffrey Chaucer*. Boston: Houghton Mifflin.

Robinson, Ian
   1971   *Chaucer's prosody: a study of the Middle English verse tradition*. London: Cambridge University Press.

Ruggiers, Paul G. (ed.)
   1979   *The Canterbury tales: a facsimile and transcription of the Hengwrt manuscript, with variants from the Ellesmere manuscript, a variorum edition of the works of Geoffrey Caucer*, vol. 1. Norman, Okla.: University of Oklahoma Press.

Saintsbury, George
   1923   *A history of English prosody*, vol. 1. Oxford: London: Macmillan.

Schipper, Jakob
   1895   *Grundriß der englischen Metrik*. Vienna.
   1910   *A history of English versification*. Oxford: Clarendon.

Skeat, Walter W.
   1894   *The complete works of Geoffrey Chaucer*. Oxford: Clarendon.

Smithers, Geoffrey Victor
   1983   "The scansion of *Havelok* and the use of ME *-en* and *-e* in *Havelok* and by Chaucer", in: Douglas Gray – E. G. Stanley (ed.), *Middle English studies presented to Norman Davis in honour of his seventieth birthday*. Oxford: Clarendon, 195-233.

Southworth, John G.
   1954   *Verses of cadence: an introduction to the prosody of Chaucer and his followers*. Oxford: Blackwell.

Tarlinskaja, Marina
    1987   *Shakespeare's verse: iambic pentameter and the poet's idiosyncrasies*. New York: Lang.
    1992   *Strict stress-meter in English poetry compared with German and Russian*. Calgary: The University of Calgary Press.

Ten Brink, Bernhard
    1885[1], 1899[2] *Chaucers Sprache und Verskunst*. Leipzig. (Trans. by M. B. Smith, *The language and metre of Chaucer*, London 1901.)

Weismiller, Edward R.
    1989   "Triple threats to duple rhythm". *Phonetics and phonology*, vol. 1: *Rhythm and meter*. New York: Academic Press.

Wild, Friedrich
    1915   *Die sprachlichen Eigentümlichkeiten der wichtigeren Chaucerhandschriften und die Sprache Chaucers*. Wien: Braumüller.

Youmans, Gilbert
    1996   "Reconsidering Chaucer's prosody", in: Christopher McCully – John M. Anderson (eds.), 185-210.

Laura Charlotte Wright (Cambridge)
**Models of language mixing:
code-switching versus semicommunication in medieval
Latin and Middle English accounts**

1. Introduction

In this paper[1] I will consider the linguistic phenomenon of medieval account keeping in Britain, which commonly mixed Medieval Latin and English, and Anglo-Norman and English, on an orderly basis, and examine the model of semicommunication, as presented by Haugen (1966) and Braunmüller (1997). In Wright (1998) I attempted to compare this scribal practice with present-day studies on the code-switching patterns of multilingual speakers. Studies of multilingual present-day speakers have not uncovered a universal rule with predictive force which would determine when switch points occur. There seem to be counter-examples from present-day languages to all the models proffered so far, and indeed I was able to find counter-examples in medieval accounts for the switch-points suggested. The great difference seems to be that multilingual present-day speakers code-switch for all sorts of pragmatic, syntactic, and semantic reasons, which no one rule can capture; whereas British medieval business accounts are, by contrast, a strictly-controlled text type, following a syntactic template, mixing no more than three languages (excluding loan-words, which could, in theory, come from any language) in an orderly manner. One of the worthwhile points of comparison between present-day code-switching speakers and medieval mixed-language accounts is the requisite competence in two or more languages. Paradoxically, whilst code-switching requires competence in two languages, multilingual speakers may be popularly deemed linguistically inadequate when they switch languages. Medieval account-keeping too has been seen as the result of a scribal want of Latin, or the general

---

[1] I should very much like to thank Ernst Håkon Jahr for introducing me to the work of Kurt Braunmüller; Stephen Miller of the Oxford University Computing Service for designing the medieval abbreviation and suspension font; David A. Trotter, editor, *Anglo-Norman Dictionary*, for alerting me to the term "pargetting"; and Dieter Kastovsky for the presentation of this paper, which is a development from an earlier draft read at the International Conference on Language and Contact in the History of English, Tulln bei Wien, 1994. My thanks also to Jonathan Hope for reading of drafts.

degeneracy of Anglo-Norman, and subsequently this text type has largely been overlooked by historical linguists.[2] In fact, such was not the case; accounts-clerks were competent Medieval Latin writers, as will be discussed below.

2. Data: the Archive of London Bridge

For the purposes of the present investigation I have limited myself to just one archive, as the constraints governing switching in a variety of regional medieval accounts have been addressed in Wright (1995a, 1998). The following data comes from the archive of London Bridge. London Bridge was both a civic and a religious perpetual institution. In the Middle Ages there was a chapel on the bridge, dedicated to St Thomas the Martyr, and travellers over the bridge would give alms to the chapel. This money maintained both the bridge and the chaplains. As time went by, local people would leave their worldly goods in their will to the chapel, and gradually the bridge built up a portfolio of property in the City, on the bridge and in Southwark. By the fourteenth century this property-holding was so great that the bridge needed to employ not only chaplains for the chapel and what were known as "tidemen" to maintain the bridge infrastructure, but also a large workforce of carpenters, masons, bricklayers, sawyers, glaziers, tilers, daubers, plumbers and other workmen to maintain all the buildings. These workmen were paid weekly, and the bridge foundation (known as "The Bridge House Estate") had to purchase materials for them to work with, and also to sell off some of its excess holdings. It is the record of these transactions, together with the details of rent paid to the bridge, that go to make up the two sets of records known as *The Bridge Masters' Yearly Rentals*, and *The Bridge Masters' Weekly Payment Books*.[3] To demonstrate the mixing of languages, I have chosen at random a few folios from the fifteenth-century *Bridge Masters' Yearly Rental*.

---

[2] For further discussion on the status of Medieval Latin see Wright (1997b).
[3] For further description and a translation of the bridge archive, see Harding — Wright (1995).

2.1. Transcription and translation of data[4]

Corporation of London Records Office MS *Bridge House Rental*, volume 3, 29 Sept 1460 – 29 Sept 1461, fo 32, fo 32v

1. *Empcō Arene*
   'Purchases of sand'
2. *Ît Johanne Brouñ p xxvij lodes ʒabuli ab iṗa empt' & expñ in factur' de morter*
   'And to Joan Broun for 27 loads of gravel from her bought and expended in making mortar'
3. *p tegulat' daubur' & piectyng muroʒ tēn de la Casteīl in wodestrete & tenement'*
   'for tiling, daubing and pargeting walls in the tenement of the Castle in Wood Street, and the tenement of the'
4. *pistrini in Graschirchestrete & alioʒ teñtoʒ hoc anno repat' xŝ Ît Davy Taillour'*
   'bakehouse in Gracechurch Street, and other tenements this year repaired, 10s. And to Davy Taillour'
5. *p vj lodes ʒabłi ab iṗo empt' & expñ in tentô pontis situat' in pochia scī Botulphi*
   'for 6 loads of gravel from him bought and expended in a bridge tenement situated in the parish of St Botulph'
6. *ex'' algate ṗc' cʒ̃ lode vḍ – ijŝ vjḍ Ît Henr' Walshe pro xlij lodes grauełl ab iṗo*
   'without Aldgate, price for each load, 5d; 2s 6d. And to Henry Walshe for 42 loads of gravel from him'
7. *empt' & expñ in pauitur' coram teñta pontis in Ciuitate londoñ hoc anno ṗc'*
   'bought and expended in paving in front of bridge tenements in the City of London this year, price'
8. *cuiuslt lode vḍ – xvijŝ vjḍ Ît p ij lodes grauełl n' Ît p ij lodes ʒabłi xijḍ expñ in*
   'of each load, 5d; 17s 6d. And for 2 loads of gravel, nil. And for 2 loads of gravel, 12d, expended in'

---

[4] My especial gratitude to the Corporation of London, for their kind permission to reproduce folios from *Bridge Masters' Rental*, vol. 3, 1460-1484. [!] means 'as in the manuscript' (it may cause confusion to place the Latin term *sic* in a Latin text), and the abbreviations and suspensions are as in the manuscript.

9. *pauitur̃ apud teñ de la Crouñ in Suthwerk Ît dcô Dauy Taillour p vj lodes grauell*
   'paving at the tenement of the Crown in Southwark. And to the said Davy Taillour for 6 loads of gravel'
10. *expñ in pauitur̃ in pauitur̃* [!] *tento% ad macell & scê Mar̃ Attehill ijŝ*
    'expended in paving tenements at the Shambles and at St Mary at Hill, 2s.'
11. *Empcō Merem*[9]
    'Purchases of timber'
12. *Ît Ricardo Lyoñ p C lodes meremij querc̃ quadrati ab ipô empt̃ & delibat̃ infra*
    'And to Richard Lyon for 100 loads of squared oak timber from him bought and delivered to'
13. *le Brughous in Suthwerk p̃c cuiust lode vjŝ viijd̃ - xxxiij ƚi vjŝ viijd̃ Ît eidm̃ Ricô*
    'the Bridge House in Southwark, price of each load, 6s 8d; £33 6s 8d. And to the same Richard'
14. *p iiijor lodes consilio meremij p̃c cuiust lode apud Croydoñ vŝ - xxŝ Ît Johi*
    'for 4 loads of similar timber, price of each load at Croydon, 5s; 20s. And to John'
15. *Kynge p xl lodes meremij querc̃ p̃c cuiust lode apud Croydoñ vŝ - x ƚi Ît Johi*
    'Kynge for 40 loads of oak timber, price of each load at Croydon, 5s; £10. And to John'
16. *Willynghurst p iij lodes meremij querc̃ xxŝ viijd̃ & C & dī^c de Shipborde ab ipô*
    'Willynghurst for 3 loads of oak timber, 20s 8d; and 100 and half a 100 of shipboard from him'
17. *empt̃ in grosß liiijŝ iiijd̃ Ît Willo Draytoñ de Totenham p viginti lodes Meremij*
    'bought, in gross 53s 4d. And to William Drayton of Tottenham for twenty loads of squared oak timber'
18. *querc̃ quadrat̃ ab ipô empt̃ & delibat̃ infra le Brughous p̃c cuiust lode vjŝ jd̃ -*
    'from him bought and delivered to the Bridge House, price of each load, 6s 1d;'
19. *vj ƚi xxd̃ Ît Henr̃ Carpent̃ de Dorking p xxiij lodes & Ricō lambale p xxviij lod̃*

'£6 20d. And to Henry Carpenter of Dorking for 23 loads and Richard Lambale for 28 loads'

20. *meremij querc̄ quadrat̄ p̄c̄ cuiuslt lode infra domū pontis delibat̄ vjŝ -xv lī vjŝ*

'of squared oak timber, price of each load delivered to the bridge house, 6s; £15 6s.'

21. *Ît Thome Cappes p iiij^or lodes mer̄ p̄c̄ cuiuslt* [fo 32v] *lode vŝ - xxŝ Ît Willo*

'And to Thomas Cappes for 4 loads of timber, price of each load 5s; 20s. And to William'

22. *Carter de Croydoñ pro ij lodes meremij querc̄ vocat̄ principall peces ab ipō*

'Carter of Croydon for 2 loads of oak timber called principal pieces, from him'

23. *empt̄ p̄c̄ apud Croydoñ xiijŝ iiijd Ît Ricardo Belamy p viij lodes meremij querc̄*

'bought, price at Croydon 13s 4d. And to Richard Bellamy for 8 loads of oak timber'

24. *ab ipō empt̄ p̄c̄ cuiuslt lode iiijŝ viijd - xxxvijŝ iiijd Ît Thome Leman p iiijx^xx vlm̂*

'from him bought, price of each load 4s 8d; 37s 4d. And to Thomas Leman for 90 elms'

25. *ab ipō empt̄ apud Bedyngtōn sine ramis xlŝ Ît Johi Bartoñ p lij vlmis ab ipō*

'from him bought at Bedyngton without branches 40s. And to John Barton for 52 elms from him'

26. *empt̄ apud* [blank] *xvjŝ viijd Ît* [blank] *Hamond pro ij fraxinis ab ipō empt̄ apud*

'bought at ... 16s 8d. And to ... Hamond for 2 ashtrees from him bought at'

27. *Chellesham vŝ Ît Simoî Terry p vij querc̄ Curvis ab ipô empt̄ apud westwode in*

'Chellesham, 5s. And to Simon Terry for 7 curved oaks from him bought at Westwood in'

28. *pochia de Leuesham p rotis molend inde fiend ixŝ Ît eidm̂ p iiij^or lodes meremij*

'the parish of Lewisham for the making of mill wheels therefrom, 9s. And to the same for 4 loads of curved timber'

29. *curvi ibm̂ empt̄ p wranges p batell inde fiend vjŝ viijd Ît expñ de dcō meremio in*

367

'there bought for wrongs for the making of boats therefrom, 6s 8d. And of the said timber'

30. *nouo edificio teñ de la Crouñ hoc anno de nouo fact̄ & parcelle teñti de la*

'used in the new building of the tenement of the Crown this year newly made and part of the tenement of the'

31. *Castell ac sursumposicōe vnius teñ apud Macella scī Nichi in tenur̄ Thome*

'Castle and in shoring up one tenement at St Nicholas Shambles in the tenure of Thomas'

32. *Johnsoñ Carnific̄ ac in repacōe & emendac̄ diu̇s̄ alioꝜ teñtoꝜ & MolendinoꝜ*

'Johnson, butcher, and in repairing and mending diverse other tenements and mills'

33. *apud Stratford & leuesham hoc anno repat̄* [blank] *lodes*

'at Stratford and Lewisham this year repaired ... loads.'

34. *Empcô tabularꝜ*

'Purchases of board'

35. *Ît Nicho Herman de Crawley ꝑ M̄ M̄ M̄ M̄ v^C dī viij peḋ de quarterborde p̄c̄ c₃*

'To Nicholas Herman of Crawley for 4,558 feet of quarterboard, price for every'

36. *C^ne ijṡ ijḋ - iiij l̄i xviijṡ viijḋ Ît sol̄ ꝑ dī C de Shipborde empt̄ & expñ in fc̄ur̄*

'hundred, 2s 2d; £4 18s 8d. And paid for half a hundred of shipboard bought and expended in making'

37. *vnius noue batelle & repacōe & emendac̄ batelle vocat̄ the Chalkebote xvjṡ viijḋ*

'one new boat and repairing and mending the boat called the Chalkboat, 16s 8d.'

38. *Ît Johi Mideltoñ aldrd̄ ꝑ dī C de Waynescote ab ipô empt̄ xxiijṡ iiijḋ Ît Henr̄*

'And to John Midelton, alderman, for half a hundred of wainscot from him bought, 23s 4d. And to Henry'

39. *Carpent̄ de Dorking ꝑ CCCCl pedes et Waltero Offold ꝑ xviij^C peḋ de*

'Carpenter of Dorking for 450 feet and Walter Offold for 1,800 feet of'

40. *plancheborde p̄c̄ c₃ C^ne apud le Brughous Wharf ijṡ - xlvṡ* [erasure] *ab ipîs empt̄*

'planchboard, price for every hundred at the Bridge House wharf, 2s; 45s from them bought'

41. *et expñ tam in fcur* hostio⹋ & fenestra⹋ ac garnysshyng & floryng dcô⹋*
'and expended both in making doors and windows and in garnishing and flooring the said'

42. *tenemento⹋ de la Crouñ & le Castell ac diu̇ɾs alio⹋ tenemento⹋ pontis ho* [!]
'tenements of the Crown and the Castle and diverse other bridge tenements this'

43. *anno Ît Ricô Lyoñ p vj lodes de hertlatth ab ipô empt* hoc anno & expñ in noua*
'year. And to Richard Lyon for 6 loads of heartlath from him bought this year and expended in newly'

44. *tegulac* teñto⹋ p̃dcô⹋ ac daubur* muro⹋ eo⹋dm̃ & alio⹋ têntó⹋ ponť hoc anno*
'tiling the aforesaid tenements and in daubing the walls of the same and other bridge tenements repaired this year,'

45. *repacť p̃c* cuiusłt lode xiijŝ iiijḋ – iiij łi Ît dicto Nicho Herman p M̃ M̃ Ciiij^{xx} peḋ*
'price of each load 13s 4d; £4. And to the said Nicholas Herman for 2,180 feet'

46. *de quartborde & planchborde C^{ne} ad xxijḋ – xxxixŝ xḋ Ît eidm̃ p C de Eueslatth*
'of quarterboard and planchboard, the hundred at 22d; 39s 10d. And to the same for 100 of eaveslath'

47. *viijŝ Ît Thome Leycetre p xxvj^C hertlatthe p̃c* C^{ne} vḋ q^a – xjŝ vjḋ*
'8s. And to Thomas Leycetre for 2,600 heartlath, price per hundred 5d farthing; 11s 6d.'

2.2. Language mixing in medieval accounts

The set of accounts from which this extract is taken is written in a mixture of Medieval Latin and Middle English. However, the distribution of the two languages is not random. Nouns, deverbal *-ing* forms and adjectives may optionally appear in English; but prepositions, conjunctions, pronouns and past participles must appear in Latin. Let us apply this to the first paragraph, *Empcō Arene* 'purchases of sand' (lines 1-10); where *lodes* (2), *morter* (2), *piectyng* (3), *grauell* (6) appear in English, and *p* (2), *apud*

(9), *de* (2), *ab* (2), *ipîs* (40), *empt'* (2), *expñ* (2), *repat'* (4), *delibat'* (12) appear in Latin.

In practice, there are two mitigating factors which blur this clear-cut distinction. Firstly, there is some matter which is common to both Medieval Latin and Middle English, such as the preposition *in*, for example; or roots which are either cognate or which English had borrowed from Latin. Inflexions too can be common to both languages, such as the plural noun marker *-is*, which marked both English plural nouns and some plural declensions of Latin nouns. Secondly, the medieval abbreviation and suspension system reproduced in the transcription above served to visually hide many of the Latin inflexions, so that roots are far more visible than appended morphemes. An example is *tenement'* (3). It is easy to ignore the abbreviated inflexion, and it doesn't matter whether the reader knew that the root entered Anglo-Norman from Medieval Latin, as it can simply be read as the English loan-word *tenement*. There was, therefore, considerable room for exploitation of the overlap of the two languages.

Scribes often wrote a lexeme in both its Latin and its English realisation in close proximity, sometimes on the same folio. In our text, the scribe wrote both *ʒabuli* (2) and *grauell* (6). It is possible that these two words denote different grades of gravel and hence are not synonymous, but the price per load is 5d for both the *ʒabli* (5) provided by Davy Taillour and the *grauell* (6) provided by Henry Walshe. The sheer number of apparent calques found in any one archive leads me to suggest that scribes did know how to translate terms, but chose to do so variably. Secondly, the base text is unmistakably Medieval Latin. The scribe knew how to decline nouns, conjugate verbs, match up preposition and case inflexion, mark adjectives for number and gender, and a great deal of later Latin vocabulary.[5] This took skill and learning.

3. The model of semicommunication

3.1. I propose to examine the Scandinavian model of semicommunication as described by Haugen (1966). This model has been applied to documents produced in Medieval Latin and Middle Low German by merchants of the Hanseatic League, by Braunmüller and his associates (see Braunmüller – Diercks (eds.) 1993, Braunmüller (ed.) 1995, Braunmüller 1997).

---

[5] See Wright (1997a) for further discussion of London-produced Medieval Latin in legal texts.

Semicommunication refers to the way in which speakers of the various Scandinavian languages are willing to surrender their own linguistic identity, and arrive at a compromise language, in order to communicate with each other. There is a partial mapping of linguistic codes so that, for example, a Dane may be understood by a Norwegian, to a certain extent. It is based upon a willingness
- not to be daunted
- to be satisfied with partial comprehension
- to allow pragmatic considerations to override linguistic ones.

How do Scandinavian speakers do this? Amongst other things:[6]
- They concentrate on nouns and verbs.
- They ignore alien grammatical markers.
- They seek out and exploit such areas as overlap, and suppress those that don't.
- They suppose word-order is equivalent to that with which they are familiar.
- They assume nominatives come first and are unmarked.
- They assume that datives and accusatives follow the order "indirect first, direct object follows".
- They assume that genitives are marked as such.
- They welcome and encourage loans.

Armed with this checklist, does any of this apply to our accounts (it should be borne in mind that Braunmüller's checklist was drawn up for Scandinavian languages, and so what is of interest is the spirit, rather than the particulars, of the model)?

3.2. Concentrate on nouns and verbs; loan-words are welcome:

By far the greatest amount of English is to be found as a noun, and there are also deverbal -*ing* forms in our text: *lodes* (OE *lād*) (1), *Shipborde* (OE *scip* + OE *bord*) (36), *wranges* (OE *wrang*) (29), *quarterborde* (AN *quarter* + OE *bord*) (35), the *Chalkebote* (OE *cealc* + OE *bāt*) (37), *Waynescote* (?MLG *wagenschot*) (38), *plancheborde* (AN *planche* + OE *bord*) (40), *Wharf* (OE *hwearf*) (40), *floryng* (OE *flor*) (41), *hertlatth* (OE *heorte* + OE *lætt*, **læþþ*) (43), *Eueslatth* (OE *efes* + OE *lætt*, **læþþ*) (46). I include here loans from Anglo-Norman into English, e.g. *morter* (AN *morter*) (1),

---

[6] I am quoting selectively here from the checklist provided by Braunmüller (1997: 369). Not all the points mentioned (e.g. those to do with prosody and intonation) are relevant to medieval written accounts.

daubur (AN *dauber*) (3), *piectyng* (AN *pargettre*) (3), *graueĩl* (AN *gravel*) (6), *tenement* (AN *tenement*) (3), *principaĩl peces* (AN *principal* + AN *peces*) (22), *garnysshyng* (AN *garniss*) (41), *bateĩl* (AN *batel*) (29). Occasionally English hyponyms have a Latin superordinate: *querc̃ vocat̃ principaĩl peces* (22), *batelle vocat̃ the Chalkebote* (37).

3.3. Suppression of alien grammatical markers: exploiting the overlap between two languages:

This is where the medieval abbreviation and suspension system plays such an important role, so that words such as *situat̃* (5), *diũs̃* (32) can be read as either Latin or English; and even roots such as *repacōe, emendac̃* (32) can be read as 'repair', 'mend', by ignoring prefixes and suffixes. The medieval abbreviation and suspension system serves to highlight roots at the expense of functional morphology. It is interesting to note that the Romance article, which declined for gender in Medieval Latin although not in English, is variable in our text: (tokens for the tenement names *Crown* and *Castle*): *la Casteĩl* (3), *la Croun̄* (9), *la Croun̄* (30), *la Casteĩl* (30-31), *la Croun̄ & le Casteĩl* (42).[7] Elsewhere on this folio (fo 32v) the scribe renders the place name of 'St. Nicholas Shambles' as *Macella scî Nic̃hi* (31), and so he could have written *la corona* and *le castrum* had he wished to (and as appears elsewhere in the volume). I suspect that the variable assignation of gender in such instances is a deliberate choice, and forms part of the characteristic quality of this text-type.

3.4. Word-order will be like that with which you are familiar:

This is not the case with adjective placement in the Latin component, where the adjective postmodifies the noun: *tenement̃ pistrini* (3-4), *ten̄ta pontis* (7), *meremij querc̃ quadrati* (12), *querc̃ Curvis* (27), *rotis molenq* (28) (with the exception of *nouo edificio* (30), *noue batelle* (37), which belongs to a small subset of adjectives which precede nouns in Romance languages). In the English component, however, the adjective precedes the noun: *principaĩl peces* (22), *Shipborde* (16), *quarterborde* (35), *Chalkebote* (37), *plancheborde* (40), *hertlatth* (43), *Eueslatth* (46). Apart from adjective placement, the word-order can certainly be read as English, as the virtually word-for-word translation shows. So this text type has both Romance-style adjective placement and Germanic-style adjective placement. Both are integrated into the text, and yet both are kept separate.

---

[7] See Wright (forthc.) for a further discussion of variable gender-assignment with regard to proper names.

3.5. Nominatives come first and are unmarked; indirect objects precede direct objects:

Our model cannot easily be applied here as our text does not display default sentences with finite verbs, subjects and objects, but has a default construction of 'to so-and-so for x amount of something bought from him/her and used in such-and-such a way'; 'to so-and-so for three days doing x', and so on. But there is a default order, as seen in *Ît Johanne Brouñ p xxvij lodes ʒabuli ab iṗa empṱ* (1) 'And to Joan Broun for 27 loads of gravel from her bought'. The person being paid, whilst not in the nominative case (and always in a marked oblique case) is always in the first position in the clause after the clause-marker *Item*. Then come the prepositional phrases denoting how much and bought from whom and how used, and finally the verb. Entries are always found in this order, which makes recognition and interpretation easier. To go with the spirit of the model, we can say that a fixed word-order aids interpretation.

3.6. Genitives are marked:

This is so for Latin lexemes: *lodes ʒabuli* (1) 'loads of gravel', *lodes meremij* (12) 'loads of timber', *piectyng muroϟ* (3) 'pargeting of walls'; but apparently not for English lexemes: *lodes graueīl* (6), where the English term *gravel* does not take a Latin genitive suffix. There are no English-lexeme possessives of the *-s* type to check in this text.

4. Conclusion

I conclude from this exercise that the Scandinavian model of semi-communication does indeed provide a useful present-day comparison for medieval mixed-language accounts. Such accounts were not the result of a dearth of knowledge of Medieval Latin; rather they were constructed by clerks who were skilled in Latin, but who understood the pragmatic function of their texts. If estrichbord was produced on the continent, then a merchant had to travel abroad to buy it – or a foreign merchant had to bring it to London. Either way, merchants had to operate in a linguistic culture that was not their own, and, most importantly, to ascertain that the price they paid or received was the correct one. To construct a mixed-language account took great skill, but to read and understand such an account to a greater or lesser degree demanded less linguistic ability on the part of the merchant. A grasp of Medieval Latin was needed, but that grasp need not encompass a full vocabulary, or a total comprehension and manipulation of grammatical morphology. So long as a merchant was willing not to be daunted, to be satisfied with partial comprehension, and to allow pragmatic considerations to override linguistic ability, then this orderly mixing of languages provided a stable and successful mercantile writing system.

References

Braunmüller, Kurt
    1997   "Communication strategies in the area of the Hanseatic League: the approach by semi-communication", *Multilingua* 16: 365-374.

Braunmüller, Kurt (ed.)
    1995   *Niederdeutsch und die skandinavischen Sprachen*, vol. 2. (Sprachgeschichte 4.) Heidelberg: Winter.

Braunmüller, Kurt — Willy Diercks (eds.)
    1993   *Niederdeutsch und die skandinavischen Sprachen*, vol. 1. (Sprachgeschichte 3.) Heidelberg: Winter.

Gooder, Eileen A.
    1961 [1979]   *Latin for local history: an introduction.* London: Longman.

Harding, Vanessa — Laura Charlotte Wright (eds.)
    1995   *London Bridge: selected accounts and rentals 1381-1538.* (London Record Society 31 (for the year 1994).) London: London Record Society.

Haugen, Einar
    1966   "Semicommunication: the language gap in Scandinavia", *Sociological Inquiry* 36: 280-279.

Kurath, Hans — Sherman S. Kuhn (eds.)
    1952-   *Middle English dictionary.* Ann Arbor, Mich.: University of Michigan Press.

Latham, Ronald E. (ed.)
    1965   *Revised medieval Latin word-list from British and Irish sources.* London: The British Academy.

Latham, Ronald E. — David R. Howlett *et al.* (eds.)
    1975-   *Dictionary of medieval Latin from British sources.* Oxford: The British Academy and Oxford University Press.

Murray, James A. H. — Henry Bradley — William Alexander Craigie — Charles T. Onions (eds.)
1884-1933 *A new English dictionary.* Oxford: Oxford University Press.
Rothwell, William — Louise W. Stone — T. B. W. Reid (eds.)
1992 *Anglo-Norman dictionary.* London: Modern Humanities Research Association.
Wright, Laura Charlotte
1992 "Macaronic writing in a London archive, 1380-1480", in: Matti Rissanen — Ossi Ihalainen — Terttu Nevalainen — Irma Taavitsainen (eds.), *History of Englishes: new methods and interpretations in historical linguistics.* (Topics in English Linguistics 10.) Berlin: Mouton de Gruyter, 762-770.
1994a "Early Modern London business English", in: Dieter Kastovsky (ed.), *Studies in Early Modern English.* (Topics in English Linguistics 13.) Berlin: Mouton de Gruyter, 449-465.
1994b "On the writing of the history of Standard English", in: Francisco Fernández — Miguel Fuster — Juan José Calvo (eds.), *English historical linguistics 1992.* (Amsterdam Studies in the Theory and History of Linguistic Science 4, 113.) Amsterdam: Benjamins, 105-115.
1995a "A hypothesis on the structure of macaronic business writing", in: Jacek Fisiak (ed.), *Medieval dialectology.* (Trends in Linguistics: Studies and Monographs 79.) Berlin: Mouton de Gruyter, 309-321.
1995b "Trade between England and the Low Countries: evidence from historical linguistics", in: Caroline Barron — Nigel Saul (eds.), *England and the Low Countries in the Late Middle Ages.* Woodbridge: Sutton, 169-179.
1995c "Middle English *-ende* and *-ing*: a possible route to grammaticalization", in: Jacek Fisiak (ed.), *Linguistic change under contact conditions.* (Trends in Linguistics: Studies and Monographs 81.) Berlin: Mouton de Gruyter, 365-382.
1996 *Sources of London English: medieval Thames vocabulary.* Oxford: Oxford University Press.

1997a "Medieval Latin, Anglo-Norman and Middle English in a civic London text: an inquisition of the River Thames, 1421", in: Steward Gregory — David A. Trotter (eds.), *De mot en mot: aspects of medieval linguistics: essays in honour of William Rothwell*. Cardiff: University of Wales Press and the Modern Humanities Research Association, 223-260.

1997b "The records of Hanseatic merchants: ignorant, sleepy or degenerate?", *Multilingua* 16: 339-350.

1998 "Mixed-language business writing: five hundred years of codeswitching", in: Ernst Håkon Jahr (ed.), *Language change: advances in historical sociolinguistics*. (Trends in Linguistics: Studies and Monographs 114.) Berlin: Mouton de Gruyter, 99-118.

forthcoming "Lexical and grammatical convergence and language shift in London mixed-language business writing, 1066-1550".

Index of subjects

A

abbreviation
   medieval ~ and suspension system 370, 372
absolute construction 171, 174, 175, 176, 180, 184, 185, 187, 259
accent
   ~ shift 132
   fixed stress ~ 132
acceptance 154, 157, 242
accommodation 99, 235, 237, 245
account 14, *363*, 364, 369, 371, 373
   ~ keeping 363
   business ~ s 311, 363
accusative case 282, 289
acquisition
   ~ device 241
   second language ~ 14, 45, 48, 50, 82, 144, 154, 233, 234, 238, 239, 240
actuation 43
adaptation 11, 126, 129, 225, *279*
address
   form of ~ 24
   term of ~ 245, 272
adjectival
   ~ paradigm 293
   ~ plural 20
adjective 20, 21, 39, 78, 80, 81, 159, 173, 203, 282, 287, 288, 289, 293, 344, 346, 369, 370, 372
administration 18, 174
adstrate 107, 235
advantage
   phonological ~ 287
adverb
   ~ial phrase 19, 321, 323
   ~ial subordinator 83
   ~ial tag 24
affrication 84
African
   ~ languages 82, 143
   South ~ English 12, 235, 240
agent
   ~ experiencer 82
Aktionsart 156
alternating
   ~ stress pattern 338
   duple ~ stress 352
alternation
   duple ~ 355
American
   ~ English . 12, 98, 101, 103, 148, 235, 236, 237
   Latin ~ 143
   North ~ dialects 37
analysis
   error ~ 241
analytic
   ~ comparison 81
   ~ future tense 79, 81
   ~ morphology 81
   ~ passive 81
   ~ past tense 81
   ~ plural 142
   ~ redundancy 49
   ~ structure 49
analyticity 49
anapaest 343, 357
~ic foot 338, 343, 351
anaphoric
   ~ referential devices 22
andative future tense 82
Anglicisation 11, 124, 140, 141, *337*
Anglo-
   ~French 17, 28, 174, 176

Index of subjects

~Norman 22, 27, 136, 137, 147, 221, 222, 223, 329, 346, 363, 370, 371
~Saxon .. 55, 100, 195, 196, 197, 203, 204, 205, 206, 209, 211, 264, 267, 268, 279, 280, 281, 282, 283, 287, 288, 290, 291, 293, 294, 295, 301
apocope 339, 344
apostrophe 85
Arabic 82
archilect 82
area
   Atlantic linguistic ~ 73
   convergence ~ 72, 74, 75, 148
areal
   ~ diffusion 75, 80, 82
   ~ linguistics 11, 72
   ~ typology 71, 72
   historical ~ linguistics 74, 88
aristocracy 18, 19, 138, 139, 275
article 77, 79, 80, 81, 201, 372
   definite ~ 13, 74, 77, 80, 201, 267, 268
   indefinite ~ 74, 77, 80
   post-positive ~ 145
aspect 142, 156, 239
   ~ual types 131, 155
   habitual ~ 142
   immediate perfective ~ 135, 142
   resultative perfective ~ 136
assertion 24, 25, 38
Atlantic linguistic area 73
Attic style 23
augmentation
   lexical ~ 102, 104
Australian English 12
autonomisation 243
auxiliary 74, 81, 82, 241, 322
   ~ verb 81
auxiliation 81
avoidance of hiatus 293

B

back construction .. 173, 174, 175, 176, 178, 179, 180, 184, 185, 186, 187
Balkans 145, 148
Baltic 132, 148
base language 307
basic clause pattern 173
Basque 133
beat
   ~-isochronous 350
   four-~ heresy 354
Bedeutungsentlehnung 199
behaviour
   bilingual speech ~ 330
Belfast English 37, 43
Benedictine reform ... 10, 198, 199, 209
bilateral influence 39
bilingual speech behaviour 330
bilingualism 18, 74, 108, 109, 134, 138, 140, 144, 155, 197, 243, 281, 283, 305, 307, 308, 317, 330
   imperfect ~ 141, 143
   instrumental ~ 140
bi-uniqueness .. 287, 288, 289, 290, 294, 295, 301
Black English 235
bleaching
   semantic ~ 49
borrowing 10, 12, 13, 22, 39, 48, 59, 62, 64, 65, 67, 71, 85, 122, 135, 171, 196, 198, 199, 221, 264, 270, 280, 281, 282, 287, 307, 319, 321, 337, 339
   ~ hypothesis 48
   cultural ~ 106, 108, 143, 159
   dialect ~ 106
   intimate ~ 107, 108
   lexical ~ 12, 13, 18, *97*, 144, 159, 305
   morphological ~ 206, 282
   post-~ polysemy 63, 64, 65
   pre-~ polysemy 62, 63, 64, 65
   semantic ~ ... 102, 104, 198, 199, 202, 203, 207
   superstrate ~ 107

Index of subjects

boundaries
    major constituent ~ 144
boundary
    switch ~ 313
British
    ~ English..... 101, 103, 139, 148, 234, 237
    northern ~ English 37
Bulgarian 47, 49, 81
Burgundian court style.................. 22
business
    ~ accounts............................311, 363
    ~ text 309, 319
    parliamentary ~ 22

C

caesura 339, 340, 341, 342, 344, 345, 348, 349, 353, 358
    ~l placement................................ 341
    epic ~ 339, 340, 341
calque .14, 19, 20, 21, 23, 24, 25, 26, 82, 85, 209, 235, 357, 370
    grammatical ~ 20
    lexical ~ 20
    pragmatic ~ 20, 24
    stylistic ~ 20, 23
    syntactic ~ 20
Cameroonian English 244
Canadian.............................. 45, 236
Caribbean creoles 142
case
    accusative ~ 282, 289
    dative ~ 282, 289, 290
    genitive ~ 280, 282, 289
    nominative ~ 82, 280, 282, 284, 289, 294, 373
    objective ~ 280
    oblique ~ 259, 261
    possessive ~ 280
Castilian Spanish 244
Catalan 82
category
    expressive ~ 49

Celtic 10, 45, 76, 82, 83, 98, 107, 131, 132, 133, 134, 135, 149, 155, 235, 305
Central French 159, 221
Chancelleries
    Latin of ~ 22
Chancery
    ~ English 174
    ~ scribes................................. 22, 86
change
    global ~ pattern..................... 44, 49
    internally motivated ~ 40
    semantic ~ 13, 102, 103, 104, 267, 269, 270
    typological ~ 135
Christianisation 10
Civil War 137
Classical Latin............................... 77
clause
    ~ break......................................144
    basic ~ pattern........................... 173
    participle ~ 172
clear triple 356
clefting 149
    it-~.. 142
clergial style 22
clitic
    ~isation 74
    pronominal ~ 74
cluster
    consonant ~ simplification 244
coda
    heavy syllable ~ 147
code
    ~ shift... 28
    ~-mixing................................... 306
code-switching 10, 11, 14, 143, 144, 145, *305*, 307, 308, 309, 310, 311, 312, 318, 324, 325, 326, 330, *363*
    flagged ~ 326
    intersentential ~ 306, 317
    syntactic ~ 144
    textual functions of ~................. 325
codification.............................. 87, 189

379

Index of subjects

cognitive schema 285
coherence 72, 177, 203, 318
cohesion 318
communication 13, 18, 64, 77, 81, 101, 107, 186, 239, 240, 282, 283, 296, 297, 300, 305, 311, 353
   oral ~ 57, 186
   semi-~ 14, *363*, 370, 373
   verbal ~ 64
communicative
   ~ competence 28
   ~ efficiency 295, 301
   ~ function 239, 279, 310
comparison.. 81
   analytic ~ 81
competence
   reading ~ 187
Complex Adaptive System ..279, 286, 295, 297, 298, 299
complication
   interlanguage ~............................ 50
compositeness
   morphotactic ~......................... 287
compound197, 200, 210, 223, 225, 245, 256
   quasi-~ 210
compression
   syntactic ~ 176
connotation...............27, 268, 270, 271
connotative
   ~ sphere 268
conservativeness........................... 236
consonant
   ~ cluster simplification............. 244
   ~ lenition 134, 146
   ~al digraph..........................217, 221
   ~al length.................................... 221
   ~al strength 293
   double ~....................................... 85
   geminate ~................................... 84
   strong ~....................................... 291
   weak ~... 291
constituent

major ~ boundaries.................... 144
constitutive
   ~ information............................268
constraint
   equivalence ~.............. 144, 321, 324
   weak ~s 48
construction
   absolute ~....171, 174, 175, 176, 180, 184, 185, 187, 259
   back ~ ..173, 174, 175, 176, 178, 179, 180, 184, 185, 186, 187
   front ~..173, 174, 175, 176, 178, 179, 180, 181, 182, 183, 184, 185, 186, 187
   gerund ~......................................173
   literate ~ 186, 189
   participle ~..172, 176, 186, 187, 188, 189
   postmodifying ~ 173, 174, 175, 176, 179, 181, 183, 184, 185, 186, 187
   transitional ~ 180
contact
   ~ situation ...9, 11, 17, 27, 44, 62, 65, 71, 78, 97, 98, 100, 106, 107, 108, 109, 110, 131, 132, 137, 154, 157, 196, 237, 238, 241, 281
   indirect language ~.....................143
   interlinguistic ~............................12
   intralinguistic ~............................12
   language ~ 11, 19, 20, 57, 97, 99, 101, 102, 103, 104, 108, 109, *131*, 132, 133, 134, 145, 245
   language internal ~9, 10, 11, 12, 13, 14, 17, 18, 27, 28, *35*, 44, 45, 46, 49, 57, 59, 62, 64, 65, 69, 71, 72, 75, 76, 77, 78, 80, 81, 82, 83, 84, 88, 118, 189, 195, 196, 199, 218, *233*, 235, 237, 238, 240, 241, 242, 243, 244, 245, 246, 251, 262, 274, 275, 281, 283, 301, 305, 308, 337, 338
   partial-~ origin..................... 11, *337*
   speaker ~ 143, 152, 155

Index of subjects

types of ~ 132, 134
contextualisation 79, 243, 245, 246
Continental
  ~ decasyllabic line 348
  ~ French 17
  ~ Germanic 10
contrast
  structural ~ 12
convergence .. 72, 74, 75, 76, 77, 78, 85, 135, 145, 148, 238
  ~ area 72, 74, 75, 148
  typological ~ 71
conversion 59, 104, 245
coordination 157, 182
copying
  pronoun ~ 245
count
  syllabic ~ 340
creole .. 9, 35, 49, 50, 142, 143, 154, 239, 241
  Caribbean ~s 142
creolisation 12, 17, 28, 50, 141, 142, 143, 234
creoloid 143
cultural
  ~ borrowing 106, 108, 143, 159
  ~ factors 13, 59
  multi~ society 28
curial
  ~ prose 174, 176
  ~ style 22, 23, 25

D

dactyl 343
Danish 38, 78, 80, 274, 281, 283
dative
  ~ case 282, 289, 290
  ~ experiencer 82, 171, 254, 259, 289
  ethical ~ 153
decasyllabic
  ~ line 340, 353
  Continental ~ line 348
decreolisation 14, 50

definite
  ~ article.13, 74, 77, 80, 201, 267, 268
deletion
  for-~ 35
demarking 46, 49
demonstrative.77, 79, 80, 81, 255, 279, 287, 288, 289
  ~ pronoun 79, 81, 109, 287, 289
demotion 345
dependent-marking 134
derivation 223, 225, 227
derivative
  prefixal ~ 210
  pseudo-~ 208
development
  internal ~ .14, 40, 48, 49, 62, 64, 133, 237
  sense ~ 63, 64, 66, 67
device
  acquisition ~ 241
diagrammatic 287, 290, 293
dialect
  ~ borrowing 106
  ~ levelling 235, 237
  ~ mixture 9, 217
  North American ~s 37
  Scottish ~s 38, 133, 148, 233, 236
dialectalisation 69
didactic
  ~ technique 206
differentiation
  semantic ~ 102, 103, 105, 270
diffusion .. 70, 71, 72, 74, 76, 83, 85, 86, 242
  areal ~ 75, 80, 82
  lexical ~ 135
diglossic 306, 308, 310
digraph 218, 221, 222
  consonantal ~ 217, 221
  vocalic ~ 220, 221
diphthongisation 70, 84
direct
  ~ transfer 134

Index of subjects

discourse 27, 28, 139, 152, 305, 307, 310, 312, 321, 325, 328
  ~ element 152
  ~ function 306, 325, 329
  technical ~ 19
divergence 69, 76, 81, 201, 238
domain
  functional ~ 189
  semantic ~ 13, 97, 105, 107, 110
Donegal Irish 148
double
  ~ modal 236
  ~ spelling 217, 220
Dravidian 145
drift 11, 49, 70, 71
  pre-dialectal ~ 70
  syntactical ~ 70
  typological ~ 70, 71, 76
Dublin
  ~ English 131, 136, 158
  ~ Irish 155
  ~ vowel shift 136
duple
  ~ alternating stress 352
  ~ alternation 355
  ~ metre 348
  ~ rhythm 337, 338, 353, 356, 357
  ~ time 349, 350, 352
  rhythmic ~ 349
Dutch..76, 78, 80, 81, 82, 83, 84, 85, 86, 87, 107, 125
dyasystem 243

E

Early
  ~ Modern English 14, 36, 64, 67, 70, 171, 172, 174, 176, 188, 225, 311, 319
efficiency
  communicative ~ 295, 301
element
  discourse ~ 152
  tag-like ~ 329

elision 340, 344, 348, 351
eloquence 28, 138
embedded
  ~ language 307, 313
emphasis.....36, 38, 48, 83, 98, 109, 342
emulation 23, 24, 28, 139
English
  American ~ 12, 98, 101, 103, 148, 235, 236, 237
  Australian ~ 12
  Belfast ~ 37, 43
  Black ~ 235
  British ~ 101, 103, 139, 148, 234, 237
  Cameroonian ~ 244
  Chancery ~ 174
  Dublin ~ 131, 136, 158
  Early Modern ~ 14, 36, 64, 67, 70, 171, 172, 174, 176, 188, 225, 311, 319
  extraterritorial ~es 12, *233*
  extraterritorial ~es as L1 234
  extraterritorial ~es as L2 234
  Hawaiian ~ 244
  Irish ~ 11, 45, *131*
  Japanese ~ 244
  Late Middle ~ 65, 174, 188
  London ~ 17, 26, 27, 50
  Middle ~....10, 11, 13, 14, 18, 24, 25, 26, *35*, 70, 75, 77, 81, 82, 84, 86, 103, 107, 159, 210, 217, 218, 219, 220, 221, 222, 223, 224, 225, 226, 227, 228, 229, 267, 268, 269, 270, 274, 275, 279, 282, 286, 287, 293, 295, 305, 306, 308, 310, 311, 317, 318, 319, 321, 326, 345, 356, 363, 364, 369, 370
  Middle ~ open-syllable-lengthening rule 84, 223, 224, 225, 226, 228, 229
  new ~es 234
  non-native ~es ... 234, 235, 238, 240, 242, 246
  northern British ~ 37

382

Old ~ 10, 11, 12, 14, 20, 38, 39, 40, 49, 55, 70, 74, 77, 98, 100, 102, 103, 104, 106, 107, 109, 133, 147, 171, *195*, 196, 197, 198, 199, 200, 202, 204, 205, 207, 208, 209, 210, 220, 221, 222, 223, 226, 227, 228, 236, *251*, 267, 271, 275, 279, 286, 287, 288, 290, 293, 294, 295, 300, 305, 308, 311, 319, 329, 341
   Ottawa Valley ~............................ 37
   Ozark ~ 37
   planter ~ 138
   South African ~ 12, 235, 240
   Southern ~......................236, 243
   Standard ~.. 35, 38, 49, 50, 176, 188, 189
   Ulster Irish ~ 148
epenthesis................................146, 147
   vowel ~ 141, 146
epic caesura......................339, 340, 341
epistolary
   ~ form 22
   ~ practice....................................... 24
equivalence constraint....144, 321, 324
error analysis...................................... 241
Estonian......................................76, 148
ethical dative..................................... 153
etymological
   ~ technique................................. 203
etymology... 21, 26, 27, 85, 97, 98, 101, 103, 206, 224, 229, 274, 275
   folk ~ 329
Europe
   Standard Average ~an.....73, 75, 77, 80, 81, 82
   Western ~an language....72, 73, 75, 76, 81
evaluative
   ~ sphere........267, 268, 269, 270, 272
evolution
   ~ theory 12
   long vowel ~ 80
experiencer

agent ~..82
dative ~.......... 82, 171, 254, 259, 289
nominative ~ 82, 280, 284, 289, 294, 373
patient ~ 82
expiratory
   ~ stress..10
explanation
   functional ~.................. 279, 286, 316
expressive
   ~ category 49
expressivity......................................48
extension
   semantic ~...................................142
extrametricality 351
extraterritorial Englishes.........12, *233*
   ~ as L1 234
   ~ as L2 234

F

field
   lexical ~ 13, *57*, 269, 270
   semantic ~..............................65, 270
figure-and-ground 287
filter
   for-to ~...50
Finnish..76
flagged
   ~ code-switching.......................326
Flemish 44, 327, 329
focussing
   front-~ structure........................142
folk etymology................................329
foot
   anapaestic ~ 338, 343, 351
   iambic ~................................351, 353
   trisyllabic ~ . 338, 340, 344, 348, 356
for
   ~ + bare infinitive 38
   ~ *to* + infinitive......13, 35, 37, 38, 41, 42, 50
   ~-deletion....................................35
   ~-insertion................... 35, 43, 49, 50

Index of subjects

~-*to* filter 50
foregrounding 26, 86, 295
foreigner talk 49
form
   ~al letters 22
   ~s of address .. 11, 12, 24, 25, 28, 86, 272, 273
      impersonal ~ 151
      poetic ~ 311, 317, 325
      progressive ~ 10, 14, 49, 82, 209, 269
      proto-~ 69
formation
   gloss-~.. 195, 196, 197, 198, 209, 211
   loan-~ 13, *195*, 197, 199, 200, 203, 205, 208, 210
   word-~ 197, 202
fossilisation 242, 243, 244
      interlanguage ~ 242
four
   ~-beat heresy 354
   ~-stress metre 354
Franco-
   ~Germanic creole 9
   ~Latin 22
Franconian 85
   Old ~ 77
French.10, 11, 12, 14, *17*, 35, 38, 39, 44, 45, 48, 50, 57, 67, 73, 74, 75, 80, 82, 83, 84, 85, 86, 87, 98, 100, 103, 104, 105, 107, 109, 118, 119, 124, 125, 128, 132, 133, 134, 135, 147, 171, 174, 217, 218, 221, 223, 224, 225, 226, 228, 270, 305, 306, 307, 309, 310, 313, 314, 315, 316, 317, 321, 327, 329, 340, 341, 345, 346, 352
   Anglo-~ 17, 28, 174, 176
   Central ~ 159, 221
   Continental ~ 17
   Old ~ 39, 133, 222, 223, 226, 228, 341
fricativisation 147
Frisian

Old ~ 77, 84
front
   ~ construction.... 173, 174, 175, 176, 178, 179, 180, 181, 182, 183, 184, 185, 186, 187
   ~-focussing structure 142
   ~ing 42, 77, 133, 149, 245
frozen
   ~ interlanguage 242
function
   ~al domain 189
   ~al explanation 279, 286, 316
   communicative ~ 239, 279, 310
   discourse ~ 306, 325, 329
   purposive ~ 50
   rhyming ~ 329
   textual ~s of code-switching 325
future tense
   analytic ~ 79, 81
   andative ~ 82

G

Gaelic
   Scottish ~ 133
geminate 221
   ~ consonant 84
generalisation
   rule ~ 46
genetic
   ~ linguistics 108, 233
   ~ selection theory 12
genitive
   ~ case 280, 282, 289
   post-posed ~ 83
genotype 295, 296
genre 176, 186, 309, 311
German 44, 45, 47, 67, 70, 73, 74, 75, 76, 78, 79, 80, 81, 82, 84, 85, 87, 98, 100, 102, 107, 119, 122, 123, 124, 125, 129, 136, 143, 148, 152, 153, 154, 156, 159, 209, 233, 235, 285, 308
   Low ~ 84, 305
   Middle Low ~ 370

384

Old High ~ ... 98, 197, 198, 200, 209, 308, 319
   Standard High ~ 101
   Swiss High ~ 101
Germanic ...9, 10, 18, 44, 45, 50, 70, 72, 74, 75, 78, 79, 80, 81, 82, 83, 84, 85, 86, 98, 102, 104, 132, 136, 205, 209, 221, 255, 372
   ~ sentence brace 136
   Continental ~ 10
   Franco-~ creole 9
   North ~ 73, 76, 81, 83, 305
   Proto-~ 70
gerund
   ~ construction 173
global
   ~ change pattern 44, 49
gloss 12, 42, 100, 159, *195*, 196, 197, 198, 199, 200, 201, 203, 204, 205, 206, 207, 208, 209, 210, 252, 253, 258, 260, 261, 309
   ~-formation 195, 196, 197, 198, 209, 211
glossary 87, 200, 272
glottalisation 147
gradient metricality 338
grammar ..19, 20, 21, 24, 38, 42, 72, 86, 100, 142, 147, 154, 204, 205, 211, 256, 257, 258, 261, 305, 319, 371, 372, 374
   Universal G~ 234
grammatical
   ~ calque 20
grammaticalisation 81, 82
Greek
   Ancient ~ 9, 23, 98, 203, 206, 222, 312
   Modern ~ 47, 74
greetings 245
group
   prestige ~ 154

H

habit
   spelling ~ 217, 218, 227
habitual
   ~ aspect 142, 155, 156
   ~ present 156
hard words 10
have-perfect 77, 81
Hawaiian English 244
head 71, 82, 134, 142, 181, 255, 323
   ~ marking 134
   ~less lines 344, 347
heavy syllable coda 147
Hebrew 82, 98
hedge-school 137
hendecasyllabic
   ~ verse 340, 341
hiatus
   avoidance of ~ 293
historical
   ~ areal linguistics 74, 88
   ~ reconstruction 69
hypothesis
   borrowing ~ 48
   nativisation ~ 46

I

iamb 343, 349
   ~ic foot 351, 353
   ~ic pentameter ... 338, 339, 340, 343, 348, 353, 356
   ~ic rhythm 11, 359
   ~ic verse 337, 359
Ibero-Romance 133
iconicity 287, 288, 293, 301
ictic
   ~ position 341, 349
ictus ..339, 341, 342, 343, 346, 349, 350, 351
idiom 152, 202
imitation
   morphological ~ 205

Index of subjects

immediate
~ perfective 135, 142, 156
~ perfective aspect 135, 142
imperfect bilingualism...........141, 143
impersonal
~ form 151
importation........................98, 100, 102
indefinite
~ article 74, 77, 80
indexes of style 19
indirect
~ address 22
~ language contact 143
Indo-Aryan.. 145
infinitive...35, 36, 37, 38, 40, 41, 42, 43, 44, 45, 46, 47, 50, 59, 145, 158, 223, 344
for + bare ~ 38
for to + ~ 13, 35, 37, 38, 41, 42, 50
plain ~... 40
prepositional ~............................ 39
to-~ 37, 38, 40, 41
inflection 20, 78, 81, 282, 287, 295
~al morphology........79, 80, 81, 288
~al system....................260, 288, 289
weakening of ~ 84
influence
bilateral ~...................................... 39
low-level ~................................... 135
informal
~ text types. 177, 178, 180, 184, 186, 188
information
constitutive ~ 268
stereotypical ~ 268
initial
~ stress 132, 346
inkhorn terms..................................... 10
innovation. 58, 64, 66, 74, 76, 106, 148, 237, 302, 341
shared ~....................................... 75
innovativeness 236
insertion

for-~ 35, 43, 49, 50
instrumental
~ bilingualism 140
intake
semantic ~...................................102
interference .18, 72, 108, 138, 154, 159, 240, 342
interjection.......................306, 317, 329
interlanguage......35, 45, 47, 48, 49, 50, 240, 241, 242, 243, 244
~ complication.............................50
~ fossilisation.............................242
~ optimalisation 49
frozen ~ 242
interlinear
~ version.............. 195, 196, 197, 207
interlinguistic
~ contact..12
internal
~ development14, 40, 48, 49, 62, 64, 133, 237
~ly motivated change..................40
interpersonal
~ role..25
intersentential 144, 306, 317, 318
~ code-switching................306, 317
intertextual
~ relation.....................................326
intimate
~ borrowing........................107, 108
intonation 132, 133, 137, 156, 157, 222, 223, 228, 341, 342, 344, 345, 346, 347, 349, 351, 356, 357
intralinguistic
~ contact..12
intrasentential.144, 306, 312, 317, 322, 326
~ switch 307, 317, 318, 319, 326
inverse
~ spelling..................... 221, 225, 227
inversion
trochaic ~....................................356
Irish 11, 37, 45, *131*

Index of subjects

~ English...........................11, 45, *131*
  Donegal ~................................. 148
  Dublin ~................................... 155
  Munster ~................................. 146
  Ulster ~............................133, 148
  Ulster ~ English........................ 148
isochrony
  beat-~.. 350
Italian .11, 39, 44, 45, 47, 73, 74, 82, 83, 84, 86, 107, 119, 120, 121, 122, 124, 233, 235, 300, 337, 340, 341, 349, 351, 352, 353, 356
it-clefting...................................... 142
i-umlaut 70, 79, 84

J

Japanese English............................ 244

K

Kassubian 148
keeping
  account ~ 363
knowledge .. 12, 25, 26, 45, 57, 80, 137, 140, 141, 149, 235, 242, 251, 330, 373
koinéisation.................................... 235

L

language
  ~ change 12, 14, 35, 46, 105, 144, 190, 195, 241, 245, 259
  ~ change typology....................... 35
  ~ maintenance 108
  ~ mixing.........................14, 309, *363*
  ~ processing 186, 187
  ~ retention 140
  ~ shift 108, 109, 155
  ~ style.. 86
  ~ variety..............................19, 86
  African ~s..........................82, 143
  base ~ 307
  embedded ~ 307, 313
  European ~s................................ *69*

indirect ~ contact 143
matrix ~ 313
noun-centred ~........................... 149
prestige ~ 73
second ~ acquisition. 14, 45, 48, 50, 233, 238, 239, 240
standard ~s 86
standard literary ~....................... 79
Western European ~s72, 73, 75, 76, 81
written ~.............. 18, 33, 76, 86, 186
Late Middle English......... 65, 174, 188
Latin9, 12, 13, 14, 22, 23, 24, 28, 57, 58, 59, 62, 63, 64, 65, 66, 67, 73, 75, 76, 77, 78, 82, 83, 85, 86, 87, 98, 100, 102, 104, 107, 108, 124, 133, 135, 143, 159, 171, 172, 174, 176, 188, 189, 195, 196, 197, 198, 199, 200, 201, 202, 203, 204, 205, 206, 208, 209, 210, 211, 217, 221, 222, 227, 228, *251*, 305, 306, 308, 309, 311, 312, 313, 314, 315, 316, 320, 321, 322, 323, 325, 327, 328, 329, *363*, 365, 369, 370, 372, 373
  ~ American 143
  ~ of Chancelleries 22
  Classical ~ 77
  Franco-~ 22
  Medieval ~ 14, 57, 77, 363, 364, 369, 370, 372, 373
  Vulgar ~ 77, 133
Latvian.. 148
Law of Swamping 236
Lehn
  ~bedeutung...................................98
  ~bildung..............................98, 103
  ~formel..................................... 202
  ~fügung.................................... 202
  ~prägung.............................98, 99
  ~wort..................................98, 99
length .14, 150, 158, 159, 220, 222, 223, 224, 225, 226, 227, 228, 258
  ~ening.................. 14, 223, 225, 228

387

Index of subjects

consonantal ~.............................. 221
Middle English open-syllable-
~ening rule84, 223, 224, 225, 226, 228, 229
open-syllable ~ening 80, 84
vocalic ~........................218, 224, 226
vowel ~ 148, 346
lenition 133, 147
consonant ~..........................134, 146
morphological ~ 147
letters
private ~ 172, 177, 179, 183, 188
levelling
dialect ~ 235, 237
paradigm ~................................. 244
lexical
~ augmentation 102, 104
~ borrowing12, 13, 18, *97*, 144, 159, 305
~ calque.. 20
~ diffusion................................... 135
~ elaboration................................. 22
~ field........................13, *57*, 269, 270
~ intake 102
~ repetition.................................... 22
~ replacement 102, 103
~ stress 341, 343, 345, 357
~ transfer 143, 159
~isation 202
lexicology...13, 98
lexicon ....*See* vocabulary
lexicosemantics 98
line
Continental decasyllabic ~....... 348
decasyllabic ~......................340, 353
headless ~............................344, 347
Lydgatian ~.........................344, 345
lingua franca.................................... 12
linguistic
~ politeness 24
areal ~s.................................11, 72
Atlantic ~ area 73
genetic ~s......................... 108, 233

historical areal ~s...................74, 88
literacy ...14, 18, 65, 134, *171*, 172, 174, 184, 185, 186, 187, 188, 189, 190, 234, 310
literary
~ language 21, 310
standard ~ language...................79
literate
~ construction...................... 186, 189
Lithuanian..148
Livonian..148
loan
~ formula 202
~ translation................................20
~blend..98
~-formation...13, *195*, 197, 199, 200, 203, 205, 208, 210
~-rendition..................................197
~shift..98
~-translation 197, 201, 202
~-word......10, 11, 63, 72, 97, 98, 102, 107, 109, 133, 222, 223, 224, 235, 239, 245, 269, 270, 363, 370, 371
semantic ~...................................198
local
~ language 17, 235, 246
long
~ vowel evolution.......................80
low
~-level influence........................135
~-level transfer 155
L~ German..........................84, 305
Middle L~ German...................370
Lydgatian lines......................344, 345

M

macaronic....14, 19, 308, 309, 311, 313, 329
~ poetry....................................311
~ style 10, 14, 19, 308, 309, 311, 313, 329
~ writing.......................................14
maintenance

388

Index of subjects

language ~ 108
malapropisms 137
marker
   stylistic ~ 24
marking
   dependent-~ 134
   head ~ 134
matrix
   ~ language 313
maximum
   stress ~ principle 357
meanings
   forms relationship 49
medieval
   ~ abbreviation and suspension
      system 370, 372
   M~ Latin ... 14, 57, 77, 363, 364, 369,
      370, 372, 373
MEOSL ..... *See* Middle English open-syllable-lengthening rule
metaphor 63, 66, 67, 70, 284, 296
metathesis 146
metre 11, 41, 42, 220, 313, 314, 316,
   317, 325, *337*, 338, 339, 342, 344,
   345, 348, 349, 350, 351, 355, 356
   duple ~ 348
   four-stress ~ 354
metrical
   ~ triplet 339
   ~ unit 338
   ~ity 356
   extra~ity 351
   gradient ~ity 338
Middle
   ~ English open-syllable-lengthening rule 84, 223, 224,
      225, 226, 228, 229
   ~ Low German 370
   Late ~ English 65, 174, 188
Mischsprache 17, 306
mix
   language ~ing 14, 309, *363*
mixture

dialect ~ 9, 217
modal
   double ~ 236
mode 17, 18, 23, 325
model
   Wave ~ 72
Modern English
   Early ~ ... 14, 36, 64, 67, 70, 171, 172,
      174, 176, 188, 225, 311, 319
modifier 255
morpho-etymology 97
morphological
   ~ borrowing 206, 282
   ~ imitation 205
   ~ lenition 147
   ~ overload 208
   ~ simplification 80
   ~ structure 97, 205
   ~ transparency 48
   ~ typology .. 97, 98, 99, 101, 102, 110
morphology 12, 13, 81, 97, 98, 134,
   145, 148, 152, 158, 246, 342, 372, 374
   analytic ~ 81
   inflectional ~ 79, 80, 81, 288
   natural ~ 286
morphosemantic
   ~ structure 287
   ~s 13, 197, 207, 287, 288, 290, 294
morphotactic
   ~ compositeness 287
   ~ transparency 288, 290, 293
motion
   verbs of ~ 82
motivation ... 24, 46, 133, 144, 147, 154,
   239, 240, 287
multilingualism .. 76, 77, 174, 243, 305,
   307, 308, 310, 363
Munster Irish 146

N

narrative
   ~ poetry 19

Index of subjects

~ prose 172, 177, 178, 179, 182, 183, 185, 188
nasalisation 147
native 21, 255, 258
nativisation
   ~ hypothesis 46
natural
   ~ morphology 286
   ~ selection 296
   ~ syntax .. 48, 49, 64, 67, 69, 76, 140, 141, 198, 200, 202, 211, 241, 245, 259, 263, 286, 287, 319, 340, 352
   ~ness 76, 284, 294, 295, 300
negation
   sentence ~ 74, 82
negative
   ~ L1 transfer 46
   ~ politeness 24
Neuerung 58, 66
neuronal
   ~ reality 285, 295
new Englishes 234
newspaper *117*
nominalisation 150
nominalised
   ~ syntax 152, 160
nominative
   ~ case 282
   ~ experiencer 82, 280, 284, 289, 294, 373
non-native
   ~ Englishes . 234, 235, 238, 240, 242, 246
norm 19, 28, 58, 177, 189, 295, 300, 346, 358
   target ~ 242
Norman
   Anglo-~ ... 22, 27, 136, 137, 147, 221, 222, 223, 289, 329, 346, 363, 370, 371
north
   ~ern British English 37
   N~ American dialects 37
   N~ Germanic 73, 76, 81, 83, 305

noun
   ~-centred language 149

O

object
   ~ pronoun 83
objective
   ~ case 280
oblique case 259, 261
octosyllabic
   ~ verse 340
Old
   ~ Franconian 77
   ~ French 39, 133, 222, 223, 226, 228, 341
   ~ Frisian 77, 84
   ~ Provencal 39
   ~ Saxon 77, 84
onset 291
   syllable ~ 291
open
   ~-syllable lengthening 80, 84
openness 222
optimalisation
   interlanguage ~ 49
optimality 284, 295, 300
optimisation 11, *279*
oral
   ~ communication 57, 186
   ~ity . 14, *171*, 172, 184, 185, 187, 188, 190
order
   word-~ 23, 28, 71, 74, 76, 81, 83, 136, 149, 155, 158, 201, 204, 255, 307, 371, 372, 373
origin
   partial-contact ~ 11, *337*
orthography .... 217, 218, 221, 280 *See* spelling
Ottawa Valley English 37
overload
   morphological ~ 208
Ozark English 37

P

palatalisation 133
paradigm
   ~ levelling 244
   adjectival ~ 293
paraphrasing 23, 328
parliamentary business 22
partial-contact
   ~ origin 11, *337*
participial
   ~ construction 14, 188
   ~ passive tense 77
participle 59, 82, 100, 135, 136, 142,
   158, 171, 172, 173, 176, 181, 186,
   187, 188, 189, 205, 344, 350, 351, 369
   ~ clause 172
   ~ construction 172, 176, 186, 187,
   188, 189
particle
   pre-verbal ~ 49, 50, 142
passive 151
   analytic ~ 81
past tense 74
patient
   ~ experiencer 82
pattern
   ~ switching 312, 317, 324, 325
   alternating stress ~ 338
   basic clause ~ 173
   global change ~ 44, 49
   syntactic switching ~ 317, 324
pejoration ... 13, 267, 268, 269, 270, 275
pentameter
   ~ verse 11, 337, 353, 356
   ~ verse *337*
   iambic ~ 338, 339, 340, 343, 348,
   353, 356
perfect
   have-~ 77, 81
perfective 155
   immediate ~ 135, 142, 156
   immediate ~ aspect 135, 142
   resultative ~ 136, 155, 156

resultative ~ aspect 136
periphrasis 142
periphrastic
   ~ tense 78, 82
personal
   ~ pronoun 11, 24, 25, 78, 83, 86, 136,
   145, 150, 158, *279*, 280, 283, 286,
   287, 289, 293, 328
petitions 22
phonemic
   ~ vowel length 148
phonological
   ~ advantage 287
phrasal
   ~ stress 341, 349
   ~ verb 210, 245
phrase 14, 19, 21, 24, 25, 26, 28, 33, 39,
   202, 234, 256, 317, 327, 329
   adverbial ~ 19, 321, 323
pidgin 35, 154, 239
   ~isation 234
placement
   caesural ~ 341
place-name 13, *117*
plain infinitive 40
planter
   ~ English 138
   ~s 138, 139
plural
   analytic ~ 142
poetic
   ~ form 311, 317, 325
poetry
   macaronic ~ 311
   narrative ~ 19
politeness 86
   linguistic ~ 24
   negative ~ 24
   positive ~ 24
polycompetence 241, 243
polysemy 61, 62, 63, 64, 65, 142
   post-borrowing ~ 63, 64, 65
   pre-borrowing ~ 62, 63, 64, 65

Index of subjects

Portuguese.........................82, 123, 340
position
  ictic ~......................................341, 349
positive
  ~ politeness 24
possessive
  ~ case.. 280
post
  ~-borrowing polysemy....63, 64, 65
  ~-posed genitive........................ 83
  ~position..........74, 76, 108, 208, 255
  ~-positive article....................... 145
postmodification....173, 176, 178, 179, 180, 183, 184
postmodify
  ~ing construction 173, 174, 175, 176, 179, 181, 183, 184, 185, 186, 187
pragmatic
  ~ calque....................................20, 24
  ~s17, 24, 86, 206, 240, 245, 246, 325, 363, 371, 373
pre
  ~-borrowing polysemy...62, 63, 64, 65
  ~-dialectal drift............................ 70
  ~-verbal particle 49, 50, 142
predicate
  raising ~... 43
preference
  universal ~s................................ 284
prefix
  ~ verb 210
  ~al derivative............................. 210
  ~ation..........................208, 209, 210
preposition ... 14, 39, 45, 48, 49, 50, 74, 76, 80, 85, 108, 150, 151, 152, 157, 173, 208, 244, 261, 282, 322, 323, 346, 369, 373
  ~al infinitive................................. 39
  ~al phrase 14, 85, *251*, 322, 323, 373
  ~al pronoun 145, 150, 151, 152, 157
  ~al serialisation......................49, 50

present
  ~ perfect tense 74
  habitual ~.................................... 156
prestige 57, 65, 73, 76, 83, 85, 154, 155, 310, 325
  ~ group.. 154
  ~ language 73
principle
  stress maximum ~......................357
printing..85, 87
private
  ~ letters........ 172, 177, 179, 183, 188
process
  psycholinguistic ~.......................233
  unmarking ~..................................48
  weakening ~ 291
processing
  language ~ 186, 187
progressive
  ~ tense 10, 14, 49, 82, 209, 269
prominence 294, 295, 301, 339, 341, 346, 354, 355
  sign ~ 287, 291, 295
pronominal
  ~ clitic 74
pronoun...11, 24, 25, 77, 78, 80, 83, 86, 145, 150, 151, 153, 157, 158, 245, 279, 280, 281, 282, 283, 284, 286, 287, 288, 289, 290, 291, 293, 294, 295, 298, 300, 301, 342, 343, 346, 369
  ~ copying 245
  ~ personal 11, *279*, 287, 293
  demonstrative ~ 79, 81, 109, 287, 289
  object ~ 83
  personal ~ 24, 25, 78, 83, 86, 136, 145, 150, 158, 279, 280, 283, 286, 289, 328
  prepositional ~ .. 145, 150, 151, 152, 157
  second person ~ 145
  subject ~ 81
prose

392

Index of subjects

~ style.. 174
curial ~.................................174, 176
narrative ~.. 172, 177, 178, 179, 182, 183, 185, 188
statutes ~ 188
prose style....................................22, 202
prosody 10, 83, 131, 132, 137, 144, 157, 224, 226, 229, 337, 341, 345, 354, 371
proto-
~form 69
P~Germanic 70
Provencal
Old ~... 39
pseudo-derivative 208
Pseudo-Mischsprache.................... 306
psycholinguistic
~ process....................................... 233
purpose .. 14, 40, 42, 43, 45, 50, 66, 158
purposive
~ function 50

Q

quasi-compound............................ 210
quotation. 119, 121, 201, 271, 272, 308, 326, 327

R

raising
~ predicate.................................... 43
rank
~ term...................................272, 273
style ~ing..................................... 243
read
~ing competence 187
reality
neuronal ~............................285, 295
reception 58, 66
reconstruction
historical ~.................................... 69
redefinition
rhythmic ~.................................. 350
reduction

~ vowel..294
redundancy
analytic ~.......................................49
reduplication.................................143
referential
~ sphere.........................13, 267, 269
anaphoric ~ devices....................22
register 295, 305
technical ~......................................86
reiteration....................................328
relation
intertextual ~ 326
thematic ~ 255, 257
relevance
subject ~ 151
Renaissance. 10, 13, 57, 85, 86, 87, 351, 357
rendition
loan-~..197
replacement
lexical ~ 102, 103
replica
structural ~ 206
resolution 147, 174, 295, 345
resultative
~ perfective.................136, 155, 156
~ perfective aspect......................136
retention
language ~ 140
shared ~...75
reversal
trochaic ~..... 342, 343, 348, 349, 353
rheme 81
rhyme. 19, 120, 220, 271, 288, 314, 329, 345, 346
rhyming
~ function....................................329
rhythm.....338, 339, 341, 342, 350, 353, 356
~ic duples...349
~ic redefinition.............................350
~ic triples 339, 349, 353, 356
~ic unit 338

393

Index of subjects

~ic variability 338
~ic variation 350
~ical triple 355
duple ~ 337, 338, 353, 356, 357
iambic ~ 11, 359
rising ~ 357, 358
triple ~ 338, 340, 342, 348, 350, 353, 354, 355, 356
rising
~ rhythm 357, 358
Romance 9, 11, 18, 33, 39, 45, 72, 74, 77, 80, 81, 82, 83, 84, 86, 103, 107, 133, 134, 135, 337, 338, 339, 340, 341, 345, 346, 347, 349, 350, 351, 357, 358, 359, 372
Ibero-~ 133
Western ~ 132, 133
Romantsch 74, 83
rule
~ generalisation 46
~ reanalysis 43
Middle English open-syllable-lengthening ~ .. 84, 223, 224, 225, 226, 228, 229
Scottish vowel length ~ 148
Russian 73, 119, 156

S

salutation 25
Saxon
Anglo-~. 55, 100, 195, 196, 197, 203, 204, 205, 206, 209, 211, 264, 267, 268, 279, 280, 281, 282, 283, 287, 288, 290, 291, 293, 294, 295, 301
Old ~ 77, 84
Scandinavian 10, 11, 13, 14, 35, 38, 39, 48, 50, 79, 80, 84, 98, 100, 103, 104, 105, 107, 109, 110, 148, 228, *267*, 270, 272, 274, 275, *279*, 280, 281, 282, 283, 284, 286, 287, 290, 291, 293, 294, 295, 300, 301, 305, 370, 371, 373

scansion ... 338, 342, 343, 344, 345, 346, 351, 352, 354, 357, 358
schema
cognitive ~ 285
Schriftsprache 255
Scottish 38, 133, 148, 233, 236
~ Gaelic 133
~ vowel length rule 148
scribes
Chancery ~ 86
second
~ language acquisition ... 14, 45, 48, 50, 82, 140, 144, 154, 155, 233, 234, 238, 239, 240
~ person pronoun 145
selection
natural ~ 296
semantic
~ bleaching 49
~ borrowing 102, 104, 198, 199, 202, 203, 207
~ change 13, 102, 103, 104, 267, 269, 270
~ differentiation . 102, 103, 105, 270
~ domain 13, 97, 105, 107, 110
~ extension 142
~ field 65, 270
~ intake 102
~ loan 198
~ shift 239
~ typology 98, 102, 105
~s 12, 13, 42, 46, 49, 50, 63, 64, 65, 66, 97, 98, 102, 103, 104, 105, 110, 137, 142, 149, 150, 152, 181, 186, 198, 201, 202, 203, 205, 206, 207, 208, 211, 239, 251, 252, 256, 258, 267, 268, 269, 270, 271, 275, 306, 307, 318, 346, 363
semi-
~communication .. 14, *363*, 370, 373
~creole 143
sense
~ development 63, 64, 66, 67

Index of subjects

~ specialisation 66
sentence
   ~ negation 74, 82
   Germanic ~ brace 136
serialisation
   prepositional ~ 49, 50
   verbal ~ 49, 50
shared
   ~ innovation 75
   ~ retention 75
shift
   accent ~ 132
   Dublin vowel ~ 136
   language ~ 108, 109, 155
   semantic ~ 239
   vowel ~ 84
shortening 14, 84, 225, 228
   ~ of vowels in closed syllables .. 84
sign prominence 287, 291, 295
signature 25
simplification
   consonant cluster ~ 244
   morphological ~ 80
situation
   contact ~ 9, 11, 17, 27, 44, 62, 65, 71,
      78, 97, 98, 100, 106, 107, 108, 109,
      110, 131, 132, 137, 154, 157, 196,
      237, 238, 241, 281
Slavic 81, 132
slurring 348
social
   ~ status 13, 21, 25, 267, 271, 272
   ~ term 13, *267*, 268, 269, 270, 271,
      272
society
   multicultural ~ 28
socio-historical
   ~ typology 98, 105, 106, 110, 189
sociolect 27
sociolinguistic
   ~ typology 108
sociolinguistic
   ~ typology 98, 106, 110

~s 11, 12, 13, 25, 110, 235, 238, 239,
   275, 282
South
   ~ African English 12, 235, 240
   ~ern English 236, 243
Spanish 39, 47, 73, 82, 133, 235, 340
   Castilian ~ 244
speaker
   ~ contact 143, 152, 155
specialisation
   sense ~ 66
speech
   ~ habits 138, 139, 155
   bilingual ~ behaviour 330
spelling 11, 14, 85, 86, 87, 118, 124,
   126, 127, 129, *217*, 218, 221, 222,
   223, 224, 225, 226, 227, 228, 280
   ~ habit 217, 218, 227
   ~ reform 85, 87
   double ~ 217, 220
   inverse ~ 221, 225, 227
sphere
   connotative ~ 268
   evaluative ~ 267, 268, 269, 270, 272
   referential ~ 13, 267, 269
Sprachbund ... 11, 47, 72, 73, 75, 76, 81,
   88
standard
   ~ languages 86
   ~ literary language 79
   ~isation 189, 190
   S~ Average European 73, 75, 77,
      80, 81, 82
   S~ English ... 35, 38, 49, 50, 176, 188,
      189
   S~ High German 101
stative
   ~ verb 156
status .. 12, 13, 18, 25, 28, 107, 131, 149,
   154, 159, 173, 176, 177, 195, 198,
   200, 202, 204, 210, 242, 267, 268,
   272, 274, 279, 281, 285, 293, 301,
   310, 329, 348, 364

Index of subjects

social ~ 13, 21, 25, 267, 271, 272
statutes 172, 174, 175, 176, 178, 183, 188, 190
    ~ prose 188
stereotypical
    ~ information 268
strength
    consonantal ~ 293
stress .. 83, 132, 133, 137, 156, 157, 222, 223, 224, 228, 229, 341, 342, 344, 345, 346, 347, 349, 351, 356, 357
    ~ maximum principle 357
    ~-timing 10
    alternating ~ pattern 338
    duple alternating ~ 352
    expiratory ~ 10
    fixed ~ accent 132
    four-~ metre 354
    initial ~ 132, 346
    lexical ~ 341, 343, 345, 357
    phrasal ~ 341, 349
strong
    ~ consonant 291
structural
    ~ contrast 12
    ~ replica 206
    ~ transfer 143
structure
    analytic ~ 49
    front-focussing ~ 142
    morphosemantic ~ 287
    syllable ~ 84
style 17, 18, 22, 23, 24, 25, 26, 27, 28, 33, 85, 86, 171, 172, 176, 189, 199, 234, 243, 245, 255, 305, 330, 338, 350, 372
    ~ ranking 243
    Attic ~ 23
    curial ~ 22, 23, 25
    indexes of ~ 19
    macaronic ~ 10
    prose ~ 22, 174, 202
stylistic

~ architecture 28
~ calque 20, 23
~ marker 24
subject
    ~ pronoun 81
    ~ relevance 151
subordinator
    adverbial ~ 83
substitution .. 23, 98, 100, 102, 227, 344
    trochaic ~ 350
substrate 10, 77, 83, 107, 133, 136, 154, 243, 244
substratum .... See substrate
superlect 75, 76, 77, 83
superstrate 107, 159
    ~ borrowing 107
superstratum .... See superstrate
suppletion 282, 287, 294
support
    ~ verb 149
    ~ive transfer 45, 158
suspension
    medieval abbreviation and ~ system 370, 372
swamping 237
    Law of S~ 236
Swedish 38, 83, 129, 308
Swiss High German 101
switch
    ~ boundary 313
    intersentential ~ 306, 317
    intrasentential ~ 307, 317, 318, 319, 326
    tag ~ 317
    tag-like ~ 306, 317
switching
    ~ pattern 312, 317, 324, 325
    syntactic ~ pattern 317, 324
syllabic
    ~ count 340
syllable
    ~ onset 291
    ~ structure 84

Index of subjects

~-timing.................................134, 350
heavy ~ coda.............................. 147
Middle English open-~-
    lengthening rule......84, 223, 224,
    225, 226, 228, 229
open-~ lengthening................80, 84
syncope....................................344, 351
synizesis....................340, 344, 348, 351
synonym..........................104, 271, 272
~ous doublet.............................. 22
syntactic
    ~ calque...................................... 20
    ~ code-switching...................... 144
    ~ compression........................... 176
    ~ group...................................... 200
    ~ switching pattern............317, 324
    ~al drift....................................... 70
syntax .10, 11, 13, 21, 28, 40, 47, 48, 49,
    66, 73, 80, 81, 103, 131, 134, 143,
    144, 148, 150, 152, 155, 156, 158,
    171, 172, 176, 180, 187, 200, 201,
    203, 204, 208, 239, 246, 251, 252,
    255, 305, 306, 307, 308, 309, 310,
    313, 317, 318, 319, 321, 324, 329,
    338, 341, 342, 348, 350, 363
    nominalised ~.....................152, 160
system
    ~atic transfer............................ 155
    Complex Adaptive S~.......279, 286,
    295, 297, 298, 299
    inflectional ~...............260, 288, 289
    medieval abbreviation and
        suspension ~...................370, 372

T

tag......................................244, 306, 317
~ switch 317
~-like element............................ 329
~-like switch.......................306, 317
~-switch..................................... 144
adverbial ~.................................. 24
talk
    foreigner ~................................... 49

target
    ~ norm..242
technical
    ~ language 19
    ~ register.....................................86
    ~ term..................... 99, 200, 203, 204
    ~ vocabulary...............................24
technique
    didactic ~...................................206
    etymological ~...........................203
telic...156
    ~ construction...........................156
template...337, 340, 341, 342, 343, 344,
    345, 348, 349, 351, 356, 357, 363
tense
    analytic future ~......................79, 81
    analytic past ~ 81
    andative future ~ 82
    participial passive ~ 77
    past ~ 74
    periphrastic ~ 78, 82
    present perfect ~ 74
    progressive ~ 10, 14, 49, 82, 209,
    269
term
    ~s of address......................245, 272
    inkhorn ~s......................................10
    rank ~ 272, 273
    social ~...*13, 267*, 268, 269, 270, 271,
    272
    technical ~............. 99, 200, 203, 204
terminology...26, 66, 98, 100, 102, 204,
    205, 206
text
    ~ type.14, 59, 83, 100, 110, 172, 176,
    177, 178, 180, 183, 187, 188, 189,
    190, 309, 311, 312, 330, 363, 372
    business ~............................309, 319
    informal ~ types 177, 178, 180, 184,
    186, 188
textual
    ~ functions of code-switching..325
thematic

Index of subjects

~ relation 255, 257
thematisation............................... 244
theme.. 81
theory
   evolution ~ 12
   genetic selection ~ 12
   Wave ~... 78
time
   duple ~..........................349, 350, 352
   triple ~ 349, 350, 353
timing
   stress-~... 10
   syllable-~ 134, 350
to-infinitive.......................37, 38, 40, 41
Tok Pisin 142
tone... 148
topicalisation............................149, 244
transfer 39, 44, 134, 136, 138, 141, 142, 145, 153, 154, 155, 157, 159, 240, 241, 337
   direct ~.. 134
   lexical ~................................143, 159
   low-level ~.................................. 155
   negative L1 ~ 46
   structural ~................................. 143
   supportive ~........................45, 158
   systematic ~ 155
transition
   ~al construction......................... 180
translation 14, 22, 23, 26, 28, 39, 48, 65, 76, 87, 100, 118, 145, 171, 195, 197, 198, 200, 202, 203, 204, 205, 206, 208, 210, 251, 253, 254, 257, 258, 259, 262, 263, 327, 328, 364, 365, 372
transparency...... 50, 154, 205, 287, 288, 289, 293, 301
   morphological ~ 48
   morphotactic ~............288, 290, 293
triple 349, 350, 352, 353, 354, 355, 356, 357, 358, 359
   ~ rhythm...... 338, 340, 342, 348, 350, 353, 354, 355, 356
   ~ time 349, 350, 353

clear ~ 356
   rhythmic ~ 339, 349, 353, 356
   rhythmical ~ 355
triplet......................... 338, 343, 350, 351
   metrical ~ 339
trisyllabic
   ~ foot............ 338, 340, 344, 348, 356
trochaic
   ~ inversion..................................356
   ~ reversal..... 342, 343, 348, 349, 353
   ~ substitution 350
trochee 343, 349
type
   ~s of contact........................132, 134
   informal text ~............................188
   informal text ~s. 177, 178, 180, 184, 186
typological
   ~ change......................................135
   ~ convergence 71
   ~ drift...............................70, 71, 76
typology .. 11, 12, 14, 47, 50, 70, 71, 72, 73, 75, 76, 80, 82, 84, *97*, *131*, 197, 198, 200, 243, 255, 264, 306
   areal ~.....................................71, 72
   language change ~.......................35
   morphological ~...... 97, 98, 99, 101, 102, 110
   semantic ~..................... 98, 102, 105
   socio-historical ~. 98, 105, 106, 110, 189
   sociolinguistic ~ ... 98, 106, 108, 110

Ü

Übernahme..58

U

Ulster
   ~ Irish...................................133, 148
umlaut.............................. 123, 125, 133
i-~ 70, 79, 84
unit

Index of subjects

metrical ~ 338
rhythmic ~ 338
universal
~ preferences 284
U~ Grammar 234
unmarking process 48

V

variability
rhythmic ~ 338
variation
rhythmic ~ 350
velarisation 133
verb
~al borrowing 59
~al communication 64
~al serialisation 49, 50
~s of motion 82
~s of verbal communication . 13, 57
~s of verbal influencing 65
auxiliary ~ 81
phrasal ~ 210, 245
prefix ~ 210
stative ~ 156
support ~ 149
vernacular
spoken ~ 18
Verschleifung 348
verse
hendecasyllabic ~ 340, 341
iambic ~ 337, 359
octosyllabic ~ 340
pentameter ~ 11, *337*
version
interlinear ~ 195, 196, 197, 207
vocabulary 10, 11, 12, 13, 18, 19, 23,
24, 26, *57*, 73, 81, *97*, 135, 138, 142,
144, 152, 159, *195*, 196, 197, 198,
199, 204, 205, 206, 207, 208, 209,
210, 239, 243, 246, 252, 253, 256,
260, 261, *267*, 268, 275, 282, 307,
309, 337, 340, 346, 359, 370, 374

vocalic
~ digraph 220, 221
~ length 218, 224, 226
vowel
~ epenthesis 141, 146
~ length 148, 346
~ reduction 294
~ shift 84
Dublin ~ shift 136
long ~ evolution 80
phonemic ~ length 148
Scottish ~ length rule 148
shortening of ~s in closed
  syllables 84
Vulgar Latin 77, 133

W

Wave
~ model 72
~ theory 78
weak
~ consonant 291
~ constraints 48
~ening of inflections 84
~ening of unstressed vowels 79
~ening process 291
Welsh 133, 134
West
~ern European languages 72, 73,
75, 76, 81
word
~ class 149
~-formation 21, 33, 81, 108, 197, 202
~-order. 23, 28, 71, 74, 76, 81, 82, 83,
136, 149, 155, 158, 201, 204, 255,
307, 371, 372, 373
~-play 329
hard ~s 10
loan-~ 10, 11, 72, 97, 98, 102, 107,
109, 133, 222, 223, 224, 235, 239,
245, 269, 270, 363, 371
written language 18, 33, 76, 86, 186

Index of names

A

Adams, George Brendan 140, 160
Adamson, Sylvia 246, 247, 248
Ælfric 100, 197, 203, 204, 205, 251, 253, 256, 257
Aitken, Adam Jack 148, 161
Alfred 251
Algeo, John 101, 110, 237, 246
Amos, Ashley Grandell 212
Andersen, Henning 46, 94
Anderson, John M. 96, 360, 362
Appel, René 134, 161
Archibald, Elizabeth 330
Ard, Josh 242, 246
Armin, Robert 177, 178, 179, 181, 182, 185, 190
Asher, Robert E. 88, 89, 95
Aspin, Isabel S. T. 314, 316, 330
Aungier, George James 29
Auwera, Johan van der .. 72, 88, 91, 92

B

Bæda .... *See* Bede, Saint
Bahner, Werner 51
Bailey, Charles-James N. 9, 15, 17, 29, 142, 161, 166
Bailey, Guy 237, 246
Ball, Christopher 210
Banfi, Emanuele 75, 88, 93
Barber, Charles 171, 192, 344, 359
Barber, Nicolas 344, 359
Barney, Stephen A. 337, 338, 344, 345, 350, 351, 353, 359
Bauer, Gero 9, 15
Baugh, Albert Croll . 18, 29, 75, 88, 97, 109, 111, 280, 302
Baum, Paul 338, 340, 348, 356, 359

Bechert, Johannes 71, 73, 74, 76, 88, 89, 95
Becker, H. 73, 89
Bede, Saint 252, 254
Behn, Aphra 172, 177, 178, 179, 183, 185, 190
Behrens, Werner 171, 178, 186, 192
Benskin, Michael ... 161, 217, 230, 306, 330
Benson, Evelyn 101, 111
Benson, Larry D. 342, 359
Benson, Morton 101, 111
Bentley, Elna 19, 29
Berk-Seligson, Susan 144, 161, 330
Berndt, Rolf 97, 109, 111, 217, 229, 310, 330
Bernini, Giuliano 73, 74, 81, 83, 88, 89, 95
Berns, J. 78, 90, 93, 95
Bertinetto, Pier Marco 90
Besch, Werner 94
Betz, Werner 98, 100, 103, 111, 198, 202, 209, 211, 215
Bhaldraithe, Tomás de 146, 161
Biber, Douglas 186, 192
Bickerton, Derek 54, 241, 243, 246
Birkmann, Thomas 205, 212
Birnbaum, Henrik 76, 89
Bischoff, Otto 338, 340, 348, 359
Björkman, Erik 275, 276
Blake, B. J. 73
Blake, Norman 86, 89, 280, 302
Blatt, Franz 171, 172, 192
Bliss, Alan J. 37, 51, 137, 138, 144, 156, 161, 162
Bloomfield, Leonard . 72, 89, 106, 107, 108, 111
Blum, Siegfried 213
Boas, Franz 72, 89

401

Index of names

Bøgholm, Niels 112, 221, 225, 227, 229
Böker, Uwe 230
Bolton, Whitney French ... 97, 101, 111
Bornstein, Diane 22, 23, 29
Bossong, Georg 82, 89
Bosworth, Joseph 198, 211, 261, 264
Bourcier, Georges 97, 112
Bradley, Henry 215, 276, 375
Braunmüller, Kurt .. 363, 370, 371, 374
Breatnach, Risteard B. 146, 161
Brekle, Herbert Ernst 112
Brett-Smith, H. F. B. 191
Brinton, Laurel J. 156, 162
Broderick, George 168, 304
Brooks, Nicholas 211, 214, 215
Brown, Penelope 24, 29
Bruce, John 191, 208
Bruch, Robert 44, 51
Brunner, Karl 17, 29, 38, 45, 51
Buchholz, Oda 75, 91
Búrca, Seán de 146, 162
Burchfield, R. W. 215
Buridant, Claude 73, 88, 89, 95
Burnley, John David 11, 12, *17*, 22, 26, 29, 174, 175, 176, 192, 274, 276, 337
Busse, Ulrich 99, 112
Busse, Wilhelm G. 112
Bynon, Theodora 84, 89
Byrne, F. J. 161
Byrne, Francis 54

C

Cable, Thomas 75, 88, 97, 109, 111, 280, 302
Calvo, Juan José 52, 375
Cameron, Angus 210, 212
Campbell, Lyle 71, 89, 279, 302
Carpenter, Charles Whitney II 209, 211
Carroll, Susanne 37, 43, 51
Carstensen, Broder 99, 112
Caxton, William ... 22, 25, 39, 177, 179, 182, 185, 190, 194

Chamberlain, John 191
Chambers, Jack K. ... 163, 164, 168, 169
Chambers, Raymond Wilson..... 22, 29
Chaucer, Geoffrey 11, 18, 20, 22, 24, 25, 26, 34, 39, 40, 41, 52, 120, 126, 218, 226, 274, 313, 337, 338, 339, 340, 341, 342, 343, 344, 345, 346, 347, 348, 349, 350, 351, 352, 353, 354, 356, 357, 358, 359
Cheshire, Jenny 248
Chomsky, Noam 35, 51, 229
Clahsen, Harald 47, 51
Clanchy, Michael Thomas .. 28, 30, 76, 86, 89
Clark Hall, John R. 198, 212
Clarkson, Iain 96
Clemoes, Peter 114
Clyne, Michael 144, 162, 331
Coetsem, Frans van 86, 90
Coleman, Evelyn Scherabon . 198, 212
Comrie, Bernard 81, 90
Congreve, William 172, 177, 178, 179, 182, 183, 185, 191
Corder, Stephen Pit 242, 246
Corrigan, Karen P. 37, 51, 141, 162
Coseriu, Eugenio 58, 66, 67
Craigie, William Alexander . 215, 276, 375
Criper, Clive 249
Crozier, Alan 162
Cruse, David Alan 66, 67
Crystal, David 227, 229

D

Dahl, Östen 81, 90, 156, 162
Dahlgreen, Kathleen 267, 268, 276
Danchev, Andrei ..5, 11, 13, *35*, 39, 40, 44, 46, 49, 50, 51, 52, 82, 90, 238, 247
Daunt, Marjorie 22, 29
Davies, Alan 249, 319, 324, 326
Davis, Alison 36, 37, 43, 52
Davis, Norman 178, 192, 231, 361
Dawkins, Richard 279, 297, 302

Index of names

Décsy, Gyula 73, 82, 90
Dekeyser, Xavier 18, 30
Delattre, Pierre 133, 162
Delepierre, Octoave 331
Deloney, Thomas ... 173, 177, 179, 183, 185, 190
Diercks, Willy 370, 374
Dietrich, Julia C. 162
Dilevski, Nikolai 53
Diller, Hans-Jürgen 10, 13, 57, 309, 331, 337
Dirven, René 91
DiSciullo, Anna-Maria 144, 162
Dolan, Terence P. 162, 163, 165
Dowling, Patrick J. 137, 162
Dressler, Wolfgang U. 279, 284, 286, 302, 303
DuCange, Charles 68
Dudley, Robert 191
Duffell, Martin J. 338, 340, 341, 343, 349, 350, 351, 352, 353, 355, 360
Dury, Richard 11, 69

E

Earle, John 264
Eastman, Carol M. 144, 162, 332
Eaton, Roger 115, 303, 304
Edmondson, Jerold A. 166
Edwards, John 155, 163
Einenkel, Eugen 38, 45, 52, 55
Ekwall, Eilert 109, 112
Ellis, Henry 19, 30
Emeneau, Murray Barnson 72, 90, 145, 163
Erickson, Jon L. 69, 95

F

Fanego, Teresa 42, 52
Farkas, Judith 38, 52
Fase, Willem 163
Fasold, Ralph W. 166
Feagin, Crawford 142, 163, 166

Ferguson, Charles A. 49, 53
Fernández, Francisco 52, 375
Ferré, Barbara M. 99, 113
Feuillet, Jack 72, 89, 90
Fillmore, Charles 166
Filppula, Markku 131, 155, 163
Finkenstaedt, Thomas 58, 68
Fischer, Andreas 12, 97, 103, 112
Fischer, Olga 36, 40, 42, 43, 50, 53, 115, 303
Fisher, Jane L. 175, 193
Fisher, John H. 86, 90, 174, 175, 193
Fisiak, Jacek ... 9, 30, 52, 68, 89, 90, 113, 164, 165, 222, 229, 265, 375
Fitikides, T. J. 47, 53
Fix, Hans 205, 212
Flasdieck, Hermann Martin 17, 30
Forcellini, Egidio 68
Foster, Robert F. 139, 163
Fowkes, Robert A. 134, 163
Frank, Roberta 210, 212
Freeman, Thomas Walter 140, 163
Fréine, Séamus de 137, 163
Fretheim, Thorstein 163
Fries, Udo ... 13, 117, 217, 221, 229, 337
Frieshammer, Johann 220, 229
Frye, Northrop 354, 356, 360
Fuster, Miguel 52, 375

G

García, Ofelia 110
Gardner, Robert C. 140, 163
Garmonsway, G. N. 264
Gasparov, M. L. 349, 360
Gass, Susan M. 246, 247, 248
Gatenby, Edward Vivian 68
Gawdy, Philip 191, 192
Gaylord, Alan 338, 360
Gell-Mann, Murray 279, 286, 298, 303
Genthe, F. W. 311, 331
Georges, Karl Ernst 63, 68
Georgiev, Emil 53
Gerritsen, Marinel 54, 95

403

Index of names

Gneuss, Helmut 98, 113, 196, 197, 198, 199, 203, 205, 207, 212
Godden, Malcolm R. 252, 255, 265
Gollancz, Israel.................... 113, 192
Göller, Karl Heinz.......................... 230
Gooder, Eileen A........................ 374
Görlach, Manfred....... 78, 90, 101, 104, 105, 113, 167, 171, 193, 236, 247
Götz, Heinrich........................ 197, 213
Gower, John........... 19, 20, 30, 217, 352
Gradon, Pamela 20, 31
Gray, Douglas 231, 361
Greenbaum, Sidney............... 193, 194
Greenberg, Joseph Harold........ 71, 91
Greene, David 156, 163
Gregory, Steward........................ 376
Grigor'ev, P. V................................ 360
Grosse, Rudolf............................. 213
Grundy, Lynne..................... 106, 115
Gumperz, John J. 28, 30, 145, 164, 247, 305, 306, 312, 325, 326, 328, 329, 331
Gusmani, Roberto................. 270, 276
Guthrie, Steven R.... 342, 357, 358, 360

H

Hagège, Claude.......................... 83, 91
Halle, Morris 229, 357, 359, 360
Halliday, Michael 86, 91
Hancock, Ian................................ 249
Harding, Vanessa 364, 374
Harington, John 191
Harley, Brilliana..... 192, 310, 316, 317, 319, 320, 322, 327, 329
Harman, Thomas... 177, 179, 181, 182, 185, 190
Harris, John 131, 135, 156, 164, 168, 247
Harris, M............................... 117, 129
Harris, Roy................................... 166
Harris, Zellig S. 70, 91
Harvey, Carol J............................ 331
Haspelmath, Martin 73, 75, 77, 80, 81, 82, 89, 91

Haugen, Einar 84, 91, 98, 100, 113, 143, 164, 189, 193, 363, 370, 374
Hawkins, John A. 255, 265, 303
Hayes-McCoy, G. A. 137, 164
Healey, Antonette diPaolo.... 212, 214
Heinzle, Joachim 213
Hellan, Lars.................................. 163
Heller, Monica 331, 332, 333
Hendrick, Randall 168
Henel, Heinrich 264
Henkel, Nikolaus................... 197, 213
Henry, Patrick Leo 152, 164
Henry, Philip................................ 192
Herzog, Marvin I..................... 241, 249
Heuser, Wilhelm 17, 30
Hickey, Raymond...... 10, 45, 113, *131*, 143, 148, 164, 165, 265
Hiltunen, Risto......................... 83, 91
Hindley, Reg 140, 165
Ho, Mian Lian...................... 245, 248
Hoccleve, Thomas 19, 20
Hock, Hans Henrich 74, 75, 76, 91
Hoenigswald, Henry M............. 71, 91
Hoffmann, Charlotte...... 144, 155, 165
Hofmann, Johann Baptist...... 200, 213
Hogg, Richard M.... 213, 251, 255, 265
Holm, John 154, 165, 167
Holmes, Janet.............................. 193
Holt, Robert.................... 291, 292, 303
Honey, John 53
Hornby, Albert Sydney 68
Hosali, Priya........................... 245, 248
Howatt, Anthony P. R. 249
Howlett, David R. 321, 332, 374
Huchon, René Louis................... 78, 92
Huebner, Thom 54
Hughes, Kathleen 114
Hunt, Tony 309, 310, 331
Hussey, S. S. 29
Hymes, Dell................ 28, 30, 164, 247

I

Ihalainen, Ossi 32, 52, 93, 249, 335, 375

404

Ilson, Robert F. 101, 111

J

Jack, George 41, 42, 50, 53, 55, 323
Jacobson, Rodolfo 331
Jahr, Ernst Håkon 376
Jake, Janice L.....................307, 319, 331
Jakobson, Roman..............48, 148, 165
Jaspaert, Koen 163
Jeayes, Isaac Herbert 191, 192
Jespersen, Otto ... 18, 30, 43, 53, 75, 83, 92, 102, 112, 113, 280, 303
Johnson, Leslie Peter 213, 368
Jones, Charles................................. 96
Jorgensen, Peter A. 99, 113
Joseph, John Earl...................77, 86, 92
Jungandreas, Wolfgang 79, 92

K

Kaartinen, Anja 41, 50, 53
Kachru, Braj B.................144, 165, 247
Kallen, Jeffrey.......... 131, 140, 156, 165
Kaplan, Theodore 18, 30
Käsmann, Hans.......................104, 113
Kastovsky, Dieter 9, 15, 30, 52, 90, 112, 207, 208, 209, 210, 213, 256, 265, 335, 363, 375
Kaufman, Terrence .. 38, 39, 40, 54, 80, 96, 97, 108, 109, 115, 139, 159, 168
Kay, Christian J. 106, 114, 115
Kerkhof, Jelle 41, 53
Kess, Joseph F............................... 111
Keyser, Samuel Jay 357, 359, 360
Kibbee, Douglas A...... 17, 30, 310, 331
Kilpiö, Matti 252, 265
Kimmens, Andrew C. 208, 213
King, J. E.. 264
Kingsmill, Allison...................208, 214
Kjellmer, Göran.....................83, 92
Klaeber, Fr. 264
Klemola, Juhani.............................. 93
Knowles, Gerry 78, 79, 80, 92

Köbler, Gerhard......................209, 213
Kohnen, Thomas 10, 14, *171*, 172, 185, 193
Kökeritz, Helge.......................344, 360
Kolb, Eduard................................112
Kolb, Herbert 215
Koopman, Willem 115, 303
Kornexl, Lucia.....10, 12, 100, *195*, 199, 200, 201, 204, 207, 210, 213
Kortmann, Bernd.......................83, 92
Koziol, Herbert . 97, 114, 186, 193, 230
Kristensson, Gillis 217, 229
Kroon, Sjaak 163
Kuhn, Sherman S....................276, 374
Kurath, Hans..........................276, 374
Kuteva, Tania.................. 73, 77, 81, 92
Kytö, Merja....11, 13, *35*, 44, 52, 82, 90, 114, 172, 177, 190, 193, 194, 230

L

Labov, William ...84, 92, 238, 241, 247, 249, 311
Laing, Margaret..... 231, 306, 311, 330, 332
Lambert, Wallace E. 140, 163
Langendoen, D. Terence...............166
Langland, William 25, 29, 308, 333
Lapidge, Michael.... 199, 206, 213, 214
Lasnik, Howard........................35, 51
Lass, Roger 79, 85, 92, 97, 111, 114, 236, 238, 247, 279, 281, 282, 285, 287, 303, 304, 360
Latham, Ronald E........... 321, 332, 374
Lauffer, Hartmut...........................215
Law, Vivien...................................246
Lazzarini, L. 332
Lazzeroni, Romano 276
Lee, Alan......................................129
Leech, Geoffrey.............................194
Leek, Frederike van der......... 115, 303
Lefebvre, Claire 39, 53
Legge, Dominica................... 18, 22, 31
Lehiste, Ilse...........................148, 165

Index of names

Lehmann, Winfred Philip . 81, 92, 249, 302
Leisi, Ernst 58, 68, 101, 114
Leith, Dick 267, 270, 276
Lekov, Ivan 53
Leuvensteijn, J. van 78, 90, 93, 95
Levinson, Stephen C 24, 29
Lewey, Ernst 73, 93
Lewis, Clive Staples 312, 332, 354, 356, 360
Lightfoot, David W .. 36, 43, 50, 53, 82, 93
Lindelöf, Uno Lorenz 78, 93
Lindsay, Wallace Martin 203, 214
Lipka, Leonhard 112
Little, David 168
Luick, Karl 15, 217, 221, 230
Lutz, Angelika 205, 210, 214

M

Machan, Tim William 276, 308, 332
Mackenzie, B. A. 30
MacNamara, John 140, 166
MacNeice, Louis 139
Mahootian, Shahrzad 332
Mairs, Jane Lowenstein 242, 248
Malkiel, Yakov 70, 91, 93, 249, 302
Manczak, Witold 65, 68
Marchand, Hans 112
Markey, Thomas L 71, 84, 93
Markus, Manfred 11, 14, *217*, 218, 224, 227, 228, 230, 337
Maroldt, Karl 9, 15, 17, 29
Marsh, Henry 269, 276
Martin, F. X. 161, 164, 308
Martinet, André 133, 166
Masser, Achim 209, 214
Mätzner, Eduard 35, 39, 53
Maurer, Friedrich 111, 211
Mayerthaler, Willi 287, 303
Mazzon, Gabriella 12, *233*, 245, 248
McArthur, Tom 101, 114, 248
McCawley, John D 156, 166

McClure Thomson, Elizabeth 191
McClure, Erica 329, 332
McClure, Malcolm 329, 332
McClure, Norman Egbert 191
McCully, Christopher 337, 360, 362
McIntosh, Angus 161, 217, 230, 330
Meillet, Antoine 69, 93
Meisel, Jürgen Michael 15, 29, 51
Meli, Marcello 75, 82, 93
Meritt, Herbert D 198, 212
Mersand, Joseph 18, 31
Mesthrie, Rajend 240, 248
Mettinger, Arthur 9
Milfull, Inge B. 207, 214
Miller, Thomas .. 25, 264, 346, 353, 354, 355, 363
Milroy, James 78, 93, 218, 230, 238, 241, 248
Milroy, Lesley 332
Minkova, Donka 11, *337*, 346, 361
Mioni, Alberto M 87, 93
Mitchell, Bruce 181, 193, 208, 214
Mocciaro, Antonia G 89, 94
Molhova, Žana 47, 53
Montgomery, Michael 166
Moody, Theodore William ... 161, 164, 167
Morani, Moreno 81, 94
More, Thomas 191
Morgan, James A 311, 332
Morris, Richard 20, 31, 55, 270, 276
Morsbach, Lorenz 17, 31
Mossé, Fernand 217, 230, 279, 280, 281, 303
Mühlhäusler, Peter 144, 166
Murray, James A. H. 215, 276, 375
Mustanoja, Tauno F. 36, 40, 41, 50, 53
Muysken, Pieter 134, 144, 161, 162, 332
Myers-Scotton, Carol M. 144, 166, 306, 307, 319, 332

N

Nagucka, Ruta............. 10, 14, *251*, 265
Nelde, Peter H.................................. 96
Nevalainen, Terttu..... 32, 52, 177, 193, 249, 335, 375
Nichols, Johanna... 29, 71, 94, 134, 166
Nihalani, Paroo 245, 248
Nixon, Graham 53
Nolan, P. E. 308, 333

O

Ó Baoil, Dónall P.... 72, 88, 91, 92, 131, 166
Ó Cuív, Brian... 140, 145, 146, 155, 166
Ó Dónaill, Niall.......................159, 167
Ó Neachtain, Seán 155
Ó Siadhail, Micheál 157, 167
O'Connell, Daniel 140
O'Muirithe, Diarmuid.................. 161
Odlin, Terence.........................240, 248
Öhmann, Suzanne 198, 215
Ojanen, Muusa 163
Onions, Charles T. 68, 215, 375
Orioles, Vicenzo........................... 88, 96
Orr, John 21, 31, 39, 54
Orton, Harold............................37, 54
Otheguy, Ricardo........................... 110

P

Page, R. I. 109, 114, 199, 215
Palander, Marjatta 163, 172, 177
Palander-Collin, Minna 172, 177, 193, 194
Panagl, Oswald 303
Partridge, Astley Cooper.......188, 194
Paston, Katherine.............178, 191, 192
Pearsall, Derek 361
Penzl, Herbert 5, 308, 333
Pfaff, Carol W. 144, 167, 306, 307, 323, 333
Pfeffer, J. Alan 100, 114
Phillips, Helen.............................. 29

Pienemann, Manfred 51
Pilch, Herbert...........................220, 230
Pinney, John................................192
Pinsker, Hans Ernst 225, 226, 228, 230
Platt, John 245, 248
Plummer, Charles............................264
Polenz, Peter von................... 75, 87, 94
Polomé, Edgar C.................... 91, 94, 96
Poplack, Shana 144, 167, 306, 307, 312, 318, 319, 321, 322, 323, 325, 326, 333
Poplack, Shona 144
Popov, Konstantin.............................53
Poussa, Patricia................................248
Pride, J. B. 193
Prince, Alan........................... 284, 303
Prins, Anton Adriaan.................39, 54
Puppel, Stanisław 113, 165, 265
Pütz, Martin 91

Q

Quirk, Randolph.......36, 41, 42, 54, 55, 173, 194

R

Ralegh, Walter 191
Ramat, Anna Giacalone. 44, 54, 90, 94
Ramat, Paolo .71, 72, 74, 75, 81, 83, 84, 89, 90, 91, 94, 95
Rasmussen, Jens 22, 31
Raumolin-Brunberg, Helena. 177, 193
Reichmann, Oskar 94
Reid, T. B. W. 375
Richardson, Malcolm...............175, 193
Richter, Michael.......... 17, 31, 310, 333
Rinelli, Gabriele............ 11, 13, 47, *267*
Ringen, John...........................279, 302
Rissanen, Matti ...32, 52, 114, 172, 177, 193, 194, 230, 246, 247, 248, 249, 334, 335, 375
Ritt, Nikolaus................. 11, 14, 16, *279*
Roberts, Jane 106, 114, 115
Robinson, Fred Norris 338, 361

Index of names

Robinson, Ian 356, 361
Robinson, M. L. 217, 231
Romaine, Suzanne . 144, 155, 167, 305, 306, 307, 308, 317, 318, 325, 333
Ross, Alan S. C. 201, 215
Ross, Garry 237, 246
Rothwell, William 17, 31, 333, 375, 376
Roy, John 142, 167
Ruggiers, Paul G. 342, 361
Runciman, Walter G. 268, 269, 276
Rupp, Heinz 111, 211
Rusakiev, Simeon 53
Rutherford, William E. 54, 55
Rynell, Alarik 103, 115

S

Saintsbury, George 345, 348, 349, 353, 361
Salmons, Joseph C. 132, 167
Samuels, Michael Louis 17, 20, 32, 83, 84, 95, 109, 115, 161, 217, 230, 231, 279, 303, 304, 330
Sanderson, S. F. 37, 54
Sandved, Arthur O. 217, 218, 231
Sankoff, David 144, 167, 319, 322, 323, 333
Sapir, Edward 11, 70, 71, 95
Schachter, Jacqueline 246, 247, 248
Schäfer, Jürgen 87, 95
Schaller, Helmut Wilhelm 145, 167
Scheler, Manfred 115
Schendl, Herbert .. 10, 11, 14, *305*, 319, 324, 333
Schildt, Joachim 51
Schipper, Jakob 338, 348, 361
Schneider, Edgar 142, 168
Schoeck, Richard J. 333
Schottmann, Hans 198, 215
Schöwerling, Rainer 230
Schwegler, Armin 83, 95
Scott, Charles T. 69, 95, 276
Scragg, Donald G. 212, 217, 231

Selinker, Larry 240, 249
Serjeantson, Mary Sidney 100, 115
Seuren, Pieter 49, 54
Shearin, Hubert Gibson . 35, 36, 45, 54
Sheridan, Richard Brinsley 138
Short, Ian 17, 27, 28, 31, 32, 310, 334
Shuy, Roger W. 166
Siebert, Eberhard 202, 215
Siewert, Ursula 230
Siewierska, Anna 72, 95
Silva-Corvalán, Carmen 46, 54
Simpson, J. M. Y. 71, 73, 95
Simpson, John A. 215, 276
Singh, Rajendran 144, 162, 168
Singleton, David 168
Sisam, Celia 208, 215
Sisam, Kenneth 208, 215
Skeat, Walter W. 338, 339, 348, 361
Skelton, John 312
Smith, Jeremy J. ...79, 95, 218, 231, 274, 276
Smith, Logan Pearsall 191
Smith, M. B. 362
Smithers, Geoffrey Victor. 20, 32, 270, 272, 276, 338, 340, 344, 348, 361
Smolensky, Paul 284, 303
Smyth, Mary W. 326, 334
Sonderegger, Stefan 94
Soravia, Giulio 89, 94
Sørensen, Knud 171, 172, 194
Southworth, John G. 354, 356, 361
Sridhar, Kamal K. 239, 249
Stanley, E. G. 231, 361
Stapleton, Thomas 264
Stein, Dieter 54, 78, 82, 95
Stein, Gabriele 87, 95
Steiner, Richard 84, 92
Stenson, Nancy 144, 168
Stenton, Doris Mary 269, 274, 277
Stockwell, Robert P. 11, *337*
Stolt, Birgit 308, 319, 333, 334
Stone, Louise W. 375

408

Index of names

Strang, Barbara Mary Hope ...53, 280, 304
Stroop 83, 95
Sullivan, Sister Carmeline 308, 313, 318, 334
Svartvik, Jan 36, 41, 42, 54, 55, 194
Sweet, Henry .. 195, 197, 198, 200, 210, 215, 264, 280, 304
Sweetser, Eve E. 63, 68
Swift, Jonathan 138, 155
Szantyr, Anton 200, 213
Szarmach, Paul E. 212
Szwedek, Aleksander 30

T

Taavitsainen, Irma 32, 52, 249, 335, 361, 375
Tarlinskaja, Marina. 348, 349, 351, 362
Ten Brink, Bernhard 338, 362
Tesch, Gerd 198, 215
Thomason, Sarah Grey .. 38, 39, 40, 54, 80, 96, 97, 108, 109, 115, 139, 159, 168
Timm, L. A 318, 319, 322, 323, 325, 334
Todd, Loreto 249
Toller, T. Northcote 198, 211, 264, 277
Tongue, R. K. 245, 248
Toorn, M. C. van den 86, 96
Toth, Karl 200, 216
Traugott, Elizabeth Closs 68, 256, 265
Trotter, David A 334, 363, 376
Trubetzkoy, Nikolaj 72, 96
Trudgill, Peter 163, 164, 168, 169, 237, 238, 249
Tyler, S. 30

U

Ullmann, Stephen 103, 115
Ureland, Per Sture 71, 73, 96, 168, 304

V

Vachek, Joseph 39, 48, 54
Varvaro, Alberto 249
Vaughan, William E. 167
Venezky, Richard L. 214
Vennemann, Theo 82, 96, 284, 291, 295, 304
Viehweger, Dieter 51
Vincent, Nigel 246
Visser, Fredericus Theodorus ... 36, 55
Vollmann-Profe, Gisela 213

W

Wagner, Heinrich 96
Wakefield, H. 68
Wartburg, Walther von . 107, 132, 168
Weber, Heidi 245, 248
Wehrle, William O. 334
Weiner, E. S. C. 215, 276
Weinreich, Uriel .72, 96, 143, 168, 233, 241, 243, 249
Weismiller, Edward R. . . 356, 357, 358, 362
Wenzel, Siegfried 309, 334
Werner, Otmar 282, 287, 304
West, Michael 58, 68
Wetzel, Claus-Dieter 205, 216
Whitelock, Dorothy 114, 268, 277
Whorf, Benjamin Lee 73, 75, 96
Widdowson, John D. A. 37, 54
Wieland, Gernot 196, 199, 216
Wild, Friedrich 217, 230, 231, 338, 362
Williams, Jeffrey P .. 142, 168, 242, 249
Williamson, Keith 231, 311, 332
Wolff, Dieter 58, 68
Wolfram, Walt 169
Woolford, Ellen 144, 169
Workman, Samuel K .. 39, 55, 171, 194
Wotton, Henry 182, 191, 192
Wright, C. E. 227, 231
Wright, Elizabeth Mary 291, 304
Wright, Joseph 291, 304

409

Index of names

Wright, Laura Charlotte 10, 11, 14, 19, 32, 309, 319, 335, *363*, 364, 370, 372, 374, 375
Wright, Susan 114, 230, 246
Wurzel, Wolfgang U. 303

Y

Yaeger, Malcah 84, 92
Yeats, William Butler 139

Youmans, Gilbert ... 338, 350, 356, 362
Young, Jean 19, 29

Z

Zobl, Helmut 46, 47, 48, 49, 54, 55, 241, 249
Zumthor, Paul 335
Zupitza, Julius 197, 203, 216, 265

Studies in English Medieval Language and Literature

Edited by Jacek Fisiak

Vol. 1 Dieter Kastovsky / Arthur Mettinger (eds.): Language Contact in the History of English. 2001.

Günther Lampert / Martina Lampert

The Conceptual Structure(s) of Modality: Essences and Ideologies

A Study in Linguistic (Meta-)Categorization

Frankfurt/M., Berlin, Bern, Bruxelles, New York, Wien, 2000. 328 pp.
ISBN 3-631-35219-0 · pb. DM 89.–*
US-ISBN 0-8204-4738-2

This study takes George Lakoff's dictum seriously that arguments about conclusions will be pointless unless there is agreement on the premises. Subscribing to the view that linguistic categorization is necessarily theory-driven, it reconstructs and assesses, on the basis of domain-independent parameters derived from General Systems Theory and Leonard Talmy's Overlapping Systems Model of Cognitive Organization, hegemonic western conceptualizations of modality. Arguing against such propositional models of modality, the book outlines, modifies, and extends conceptualizations from a Cognitive Linguistics point of view, grounding "modality" especially on Talmy's Imaging System of Force Dynamics. The authors conceive of modality as an experientially and perceptually motivated radial category of Image Schemata, ultimately challenging the categorial status of modality as a cognitively motivated category.

Contents: Theories of Categorization · Hegemonic Conceptualizations of Modality: Functionalist, Logical, National Models · Cognitive Linguistics Models of Modality · Force Dynamics: Extensions and Modifications · Challenging the Categorial Status of Modality

Frankfurt/M · Berlin · Bern · Bruxelles · New York · Oxford · Wien
Distribution: Verlag Peter Lang AG
Jupiterstr. 15, CH-3000 Bern 15
Fax (004131) 9402131
*incl. value added tax
Prices are subject to change without notice.